Sustainability or Collapse?

An Integrated History and Future of People on Earth

Goal for this Dahlem Workshop:

To understand better the dynamic interactions between human societies and their environment by linking various forms of knowledge on human history and environmental change at multiple temporal scales (millennial, centennial, decadal, and future scenarios).

Report of the 96th Dahlem Workshop on
Integrated History and future Of People on Earth (IHOPE)
Berlin, June 12–17, 2005

Held and published on behalf of the
President, Freie Universität Berlin

Supported by:

Freie Universität Berlin
AIMES (IGBP): Analysis, Integration and Modeling of the Earth (International Geosphere–Biosphere Programme)
QUEST: Quantifying and Understanding the Earth System, U.K. Natural Environment Research Council

Sustainability or Collapse?

An Integrated History and Future of People on Earth

Edited by

Robert Costanza, Lisa J. Graumlich, and Will Steffen

Program Advisory Committee:

Robert Costanza, Lisa J. Graumlich, and
Will Steffen, Chairpersons
Carole L. Crumley, John A. Dearing, Eric F. Lambin,
Rik Leemans, and Frank Riedel

The MIT Press
Cambridge, Massachusetts
London, U.K.

in cooperation with Dahlem University Press

MIT Press books may be purchased at special quantity discounts for business or sales promotional use. For information, please email special_sales@mitpress.mit.edu or write to Special Sales Department, The MIT Press, 55 Hayward Street, Cambridge, MA 02142.

This book was set in TimesNewRoman by BerlinScienceWorks.

Printed and bound in the United States of America.

Library of Congress Cataloging-in-Publication Data
Dahlem Workshop on Integrated History and Future of People on Earth (2005 : Berlin, Germany) Sustainbility or collapse? An integrated history and future of people on earth / edited by R. Costanza, L. J. Graumlich, and W. Steffen ; Program Advisory Committee, R. Costanza … [et al.].
p. cm.
"Report of the 96th Dahlem Workshop on Integrated History and Future of People on Earth (IHOPE) Berlin, June 12–17, 2005."
Includes bibliographical references and indexes.
ISBN-13: 978-0-262-03366-4 (hardcover : alk. paper)
ISBN-10: 0-262-03366-6 (hardcover : alk. paper)
1. Human ecology—History—Congresses. I. Costanza, Robert. II. Graumlich, Lisa. III. Steffen, W. L. (William L.), 1947– IV. Title.
GF13.D35 2005
304.2—dc22
2006023346

10 9 8 7 6 5 4 3 2

Contents

Dahlem Workshops

History

During the last half of the twentieth century, specialization in science greatly increased in response to advances achieved in technology and methodology. This trend, although positive in many respects, created barriers between disciplines, which could have inhibited progress if left unchecked. Understanding the concepts and methodologies of related disciplines became a necessity. Reuniting the disciplines to obtain a broader view of an issue became imperative, for problems rarely fall conveniently into the purview of a single scientific area. Interdisciplinary communication and innovative problem-solving within a conducive environment were perceived as integral yet lacking to this process.

In 1971, an initiative to create such an environment began within Germany's scientific community. In discussions between the *Deutsche Forschungsgemeinschaft* (German Science Foundation) and the *Stifterverband für die Deutsche Wissenschaft* (Association for the Promotion of Science Research in Germany), researchers were consulted to compare the needs of the scientific community with existing approaches. It became apparent that something new was required: an approach that began with state-of-the-art knowledge and proceeded onward to challenge the boundaries of current understanding; a form truly interdisciplinary in its problem-solving approach.

As a result, the *Stifterverband* established *Dahlem Konferenzen* (the Dahlem Workshops) in cooperation with the *Deutsche Forschungsgemeinschaft* in 1974. Silke Bernhard, formerly associated with the Schering Symposia, was

Figure adapted from *L'Atmosphère: Météorologie Populaire*, Camille Flammarion. Paris: Librairie Hachette et Cie., 1888.

engaged to lead the conference team and was instrumental in implementing this unique approach.

The Dahlem Workshops are named after a district of Berlin known for its strong historic connections to science. In the early 1900s, Dahlem was the seat of the Kaiser Wilhelm Institutes where, for example, Albert Einstein, Lise Meitner, Fritz Haber, and Otto Hahn conducted research. Today the district is home to several Max Planck Institutes, the *Freie Universität Berlin*, the *Wissenschaftskolleg*, and the Konrad Zuse Center.

In its formative years, the Dahlem Workshops evolved in response to the needs of science. They soon became firmly embedded within the international scientific community and were recognized as an indispensable tool for advancement in research. To secure its long-term institutional stability, *Dahlem Konferenzen* was integrated into the *Freie Universität Berlin* in 1990.

Aim

The aim of the Dahlem Workshops is to promote an international, interdisciplinary exchange of scientific information and ideas, to stimulate international cooperation in research, and to develop and test new models conducive to more effective communication between scientists.

Concept

The Dahlem Workshops were conceived to be more than just another a conference venue. Anchored within the philosophy of scientific enquiry, the Dahlem Workshops represent an independently driven quest for knowledge: one created, nurtured, and carefully guided by representatives of the scientific community itself. Each Dahlem Workshop is an interdisciplinary communication process aimed at expanding the boundaries of current knowledge. This dynamic process, which spans more than two years, gives researchers the opportunity to address problems that are of high-priority interest, in an effort to identify gaps in knowledge, to pose questions aimed at directing future inquiry, and to suggest innovative ways of approaching controversial issues. The overall goal is not necessarily to exact consensus but to search for new perspectives, for these will help direct the international research agenda.

Governance

The Dahlem Workshops are guided by a Scientific Advisory Board, composed of representatives from the international scientific community. The board is responsible for the scientific content and future directions of the Dahlem Workshops and meets biannually to review and approve all workshop proposals.

Workshop Topics

Workshop topics are problem-oriented, interdisciplinary by nature, of high-priority interest to the disciplines involved, and timely to the advancement of science. Scientists who submit workshop proposals, and chair the workshops, are internationally recognized experts active in the field.

Program Advisory Committee

Once a proposal has been approved, a Program Advisory Committee is formed for each workshop. Composed of 6–7 scientists representing the various scientific disciplines involved, the committee meets approximately one year before the Dahlem Workshop to develop the scientific program of the meeting. The committee selects the invitees, determines the topics that will be covered by the pre-workshop papers, and assigns each participant a specific role. Participants are invited on the basis of their international scientific reputation alone. The integration of young German scientists is promoted through special invitations.

Dahlem Workshop Model

A Dahlem Workshop can best be envisioned as a week-long intellectual retreat. Participation is strictly limited to forty participants to optimize the interaction and communication process.

Participants work in four interdisciplinary discussion groups, each organized around one of four key questions. There are no lectures or formal presentations at a Dahlem Workshop. Instead, concentrated discussion—within and between groups—is the means by which maximum communication is achieved.

To enable such an exchange, participants must come prepared to the workshop. This is facilitated through a carefully coordinated pre-workshop dialog: Discussion themes are presented through "background papers," which review a particular aspect of the group's topic and introduce controversies as well as unresolved problem areas for discussion. These papers are circulated in advance, and everyone is requested to submit comments and questions, which are then compiled and distributed. By the time everyone arrives in Berlin, issues have been presented, questions have been raised, and the Dahlem Workshop is ready to begin.

The discussion unfolds in moderated sessions as well as during informal times of interaction. Cross-fertilization between groups is both stressed and encouraged. By the end of the week, through a collective effort directed that is directed by a rapporteur, each group has prepared a draft report of the ideas, opinions, and contentious issues raised by the group. Directions for future research are highlighted, as are problem areas still in need of resolution. The results of the draft reports are discussed in a plenary session on the final day and colleagues from the Berlin–Brandenburg area are invited to participate.

Dahlem Workshop Reports

After the workshop, attention is directed toward the necessity of communicating the perspectives and ideas gained to a wider audience. A two-tier review process guides the revision of the background papers, and discussion continues to finalize the group reports. The chapters are carefully edited to highlight the perspectives, controversies, gaps in knowledge, and future research directions.

The publication of the workshop results in book form completes the process of a Dahlem Workshop, as it turns over the insights gained to the broad scientific community for consideration and implementation. Each volume in the Dahlem Workshop Report series contains the revised background papers and group reports as well as an introduction to the workshop themes. The series is published in partnership with The MIT Press.

Julia Lupp, Program Director and Series Editor
Dahlem Konferenzen der Freien Universität Berlin
Thielallee 50, 14195 Berlin, Germany

List of Participants

Roelof M. J. Boumans Gund Institute for Ecological Economics, University of Vermont, 590 Maine Street, Burlington, VT 05405, U.S.A.
Global simulations of human welfare, ecosystem services, and ecosystem service values

Robert Costanza Gund Institute for Ecological Economics, Rubenstein School of Environment and Natural Resources, University of Vermont, 590 Main St., Burlington, VT 05405–1708, U.S.A.
Ecological economics, systems ecology, environmental policy, landscape ecology, spatial, dynamic, ecological modeling, social traps, incentive structures and institutions

Carole L. Crumley Department of Anthropology, University of North Carolina, 301 Alumni Building, Chapel Hill, NC 27599–3115, U.S.A.
Archaeology, ethnohistory, historical ecology and climatology, contemporary climate change and agriculture, complex systems theory

Paul J. Crutzen Abteilung Atmosphärenchemie, Max-Planck-Institut für Chemie, Postfach 3060, 55020 Mainz, Germany
The role of atmospheric chemistry in biogeochemical cycles and climate

John A. Dearing Department of Geography, University of Liverpool, Roxby Building, Liverpool L69 7ZT, U.K.
Human–environment interactions; reconstructing past environments; simulating complex systems

Bert J. M. de Vries Netherlands Environmental Assessment Agency (MNP), P.O. Box 303, 3720 AH Bilthoven, The Netherlands, and Copernicus Institute for Sustainable Development and Innovation, Utrecht University, Heidelberglaan 2, P.O. Box 80.115, 3508 TC Utrecht, The Netherlands
Sustainable development concepts/modeling; energy and climate modeling/policy; historical socioecologial developments

John Finnigan CSIRO Centre for Complex Systems Science, Pye Laboratory, G.P.O. Box 1666, Canberra ACT 2601, Australia
Complex systems science; Earth system science; terrestrial carbon cycle; turbulent exchange between atmosphere and biosphere

Lisa J. Graumlich Big Sky Institute, 106 AJM Johnson Hall, Montana State University, Bozeman, MT 59717, U.S.A.
Climate variability on decade-to-century timescales and its impacts on ecosystem dynamics and services

Richard H. Grove Resource Management in the Asia Pacific Program, Research School of Pacific and Asian Studies, Australian National University, Canberra ACT 0200, Australia
Environmental history, history of science, forest history, history of witchcraft and crisis; extreme climate events and Dark Ages; socioeconomic crises in world history; South Asian environmental history; island environmental histories; African environmental history

Arnulf Grübler IIASA, Schlossplatz 1, 2361 Laxenburg, Austria, and School of Forestry and Environmental Studies, Yale University, New Haven, CT 06511, U.S.A.
Long-term history and future of technology with focus on energy, transport, and communication systems

Helmut Haberl Institute of Social Ecology, IFF Vienna, Klagenfurt University, Schottenfeldgasse 29, 1070 Vienna, Austria
Long-term changes in society–nature interaction, integrated analysis of socioecological systems, human appropriation of net primary production (HANPP), including its causes and consequences (e.g., on biodiversity and on the carbon household), long-term socioecological research (LTSER)

Fekri A. Hassan Institute of Archaeology, University College London, 31–34 Gordon Square, London WC1H 0PY, U.K.
Implications of water history as revealed by archaeological and historical sources for the resolution of current world water problems; study of the past as a means for coping with the present and developing strategies for a better future world

Kathy A. Hibbard AIMES International Project Office, Climate and Global Dynamics Division, National Center for Atmospheric Research, P.O. Box 3000, Boulder, CO 80307–3000, U.S.A.
Effects of management practices (fire suppression, heavy grazing) and disturbance on the carbon cycles of savanna and forested ecosystems in the context of altered biogeochemistry and successional dynamics through the integration of field observations and ecosystem modeling; the international Global Carbon Project and the IGBP Analysis, Integration and Modeling of the Earth System Project

Frank Hole Department of Anthropology, Yale University, Box 8277, New Haven, CT 06520–8277, U.S.A.
Near East, archaeology, climate history, land use, agriculture, sustainability

Eric F. Lambin Department of Geography, University of Louvain, 3, place Pasteur, 1348 Louvain-la-Neuve, Belgium
Land-use/cover change, remote sensing of land

Rik Leemans Environmental Systems Analysis Group, Wageningen University, P.O. Box 47, 6700 AA Wageningen, The Netherlands
Ecology, integrated assessment, consequences of land-use change; ecosystem services; global and regional models

Diana M. Liverman Environmental Change Institute, Oxford University Centre for the Environment, Dyson Perrins Building, South Parks Road, Oxford OX1 3QY, U.K.
Human dimensions of global change; environmental policy in the Americas

Nathan J. Mantua Climate Impacts Group, University of Washington, Box 354235, Seattle, WA 98195–4235, U.S.A.
Causes for year-to-year, decade-to-decade, and multi-decadal climate variations; climate predictability and prediction; climate impacts on ecosystems and society; paleoclimate reconstructions; use of climate information in resource management

John R. McNeill History Department, Georgetown University, Washington, D.C. 20057, U.S.A.
Environmental history; energy history

Dennis L. Meadows Laboratory for Interactive Learning, P.O. Box 844, Durham, NH 03824, U.S.A.
Innovative educational methods for helping people understand the behavior of complex systems; social, economic, and political implcations of limits to growth

Bruno Messerli Institute of Geography, University of Bern, Hallerstrasse 12, 3012 Bern, Switzerland
Millennial scale: climate change and human history (e.g., African mountains, Andes). Centennial/decadal scale: water resources/floods/droughts and human population (e.g., Himalayas and Bangladesh). Decadal/future scale: human–environment systems (concepts)

João M. F. Morais International Geosphere–Biosphere Programme, Deputy Director, Social Sciences, Royal Swedish Academy of Sciences, Box 50005, 10405 Stockholm, Sweden
Earth system science; environmental archaeology

Christian Pfister Section of Economic, Social, and Environmental History, Institute of History, University of Bern, Erlachstr. 9a, 3000 Bern 9, Switzerland
Climatic change (Europe, last millennium), demographic and economic impacts on societies, buffering strategies and innovations

Charles L. Redman Global Institute of Sustainability, Arizona State University, P.O. Box 873211, Tempe, AZ 85287–3211, U.S.A.
Human impacts on ancient environments, urban ecology, integration of social and life sciences

Frank Riedel Interdisciplinary Centre for Ecosystem Dynamics in Central Asia, Freie Universität Berlin, Malteserstr. 74–100, Haus D, 12249 Berlin , Germany
Environmental and human dynamics in northwestern China during the late Quaternary; ecosystem dynamics in Central Asia; palaeoclimate

Vernon L. Scarborough Dept. of Anthropology, University of Cincinnati, P.O. Box 210380, Cincinnati, OH 45221–0380, U.S.A.
Archaeology, anthropology, tropical ecosystems, landscapes, water management, civilization, Maya

Will Steffen Director, CRES and ANU Institute for Environment, Centre for Resource and Environmental Studies (CRES), Australian National University, W.K. Hancock Building (43), Canberra ACT 0200, Australia
Earth system science in general, with more specific interest in (i) the global carbon cycle, (ii) abrupt changes in Earth system functioning, and (iii) evolution of the human–environment relationship in an Earth system context

Uno Svedin FORMAS, The Swedish Research Council for Environment, Agricultural Sciences, and Spatial Planning, Box 1206, 111 82 Stockholm, Sweden
Research policy, especially environment and sustainable development, connected systems analysis, connected governance, connection between socioeconomic and biogeophysical aspects of environmental challenges, societal risk, cultural connotations of humans–environmental nexus, micro–macro phenomena relations

Joseph A. Tainter Global Institute of Sustainability and School of Human Evolution and Social Change, Arizona State University, P.O. Box 873211, Tempe, AZ 85287, U.S.A.
Sustainability; evolution of complexity

Peter Turchin Department of Ecology and Evolutionary Biology, University of Connecticut, 75 N. Eagleville Road, U–43, Storrs, CT 06269–3042, U.S.A.
Population dynamics, historical demography, complex dynamics in social and economic structures of historical societies

Sander E. van der Leeuw School of Human Evolution and Social Change, Arizona State University, P.O. Box 872402, Tempe, AZ 85287–2402, U.S.A.
Archaeological method and theory, long-term relations between society and environment, land use and cover change, modeling

Yoshinori Yasuda International Research Center for Japanese Studies, Nishikyoku, 3–2 , Oeyama-cho, Kyoto 610–1192, Japan
Environment archaeology, human–environment interaction, especially deforestation; climate change and the rise and fall of civilization, Jomon and Yangtze River civilization; high-resolution chronology by analysis of annually laminated sediments

Marianne N. Young Marianne Young Planning, 216 Gilles Street, Adelaide SA 5000, Australia
Urban and regional town planning; social science—communication

Michael D. Young CSIRO Land and Water, Private Bag 2, Glen Osmond 5064 , Australia, and Water Economics and Management, School of Earth and Environmental Sciences, The University of Adelaide, Adelaide 5005, Australia
Market-based instruments; resource accounting; ecological economics; policy review and development

Foreword

The Mirror of Galadriel

Hans Joachim Schellnhuber

Potsdam Institute for Climate Impact Research (PIK), Postfach 60 12 03,
14412 Potsdam, Germany

Sometime in the summer of 2002, I moderated the Program Advisory Committee meeting charged with the design of the Dahlem Workshop "Earth System Analysis for Sustainability." I recall that toward the end of our incredibly intense and exhausting deliberations Paul Crutzen, a Nobel laureate and one of my designated workshop co-chairs, exclaimed: "This is going to be the most ambitious of all Dahlem events so far! I really don't know whether it will work."

This was a well-justified concern indeed since the workshop topic was nothing less than the generic mode of operation of the planetary machinery under qualitatively different circumstances—for instance, in response to massive volcano eruptions, under asteroid bombardment, with different distributions of continental masses, after the great oxidation as caused by primitive life, with and without strong anthropogenic interference, driven by the blind expansion of globalized business-as-usual economy or wisely steered by sophisticated supranational institutions...

I was as worried as Paul, yet the workshop—held during a glorious week in May 2003—developed into a major success as documented in a recent MIT publication (Schellnhuber et al. 2004). Only two years later, another Dahlem event dedicated to global long-term sustainability took place and gave birth to this book. Its focus, "An Integrated History and future Of People on Earth (IHOPE)" is not just ambitious, it is aspirational! I will have to elaborate a bit on why this is so and why the aspirations are nevertheless warranted. Prior to that, let me mention that Paul Crutzen undauntedly did it again by participating in the IHOPE workshop. I was actually supposed to serve as a co-chair, but was eventually denied this role by urgent conflicting obligations. The Dahlem organizers took full revenge, however, by asking me to produce a preface to this report. So here we are.

> With water from the stream Galadriel filled the basin to the brim, and breathed on it, and when the water was still again she spoke. "Here is the Mirror of Galadriel," she said. "I have brought you here so that you may look in it, if you will."

> The air was very still, and the dell was dark, and the Elf-lady beside him was tall and pale. "What shall we look for, and what shall we see?" asked Frodo, filled with awe (Tolkien 1954).

I am citing Tolkien here because the IHOPE protagonists are in search of something that reminds me of Elvish magic: they set out to tell the grand story of humankind's journey through its lifetime on Earth. This is not meant to be a traditional (hi)story of kings and towers and battles, but a scientific narrative describing and explaining how human civilization has developed as a part of nature and, therefore, in perpetual material, sensual, and spiritual interaction with its life-supporting environment. Scholars have previously tried to provide "holistic" accounts of important episodes in cultural evolution—the most shining example may be Fernand Braudel's masterpiece about the Mediterranean in the times of Philip II of Spain, where the French genius interweaves political, social, economic, technical, and natural dimensions into a superb portrait of the 16th century (Braudel 1949). Yet Braudel's approach is focused on a fairly limited spatiotemporal window, and his investigation remains qualitative, in spite of an impressive array of figures underpinning the reasoning. By way of contrast, IHOPE is virtually unlimited in its goals: the project aims at a global panorama of coupled human–environment history since the dawn of civilization, derived from quantitative insights into archetypical relationships and processes.

Humankind is distinct from all other species on Earth in its superior way of processing information and, as a consequence, its ability to build instruments, infrastructures, and institutions. Like termites compiling their conspicuous nest, *Homo sapiens* are thus generating the *anthroposphere*, which is about to become the dominant factor of planetary dynamics in the not-too-distant future. Quite remarkably, IHOPE claims to be able to tell some story of that future, too.

How can such an intellectual bravado be justified? It all depends on *what we look for* and *what we expect to see*, as Frodo so aptly put it in Tolkien's tale (see above). First, it is of utmost importance to specify the variables in which one is interested. These cannot be simple, measurable quantities since there is no way whatsoever to measure the future in advance. The pertinent IHOPE variables must rather be "observables" in the sense of quantum theory, i.e., entities derived from intricate intellectual constructions involving whole sets of numbers instead of single figures. This can be nicely illustrated by climate analysis, which is a relevant discipline in our context anyway:

"Predictions are notoriously difficult, particularly when they concern the future!" This corny joke was fairly popular, a while ago, among climate modelers who are wrongly expected to be able to anticipate the *weather* (e.g., the cloudiness over Oxfordshire in the afternoon of the 27th of May, 2081). A scientific first-principles explanation as to why that expectation is nonsensical has to do with nonlinear dynamics, multiscale stochasticity, etc., and so could easily fill a 500-page monograph. Things look quite differently, however, if climate scientists are asked to do what they have been trained for, namely, anticipating the *climate*. The latter is not a real entity but a human construction through the definition of a set of macrovariables which are evaluated by massive standardized spatiotemporal averaging over complicated atmospheric microdynamics. Take, for instance, the lead parameter in the current environmental debate, namely global mean temperature, T. It can neither be felt nor measured directly; there is not even a practically unambiguous way to determine it! Our knowledge about this aggregate variable is characterized in the following, heavily stylized, diagram (see Figure 1).

The figure indicates, first of all, that even the present planetary temperature comes with a nonnegligible error bar. When we look back in time, the uncertainty in T generally grows with the distance from the present, yet not necessarily in a monotonic way. This has to do with proxy data availability that varies wildly along the timeline, but which clearly deteriorates in the deep geological past. As a consequence, there is a rough time symmetry in our ignorance about global mean temperature: the paleoreality is evidently distinguished from virtual future by its factual uniqueness, yet this uniqueness cannot be translated into a unique reconstruction (NRC 2006).

Yet even the delineation of an uncertainty band for the paleovariation of T may reveal important characteristics of this planetary parameter. One may discern, for instance, conspicuous seesaw patterns, as recorded in the famous Vostok ice cores and similar Earth System archives. Figure 1 alludes, in a cartoon way, to such a behavior. It is not unreasonable to assume that qualitative patterns of that kind *can* repeat in time, and that the range of past variations *might* provide certain quantitative constraints on future dynamics. Note, however, that this by no means deserves to be called a "prediction."

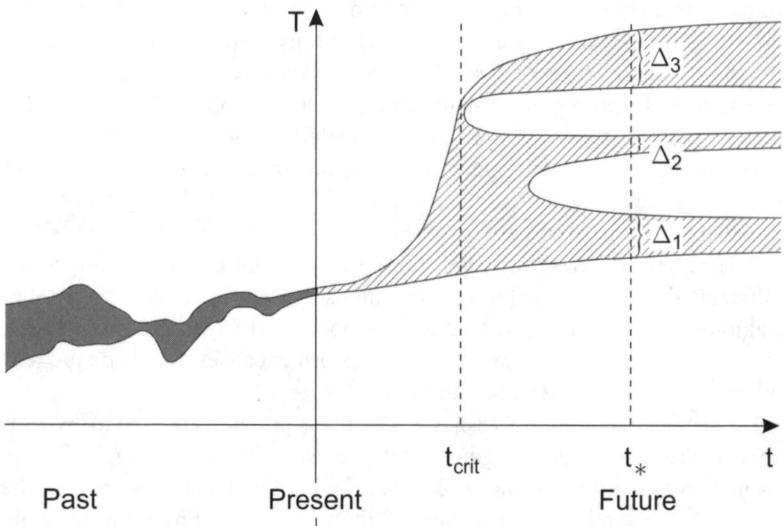

Figure 1 Global mean temperature—never ever certain…

> "Many things I can command the Mirror to reveal," she answered, "and to some I can show what they desire to see. But the Mirror will also show things unbidden, and those are often stranger and more profitable than things which we wish to behold. What you will see, if you leave the Mirror free to work, I cannot tell. For it shows things that were, and things that are, and things that yet may be. But which it is that he sees, even the wisest cannot always tell. Do you wish to look?" (Tolkien 1954).

The prescience of climate science regarding the future development of global mean temperature is fairly limited, even if the socioeconomic driving forces of planetary warming were precisely known (see below). This is epitomized by the notoriously vague T-projections for 2100 by the IPCC, where the error bars obstinately refuse to shrink. Yet there is more to be said about the temperature in the future than just asserting its perpetual rise: On the one hand, fundamental geophysics, taking into account systems inertia and process timescales (like the ones associated with heat overturning in the oceans), defines rather crisp lower limits to that rise. On the other hand, the topology of climate possibility space may be nontrivial, as sketched on the right-hand side of Figure 1. For example, incessant human forcing may drive the Earth System through a series of "tipping points," where discontinuities in T might occur. In our cartoon, one such discontinuity may—or may not—happen around the time t_{crit}, triggered perhaps by the collapse of the Amazon rainforest, an event not unlikely if the planet keeps on wandering in the high-temperature domain throughout this century. Figure 1 suggests that some time after that event, T might be confined to an ultra-hot set of values, separated from the traditional interglacial range by an inaccessible zone. Due to intricate, perhaps unprecedented feedback dynamics, the interglacial band might be split up again later to bring about even more fragmentation of temperature possibility space. Thus, at an arbitrary inspection time t_* in the future, the potential values of T would be confined to the sets Δ_1, Δ_2, and Δ_3.

It is precisely this type of set-valued projections for appropriately aggregated variables that IHOPEian approaches aspire to derive from whole-systems analyses of the past and present operation of the nature–civilization complex. The pertinent variables are even less sensuous and more synthetic than global mean temperature; the narratives will have to deal with entities like resource availability, disaster preparedness, or social cohesion. The IHOPE team claims that such entities are driven by a combination of sluggishly evolving environmental–cultural forcing and incalculable contingency noise. So there is a chance to anticipate, *grosso modo*, at least the effects of the former causes. Read and judge for yourself whether that claim is warranted.

The second major point I wish to make in this preface is related to "what we expect to see" and requires arguments at an even higher level of sophistication. The main reason for this is the aspect of self-referentiality always present when science ponders relevant problems of the real world. From the reasoning sketched thus far, one may conclude that the future of a dynamical system such as the global human–environment complex cannot be predicted even if we were able to constrain possibility space to a nontrivial domain by ingenious analyses and techniques. Predictive power appears to be an even more elusive skill if one takes into account the observation that any attempt of anticipation is generally fed back into the system (e.g., through the reactions of important stakeholders) and thus modifies the very assumptions, conditions, and processes that were involved in the original projections. In other words, we are faced here with the dilemma of self-destroying prophecy! If a scientific study were to predict, for instance, that technological innovation as generated by pure market forces would suffice to stabilize atmospheric CO_2 at subdangerous levels, then each and every climate policymaker could complacently lean back only to miss the opportunities for the strategic induction of the pertinent innovations. And vice versa…So IHOPE is a hopeless endeavor, right?

Not necessarily: The same actors that can destroy a prophecy can also help to realize it. In fact, the best way of anticipating the future is *by construction*. Nobody is able to predict the precise position of a given dozen of individuals a week in advance under normal circumstances; however, the same task becomes fairly simple if one organizes a get-together with them at a certain location at the time in question. This observation is much less trivial than it appears to be at first glance. I tried to elaborate on the main arguments some years ago in an article where the concept of fuzzy control plays a prominent role (Schellnhuber 1998). It is interesting to see how the IHOPE authors revisit that approach and what conclusions they try to draw.

The factual making of a (desired/expected/abhorred) future might be simulated, to some extent, through a sophisticated "hybrid" modeling technique: The basic idea—also reflected in this book—is to couple directly electronic simulators forecasting the dynamics of calculable macrovariables with representative stakeholder cohorts mimicking human microbehavior in crucial decision making under uncertainty. This way, the best Earth System model might be combined with, say, 1000 leading stakeholders in a grand co-animation experiment that produces a self-consistent virtual realization of the future. When I mentioned this concept a while ago in an essay for an IGBP book (Schellnhuber 2002), introducing the term "hyberspace simulation," the editors amusingly "corrected" it to the more familiar expression "cyberspace simulation."

These are just a few thoughts on the IHOPE enterprise, which deserves to be called one of the most exciting intellectual journeys in contemporary science. We do not know at all where that journey is heading, let alone, where it might end. Yet it is worthwhile to embark on it—with painstaking care.

> "Like as not," said the Lady with a gentle laugh. "But come, you shall look and see what you may. Do not touch the water!" (Tolkien 1954).

REFERENCES

Braudel, F. 1949. La Méditerranée et le Monde Méditerranéen a l'époque de Philippe II. Paris: Colin.

NRC (National Research Council). 2006. Surface Temperature Reconstructions for the Last 2,000 Years. Washington, D.C.: Natl. Acad. Press.

Schellnhuber, H.J. 1998. Earth system analysis: The scope of the challenge. In: Earth System Analysis, ed. H.J. Schellnhuber and V. Wenzel, pp. 3–195. Berlin: Springer.

Schellnhuber, H.J. 2002. Coping with Earth system complexity and irregularity. In: Challenges of a Changing Earth, ed. W. Steffen, J. Jäger, D.J. Carson, and C. Bradshaw, pp. 151–156. Berlin: Springer.

Schellnhuber, H.J., P.J. Crutzen, W.C. Clark, M. Claussen, and H. Held, eds. 2004. Earth System Analysis for Sustainability. Dahlem Workshop Report 91. Cambridge, MA: MIT Press.

Tolkien, J.R.R. 1954. The Fellowship of the Ring. Chapt. VII: The Mirror of Galadriel (reset ed., 1999), pp. 474–475. London: HarperCollins.

Introduction

1

Sustainability or Collapse

Lessons from Integrating the History of Humans and the Rest of Nature

Robert Costanza,[1] Lisa J. Graumlich,[2] and Will Steffen[3]

[1]Gund Institute for Ecological Economics, Rubenstein School of Environment and Natural Resources, University of Vermont, Burlington, VT 05405–1708, U.S.A.
[2]Big Sky Institute, Montana State University, Bozeman, MT 59717, U.S.A.
[3]Centre for Resource and Environmental Studies, Australian National University, Canberra ACT 0200, Australia

INTRODUCTION

What is the most critical problem facing humanity at the beginning of the 21st century? Global pandemics, including AIDS? Global warming? Meeting global energy demands? Worldwide financial collapse? International terrorism? The answer is all of these and more. Most of us live in an increasingly global system in which our most critical problems span national borders, cover continents, or are truly global. When past civilizations collapsed, they were isolated from other parts of the world. The socioeconomic and natural drivers of these collapses were local and regional. Today in our interconnected global civilization, massive social failure in one region can threaten the stability of the entire global system. Can the current global civilization adapt and survive the accumulating, highly interconnected problems it now faces? Or will it collapse like Easter Island, the Classic Maya, the Roman Empire, and other past civilizations, but on a larger scale? What can we learn from these past civilizations (and especially the ones that did *not* collapse) to help guide our current civilization toward sustainability?

To answer this question requires a new, more integrated understanding of how humans interact with each other, with resources, with other species and with the environment. The essence is thus to understand the interaction of human systems with the rest of nature. Our phrasing of the previous sentence is quite deliberate. "Humans and nature" implies that humans are separate from nature, whereas "humans and the rest of nature" implies that humans are a part of nature, not separate from it. We need to understand better how humans have interacted with the rest of nature in the past, how we currently interact, and what the options are for future interactions. Based on this, we can attempt to create a sustainable and desirable future for our species. If we continue to operate in ignorance or denial of this integrated understanding, we run the very real risk of going the way of the Easter Islanders and others, but on a much larger scale.

This book is devoted to the first steps in developing a fully integrated history of humans and the rest of nature and thus serves as a foundation for the ongoing Integrated History and future Of People on Earth (IHOPE) project.

INTEGRATING HUMAN AND NATURAL HISTORY

Human history has traditionally been cast in terms of the rise and fall of great civilizations, wars, and specific human achievements. This history leaves out, however, the important ecological and climate context which shaped and mediated these events. The capability to integrate human history with new data about the natural history of the Earth at global scales and over centuries to millennia has only recently become possible. This integrated history could not have been accomplished even ten years ago and is a critical missing link that will provide a much richer picture of how (and why) the planet has changed in historical times. When compiled, such an integrated history will advance research from various perspectives of the Earth's history and possible futures. Finally, it will be used as a critical data set to test integrated models of humans in natural systems.

Human–environment systems are intimately linked in ways that we are only beginning to appreciate (Steffen et al. 2004; Diamond 2005; Kirch 2005). To achieve the ambitious goals of IHOPE, multiple scientific challenges must be met. To understand the integrated history of the Earth it is necessary to integrate the different perspectives, theories, tools, and knowledge of multiple disciplines across the full spectrum of social and natural sciences and the humanities.

LONG-TERM GOALS OF THE IHOPE PROJECT

The IHOPE project has three long-term goals:

1. Map the integrated record of biophysical and human system change on the Earth over the last several thousand millennia, with higher temporal and spatial resolution over the last 1000 and the last 100 years. The longest time frame of analysis will depend on the region. For example, Australian history might span the last 60,000 years, whereas in southern Europe, the last 20,000 years would capture initial colonization since the Last Glacial Maximum (LGM).
2. Understand the connections and dynamics of human and Earth history by testing humans-in-environment systems models against the integrated history. For example, how well do various models of the relationships between climate, agriculture, technology, disease, language, culture, war and other variables explain the historical patterns of human settlement, population, energy use, and Earth system cycles such as global biogeochemistry?

3. Project with much more confidence and skill options for the future of humanity and Earth system dynamics, based on models and understanding that has been tested against the integrated history and with participation from the full range of stakeholders.

A first step toward the development of such an integrated history and future took place at the 96th Dahlem Workshop, which was convened in Berlin, Germany, from June 12–17, 2005. This workshop assembled an interdisciplinary group of 40 top scientists from a range of natural and social science disciplines with the goal of identifying mechanisms and generalizations of how humans have responded to and affected their environment over millennial (up to 10,000 years ago), centennial (up to 1000 years ago), and decadal (up to 100 years ago) timescales as well as a glimpse of the future of the human–environment system. The Dahlem Workshop was the kickoff event for a series of coordinated interdisciplinary research projects around the world that will allow us to learn about the future from the past.

The overall conclusion from the workshop was that human societies respond to environmental (e.g., climate) signals through multiple pathways including collapse or failure, migration, and creative mitigation strategies. Extreme drought, for instance, has triggered both social collapse and ingenious management of water through irrigation. Future response and feedbacks with the human–environment system will depend on our understanding of the past and adaptation to future surprises.

OVERVIEW OF THE BOOK

This volume is divided into five sections, with the overall organizational principle being the timescale at which the analyses are conducted. The approach was to address the collection, integration, interpretation, and analysis of knowledge on human history and environmental change at three complimentary temporal scales for the past—millennial, centennial, decadal—and to bring the same tools to a consideration of the future.

The book begins with an initial section that provides background information on important cross-cutting issues that apply to all of the timescales considered in the volume. Dearing (Chapter 2) presents an overview of the ways in which information is generated, integrated, and analyzed to provide a better understanding of human–environment interactions. In Chapter 3, Costanza addresses data quality as a key cross-cutting methodological issue that is especially important for interdisciplinary studies.

The foci of the other four sections are the three historical time periods and the future. Each section contains the background papers and a group report that synthesizes the findings for the section. The background papers were initially prepared before the meeting to initiate discussion in Berlin; they were subsequently peer reviewed and revised. The group reports are a product of the intense discussions during the workshop as well as the dialog that has continued ever since. Below we briefly summarize the background papers and the conclusions of each of the working groups.

Group 1: Millennial-scale Dynamics

Redman et al. (Chapter 9) sought to go beyond a simple comparison of environmental change data and human activities, to address more fundamental characteristics of their mutual interactions. In other words, what were the contributing conditions or circumstances leading to effective or ineffective responses to climatic change?

We are aware that different cultural elements (social and political structure, traditional practices, and beliefs, to name a few) enable or constrain social responses to the environment. We are also aware that even global-scale incidents (e.g., climate change, major volcanic activity) do not affect all regions equally with regard to either timing or intensity of events. Thus the group began to develop a conceptual model to test how extant societal characteristics and environmental conditions affect societies' ability to cope with climatic change.

These issues are addressed from a number of perspectives. Friedman's analysis (Chapter 8) of hegemonic decline in global systems develops a model describing a long-term process of cyclical expansion/contraction and geographical shift in the center of accumulation with periodic declines and "dark ages" when external limits to social reproduction are reached. The case study of the southern lowland Maya of the Yucatán Peninsula, presented by Scarborough (Chapter 4), is a classic example of the evolution of a human–environment system that was initially highly adaptive and successful but eventually collapsed or failed. The roles of self-organization and heterarchical networks are considered to be especially important in the context of social complexity.

The importance of the nature of the response of a society to both internal and external (environmental) stresses is described by Tainter and Crumley (Chapter 5) in their treatment of the dynamics of the Roman Empire. They particularly emphasize the development of complexity, costliness, and ineffectiveness in problem-solving as a major element in the eventual collapse of the Empire. Hole's analysis (Chapter 6) of the socioecological system of the Khabur River Basin in northeastern Syria shows how even small changes in the environment, such as the timing and amount of precipitation, can have a major impact on societies, while the application of irrigation and fertilization, which buffer against environmental variability, may now strain the resilience of the socioecological system. The exceptionally long trajectory of human–environment evolution in Australia, presented by Flannery, notes a period of great stability in the socioecological system between 45,000 and 5,000 years ago despite the rapid climate shifts of the LGM and the transition to the Holocene. However, this analysis is based on very sparse data and emphasizes the need for much more information to be able to understand better the long human interaction with the Australian environment.

For discussion and analysis, two periods in climate history were chosen for which researchers agree that evidence for regional climate shifts is strong:

1. The 4.2 K Event, a drier and cooler period that occurred between 2200 and 1800 B.C.;
2. The 7th- to 10th-Century Episode, a warmer and drier period, which began in the century before A.D. 1000.

For both periods, discussion centered on societies with detailed archaeological and historical information, including China, India, Egypt, the Near East, Europe, and the Americas. Some of these societies prospered in changed circumstances, some struggled and survived, while others collapsed. By analyzing both the impact on the environment/natural resources and on society (e.g., population, social and political organization, agriculture, trade, technology, religion), initial analyses can begin to discern how structure, practices, and attitudes may respond to ongoing and future environmental change.

The group concluded that a simple, deterministic relationship between environmental stress, for example, a climatic event, and social change cannot be supported. Redman et al. (Chapter 9) note that there are organizational, technological, and perceptual mechanisms that mediate the responses of societies to environmental stress, and that there may also be time-series sequences and lags to societal responses. Despite the apparent complexity of the relationship between environmental stress and societal response, the group concluded that a case-study approach could tease out some useful regularities or parallels in the evolution of the human–environment relationship.

Group 2: Centennial-scale Dynamics

Both globalization and global environmental change have deep roots in humanity's relationship with nature over the past millennium. While we often associate the term "global change" with the greenhouse gas warming evident in the last decade, changes of a global scale were put in motion over the past 1000 years. Historians, archeologists, and ecologists collaborated in this group to examine long-term patterns and trends in society's relationship with the environment. The last 1000 years was examined in detail because of the remarkable extension of the footprint of humanity on the Earth during that period. Important phenomena included a rise in human population, the strengthening of nation-states, the global transfer of European inventions and values, the beginning of industrialization, and the rise of global communications. The last 1000 years is also particularly interesting because this was a time when broad swings in temperature as well as clusters of extreme weather events arguably changed the trajectory of history. For example, the 14th century in Europe saw the end of the Medieval Warm Period; particularly during the period from A.D. 1315–1317 western Europe experienced a combination of rainy autumns, cold springs, and wet summers that led to crop failures and a dramatic slowdown in urban expansion. These early Europeans were further subjected to the last major locust invasion (1338), the "millennium flood" (1342), and the coldest summer of the millennium in 1347. From 1347 to 1350 the "Black Death" devastated populations. Dearing et al. (Chapter 14) suggest that the clustering of extreme events in the 14th century fundamentally undermined social order and was a key factor in a major wave of anti-Semitic pogroms and systematic discrimination.

The background papers for this section highlight more detailed examples of the interplay between environmental variability and human societies. Grove (Chapter 10) focuses on the fascinating effects of the exceptional 1788–1795 El Niño event, which reverberated around the world in places as far apart as the first British colonial settlement in Australia, the Indian monsoon region, Mexico, and western Europe. An even more detailed exploration of the interplay between environmental stress and human response is found in Pfister's exploration of the impacts of the Little Ice Age (A.D. 1385 to 1850) on food vulnerability in the Bern region of Switzerland (Chapter 12). His analysis is particularly relevant for the issue of climate change and contemporary society as Pfister raises the concepts of adaptive and buffering strategies as important ways through which humans attempt to mitigate social vulnerability.

The contributions by Hassan (Chapter 11) and van der Leeuw (Chapter 13) address more general aspects of the human–environment relationship at the centennial timescale. Hassan discusses the tension between the modern nation-state and the emergence of multinational corporations and international political institutions. He argues that understanding the present suite of environmental and societal problems must be based on a careful analysis of the history of the past several centuries. An innovative approach is taken by van der Leeuw to describe the past millennium. Rather than focusing on historical detail, he treats the underlying socionatural dynamics and attempts to reconstruct the past from an understanding of present-day dynamics.

The discussions of Dearing et al. (Chapter 14) were open-ended and addressed perceptions, open questions, and controversies within the centennial time frame. The effort was aimed at critical and generic issues of methodology and understanding rather than historical review. Two main conclusions emerged. First, the present nature and complexity of socioecological systems are heavily contingent on the past; we cannot understand the present condition without going back centuries or even millennia. An important implication is that societal actions today will reverberate—in climatic and many other ways—for centuries into the future. Second, the records of the dynamics of past socioecological systems are immensely rich and will provide an excellent base for exploring the contemporary phenomenon of global change.

Group 3: Decadal-scale Dynamics

Hibbard et al. (Chapter 18) addressed the rapid changes in population, economic growth, technology, communication, transportation, and other features of the "human enterprise" that typified the 20th century. The past century also witnessed several sharp discontinuities in the evolution of socioecological systems: two global-scale wars and the Great Depression. The group considered the growing imprint that human numbers and activities were having on the environment, with the clear impacts at the global scale being an important feature of this time period. This time period is also typified by a vast array of socioeconomic and biophysical data; it marks the first century for which instrumental records of many environmental parameters have become available and for which detailed statistical records of many human activities have also been collected. Thus, there is a rich and growing array of data and information to underpin analyses of the rapidly changed human–environment relationship.

The background paper by Mantua (Chapter 15) outlines the major changes that occurred in Earth's natural systems through the 20th century. These include not only changes in atmospheric composition and consequent changes to climate, but also the growing human imprint on the cycling of critical elements through the planetary system and on the structure and composition of the terrestrial and marine biospheres. McNeill (Chapter 16) provides a complementary account of the dynamics of global political economy that helped to drive the environmental change documented in Mantua's paper. An intriguing element of McNeill's paper is the description of the global struggle between efforts to build centralized, imperial economies (e.g., Germany, Japan, the former Soviet Union) and those to build an integrated, international economy (e.g., United Kingdom, United States).

The issue of desertification (i.e., the widespread reduction of productivity in arid and semiarid lands caused or exacerbated by human activities) provides a contemporary example of an integrated human–environment system undergoing significant change. Lambin and colleagues (Chapter 17) outline the biophysical and socioeconomic linkages that lead to dryland degradation and the far-reaching impacts that this degradation has on human welfare. They describe a new paradigm for understanding desertification, a framework consistent with the IHOPE goals of building a more integrated understanding of how human–environment systems evolve in highly interactive ways.

The group's discussions (Chapter 18) highlighted the most remarkable phenomenon of the 20th century: the "Great Acceleration," that is, the sharp increase in human population, economic activity, resource use, transport, communication and knowledge–science–technology that was triggered in many parts of the world (North America, Western Europe, Japan, and Australia/New Zealand) following World War II and which has continued into this century. Other parts of the world, especially the monsoon Asia region, are now also in the midst of their own Great Acceleration. The "engine" of the Great Acceleration is an interlinked system consisting of population increase, rising consumption, abundant energy, and liberalizing political economies. Globalization, especially an exploding knowledge base and rapidly expanding connectivity and information flow, acts as a strong accelerator of the system. The environmental effects of the Great Acceleration are clearly visible at the global scale: changing atmospheric chemistry and climate, degradation of many ecosystem services (e.g., provision of freshwater, biological diversity), and homogenization of the biotic fabric of the planet. The Great Acceleration is arguably the most profound and rapid shift in the human–environment relationship that the Earth has experienced.

Toward the end of the 20th century, there were signs that the Great Acceleration could not continue in its present form without increasing the risk of crossing thresholds and triggering abrupt changes. Transitions to new energy systems are required. There is a growing disparity between wealthy and poor, and, through modern communication, a growing awareness by the poor of this gap, which has created a potentially explosive situation. Many of the ecosystem services upon which human well-being depends are degrading, with possible rapid changes when thresholds are crossed. The climate may be more sensitive to increases in carbon dioxide and may have more in-built momentum than earlier thought, raising concerns of abrupt and irreversible changes in the planetary environment as a whole.

Group 4: Anticipating the Future

What we know from investigations of the past is that there are circumstances when a society is resilient to perturbations (i.e., climate change) and others when a society is so vulnerable to perturbations that it will be unable to cope and may be severely affected or even collapse (Diamond 2005). To use this information to meet the challenges of the future, we need to construct a framework to help us understand the full range of human–environment interactions and how they affect societal development and resilience. We now have the capacity to develop this framework in the form of more comprehensive integrated models, using different but complementary scientific approaches, ranging from systems dynamics models to agent-based models to simulation games to scenario analysis. This will allow us to increase our understanding of the major components and behavioral characteristics of both past and present human–environment interactions. Although the future will differ from the past, insights from modeling and analysis of the rich array of well-documented integrated historic events can be used to structure, test, and further develop these models.

The background papers to this section present an overview of the variety of tools that can provide insights into the future. A survey of the development of Integrated Global Models is given by Costanza et al. (Chapter 21),who analyze seven such models in terms of characteristics, performance, and limitations. In

Chapter 19, de Vries focuses on the use of scenarios for gaining insights into the future. He reviews two important sets of scenarios involved with large global change projects and suggests ways in which the next set of scenarios can be improved. One of the first model-based projections of the future was the Limits to Growth project (Meadows et al. 1972). In Chapter 20, Dennis Meadows, one of the original members of the team, compares the original 1972 projections to what is observed now, drawing on the recent report of the Millennium Ecosystem Assessment (MEA 2005). He also provides a retrospective analysis of some of the early criticisms of the Limits to Growth project.

The group's discussions (Young, Leemans et al., Chapter 22) centered around the fundamental question of how historical narratives and models of human–environment systems can generate plausible insights about the future. In an attempt to gain insights from the past, the group considered the interplay between regularities in the behavior of the Earth system and contingent events. Regularities in the functioning of the Earth system (e.g., the laws of thermodynamics and conservation of mass) allow some level of predictability of the future. However, when considering complex socioecological systems, the role of contingent events becomes important; these "chance events" limit the precision with which the future can be predicted. Nevertheless, an array of different modeling approaches—some focused strongly on the biophysical aspects of the Earth system (e.g., General Circulation Models of climate) and others centered on socioeconomic aspects (e.g., models of the global economy)—have been developed for projecting Earth system behavior into the future. The group concluded that a comparison and synthesis of results from different modeling approaches may provide a more robust strategy than the reliance on a single approach. Alternatively, various modeling approaches could be integrated into a single hybrid simulation framework.

SYNTHESIS OF WORKING GROUP REPORTS

Each of the discussion groups provided fascinating information on and many useful insights into the evolution of socioecological systems through time (see Chapters 9, 14, 18 and 22). Here we offer a synthesis of the group reports toward (a) achieving a deeper understanding of human–environment interactions by considering various timescales, (b) developing common themes across all timescales, and (c) defining some of the most important research questions for IHOPE to tackle in the future. We use the term *socioecological system* to refer to human societies embedded in and interacting with the natural world around them. We often differentiate the two major aspects of socioecological systems and use a variety of terms (e.g., *human enterprise, society, civilization*) to refer to the human part and another set of terms (e.g., *nature, natural world, environment*) to refer to the rest of such systems.

Socioecological Systems from Different Time Perspectives

The analysis of socioecological systems around a range of timescales—from millennial through centennial and decadal and into the future—provided a rich basis for a deeper understanding of human–environment interactions. For example, over the past several millennia, humans moved from hunter–gatherers to agriculture and civilizations, and developed a stronger capacity to manipulate nature, at least at the local and regional level. However, impacts in the reverse direction of this human–environment relationship (i.e., impacts of natural environmental variability and change on human societies) were stronger and, for the most part, dominated the relationship before the Industrial Revolution. Over the past several centuries, the two-way interactions between humans and the natural world, especially at larger spatial scales, have become more balanced. The imprint of humans at large regional scales became clearer and the first signs of significant global impact appeared. The Great Acceleration carried this trend dramatically forward. Humans are now a global force that rivals the great geophysical forces of nature in many aspects. A feature of the Great Acceleration that points toward the future evolution of socioecological systems is the fundamental role of technology in mediating the interactions between humans and the rest of the natural world.

Another way of looking at these trends in the human–nature relationship is to contrast the connectivity of humans to nature with the size and power of the human enterprise. One end point is represented by hunter–gatherers, who are strongly connected to nature but are small in numbers and have a weak capacity to impact the natural world at large scales. Agrarian societies evolving into the early civilizations represent an interesting midpoint, in which the human enterprise had become large and clever enough to impact the natural world significantly at more than local scales. On the other hand, early human civilizations still retained a strong connection to the natural world through their direct and visible reliance on ecosystem services for their success and well-being. The other end point is the current highly technological, globalizing society, which is less overtly (or obviously?) connected to nature than ever before but also more numerous and economically powerful than ever before. The human enterprise has grown to enormous size and strength. It can (and does) insulate people from both the direct knowledge and experience of the ecosystem services on which we all ultimately still depend and from the many global-scale impacts of the burgeoning human enterprise on the natural world.

Insights can also be obtained from examining the evolution of socioecological systems from a particular time perspective, but in a broader context. For example, a particular strength of the millennial-scale analysis is that it addresses the importance of the long-term evolution of societies. The analysis is able to go beyond shorter-term historical cycles to multiple completed cycles of the rise, spread, and eventual decline of civilizations. This raises some intriguing questions that would not necessarily arise from examining shorter timescales. How do societies reorganize after a decline or collapse? What are some of the more important "slow processes" (cf. resilience perspective in the next section) that are barely discernible at shorter timescales but can dramatically affect the success or failure of socioecological systems? Are there particular points in the evolution of socioecological systems at which slow processes flip from being adaptive to being destabilizing?

Finally, examining socioecological systems across multiple timescales can identify the antecedents further back in time of major phenomena that occur in a particular era or time. A good example is the Great Acceleration (ca. 1950 to the present). The phenomenon is well described from a decadal perspective but the antecedents, especially in the socioeconomic sphere (e.g., globalization, fossil fuel use, increased information flow), go well back into the previous centuries. Examining the Great Acceleration from a longer time perspective also uncovers the "stillborn" Great Acceleration of the late 19th and early 20th centuries. Most of the ingredients for an acceleration of the human enterprise were apparent, but the decline and collapse of many countries and regions during the period of 1915–1945, due to economic depression and two world wars, delayed the phenomenon for a half-century. On the other hand, this could also be interpreted from a resilience perspective as two adaptive cycles of the modern, globalized socioecological system.

Common Themes across Timescales

Although the four working groups approached their analyses of socioecological systems from very different perspectives, several common themes emerged from the group reports. The most important of these are described below.

1. There is a general movement away from simple cause-and-effect paradigms as a credible explanatory framework. There is a strong consensus that we are dealing with complex, adaptive, integrated socioecological systems that often defy simple cause–effect logic in their behavior. Complex systems may exhibit multiple interactions between apparent drivers and responses where the direction and strength of interaction are not necessarily explicable in terms of simple, direct, and linear causative links; there may be internal dynamics that drive system changes. IHOPE studies, therefore, will need to encourage the use of concepts from complexity science, including linear and nonlinear dynamics, feedback, thresholds, emergence, historical contingency, and path dependence as well as the application of nonlinear simulation tools, spatially explicit and agent-based models to simulate relevant phenomena (cf. Young, Leemans et al., Chapter 22).
2. A dichotomy often arises between explanatory power and predictive success. Could anyone have predicted the collapse of the Classic Maya

civilization a century before it occurred? Could anyone in 1900 have predicted the evolution of human societies, especially their relationship to the natural world, through the 20th century? In both of these (and other) cases, we have impressive explanatory power in describing what unfolded, but that does not yet translate into an ability to predict the future trajectories of complex socioecological systems. The ability to influence the future comes with a loss of ability to predict it. A better way to look at it is that IHOPE can use a deeper understanding of the past to help us create a better future, rather than to predict the future.

3. Resilience theory, particularly that aspect which focuses on adaptive cycles, played a strong role in the discussions of all working groups. For example, Dearing et al. (Chapter 14) used the concept of "risk spirals" to describe an inadvertent loss of resilience through time. They defined a risk spiral as being derived from "…a transformation of environmental complexity into social complexity. The key point is that while human actions often succeed in reducing specific risks, these efforts also create qualitatively new risks at a larger spatial scale and/or a longer time frame." The notion of risk spirals points to a dangerous positive feedback loop. As human societies become more complex, they are less able to withstand shocks from the natural world and, ironically, in the process of making themselves more complex, societies inadvertently and (often) unknowingly change natural systems in ways that make these systems more prone to abrupt changes or extreme events!
4. A critical aspect of any society is the trade-off between short-term production and long-term resilience or sustainability. These values are often in conflict. In general, there is a need to keep production systems well below theoretical carrying capacity to avoid a severe drop in resilience. Cultural traditions have played an important role in building long-term resilience by acting as a brake on short-term production that would damage or diminish resilience and long-term sustainability. During the Great Acceleration, many of these cultural traditions dissipated such that resilience and long-term sustainability may be adversely affected.
5. The role of feedback processes is crucial in complex socioecological systems (and a big reason why simple cause-and-effect paradigms often have little explanatory power). A potentially dangerous positive feedback loop was mentioned above. Are there, however, counteracting negative feedback loops that can generate increased resilience in socioecological systems? For example, is there a general self-regulating feature in human civilizations that acts to lessen environmental stresses when they become apparent? Are the "decelerating trends" we see now in some aspects of the contemporary human enterprise part of a self-regulating feature that will slow the Great Acceleration?
6. Finally, the group reports point to a number of phenomena that are difficult to model or project but are nevertheless extremely important:

- Temporal dynamics, especially rates of change in critical phenomena. This includes thresholds, nonlinearities, and abrupt or extreme events (in both human and natural parts of the system). Are we approaching global-scale thresholds in contemporary socioecological systems, especially in either the natural or the human part of the system? Is the Earth system shifting to another state? Can increasing resource scarcity and environmental impacts trigger a collapse of the global economic system?
- Contingencies or contingent events—chance events can strongly affect the trajectory of a socioecological system—and legacies from the past (or path dependencies) are very important. An example of the latter is the contemporary energy system, which cannot be changed immediately in response to climate change.
- The phenomenon of "collapse." This is a central concept in IHOPE and probably the most critical question facing current society, but it needs to be defined and used carefully. What do we mean by collapse and what can we learn from past collapses?

Research Challenges

The group reports individually set out a number of research questions relevant to their particular timescales. Here we explore common threads among those questions to develop a single set of IHOPE research challenges that will need to be met regardless of the timescale or particular aspect of IHOPE of interest.

1. Data on the behavior of socioecological systems are critically important for IHOPE but vary enormously in quality, selection, interpretation, resolution, dating/chronologies, and unevenness (cf. Costanza, Chapter 3). The amount of data rises dramatically as we approach the present, and this could easily distort analyses.
2. There is an issue regarding the comparability of social and environmental data. On long timescales, proxy data are used to describe important social and environmental characteristics. For example, artifact assemblages are used to reconstruct trade relationships, and pollen in lake sediments are used to reconstruct vegetation and climate history. Although interpretation protocols within a discipline lend rigor to analyses of a single aspect of the socioecological system, we need better theory and models to integrate different proxy measurements that vary in spatial and temporal resolutions.
3. There is often a dichotomy in research approaches (reductionist vs. systems-oriented) that can lead to tension within research teams and thus pose major challenges to interdisciplinary research projects. IHOPE studies need to adopt a range of alternative explanatory frameworks, embracing conventional scientific positivist approaches as well as

discipline-specific protocols and more systems-oriented approaches. However, a key issue for IHOPE is the evaluation of explanations and the realistic appreciation of uncertainty. How we learn from the past takes different forms (cf. Dearing, Chapter 2): the type and range of data sources, the different disciplinary conventions, and the nature of conceptual and predictive models used imply that there is no single method to determine the quality and certainty of explanations. In some contexts, it may be possible to utilize a hypothesis-testing approach, whereas in others the ability to falsify hypotheses may be severely restricted. In many historical studies, the use of approaches that argue from the perspective of mutual internal consistency or weight of evidence may be more appropriate. For some disciplines, it may be necessary to construct a set of interpretative protocols for IHOPE studies.

4. In analyzing socioecological systems or simulating their behavior into the future, biophysical laws governing aspects of nature can provide an "envelope of regularities" in projections or analyses (but complex natural systems can also have strong nonlinearities). This broad envelope of regularities can define the "environmental space" within which human societies operate, but contingent events, which are difficult or impossible to predict, often determine the trajectories of socioecological systems within that space and are thus crucially important to how the future will actually unfold. We need to know what the range of possibilities is, as we continue to create the future.
5. Comprehensive models of the integrated Earth system are still in their infancy (cf. Costanza et al., Chapter 21). Nearly all models begin with a strong emphasis on either the natural or the human part of socioecological systems. There is a strong need for more balanced, hybrid approaches that can take on the research challenges outlined above. The insight, data, and models generated from the IHOPE activity through the close collaboration between environmental historians, archeologists, paleoenvironmentalists, ecologists, modelers, and many others will allow the construction and testing of new ideas about humans' relationship with the rest of nature. It will also allow the calibration and testing of a new generation of integrated global Earth system models (cf. Young, Leemans et al., Chapter 22) that contain a range of embedded hypotheses about human–environment interactions.

BIG QUESTIONS

IHOPE is poised to address a number of critical research and policy questions affecting the life of all humans on Earth. This book takes a "first cut" at those questions. It is thus fitting to conclude not with answers, but with questions. The big, general questions for IHOPE (consistent with the long-term goals stated earlier) can be summarized as follows:

- What are the complex and interacting mechanisms and processes resulting in the emergence, sustainability, or collapse of coupled socioecological systems?
- What are the pathways to developing and evaluating alternative explanatory frameworks, specific explanations, and models (including complex systems models) with and against observations of highly variable quality and coverage?
- How can we use this integrated knowledge of human perceptions of and behaviors in the environment in the past to understand and *create* the future?

This Dahlem Workshop Report addresses these questions from a number of perspectives and time frames, but it obviously represents just the beginning.

It has been said that if one fails to understand the past, one is doomed to repeat it. IHOPE takes a much more "hopeful" and positive attitude. If we can *really* understand the past, we can *create* a better, more sustainable and desirable future.

REFERENCES

Diamond, J.M. 2005. Collapse: How Societies Choose to Fail or Succeed. New York: Viking.

Kirch, P.V. 2005. Archeology and global change: The Holocene record. *Ann. Rev. Env. Resour.* **30**:409–440.

Meadows, D.H., D.L. Meadows, J. Randers, and W.W. Behrens. 1972. The Limits to Growth. New York: Universe.

MEA (Millennium Ecosystem Assessment). 2005. Millennium Ecosystem Assessment Synthesis Report: A Report of the Millennium Ecosystem Assessment. Washington, D.C.: Island. http://www.millenniumassessment.org/

Steffen, W., A. Sanderson, P. Tyson et al. 2004. Global Change and the Earth System: A Planet Under Pressure. IGBP Global Change Series. Heidelberg: Springer.

2

Human–Environment Interactions

Learning from the Past

John A. Dearing

Department of Geography, University of Liverpool, Liverpool L69 7ZT, U.K.

ABSTRACT

Largely from the perspective of paleoenvironmental science, this chapter addresses the issue of how past records of human–environment interactions can provide valuable information for deriving strategies for sustainable management of human-dominated landscapes. It contrasts the different approaches to learning from the past in the sciences and humanities and suggests a simple typology of the different types of learning: trajectories and baselines; spatiotemporal variability and scaling; process responses; and complex system behavior. It argues that there are three research priorities requiring further effort and international organization: (a) the development and testing of theory that pertains to human–environment interactions; (b) the integration and regionalization of case studies and time series; and (c) the simulation of future human–environmental interactions using tools and frameworks that allow testing against historical records. Key questions are identified and shown at the end of each subsection.

> Without a knowledge of our history, we cannot understand our present society, nor plan intelligently for the future (McCullagh 1998, p. 309).

INTRODUCTION

This discussion paper[1] attempts a brief review of the ways in which useful information about human–environment interactions can be gained by studying the past. It is essentially the personal view of an environmental scientist whose perspective has evolved through a career dealing with the reconstruction of past environments from the analysis of sediments. Thus, while it attempts to cover

[1] Parts of the paper are drawn from Dearing (2006) and Dearing et al. (2006a, b).

diverse approaches to the study of Earth and world systems, it is biased toward the physical sciences. Its primary aim is to draw out a few categories of "learning from the past" for discussion, focusing particularly on common and contrasting modes of learning across disciplines. Key questions are identified at the end of each section. My starting point is to consider the different approaches taken by the humanities and natural sciences in terms of dealing with history, and hence past human–environment interactions.

Nature of Truth

The natural scientist reading essays and accounts of the philosophy of history cannot fail to be impressed with the tradition of intense debate about the accuracy and completeness of historical information, and the striking influence that certain historical theories have had on culture and politics. For some world philosophies, such as Marxist and Popperian, the central tenet is the value that can be placed on historical knowledge itself. In contrast, the philosophical debate about the development of the physical world appears to be far less. The environmental equivalent of the sociopolitical Grand Theories might exist in the form of Darwinian evolution and Milankovitch's orbital cycles, but these are today far less contested. Does this difference essentially stem from the perceived subjectivity or intractability of human views and actions contrasted with the objectivity of factual records of past environments? Is historical information about human actions intrinsically more unreliable and, thus, debatable? From the perspective of the humanities, McCullagh (1998) reviews methods and attitudes of assessing the truth of historical information, considering the constraints of evidence, culture and language, cultural relativism, and postmodern insights. In some ways the similarities between disciplines are clear. Both the historian and the paleoecologist have to interpret raw materials: both need to know the contexts; there may be alternative interpretations. Further, each may argue that the material should not be viewed as a literal record but one that is presented according to the authors' beliefs, data and information sampling and availability, and data processing. What perhaps is different, and surprising to some environmental scientists, is the degree of theorizing and philosophizing of approaches. For example, Collingwood's constructionist theory of history (in Gardiner 1959), in which the inadequacy of historical information demands (re)constructions rather than descriptions, is viewed as not so much a theory in the paleosciences but as a logical, rational, and dominant *modus operandi*. Clearly, there is general acceptance that sometimes there is a need for different treatments of truth, for example as "coherence with existing beliefs" in the humanities or as "consensus reached by rational enquirers" in both the humanities and environmental sciences (McCullagh 1998). Are there, however, other reasons that divide attitudes to history than simply the different levels of enthusiasm for philosophical argument?

Reductionism and Laws

One issue that may divide the disciplines is the level at which reductionism has played a part in explaining phenomena and formulating laws. If the natural scientist believes that factual records of past natural environments are more reliable, it may be because there are generally accepted laws for the movements of particles, matter, and energy that allow coherence between findings and theory to be established across a broad range of scales. Following Wilson (1998), we may ask whether the problems of explanation in the humanities lie with the inability to seek explanations of human actions through reductionism to the same low level as in the natural sciences (Figure 2.1). This is not to say that historically the humanities have not considered the possibility of explaining human affairs through discoverable laws—as exemplified by Hobbes' *Leviathan*,

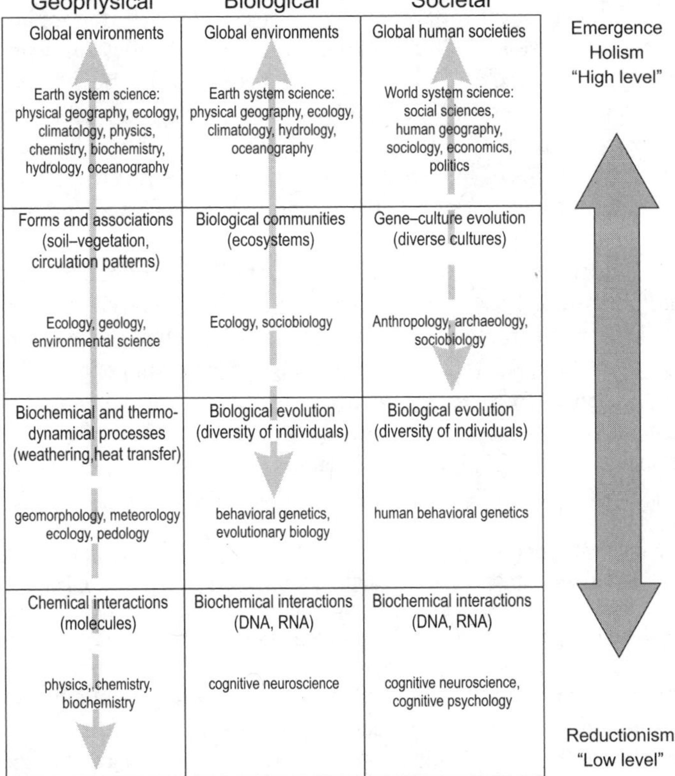

Figure 2.1 Hierarchies of explanatory rules in geophysical, biological, and social systems arranged in three columns from low level (reductionism) to high level rules (emergence/holism). The vertical arrows show the generally accepted span of explanatory rules for each system with dotted lines suggesting possible extensions in the foreseeable future (developed from Wilson [1998] and extended by the author).

Condorcet's *Sketch*, or Tolstoy's belief that laws derive from the individual tendencies of humans—but to observe that this has not been generally successful. What is apparent is that the more successful or common use of history in terms of application has been to suppose that history has "meaning" in the sense of preordination or a hidden hand (i.e., historicism): in the sense of Hegel and Marx, to observe a certain trajectory and to speculate upon its continuation into the future (cf. Gardiner 1959). "Marxism" and "hidden hands" may be considered as outmoded concepts but, as considered below, the science of complexity suggests that we should not automatically dismiss either the opportunity to find "laws of society" or the value of studying repeated patterns of emergent phenomena, such as trajectories of civilizations (cf. Friedman, this volume).

Application of Understanding

We might also analyze the difference in influence achieved by the application of theories based on history. Political, social, and cultural theories based on history have clearly affected the structure and governance of nations, but what about the impact of scientific theory? One could argue that fewer people have been directly affected by Darwin's biological theory of evolution than by the indirect political ramifications of the derived social Darwinism. It certainly seems the case that current projections of global climate change represent the first use of scientific theory based on historical analysis and testing that engages, largely via the media, directly with the lives of a major proportion of the modern world population. Perhaps it is no coincidence that one of the most influential aspects of the climate change argument is the "hockey-stick" graph of reconstructed temperatures over the past few centuries: utilizing the power of perspective to educate and influence. However, arguments for the value of *learning from the past*, as opposed to merely *knowing the past*, are often not as clear as those pertaining to the "hockey-stick" graph or have been ignored. For example, McCullagh's (1998, p. 304) statement:

> The unique value of history lies in explaining the origin and value of all social institutions, cultural practices and technological advances we have inherited…in the past, it is indeed vital to recognize the conditions which enable them [institutions] to function as they did, in case those conditions exist today or have changed.

implies that a full description and explanation of the past (i.e., knowledge of the past) is sufficient in itself. Much research, from social history to paleoecology, has been driven by the disciplinary debates—appropriate methodologies, new techniques, and the alternative explanations—rather than the development of theory about how humans interact with their environment. In fact, we may have devoted more time and effort to describing the past than analyzing it for the lessons to be gained. Where theory has emerged, it has tended to take either a predominantly cultural or physical line with little attempt to understand fully the true nature of interaction. Moreover, social and physical sciences have now embraced the implications of complexity science. As a result, theories like environmental determinism seem outmoded oversimplifications of reality.

Current global change shows accelerating trends in many social and physical phenomena driven by demography, technology, culture, and climate. At every point on the world's surface these drivers interact, usually in complex ways. As a global scientific community we strive to provide realistic advice and guidelines as to the optimal strategies for adaptation and sustainable management. What follows is a discussion about how we can learn about current and future human–environment interactions from the past by adopting frameworks and approaches based on historical ecology (Crumley 2006). It does not follow that understanding and explaining the past means that we can predict the future, but it does mean that we might be able to identify, justify, and rank alternative futures for humanity to work toward. Below I briefly review and exemplify different ways that this might be done. While the following sections represent epistemological categories, they are mainly for convenience: in practice, they are often combined.

TRAJECTORIES AND BASELINES

Our knowledge of world and Earth system history is highly variable in time and space. All documentary, reconstructed, and instrumental records are, to different degrees, incomplete, discontinuous, and inaccurate. For Earth systems, the growth of modern science has not been matched by the monitoring of those environmental processes and conditions that are now seen as essential for generating strategies for sustainable environmental management. Meteorological records for major regional stations and hydrological records for the largest rivers are often available for the last 100 years but more locally, and for time series of other conditions such as vegetation cover, biodiversity, biogeochemical cycles, phytoplankton populations, and atmospheric pollution, records are often nonexistent or significantly shorter. Some long documentary records provide dates of events, such as the famous phenological series from China, or semi-quantitative information such as the Nile River flood height, stretching back into antiquity, but these are exceptional. Environmental reconstruction of processes and conditions can substitute for and extend many of these records (Oldfield and Dearing 2003), but clearly, as in the case of crop yields, not all. The quality of our documented and archaeological histories of societies and culture is similar, usually becoming more generalized and more speculative as we reach back in time.

Where the issue is about the sustainability of ecosystem processes and services in the face of human pressures, past records are already being utilized to good effect in order to demonstrate antecedent change. For example, Steffen et al. (2004) summarize the acceleration of 20th-century changes in several sets of human activities and impacts on the Earth system. This analysis has been extended through the Syndrome Approach (Schellnhuber et al. 1997; Lüdeke et al. 2004) to defining functional patterns of regional human–environment interactions, such as the Sahel, Dust Bowl, and Green Revolution syndromes. For specific processes, particularly for those that are important locally rather than globally, a longer timescale may reveal strongly contrasting trajectories. For example, reconstructed erosion records over the past few hundreds of years show a wide range of curve shapes: accelerating in Papua New Guinea, declining in southern Yucatán, and stationary following initial sharp rises in Michigan (Dearing et al. 2006a). These records in themselves provide a basis for defining a typology of current trends (in this case, for soil erosion) that can contribute to any evaluation of modern sustainable land-use practices. The reconstructed trajectories for a single region, southern Sweden (Figure 2.2), show the diversity of human and environmental "parallel histories" available from a rigorous analysis of documentary, archaeological, instrumental, and sedimentary records (Berglund 1991).

Perhaps the simplest application of studying trajectories is to use past conditions as a goal for the management of the present. This type of analysis has become an increasingly common part of environmental regulation, where there is often a demand to identify and describe a "baseline" or "pre-impact" condition that can be used as a reference condition or rehabilitation target. Such demands commonly exist for air pollution, nature conservation, biodiversity loss, forest management, fire suppression, and water quality (e.g., EC Water Framework Directive). The concept of "reference conditions" is now particularly well-developed in studies of lake water quality where the chemical and biological status of a lake prior to recent human impact can be inferred from the lake sediment record (Battarbee 1999). This approach is more difficult to apply in terrestrial ecosystems. For example, Bradshaw et al. (2003) review the paleoenvironmental evidence for the role of grazing mammals on forest structure and conclude that no pre-impact baseline for contemporary management targets actually exists within the Holocene period. One common, and sometimes controversial, conclusion from this kind of analysis is that selecting a pre-impact or natural condition is not straightforward; it may even be unrealistic.

Key question: Can we characterize the nature of change in a region by using the trajectories of "parallel histories" to generate typologies of change in human–environmental states?

SPATIOTEMPORAL VARIABILITY AND SCALING

Ideally, reference to historical points should not assume static environments but rather dynamic systems. Thus, one important type of analysis is to define an envelope of spatial and temporal variability. The paleoenvironmental sciences routinely reconstruct past frequency and magnitude time series to compare with

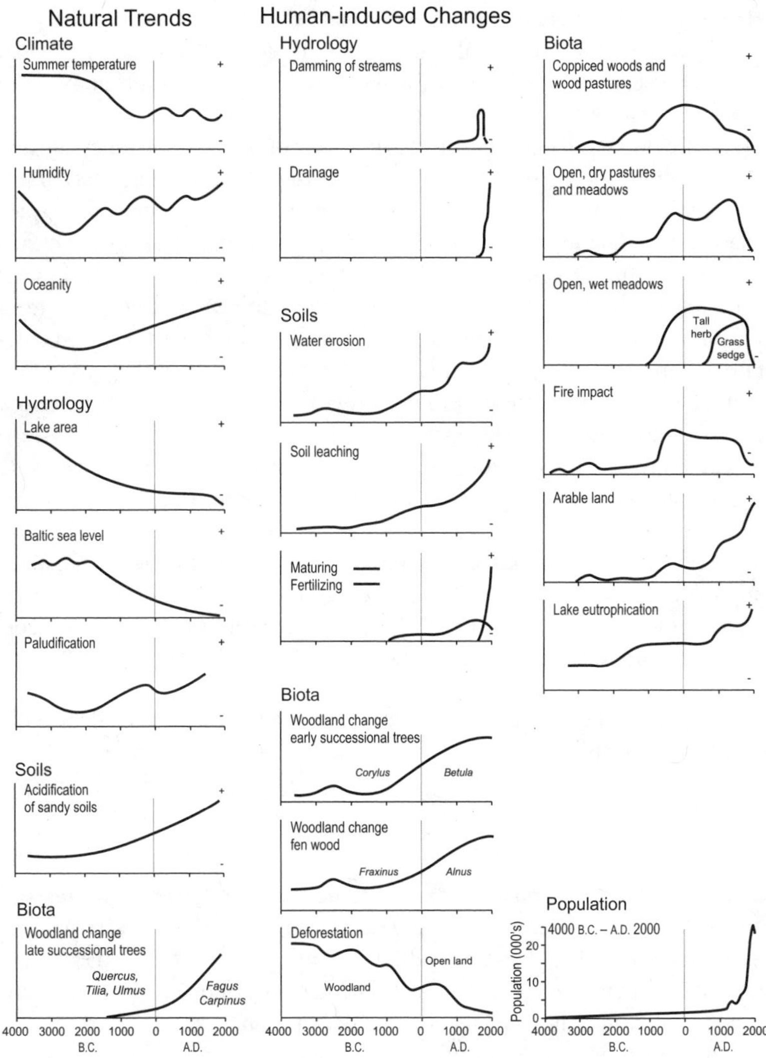

Figure 2.2 Parallel histories: trajectories of human actions and environmental conditions over the past 6000 years for southern Sweden (Berglund 1991).

modern conditions. For example, Nott and Hayne (2001) demonstrate that the recurrence interval of "super-cyclones" along the Great Barrier Reef is an order of magnitude shorter than had previously been calculated, using the period of instrumental measurements. Compiling separate time series from different sites provides an alternative way of observing spatiotemporal variability. For example, historically reconstructed fire data are now routinely used to define optimum fire suppression strategies (Swetnam et al. 1999).

However, the problem of scaling is one that lies central to linking local case studies to global processes. Ecological variability tends to increase as spatial and temporal scales become smaller, and our understanding of the controlling factors on the variability is often significantly modified by the scale of observation (e.g., Levin 1999). For time, there is the issue of defining the timescale that is relevant to the problem of concern. Over what timescales are the effects of soil conservation measures observed? Which particular flood frequency in the past resonates with climatic variation and which with the history of deforestation? In terms of space, the upscaling of cumulative local changes to the global system and the downscaling of projected impacts at a continental scale (e.g., from global climate models to local environments) present some of the greatest challenges to Earth system science. Most of our knowledge about the past comes from case studies with little uniformity in terms of spatial scale. It therefore seems sensible to promote the integration of human–environmental responses to "uniform" impacts in case studies across spatial gradients in order to generate new understanding about spatial scaling. For example, Dearing and Jones (2003) compiled past lake sediment accumulation rates in a number of catchments to calculate the effects of catchment size on the magnitude of erosional response to disturbance. Their data exhibit a spatial scaling control that seems to transcend other environmental factors, like climate. Still, examples of this sort of spatiotemporal scaling using paleodata are uncommon.

Key question: How best to integrate case studies within a region in order to gain new metadata for spatial and temporal controls on process responses?

PROCESS RESPONSES

Causation, explanation, and insight are often derived through inductive reasoning using corroborative, correlative, and converging lines of evidence from parallel sets of records. This may involve, for example, the use of instrumental and documentary records to provide independent data for external forcings, like climate and human activities, and the use of paleoenvironmental or historical data for response records. This also applies to postulated human–environment interactions from local to global scales. For example, the strength of Ruddiman's recent theory (2003) that global climate was affected by early human impact rests to a large extent on visual correlations between independent data for forest regrowth driven by epidemics and minima in the CO_2 ice record. Learning from the past in this context is often implicit: through past records we learn about the functioning of the system in question for which the present is simply the latest point in time. An exception is the use of analogs, where it is assumed that a past set of conditions closely resembles a present state, or projected future state. Deevey (1969, p. 40) stated that "where time is required for an experiment there's no substitute for history," arguing that the power of historical perspectives included learning from analogs of modern conditions. This line of argument has also been convincingly used by archaeologists and anthropologists to demonstrate the multidirectional nature of human–environment interactions: the vulnerability of past human societies and civilizations to natural climate change or events contrasted with the self-imposed impacts on support systems arising from unsustainable practices and positive feedback (e.g., Redman 1999; Diamond 2005). Such case studies clearly demonstrate the interrelatedness of human actions and biogeophysical processes, and can serve to dismiss the notion of absolute environmental determinism. They are strong conveyors of messages about unsustainable practices and the vulnerability of human society. However, we should be cautious in using them as analogs to inform the construction of mitigation or adaptation strategies to current and future stresses because the decision-making processes in past case studies can usually only serve as a basis for speculation. In this sense, Collingwood (in Gardiner 1959) saw history as a sequence of actions where the job of the historian was the study of the "thoughts" behind the actions. May (1973) took this idea further by analyzing the role of history on 20th-century U.S. foreign policy from the documented viewpoints of the crucial actors. He showed that foreign policy is often influenced by what history apparently teaches or portends, but that it makes wrong decisions because the past is an inappropriate analog for the present. Analogs are also used erroneously, as when trajectories are extrapolated into the future without qualification, or used selectively to support a moral judgment.

Overall, the task of understanding human–environment interactions through an inductive cause-and-effect paradigm may not be realistic simply because of the inability to understand the cognitive processes behind individual human actions (Wilson 1998). In this sense, the real value of inductive cause–effect "explanations" based largely on correlation lies with their generation of testable hypotheses.

Key question: How can we maximize our understanding of human–environment interactions through analysis of parallel historical records?

COMPLEX SYSTEM BEHAVIOR

Although cause–effect explanations remain a dominant mode, the view from complexity science argues against simple causative explanation. Open, dynamic systems are expected to behave nonlinearly with respect to external forcings and their internal organization (e.g., Phillips 1998; Levin 1999; Scheffer et al. 2001). External forcings may exert their influence through the transgression of thresholds, there may be time lags in a process response, and perhaps most importantly a modern system is not separated easily from its past: we should expect that it has been conditioned or sensitized by past events, or bears the legacy of past forcings and responses. Complexity science also predicts that systems may exhibit emergent phenomena: forms and structures that have evolved merely through a network of process interaction within a set of boundary conditions. Understanding the complexity of current systems in these terms is a high priority if we are to avoid environmental surprises at local and global levels (e.g., Amsterdam Declaration 2002).

If the formalization of complexity through mathematics is relatively new, the ideas are certainly not. Throughout the history of philosophy, one common observation from critics of historicism is their frequent allusion to the need to understand interactions between individuals, thus rejecting holism. Popper (1957, p. 18) argues that holistic studies of groups do not lead to an understanding of culture, "for if social structures...cannot be explained as combinations of their parts or members, then clearly it must be impossible to explain new structures by this method." Similarly, Tolstoy states: "Only by taking infinitesimally small units for observation (the differential of history, that is, the individual tendencies of men) and attaining to the art of integrating them (that is, finding the sum of these infinitesimals) can we hope to arrive at the laws of history" (in Gardiner 1959, p. 174). This raises the issue of how to integrate Earth and world systems. Essentially, do we have appropriate methodologies that can combine the natural laws of the physical world with approaches to the study of society that have largely excluded "historicism" as a mode of explanation? One approach may be to embrace more fully the new "physics of society" (Ball 2004) and utilize historical records more imaginatively to help define Tolstoy's "laws of history." In this sense Ball (2004) presents an optimistic view on the application of network and complexity theory to understanding social change—from the aggregation of individual actions to produce group behavior, through the emergence of scale-free societal properties, to the modeling of colonization and political action by national powers. The central point to be made is that long timescales of observation often enable, uniquely, complex phenomena and nonlinearities to be identified—certainly for environmental systems perturbed by human actions (e.g., Tainter 2000). In some cases high-resolution environmental time series (which include implicitly the actions of humans) may be amenable to mathematical tools that identify certain kinds of system behavior, like self-organized criticality (e.g., Dearing and Zolitschka 1999).

The idea of historical contingency has also been a common and long-running theme in the humanities and natural sciences. Whether it is Tolstoy's first method of history, whereby a series of continuous events is selected and examined ("even though there can be no beginning to any event"), Stephen J. Gould's impassioned view on the uniqueness of evolutionary paths, or the current web site (http://www.cooperative research.org/index.jsp) which describes the perceived timeline of actions and events that led to the 9/11 terrorists attacks (in the view of the web compiler now stretching back to the Russian invasion of Afghanistan in 1979), the idea that the present is conditioned by the past is an enduring one. However, while the potential value of history in defining the importance and existence of contingent processes is self-evident, the approach to be taken is not. Certainly, it seems sensible that we should not follow Churchill's view that "the farther backward you can look, the farther forward you are likely to see." Otherwise we should fall into the trap posited by Bertrand Russell in his tongue-in-cheek argument (Russell 1934) for the cause of the Industrial Revolution in terms of the chain of world events that starts with the migration of the Turks out of a desiccating Central Asia, and the fall of Constantinople. But how far back do we look?

For recent studies of ecological systems in North America, Foster et al. (2003) provide many examples of how modern ecosystems are a product of past cultural history. In some, human actions from decades past still reverberate into the present system; in others, the sensitivity of the present system to current forcings has increased because of past impacts since the times of the European pioneers. Three aspects of contingency should be highlighted here. First, the concept of *inertia*, which describes a process that once underway will not be halted without conditions changing, like demographic growth, the atmosphere–ocean system, or forest succession. Second, *emergence*, describing the appearance of a macroscale form from coevolving interactions operating at a microscale—from local cultural landscapes, to regional and world social structures such as Friedman's (2006) cyclical hegemonies, and Tainter's (2000) organizational problem-solving. Third, *conditioning*, where a past change to a system makes a particular impact more likely (e.g., deforesting land makes the fluvial system more sensitive to the same amount of rainfall than it was previously). An ability to distinguish between these facets of contingency and to define them for key environmental situations seems highly desirable.

Key question: How do we determine how far back in time our studies should cover in order to capture the important elements of contingency and emergence that are relevant to understanding today's socioenvironmental systems?

BEYOND MARX AND MILANKOVITCH: DEVELOPING AND TESTING THEORY

Learning from the past should include the development of theory, as already mentioned, but this seems quite deficient with respect to human–environment interactions. It might be argued that separate elements and processes contained within human–environment interactions, such as culture, economics, climate and ecology, are already relatively well founded on theory. However, the opposing argument made here is that there is a lack of fundamental theory (i.e., that which generates laws or axioms) pertaining to the complexity of multidirectional interactions between human spheres and the physical environment at all scales. What should these theories encapsulate and enable? Well, the whole issue of defining sustainable management, including system sensitivity, impact assessments, and societal vulnerability seems to be a prime candidate. How do the common properties and dynamics of real socioenvironmental systems translate to the languages of energetics, complex system dynamics, and "the physics of society"? How does the sensitivity of a socioenvironmental system change with spatial scale? How does the pattern of networked interactions define stability and resilience? What are the relationships between real systems, operating far from "natural" or "equilibrium" states, and sensitivity to perturbations? How do we embed the value of common property regimes for sustainability in theoretical terms? In developing new socioenvironmental theory, can we build on and reconcile current research trends: the social theory of adaptive capacity and vulnerability (e.g., Pelling 2003); world system analysis (Hornborg and Crumley 2006); ecological dynamics (e.g., Levin 1999; Pahl-Wostl 1995) and the formal mathematical approach advocated in Earth system analysis (e.g., Schellnhuber and Wenzel 1998)?

Historical information may provide the vital perspective and insight that inspires new theory, but it also serves to test theory and hypotheses. Therefore, advancing testable theory about balanced human–environment interactions rather than about either biophysical or social phenomena should not only be viewed as a scientific priority, but may also be the route to reducing the constraints imposed by methodological differences. Where paleoenvironmentalists have worked together with environmental historians within an historical ecology framework, the potential to support or refute conjectures about the causes of environmental change is clear. Reconstructing parallel histories of social, climate, and natural environmental change provides a methodology in which circular argument is minimized and deductive hypothesis-testing maximized. One example of its success is in understanding the anthropogenic causes of surface water acidification. Surface water acidification was recognized as a major problem in the U.K. and elsewhere from the early 1980s. A lack of long-term instrumental data for precipitation acidity and water quality meant that there were a number of alternative theories as to its causes. These included industrial emissions, but also the effects of forestry and long-term natural biogeochemical cycling. Different lake records were compiled (Battarbee et al. 1985), which allowed post hoc scientific control for certain variables, such as geology and the absence or presence of coniferous plantations. These records showed that increased precipitation acidity caused by industrial emissions of sulfur and nitrogen oxide gases over 100–200 years was the only plausible explanation. These findings contributed significantly to government decisions in the U.K. and elsewhere to introduce sulfur emission reduction policies.

The improved development and testing of theory probably requires two new initiatives: (a) the compilation, integration, and regionalization of existing knowledge and data and (b) the continued development of dynamic models for the simulation of human–environment interactions. These are considered in the final two sections.

Key question: How can we develop new testable theory for the behavior of socioenvironmental systems that helps guide sustainable management?

INTEGRATION AND REGIONALIZATION

There are two dominant models for the integration of human–environment interactions. The first comes from the environmental sciences and emphasizes integrative studies across natural systems. This approach (cf. Swetnam et al. 1999) tries to encompass the full set of multidirectional interactions between human activities and fluvial, ecological, geomorphic, and climatic systems; effectively treating human actions, like deforestation and drainage, as *stressors* on a natural environment, not unlike climate. The objectives seek to find explanations of human actions in terms of the wider political and economic climate, but the emphasis is on the description and reconstruction of parallel histories. Less emphasis is placed on the changing nature of social and political organization, and the role of distal economic drivers, technology, disease, and climate feedback (e.g., drought and extreme cold) are essentially implicit or speculative. The Ystad Project (Berglund 1991) exemplifies this approach, describing the cultural landscape in southern Sweden over the past 6000 years through historical and scientific reconstructions at a number of sites (Figure 2.2). It describes changes in society and the landscape in order to understand human–environment interactions better through time and to provide a sound foundation for the management of the natural environment, cultural landscapes, and ancient monuments. It poses questions about the effects and spatial patterns of human influence on vegetation change set within a broad hypothesis that argues for the development of agrarian landscapes driven by technology, population, and environmental carrying capacity. The second approach treats humans in past natural environments, explicitly, as *actors* rather than stressors. This type of integration is implicit within the aims of IGBP Core Projects (e.g., LAND and LUCC) and the wider Earth System Science Partnership, but entails more ambitious integration that bridges the gaps between world systems, social science, historical ecology, and Earth system science. In this respect, the Mappae Mundi project (de Vries and Goudsblom 2003) provides a narrative that places the sustainability of humans and their habitats in a long-term socioecological perspective, as well as a foundation for future studies.

A considerable amount of historical and paleoenvironmental information already exists for many parts of the world, yet rarely is it compiled and analyzed in a form that maximizes our learning of human–environment interactions beyond the level of the case study (for an exception, see van der Leeuw 2005). One major task is therefore to produce syntheses at either national levels or for common ecosystems and landscapes that capture the current understanding of long term (100–102 years) ecosystem dynamics. A new initiative in the IGBP Core Project "Past Global Changes" (PAGES) will attempt to do this (http://www.liv.ac.uk/

geography/PAGESFocus5/). PAGES Focus 5 encourages paleoscience and environmental history communities to interact more effectively in order to provide a fuller understanding of landscapes and environmental systems. These integrative syntheses will act as inventories of information that can help inform contemporary studies of these ecosystems (ideally linked to other IGBP Core Projects, such as LAND, or the Long Term Ecological Research Network). A draft scheme for organizing regional syntheses shows a two-dimensional matrix defined by zonal and azonal geographical regions, and simple measures of the intensity and duration of past human impact (Figure 2.3). Such a scheme will allow us to catalog regions where sufficient information and data already exist, and to prioritize new regions where new records and syntheses are required (e.g., "fragile human landscapes," "threatened human landscapes," and "highly valued ecosystems").

Ecosystem type		← Human land-use impact →		
		Low	Medium–High	
			Recent (last 1–2 ka)	*Ancient (last 1–2)*
Zonal	Temperate mixed forest			Rhine / Eifel
	Mediterranean			SW Turkey
	Temperate grassland		Upper Midwest U.S.A.	
	Tropical moist forest			Mesoamerica
	Boreal forest	Peace River, Canada		
Azonal	Large oceanic islands		North Island, New Zealand	
	Mountains			W Alps
	Large river floodplains		Murray Darling	Lower Yangtze
	Coastal zone, peatlands, etc.			Netherlands
	Lake systems		SW Scotland	

Figure 2.3 An example of an organizational matrix for the regionalization of global case studies within PAGES Focus 5. Each cell represents a zonal region or azonal system for which high-quality (well-dated, high-resolution) multi- and interdisciplinary paleoenvironmental data (including sedimentary, archaeological, instrument, and documentary data as appropriate/available) already exist and where synthesis of information for different environmental systems (e.g., lakes, fluvial) and/or at different scales is feasible. Blank cells could be targeted for new studies, with priorities set by criteria such as high biodiversity status; fragile and/or degraded regions; projected climate and /or human impacts; pollution loadings; and regions coincident with other IGBP Core Projects (Dearing 2005; Dearing et al. 2006b).

Two further aspects of international environmental change research would be addressed by these syntheses. First, a full inventory of past environmental processes and human–environment interactions within a region could make major contributions toward ranking subsystem sensitivities to particular combinations of past climate and human impact, and help to underpin other attempts to characterize functional human–environment units (Lüdeke et al. 2004; Lambin et al. 2001) where crucial long-term trajectories may be lacking. Success may require new methods for ranking the sensitivities of modern ecosystems based on long-term histories, utilizing, for example, system energetics, "distances" from pre-impact states, and rates of change in key process variables (e.g., Dodson and Mooney 2002). Second, improved ability to scale-up local case studies through coordinated regionalization will allow generalization or transfer of findings across larger geographical areas and ecosystems, giving compatibility with the scale of real and modeled environmental drivers (e.g., administrative areas, downscaled GCM outputs). An example of where this has already been attempted is the biomization of pollen diagrams (Prentice et al. 1996) used to produce global vegetation/biomass maps for chosen time periods (e.g., BIOME 6000). For some processes, it may provide the means to upscale to the global scale in order to compute new global process records, such as a Holocene record of global deforestation or sediment flux to the global coastline.

Key questions: How do we prioritize which regions or ecosystems need new and dedicated research programs to establish historical perspectives? How do we move from viewing humans as "stressors" to viewing them as "actors" in reconstructed environments?

SIMULATING FUTURE HUMAN–ENVIRONMENTAL INTERACTIONS

However powerful the insights gained from history, there will always remain gaps in the record and uncertainty with regard to narrative description and explanations. However detailed and penetrating, a full analysis of available past records will not be able to generate alternative and testable strategies for sustainable management. Enhanced levels of confidence in understanding human–environment system behavior are therefore most likely to come through mathematical simulation modeling. A key measure of the quality of our theoretical understanding of socioenvironmental systems has to be the extent to which we can simulate reality. Simulation modeling is therefore a key complement to empirical studies of human–environment interactions and may be used together with historical and paleoenvironmental data in different ways. For example, model–data comparisons are often used to isolate an individual forcing by controlling for other variables. This is a particularly valuable approach in human-interaction studies where a common issue is how to "isolate" the effect of land-use or land-cover change, forced by human actions, from the impact of climate change. However, sufficient empirical evidence now exists to show that human–environment interactions are complex and essentially nonlinear, characterized by the growth of relatively long-lived emergent phenomena at all scales: social institutions, social structures, ecosystems, and geomorphic forms. Thus, ideally, new simulation models should allow complex and macroscale emergent phenomena to arise from microscale interactions within an evolutionary framework. Such models would be run forward from the past and be validated against historical time series before simulating future systems under different scenarios of climate, environmental, and societal change: a methodology utilized in disentangling the individual and combined roles of alternative climate drivers of 20th-century global warming.

One promising approach would be to build on recent developments in spatially explicit cellular automata-type models (Dearing 2006). These models can be classified according to the level of functional rules used, the means by which and the timescales over which the model is validated, and the extent to which the activities of human agents and decision making are made explicit. As with integrating case studies, there is a logical dichotomy of approaches depending on how human actions are captured. For example, biophysical cellular models in catchment hydrology use low-level rules (Figure 2.1), long timescales ranging from decades to millennia, but with limited inclusion of human agents. Environmental changes are expressed as sequential maps or as time series of outputs from the whole catchment. In such examples, human agents are brought into play mainly as stressors to set future scenarios for hard engineering options or land-use change. In contrast, the inclusion of humans as agents makes use of high-level rules and often a restricted history. Limitations of cellular automata modeling include the constraints imposed by the simplicity of cellular models and how this simplicity has to be compromised to accommodate action-at-a-distance social processes. Beyond these problems, there are ongoing developments that are likely to see improved cellular-based modeling, through integration with GIS, macrolevel models and, in ecology, developing individual-based approaches. A recent variant of the cellular automaton approach provides a compelling spatiotemporal simulation of the global population through the Neolithic transition (Wirtz and Lemmen 2003), with validation through the archaeological record. Perhaps most headway toward the development of integrated socioenvironmental models has been gained through the development of agent-based models (ABMs), particularly among the international land-cover and land-use community (e.g., Parker et al. 2001). The emphasis in ABMs tends to be on social and economic drivers of land use rather than the coevolution of interactions between humans and environmental processes, and validation has largely come through sequential maps of land cover derived from satellite imagery since the 1960s. For example, projections of global land use for different socioenvironment scenarios by the Millennium Ecosystem Assessment (2005) utilize observed changes in global crop and forest areas since 1970 with modeled socioenvironmental scenarios until 2050. Thus, while these approaches are of great value in strategic planning, they have yet to exploit the fully reconstructed history of human–environment interactions that is often available.

Key question: How do we improve the integration of socioeconomic and biogeochemical processes within the same dynamic simulation modeling framework?

ACKNOWLEDGMENTS

I would like to acknowledge wide-ranging discussions with colleagues at the Dahlem Workshop and within the PAGES Focus 5 leadership, and the very useful comments made by reviewers of the paper.

REFERENCES

Ball, P. 2004. Critical Mass. London: Heinemann.

Battarbee, R.W. 1999. The importance of palaeolimnology to lake restoration. *Hydrobiologia* **395/396**:149–159.

Battarbee, R.W., R.J. Flower, A.C. Stevenson, and B. Rippey. 1985. Lake acidification in Galloway: A palaeoecological test of competing hypotheses. *Nature* **314**:350–352.

Berglund, B.E., ed. 1991. The Cultural Landscape during 6000 Years in Southern Sweden. Ecological Bulletin 41. Oxford: Blackwell.

Bradshaw, R.H.W., G.E. Hannon, and A.M. Lister. 2003. A long-term perspective on ungulate-vegetation interactions. *Forest Ecol. Manag.* **181**:267–280.

Crumley, C. 2006. Historical ecology: Integrated thinking at mutiple temporal and spatial scales. In: The World System and the Earth System, ed. A. Hornborg, and C.L. Crumley. Santa Barbara, CA: Left Coast Books, in press.

Dearing, J.A. 2005. Past ecosystem processes and human–environment interactions. *PAGES Newsl.* **13**:23. http://www.pages-igbp.org

Dearing, J.A. 2006. Integration of World and Earth systems: Heritage and foresight. In: The World System and the Earth System, ed. A. Hornborg and C.L. Crumley. Santa Barbara, CA: Left Coast Books, in press.

Dearing, J.A., R.W. Battarbee, R. Dikau, I. Larocque, and F. Oldfield. 2006a. Human–environment interactions: Learning from the past. *Reg. Env. Change* **6**:1–16.

Dearing, J.A., R.W. Battarbee, R. Dikau, I. Larocque, and F. Oldfield. 2006b. Human–environment interactions: Towards synthesis and simulation. *Reg. Env. Change* **6**:115–123.

Dearing, J.A., and R.T. Jones. 2003. Coupling temporal and spatial dimensions of global sediment flux through lake and marine sediment records. *Glob. Planet. Change* **39**:147–168.

Dearing, J.A., and B. Zolitschka. 1999. System dynamics and environmental change: An exploratory study of Holocene lake sediments at Holzmaar, Germany. *Holocene* **9**:531–540.

Deevey, E.S. 1969. Coaxing history to conduct experiments. *BioScience* **19**:40–43.

De Vries, B., and J. Goudsblom, eds. 2003. Mappae Mundi: Humans and their Habitats in a Long-Term Socio-Ecological Perspective. Myths, Maps and Models. Amsterdam: Amsterdam Univ. Press.

Diamond, J.M. 2005. Collapse: How Societies Choose to Fail or Succeed. New York: Viking.

Dodson, J.R., and S.D. Mooney. 2002. An assessment of historic human impact on south-eastern Australian environmental systems using late Holocene rates of environmental change. *Austral. J. Bot.* **50**:455–464.

Foster, D.R., F. Swanson, J. Aber et al. 2003. The importance of land-use legacies to ecology and conservation. *BioScience* **53**:77–88.

Friedman, J. 2006. Plus ça change: On not learning from history. In: The World System and the Earth System ed. A. Hornborg and C.L. Crumley. Santa Barbara, CA: Left Coast Books, in press.

Gardiner, P., ed. 1959. Theories of History: Readings from Classical and Contemporary Sources. Glencoe, IL: Free Press.

Hornborg, A. and C.L. Crumley, eds. 2006. The World System and the Earth System. Santa Barbara, CA: Left Coast Books, in press.

Lambin, E.F., B.L. Turner II, H.J. Geist et al. 2001 The causes of land-use and -cover change: Moving beyond the myths. *Glob. Env. Change* **11**:261–269.

Levin, S.A. 1999. Fragile Dominion: Complexity and the Commons. Cambridge, MA: Perseus.

Lüdeke, M.K.B., G. Petschel-Held, and H.J. Schellnhuber. 2004. Syndromes of global change: The first panoramic view. *GAIA* **13**:42–49.

May, E.R. 1973. "Lessons" of the Past. New York: Oxford Univ. Press.

McCullagh, C.B. 1998. The Truth of History. London: Routledge.

Millennium Ecosystem Assessment. 2005. Millennium Ecosystem Assessment Synthesis Report: A Report of the Millennium Ecosystem Assessment. Washington, D.C.: Island. http://www.millenniumassessment.org/.

Nott, J., and M. Hayne. 2001. High frequency of "super-cyclones" along the Great Barrier Reef over the past 5,000 years. *Nature* **413**:508–512.

Oldfield, F., and J.A. Dearing. 2003. The role of human activities in past environmental change. In: Paleoclimate, Global Change and the Future, ed. K.D. Alverson, R.S. Bradley, and T.F. Pedersen, pp. 143–162. Berlin: Springer.

Pahl-Wostl, C. 1995. The Dynamic Nature of Ecosystems. Chichester: Wiley.

Parker, D.C., T. Berger, and S.M. Manson, eds. Agent-based models of land use and land cover change. Report and Review of an Intl. Workshop, Oct. 4–7, 2001. LUCC Report Series 6. Bloomington, IN: LUCC Focus 1 Office, Indiana Univ.

Pelling, M. 2003. The Vulnerability of Cities. London: Earthscan.

Phillips, J.D. 1998. Earth Surface Systems: Complexity, Order and Scale. Oxford: Blackwell.

Popper, K. 1957.The Poverty of Historicism. London: Routledge.

Prentice, I.C., J. Guit, B. Huntley, D. Jolly, and R. Cheddadi. 1996. Reconstructing biomes from palaeoecological data: A general method and its application to European pollen data at 0 and 6 ka. *Clim. Dyn.* **12**:185–194.

Redman, C.L. 1999. Human Impact on Ancient Environments. Tucson: Univ. of Arizona Press.

Ruddiman, W.F. 2003. The anthropogenic greenhouse era began thousands of years ago. *Clim. Change* **61**:261–293.

Russell, B. 1934. Freedom and organization, 1814–1914. London. (Reproduced in: Gardiner, P., ed. 1959. Theories of History: Readings from Classical and Contemporary Sources. Glencoe, IL: Free Press, chap. XVIII).

Scheffer, M., S. Carpenter, J.A. Foley, C. Folke, and B. Walker. 2001. Catastrophic shifts in ecosystems. *Nature* **413**:591–596.

Schellnhuber, H.-J., A. Block, M. Cassel-Gintz et al. 1997. Syndromes of global change. *GAIA* **6**:19–34.

Schellnhuber, H.-J., and V. Wenzel, eds. 1998. Earth System Analysis. Berlin: Springer.

Steffen, W., A. Sanderson, P.D. Tyson et al. 2004. Global Change and the Earth System: A Planet under Pressure. Berlin: Springer.

Swetnam, T.W., C.D. Allen, and J.L. Betancourt. 1999. Applied historical ecology: Using the past to manage for the future. *Ecol. Appl.* **9**:1189–1206.

Tainter, J.A. 2000. Problem solving: Complexity, history, sustainability. *Pop. Env.* **22**:3–41.

van der Leeuw, S.E. 2005. Climate, hydrology, land use and environmental degradation in the lower Rhone valley during the Roman period. *C.R. Geoscience* **337**:9–27.

Wilson, E.O. 1998. Consilience: The Unity of Knowledge. London: Little Brown.

Wirtz, K.W., and C. Lemmen. 2003. A global dynamic model for the Neolithic Transition. *Clim. Change* **59**:333–367.

3

Assessing and Communicating Data Quality

Toward a System of Data Quality Grading

Robert Costanza

Gund Institute for Ecological Economics, Rubenstein School of Environment and Natural Resources, The University of Vermont, Burlington, VT 05405–1708, U.S.A.

ABSTRACT

IHOPE will require the integration and synthesis of data from a huge range of sources of highly variable quality. Although experts in a field of study usually have a good working understanding of the quality constraints on their data, this understanding is not often or easily communicated across fields. What we need for the IHOPE effort is a system to communicate the full range of data quality: from statistically valid estimates to informed guesses, from historical narratives to the results of computer simulations. Communicating data quality is a prerequisite to effectively integrating the full range of information we hope to assemble. One can think of this process as *grading* data. A grading scheme for communicating the "degree of goodness" associated with data has high potential utility. If consistently applied, it can provide nonexperts with greater competence in interpreting the degree of uncertainty associated with complex estimates. In modeling and analysis, it will provide a much needed input to help assess the overall uncertainty of results, based on the quality of the input data, combined with information on the structure and quality of the models used to process the data.

INTRODUCTION

There are three principle sources of uncertainty in scientific analysis. Gaps in knowledge or understanding can arise from any or all of these sources.

1. Parameter uncertainty: the uncertainty associated with model parameters. This is also known as "within model" uncertainty. The usual way to communicate this uncertainty is through statistics and sensitivity analysis of various kinds.

2. Model uncertainty: the uncertainty associated with the choice of model or underlying assumptions. This is also known as "between model" uncertainty. The usual way to communicate this uncertainty is to display the results of alternative models or sets of assumptions. For example, the global change community supports the development of multiple global climate models, and a rigorous intercomparison of model results has helped to communicate "between model" uncertainty.
3. Data quality: the uncertainty associated with the quality of the data going into the models and analysis. The famous saying, "garbage in—garbage out," captures this situation. The methods to communicate this source of uncertainty have not been adequately worked out or accepted. Because of this, data quality is often either ignored completely or oversimplified into "good" versus "bad" data.

In this chapter I explore the underlying issues of data quality, the ways in which the full range of data quality can be communicated adequately, and the influence of data quality on the overall uncertainty of scientific analysis. A grading system for data to assess and communicate data quality is proposed.

THE PROBLEM OF DATA QUALITY IN INTEGRATED ASSESSMENT

In scientific research, as in any other sphere of activity, the maintenance of the quality of products is critical for their effective use. In mature fields of traditional science, quality control is exercised informally by competent practitioners (Ravetz 1971). In most scientific studies, the scientists actually doing the analysis have a good working understanding of the inherent quality of their measurements and results. However, there is no accepted method to communicate this knowledge of data quality to potential users of the information in other fields. When research results are used as inputs to an integrated, interdisciplinary assessment (as we intend to do in the IHOPE project), the users of this information must either be knowledgeable in the details of the research methods or accept the results with no idea of their quality. Usually, they lack the knowledge for performing their own assessment of quality. In the absence of a quality assessment system these deficiencies are largely unrecognized, and their consequences are difficult to estimate. Grades for quality are routinely assigned in innumerable spheres of activity in our society (e.g., academic performance or quality of meat and eggs). Yet in the case of information, one of the most sensitive products we have, there are no standard systems for grading and hence no means for a socially effective system of quality control. In this chapter, I present some ideas aimed at rectifying this situation. If the IHOPE project is to be a success, we ultimately will have to address this issue.

Statistics Is Not the Answer

The standard techniques of statistics were developed to handle a particular aspect of uncertainty. They are based on the assumption that uncertainty is due to real, precisely measured variation in the populations being sampled. They generally assume that we have a probability distribution with which to work, without asking how well we know that distribution (Mosleh and Bier 1996; Kuhn 1997). Such assumptions are frequently justifiable in the case of the traditional experimental or field sciences, but the data available in integrative research is frequently so scattered and coarse that refined mathematical manipulations do not possess much genuine meaning when applied to them. Here I wish to concentrate on the issue of how well we know the data distribution (i.e., quality), and how we may communicate this knowledge. The more basic problem is the problem of errors in measurement and the partitioning of the uncertainty into that which is caused by real variation in the population (statistical uncertainty) versus that due to errors in measurement (data quality).

The achievement of scientific work of high quality requires the deployment of sophisticated craft skills, as well as the motivating force of commitment and morale. A full specification of quality would therefore be as complex and subtle a task as the research itself. Fortunately, from the nonexpert user's point of view, the relevant aspects of quality depend more on the product than on the process. Since the product of research is information that has certain knowledge as its ideal, we can assess quality by that yardstick. The incompleteness of certainty (or the inevitable uncertainty) of scientific information can be used to define a system of quantitative estimates and qualitative grades. By this method, the various aspects of the uncertainty, and hence of the quality, of scientific information can be described.

No scientific activity is free from uncertainty; it may be said that the key to a science being "mature" lies in its ability to recognize, communicate, and control the various sorts of uncertainty that affect its results and predictions. These include inexactness (as expressed by significant digits), unreliability (as expressed in systematic error), epistemic uncertainty, linguistic uncertainty, and others (Regan et al. 2002). No amount of sophisticated apparatus and computer power can replace theoretical understanding of the problems of uncertainty or the practical skills of controlling and communicating it.

When quantitative information is used to provide inputs for synthesis across broad disciplinary gulfs, as in the case of the IHOPE project, the scientist's problems of management and communication of uncertainty are severe. First, original data are rarely as well controlled as in the laboratory. Well-structured theories, normally expected to be available in basic or applied science, are conspicuously absent in integrative research. Furthermore, since such research is inherently interdisciplinary, it involves fields of varying states of maturity and with very different sorts of practice in its theoretical, experimental, and social dimensions. Scientists must use inputs from fields they do not know intimately, and thus they cannot make the same sensitive judgments of quality that they would in their own subject. The result is that quality control of the research process is diluted; quality assurance of results is weaker; and the results command less confidence among users.

For example, a principal challenge to understanding and modeling the coupled effects of humans in natural systems is that social data and biophysical data are collected and organized on very different time and space scales and have very different quality characteristics. Data on human systems (GDP, literacy, energy use, press freedom) are organized by political boundaries (local, national, regional). Data on natural systems are increasingly derived from remote sensing and organized on grid systems of varying scales. Examples include data on vegetation, precipitation, desertification, deforestation, soils, land use/land cover, elevation, to name just a few. Commensuration of these disparate forms and scales of data and communication of their relative quality is a fundamental prerequisite to posing relevant questions about the relation between human development and the environment.

TOWARD A GENERAL SYSTEM OF DATA QUALITY ASSESSMENT

There have been relatively few prior proposals for solutions to the above problems. One attempt (Costanza et al. 1992) employed a system known as NUSAP (numeral, unit, spread, assessment, pedigree). It allows the more quantitative and qualitative aspects of data uncertainty to be managed separately. The NUSAP approach illustrates the major sources of uncertainty related to data quality and can guide new research aimed at the improvement of the quality of outputs and the efficiency of the procedures. A brief description of the NUSAP system is given below (for further details see Costanza et al. 1992).

The NUSAP Notational System

Every set of data has a *spread*, which is an attribute of any quantity, however derived. Spread may be considered in some contexts as a degree of precision, as a tolerance, or as a random error in a calculated measurement. It is the kind of uncertainty that relates most directly to the quantity as stated and is most familiar to students and even the lay public.

A more complex sort of uncertainty relates to the level of confidence to be placed in a quantitative statement, which relates to the "accuracy," in contrast to precision. In statistical practice, this is usually represented by the confidence limits (at, say, 95% or 99%). In practice, such judgments are quite diverse; thus safety and reliability estimates are given as "conservative by a factor of *n*."

Alternatively, in risk analyses and futures scenarios, estimates are qualified as "optimistic" or "pessimistic." In laboratory practice, the systematic error in physical quantities, as distinct from the random error or spread, is estimated on an historic basis. Thus, it provides a kind of *assessment* to act as a qualifier on the number, or alternatively (if desired) on the spread; it is this that we express as the *grade* when we wish to convey the qualitative "degree of goodness" of a number.

This assessment is one level up from spread, both in its sophistication and variety. We may imagine spread as representing *inexactness*, and assessment or grade as expressing *unreliability* (or *degree of reliability* as appropriate). It can also be seen as expressing the "strength" of the number. *Our knowledge of the behavior of the data gives us the spread; our knowledge of its production or intended use gives us assessment or grade.* But there is something more. No process—in the field or in the lab—is completely known. Even the successive accepted values of familiar physical constants tend to vary in ways that could not have been predicted, and by amounts that lie outside their "error bars," until they eventually settle down (Henrion and Fischhoff 1986). This is the realm of our *ignorance*. It includes all of the different sorts of gaps in our knowledge that are not encompassed in the previous two sorts of uncertainty. This ignorance may merely be of what is significant, as when anomalies in experiments are discounted or neglected, only to be discovered when a new and strongly different value for a physical constant is obtained. It may also be deeper, as is appreciated retrospectively when great new theoretical advances are made in science, and things which had been scarcely imaginable become commonplace.

Can we say anything useful about that of which we are ignorant? It would seem that by the very definition of ignorance, we cannot. But the boundless sea of ignorance has shores that we can stand on and map. Let us think of a *boundary with ignorance* as the last sort of uncertainty that we can now effectively control in practical scientific work. To map this boundary, we describe the state-of-the-art in the field of practice in which our quantity is produced. This is done by an evaluative analytical accounting we call the *pedigree* of the quantity. By means of a matrix it shows the boundary with ignorance by displaying the degrees of strength of crucial theoretical, empirical, and social components of the process.

The nature of the boundary, with its crucial components, will depend on the sorts of operations involved. The theoretical, empirical, and social phases (or crucial components) are *quality of models*, *quality of data*, and *degree of acceptance* (Table 3.1). If we qualify the theoretical phase of the production process of the information as *computational model,* we are implicitly stating that we do not have a *theoretical model*, and thus we record the absence of an effective theory. Similarly, if the empirical phase is not *experimental*, it can be at best *historical or field data*, as in most environmental and historical research. In the latter case, data are inherently less capable of control, and so it is less effective as an input and check on the quality of the model. The components on the social side

Table 3.1 The numerical estimate pedigree matrix.

Score	*Theoretical* Quality of Model	*Empirical* Quality of Data	*Social* Degree of Acceptance
4	Established theory • many validation tests • causal mechanisms understood	Experimental data • statistically valid samples • controlled experiments	Total • all but cranks
3	Theoretical model • few validation tests • causal mechanisms hypothesized	Historical/field data • some direct measurements • uncontrolled experiments	High • all but rebels
2	Computational model • engineering approximations • causal mechanisms approximated	Calculated data • indirect measurements • handbook estimates	Medium • competing schools
1	Statistical processing • simple correlations • no causal mechanisms	Educated guesses • very indirect approximations • "rule of thumb" estimates	Low • embryonic field
0	Definitions/assertions	Pure guesses	None

describe the evaluation of the information in its particular context. *Degree of acceptance* of a result will be straightforward in a fully matured field where criteria of quality are agreed; a rough approximation to this is the referee's judgment on the research paper.

In the NUSAP system the last three letters in the acronym refer to the spread, assessment, and pedigree already discussed. The first two refer to *numeral* and *unit*. The first category encompasses the arithmetical system, and the second, the base in which it is appropriately expressed. In a full NUSAP expression there is a balance between all the elements. Thus the number of "significant digits" in the numeral place, when combined with the scaling-factor in the units place, will be coherent with the inexactness described under spread.

There is another connection between the different categories of the NUSAP system that can be used to establish the *grade* for the strength of the information. In the absence of other information, whereby an appropriate judgment can be made for the assessment category, one can use the set of entries of the pedigree matrix. These are coded on an ordinal scale of 0–4; their average, normalized on the scale 0–1, provides a convenient measure. It should be clear that this scale provides a simple and suggestive index and not a measured quantity. Provided that it is used with that awareness and is not embedded in complex, hyper-precise mathematical manipulations, it will function as a useful tool in the evaluation of scientific information. Thus the *pedigree*, in exhibiting the limits of the state-of-the-art of the field in which the information was produced, provides us with a gauge for an *assessment* of the strength of that information, or its *grade*.

The full NUSAP form, as given above, is the most general framework for such expressions. In it, the assessment box may be used to constitute the "grade" or "degree of goodness." For many purposes it may be sufficient to use an abbreviated form, the pair (N, A), where N (the numeral) is a representative number and A (the assessment) is a code for the grade which describes the "degree of goodness" of the number, as distinct from its spread.

An Example: The Valuation of Wetland Ecosystem Services

To demonstrate the usefulness of the proposed system, let us apply it to the example case of ecosystem valuation. We use a well-documented study of the economic value of wetlands in Louisiana (Farber and Costanza 1987; Costanza et al. 1989), which employed a number of different models and methods to arrive at an estimate of the total value of the wetland's ecosystem services. The results from the original study are reproduced in Table 3.2.

Table 3.2 Summary of wetland value estimates (1983 U.S. dollars) for various components of wetlands contributing to their economic value, using two competing models. WTP: willingness to pay; EA: energy analysis. From Costanza et al. (1989).

Method	Annual value per acre	Present value per acre at specified discount rate	
		8%	3%
WTP based			
Shrimp	10.85	136	362
Menhaden	5.80	73	193
Oyster	8.04	100	268
Blue crab	0.67	8	22
Total commercial fishery	25.37	317	846
Trapping	12.04	151	401
Recreation	3.07	46	181
Storm protection	128.30	1915	7549
Subtotal	$168.78	$2429	$8977
Option and existence values	?	?	?
EA based			
GPP conversion	$509–847	$6,400–10,600	$17,000–28,200
"Best estimate"	$169–509	$2,429–6,400	$8,977–17,000

Two overall methods results are presented in Table 3.2. The "willingness to pay" method enumerates the various ecosystem services and derives an independent estimate for each one. These components are then added to yield the total value (e.g., shrimp production value was estimated as \$10.85/ac/yr and storm protection value as \$128.30/ac/yr). Option and existence value are known to be important components of the total but no direct estimate was made for them for this ecosystem.

A second method ("energy analysis") uses the total solar energy captured by the ecosystem as an indicator of its economic value. It is more comprehensive in that it does not require summing individually measured components to arrive at the total. However, the connection between energy captured and economic value is controversial and uncertain.

Finally, the "present value" of the ecosystem services is calculated using various discount rates based on the assumption that the ecosystems provide a constant stream of benefits into the indefinite future. In this case: present value = annual value/discount rate. The appropriate discount rate to use in such a situation is, however, highly uncertain.

Table 3.3 recasts these results into the NUSAP system. Here the numerical results are given only to the appropriate degree of precision and the spreads on each number are shown, using only 10% increments (except for 25% and 75%). The pedigree for each number is given, based on an analysis of the individual models and methods used (for a complete description, see Costanza et al. 1989). They are coded using the 0–4 system in Table 3.1. For example, the shrimp production estimate was based on a theoretical model relating wetland area to shrimp catch (score = 3) using historical/field data from National Marine Fisheries shrimp catch statistics and measured wetland area (score = 3) in a procedure (regression analysis) that has high, but not total, peer acceptance for the intended purpose (score = 3). Finally, the grade for each estimate is given based on the average scores in the pedigree $\frac{3+3+3}{12} = .6$. Note that grades are rounded to one digit.

Several quantities are calculated in the table using the NUSAP arithmetric. These are shown in bold. The total commercial fishery value is the sum of four components. Its spread is the weighted average of the percentage spreads of the components:

$$\frac{1E1*.1+6E0*.2+8E0*.3+1E0*.4}{2.5E1} = .2.$$

Its grade is the weighted average of its component grades:

$$\frac{1E1*.7+6E0*.5+8E0*.6+1E0*.6}{2.5E1} = .6.$$

An estimate for option and existence value is given based on studies of other areas. However, as its spread and grade indicate for this application, it is definitely an "order of magnitude" estimate. The total WTP-based value reflects the

Table 3.3 NUSAP scores and summary grades for the elements of the wetland valuation problem.

Element	Numeral N	Unit U	Spread S	Assessment Pedigree	Assessment Grade
WTP-based estimates					
Shrimp	1 E1	$/ac/yr	±10%	(3,3,3)	.7
Menhaden	6 E0	$/ac/yr	±20%	(2,2,2)	.5
Oyster	8 E0	$/ac/yr	±30%	(2,3,2)	.6
Blue crab	1 E0	$/ac/yr	±40%	(3,2,3)	.6
Total commercial fishery	2.5 E1	$/ac/yr	±20%		.6
Trapping	1.2 E1	$/ac/yr	±30%	(2,2,2)	.5
Recreation	3 E0	$/ac/yr	±10%	(3,4,3)	.8
Storm protection	1.3 E2	$/ac/yr	±20%	(2,3,2)	.6
Subtotal	1.7 E2	$/ac/yr	±20%		.6
Option and Existence Values	5 E2	$/ac/yr	±50%	(1,0,1)	.2
Total WTP	7 E2	$/ac/yr	±40%		.3
EA based					
GPP conversion	7 E2	$/ac/yr	±25%	(3,2,1)	.5
Average of two methods	7 E2	$/ac/yr	±30%		.6
Discount Rate	5 E0	%	±50%	(1,3,1)	.4
Present Value	15 E3	$/ac	±80%		.4

quantitative importance of option and existence values and their relatively low quality. We end with a spread of ±40% and a grade of .3 for this estimate.

The EA-based estimate yielded a very similar quantity estimate to the WTP estimate, and this is taken as corroborating evidence since the likelihood that this would occur by chance is small. The average of the two methods is therefore of higher grade than either of the inputs (.6 vs. [.5 and .3]), and we are left with a reasonably high quality estimate of the total annual value of wetland production (7 E2 $/ac/yr ± 30% [.6]).

Converting this to present value significantly reduces the data quality, however, because of the high uncertainty about the discount rate. The spread on the present value goes to ±80% and the grade goes down to .4.

The NUSAP representation of the series of calculations that went into the estimation of the value of wetlands offers a clear picture of the data quality. It also allows the uncertainty in the final estimate to be more easily communicated and directs research to those areas most likely to improve the quality of the final estimate.

CONCLUSIONS FOR IHOPE

The IHOPE project intends to integrate data from a very broad range of disciplines and time and space scales. This data will vary significantly in nature, statistical characteristics, and quality. Developing and implementing a system to grade the data according to their quality will therefore be essential in this effort. We can build on suggestions made previously (i.e., NUSAP) or develop our own system as part of the project.

Whichever option we choose, we need to address this issue. This chapter was intended to provide background to help us think about and discuss the issue, ultimately resulting in a system that may be implemented during data collection phases of the project.

REFERENCES

Costanza, R., S.C. Farber, and J. Maxwell. 1989. The valuation and management of wetland ecosystems. *Ecol. Econ.* **1**:335–361.

Costanza, R., S.O. Funtowicz, and J.R. Ravetz. 1992. Assessing and communicating data quality in policy-relevant research. *Env. Manag.* **16**:121–131.

Farber, S., and R. Costanza. 1987. The economic value of wetlands systems. *J. Env. Manag.* **24**:41–51.

Henrion, M., and B. Fischhoff. 1986. Assessing uncertainty in physical constants. *Am. J. Physics* **54**:791–797.

Kuhn, K.M. 1997. Communicating uncertainty: Framing effects on responses to vague probabilities. *Org. Behav. Hum. Dec. Proc.* **71**:55–83.

Mosleh, A., and V.M. Bier. 1996. Uncertainty about probability: A reconciliation with the subjectivist viewpoint. *IEEE Trans. Syst. Man Cyber. A: Syst. Hum.* **26**:303–310.

Ravetz, J.R. 1971. Scientific Knowledge and its Social Problems. Oxford: Oxford Univ. Press.

Regan, H.M., M. Colyvan, and M.A. Burgman. 2002. A taxonomy and treatment of uncertainty for ecology and conservation biology. *Ecol. Appl.* **12**:618–628.

The Millennial Timescale: Up to 10,000 Years Ago

4

The Rise and Fall of the Ancient Maya

A Case Study in Political Ecology

Vernon L. Scarborough

Department of Anthropology, University of Cincinnati,
Cincinnati, OH 45221–0380, U.S.A.

ABSTRACT

The ancient southern lowland Maya of the Yucatán Peninsula provide a case-study example of the complex and ever-changing relationships between humans and their environments from the specific vantage of a fragile biophysical setting that was engineered into a highly resilient and productive landscape. Their semitropical ecosystems, like most landscapes, have been significantly affected by humans both today and in the ancient past. However, the rate and process of ecosystem engineering is different in tropical settings when compared with semiarid or temperate regions. Because of shallow and frequently poorly drained soils, coupled with a biogeography of species-rich diversity but one associated with a sparse number of any one species within any specific patch, early sedentary adaptations by humans were challenging. Like other members of their biophysical world, humans practiced dispersed living strategies to harvest the environs. This chapter assesses the socioeconomic and sociopolitical potential for sustainability and collapse in a tropical ecosystem given the biophysical constraints. The significant roles of self-organization and heterarchical networks are examined in the context of social complexity.

INTRODUCTION

The ancient Maya of karstic lowland Central America occupied a geographic region of 250,000 km^2 with an uninterrupted cultural legacy of at least 1500 years. Although much is made of their well-known, if little understood, collapse in the 9th century A.D. (most recently popularized by Diamond 2005), the long-lived success of the Maya within a difficult and frequently inhospitable semitropical environment warrants greater attention. As a primary civilization, or a highly complex social order unlike any preceding it, and the only such "state" from a tropical regime, the Maya are best known for their towering pyramids, elaborate ball courts, developed art forms, and a writing system unparalleled elsewhere in the pre-Hispanic Americas. So how is it that a primary civilization without the wheel, sail, metal tools, beasts of burden, or navigable rivers was capable of supporting an estimated 10 million people by A.D. 700 (Rice and Culbert 1990)?

ECOLOGICAL BACKGROUND

The critical data in addressing this multifaceted quandary are a set of interdependencies within and between early Maya communities and the environments they selected to alter (Figure 4.1). Unlike the first great riverine states in the semiarid Old World—Sumeria, Pharaonic Egypt, the Indus, and the ancient Shang/Erlitou—often associated with major canalization efforts, the Maya occupied a wet–dry tropical forest resting on limestone bedrock and associated with limited

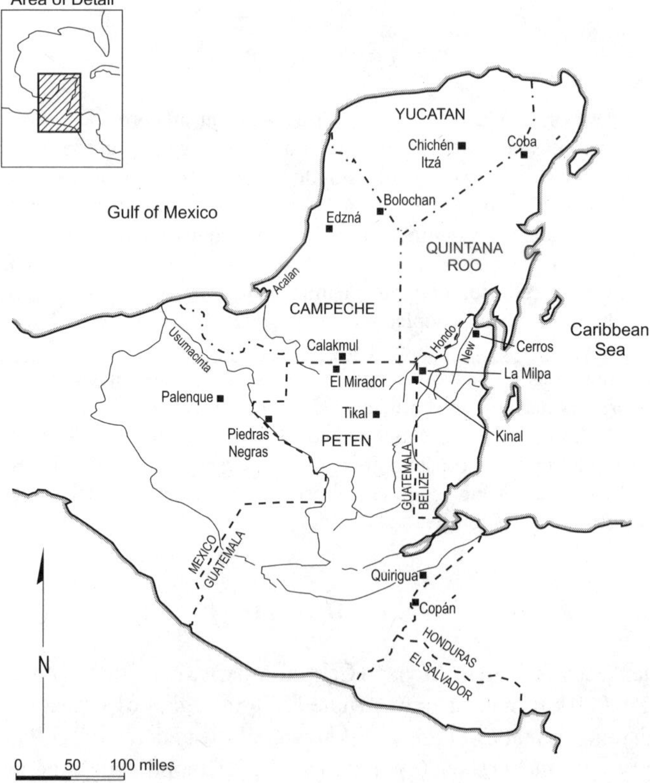

Figure 4.1 Map of the Maya area showing several of the most significant archaeological sites.

surface drainage. Two principal environmental constraints affect our assessment of the ancient Maya. The first is precipitation. Because of the marked seasonality of rainfall, four to six months of the year are drought-like. In the southern Maya Lowlands, the heart of significant cultural development, over 90% of the precipitation falls during the 7–8 month-long rainy season. Because of the pocked and fissured character of the limestone, most surface water rapidly percolates into the karstic substratum, out of reach to a stone-age technology. Although climatic modeling suggests some changes in the timing and duration of seasonal rainfall in the past (Gill 2000; Gunn et al. 1995; Hoddell et al. 1995), the setting is unlikely to have received any more rainfall than it does today.

A second ecological constraint characteristic of many tropical environments is the natural distribution of potential food resources. Although tropical ecosystems are renowned for their tremendous diversity of species when compared with temperate and semiarid settings, they are also noted for a limited number of any one species within any specific microhabitat or patch. For conventional views of complex society based on resource concentrations, stored surpluses, and dense aggregates of people (i.e., towns and cities), the natural environment poses daunting challenges. When coupled with elevated temperatures and humidity, stored surplus foodstuffs are prone to accelerated decomposition when compared with other primary states in semiarid settings.

CULTURE HISTORY

The culture history of the ancient Maya suggests that sedentary pioneer populations migrated from the Archaic Period (7000–2500 B.C.) estuary margins of the western Caribbean and southern Gulf of Mexico coasts into the interior of the Yucatán Peninsula by 1000 B.C., though early highland sedentists subsisting on maize and other domesticates were surely influential (Scarborough 1994). Movement of these populations into the dense lowland vegetation was dependent on water access from shallow internally draining lakes and seasonal swamp-like settings. Evidence suggests that by at least A.D. 1, some of the large seasonal swamps (bajos) now comprising more than a third of the landscape and widely recognized as unusable terrain were perennial lakes (Dunning et al. 2002). By the Late Preclassic Period (400 B.C. to A.D. 250), a large sedentary population had dispersed across the entire Yucatán Peninsula, most positioned in proximity to a present-day bajo or internally draining karstic depression. This population represents the first experiment in state-like complexity and is associated with all the material trappings of subsequent Classic Maya statecraft (A.D. 250–800). In addition to pyramids and ball courts, Late Preclassic communities significantly modified their landscapes to best centralize their water needs. They created "concave microwatersheds" (Scarborough 1993, 1994, 2003a) (Figure 4.2) at the margins of natural karstic sinks (aguadas) or larger depressions (bajos or polje) that contained water year around. Their communities were often

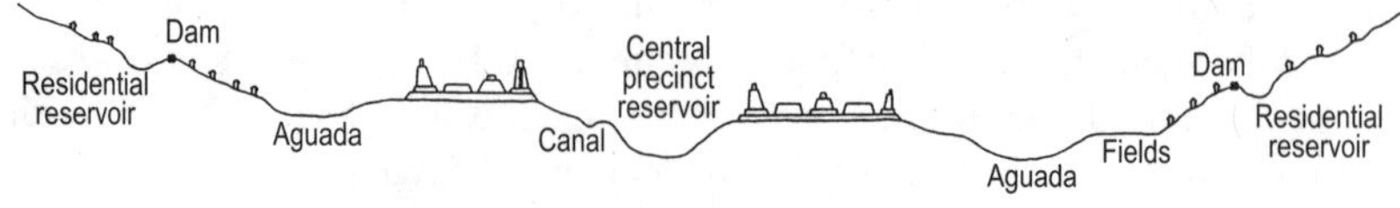

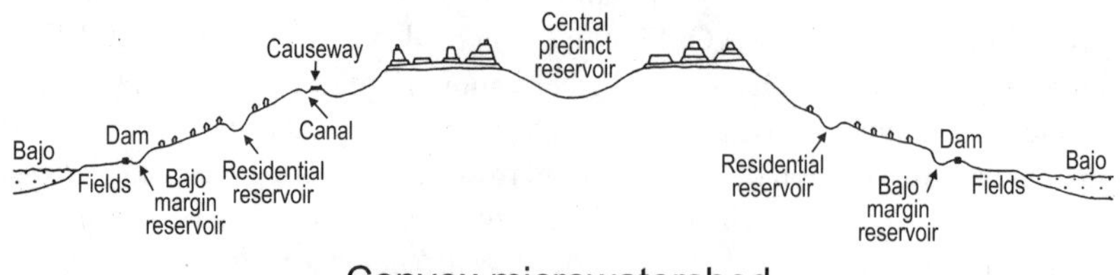

Figure 4.2 Microwatersheds in the Maya area.

carefully altered to capture the seasonal rainfall efficiently and channel it to several localities for subsequent use. Although planned and coordinated, the water system was a passive adaptation in that the natural terrain was only modified enough to accommodate the flow of water across the low-lying depression space already present. Nevertheless, some of the largest communities in Maya history were constructed at this temporal juncture (e.g., El Mirador [Hansen et al. 2002; Matheny 1986]) based on a passive concave microwatershed landscape design.

By the Early Classic Period (A.D. 250–550), fundamental landscape changes were underway. Throughout the Maya Lowlands, major centers located/relocated to the summits of hillocks or ridges quarrying limestone to construct their pyramids and huge paved plazas. Like their predecessors, but in a more active manner, these architects reinvented their environs by constructing "convex microwatersheds" (Scarborough 1993, 1994, 2003a) (Figure 4.2). Here, the elevated landscape was quarried as much for the resulting cavities that were rapidly converted into clay-sealed reservoirs as for the building stone to construct pyramids and palace structures. In addition, the carefully planed plaza surfaces were slightly canted to shed seasonal precipitation into these former large quarry scars. Thick layers of plaster were used to cover the natural pocked and pitted limestone surfaces to prevent premature water loss and to promote sizable amounts of runoff into the elevated reservoirs. Following the rainy season, potable water was then released as needed by gravity flow from the summit reservoirs to the densest portion of the population occupying a Classic Period site on the flanking slopes of the hillock. In the case of Tikal, gray water was captured in large swamp-margin tanks used for subsequent agricultural ends after passing through the densest residential zones (Scarborough and Gallopin 1991). The convex microwatershed was a creative and extremely well-designed water system in a water-stressed environment, but why did the transition from a concave to convex microwatershed system occur?

The cultural transition from the Preclassic to the Classic Period now indicates that the successes of the Late Preclassic towns were based on the intensification of those lake margin settings for agricultural productivity. Raised or ditched fields are well documented at this time along the few perennial rivers in northern Belize (Pohl et al. 1990) and are conjectured within several of the ancient bajos. Moreover, terrace construction is implemented for the first time, suggesting both the necessity for more food to support the successes of a growing population but also to control soil erosion from the deforested slopes leading into the shallow lakes and low-lying communities (Dunning et al. 2002; Hansen et al. 2002). Nevertheless, by the Early Classic Period tons of sediment eroded into the former shallow lakes displacing water sources and disrupting the ancient natural seals holding lake waters. Evidence from climatic modeling further suggests the onset of a serious drying trend which only exacerbated the water deficit (Gill 2000; Gunn et al. 1995; Hoddell et al. 1995). Late Preclassic agricultural overproductivity when coupled with the degraded fragility of the semitropical environment actually stimulated the subsequent Classic Period florescence and a political ecology associated with a centralized and controllable tank system based on the "convex microwatershed."

AN EXPLANATION

There is, however, much more to the story (Scarborough 2003b). Drawing heavily on the scholarship of Lansing (1991), Scarborough et al. (1999), and Crumley (2003), a model is proposed here grounded on the environmental changes noted above but further determined by the interplay between social decision making and the accretional pace at which changes to the landscape occurred. The concept of self-organization is used to examine the semitropical ecology occupied by the ancient Maya and their constant need to monitor and frequently substitute usable plants, soils, and some animals in attempting to mimic the natural pathways and tempos set by the many variable patches harvested. Because the biophysical environments were seldom clearly circumscribed or naturally bounded in a manner identifiable by some other nontropical settings, the ecological backdrop for the Maya was ever-changing. When attempting to make this environment a human-made setting, only slow, incremental modifications were possible without significantly altering existing pathways and disrupting the natural flow of nutrients and energy. Given the fragility of many of the microhabitats in the semitropical setting, rapid landscape change involving monocropping or clear-cutting large jungle tracts could lead to catastrophic environmental degradation and resulting population reorganization and/or relocation. Because of these constraints, the ancient Maya managed their landscape in a different way than that of several other early civilizations (Scarborough 2000, 2003a). Although more precarious than some other environments (Scarborough 2005), the Maya adapted to their semitropical ecology by dispersing their populations and attempting to replicate its natural tendencies. There were several early efforts to concentrate populations and resources during the Late Preclassic Period (as noted above) in a manner typifying the precipitous rise of several early semiarid states, but they failed. These radical attempts at centralization were initiated too rapidly and without the careful monitoring and assessing of the environment necessary to establish a sustainable long-term adaptation on the landscape. Nevertheless, by the Classic Period several socioeconomic and sociopolitical organizational adaptations were in place reflecting the self-organization of society into a highly complex human-induced environment. The uniqueness of this evolved experiment in statecraft rests in a realization by the ancient Maya that interdependencies within and between groups as well as, in turn, within and between themselves and their environs were fundamental. This was not a frequently chosen option for the short-term, radical alterations performed by early semiarid states, as most of these attempts at self-organizing onto a landscape were interrupted or redirected by significant spikes in population concentrations (true cities) induced, in part, by precipitous surplus production often based on novel breakthroughs in technology—techno-tasking (Scarborough 2003a). Furthermore, resource concentrations were frequently challenged by other groups leading to chronic and sometimes acute warfare disruptions subverting the kinds of careful microenvironmental evaluations necessary for self-organization to mature productively.

In many ways, self-organization provides the explanatory basis for the successful and lasting interdependencies identifiable for a semitropical state. Within this context, the early state became more heterarchical than hierarchical, the latter usually associated with highly stratified, centralized urban states often associated with periods of hegemonic control. Heterarchy emphasizes the network of alliances and exchanges within a system; it is less about the rigidity of class distinctions or definitions of control. Power and control remain formidable factors in any discussion of statecraft, but in the case of the ancient Maya it is the evolving set of pathways by which information and associated goods are moved that is emphasized. This is not surprising, given the agricultural implications of the seasonal and frequently erratic rainfall from zone to adjacent zone and the perishable nature of foodstuffs in a humid setting. Highly scheduled yet flexible movements of goods and services are predictable given consumption needs, a characteristic now apparent from the number of roads identifiable in the Maya Lowlands (Shaw 2001) as well as the often cited preoccupation of the Maya with time.

During the Classic Period, the Maya developed an economic adaptation that stressed a degree of specialization at the community level. This adaptation was a self-organizing adjustment to harness the diversity in the lowlands by opting to specialize in one or two resources produced or extracted in sizable quantities in excess of community-wide consumption. At the hamlet-, small village-, or town-level of production, these limited and specialized resources did not over-extend the harvesting of forest and jungle resources. In this model, the ancient Maya coordinated the distribution and exchange of goods from several large civic centers—"cities"—positioned within a region. Socioeconomic and sociopolitical regions varied in size with time and environment, though many different goods and services were brought to these civic nodes to satisfy consumption demand by all support villagers and townspeople of a region. The huge plaza spaces noted by most Mayanists—multifunctionally designed as rainfall catchments—at the principal "cities" during the Classic Period strongly suggest the forums for necessary exchange. Control and power developed with the elite manipulation of the political economy from these socioeconomic and sociopolitical nodes. Elsewhere (Scarborough and Valdez 2003), the underlying agents supporting this system are referred to as "resource-specialized communities."

During the Late Classic Period, several authors (Martin and Grube 1995; Schele and Mathews 1998) indicate that two "super-states" occupied the heartland of the Maya Lowlands. Tikal and Calakmul have different histories, but are posited to have been the principal capitals for two competing states immediately prior to the great Maya Collapse. Flannery's (1972) brilliant article argues that precipitous collapse in the case of the Classic Maya as well as most other less dramatic terminations of the early state was a result of "hypercoherence," or that condition in which the movement of information within and between levels of class, occupational specialization, and control was so unregulated and open that chaos ensued. Although written more than a generation ago, this thoughtful perspective continues to influence interpretations of early state demise. Nevertheless, the model may well be most applicable to the trajectory of the highly centralized, hegemonic state associated with much of the Old World literature than an accurate characterization of the Maya. Elsewhere (Scarborough and Valdez 2003), it is suggested that the kind of "coherence" suggested by Flannery's model is the actual glue that held the flexible and ever-adapting Maya socioeconomic and sociopolitical system together. Although there were deliberate and highly directional communication pathways in the Maya system from the outset, network interactions could rapidly change if social or environmental conditions warranted it. Given these circumstances, the great collapse of the Classic Period Maya was then a consequence of a lack of "hypercoherence," a situation in which two huge urban centers grew to dimensions and levels of centralized control that were unwieldy for the semitropical state derived from self-organized monitoring of resources, heterarchical interdependencies, and "resource-specialized communities." Tikal and Calakmul developed in isolation from their immediate sustaining population and cultivated the hubris of their rulers. Without the constant and uninterrupted evaluation of their social and biophysical environs undergirding their support in a highly evolved engineered landscape, these two huge cities began the radical devolution to collapse. Rapid and flexible movements of goods and services throughout the hinterlands carried the early semitropical state forward. Isolation, excessive centralization, and elite hubris spelled the end to Period Maya.

CONCLUSION

This chapter has not emphasized the role of climatic change in tipping sociopolitical realignments or collapse. There is little doubt that climatic conditions have fluctuated in the recent past, though the precise magnitude of those oscillations remains subject to debate. Regardless, the fundamental organizing parameters of a society are rooted in deeply convoluted histories as well as in their adaptations to an evolving landscape, histories and landscapes that determine the degree of social disruption wrecked by external changes. Given the environmental and social elasticity and longevity of the ancient Maya, other economic, political, and ideological variables internal to the underpinnings of society may weigh more heavily.

REFERENCES

Crumley, C. 2003. Alternative forms of social order. In: Heterarchy, Political Economy, and the Ancient Maya, ed. V. Scarborough, F. Valdez, and N. Dunning, pp. 136–145. Tucson: Univ. of Arizona Press.

Diamond, J.M. 2005. Collapse: How Societies Choose to Fail or Succeed. New York: Viking.

Dunning, N., S. Luzzadder-Beach, T. Beach et al. 2002. Arising from the bajos: Anthropogenic transformation of wetlands and the rise of Maya civilization. *Ann. Assn. Am. Geogr.* **92**:267–283.

Flannery, K. 1972. The cultural evolution of civilizations. *Ann. Rev. Ecol. Syst.* **3**:399–426.

Gill, R. 2000. The Great Maya Drought. Albuquerque: Univ. of New Mexico Press.

Gunn, J., W. Folan, and H. Robichaux. 1995. A landscape analysis of the Candelaria watershed in Mexico: Insights into paleoclimates affecting upland horticulture in the southern Yucatán peninsula semi-karst. *Geoarchaeology* **10**:3–42.

Hansen, R., S. Bozarth, J. Jacob, D. Wahl, and T. Schreiner. 2002. Climatic and environmental variability in the rise of Maya cvilization: A preliminary perspective from northern Peten. *Ancient Mesoamerica* **13**:273–295.

Hoddell, D., J. Curtis, and M. Brenner. 1995. Possible role of climate in the collapse of classic Maya civilization. *Nature* **375**:391–394.

Lansing, J. 1991. Priests and Programmers: Technologies of Power in the Engineered Landscape of Bali. Princeton, NJ: Princeton Univ. Press.

Martin, S., and N. Grube. 1995. Maya superstates. *Archaeology* **48**:41–43.

Matheny, R. 1986. Investigations at El Mirador, Peten, Guatemala. *Natl. Geogr. Res. Expl.* **2**:332–353.

Pohl, M., P. Bloom, and K. Pope. 1990. Interpretation of wetland farming in northern Belize: Excavations at San Antonio, Rio Hondo. In: Ancient Maya Wetlands Agriculture, ed. M. Pohl, pp. 187–254. Boulder, CO: Westview Press.

Rice, D., and T. Culbert. 1990. Population size and population change in the central Peten lakes region, Guatemala. In: Precolumbian Population History in the Maya Lowlands, ed. T. Culbert and D. Rice, pp. 123–148. Albuquerque: Univ. of New Mexico Press.

Scarborough, V.L. 1993. Water management systems in the southern Maya lowlands: An accretive model for the engineered landscape. In: Economic Aspects of Water Management in the Prehispanic New World, ed. V. Scarborough and B. Isaac, pp. 17–69. Greenwich, CT: JAI Press.

Scarborough, V.L. 1994. Maya water management. *Natl. Geogr. Res. Expl.* **10**:184–199.

Scarborough, V.L. 2000. Resilience, resource use, and socioeconomic organization: A Mesoamerican pathway. In: Environmental Disaster and the Archaeology of Human Response, ed. G. Bawden and R. Reycraft, pp.195–212. Albuquerque: Univ. of New Mexico Press.

Scarborough, V.L. 2003a. The Flow of Power: Ancient Water Systems and Landscapes. Santa Fe: School of American Research Press.

Scarborough, V.L. 2003b. How to interpret an ancient landscape. *Proc. Natl. Acad. Sci.* **100**:4366–4368.

Scarborough, V.L. 2005. The power of landscapes. In: A Catalyst for Ideas: Anthropological Archaeology and the Legacy of Douglas W. Schwartz, ed. V.L. Scarborough, pp. 209–228. Santa Fe: School of American Research Press.

Scarborough, V.L., and G. Gallopin. 1991. A water storage adaptation in the Maya lowlands. *Science* **251**:658–662.

Scarborough, V.L., J. Schoenfelder, and J. Lansing. 1999. Early statecraft on Bali: The water temple complex and the decentralization of the political economy. *Res. Econ. Anthro.* **20**:299–330.

Scarborough, V.L., and F. Valdez. 2003. The engineered environment and political economy of the three rivers region. In: Heterarchy, Political Economy, and the Ancient Maya, ed. V. Scarborough, F. Valdez, and N. Dunning, pp. 1–13. Tucson: Univ. of Arizona Press.

Schele, L., and P. Mathews. 1998. The Code of Kings. New York: Scribner.

Shaw, J. 2001. Maya sacbeob: Form and function. *Ancient Mesoamerica* **12**:261–272.

5

Climate, Complexity, and Problem Solving in the Roman Empire

Joseph A. Tainter[1] and Carole L. Crumley[2]

[1]Global Institute of Sustainability and School of Human Evolution and Social Change, Arizona State University, Tempe, AZ 85287, U.S.A.
[2]Department of Anthropology, University of North Carolina, Chapel Hill, NC 27514, U.S.A.

ABSTRACT

The Roman Empire was established in northwestern Europe in the last two centuries B.C. and the first century A.D. during a warm, dry era known as the Roman Warm Period or the Roman Climatic Optimum. In northwestern Europe the Romans disrupted earlier systems of production, exchange, and political relations to establish Mediterranean production systems oriented toward markets and government revenues. Being based on solar energy, the Empire as a whole ran on a very thin fiscal margin. The end of the Roman Warm Period would have introduced uncertainty into agricultural yields just as the Empire was experiencing a concatenation of crises during the third century A.D. The Roman response to these crises was to increase the complexity and costliness of the government and army, and to increase taxes to pay for the new expenditures. This undermined the well-being of the population of peasant agriculturalists, leading to a reduction in the government's ability to address continuing problems. The Western Roman Empire collapsed while in the process of consuming its capital resources: productive land and peasant population. The experience of the Roman Empire has implications for the IHOPE project, and for problem solving in general, in two areas: (a) the relationship of hierarchy to heterarchy, and local to global, in addressing environmental and social problems, and (b) the development of complexity, costliness, and ineffectiveness in problem solving.

INTRODUCTION

The Roman Empire has been studied for centuries by those who see in it lessons for their own time. We are among those who perceive in the Empire a case study whose value is timeless. Where once the Roman Empire was studied to draw political or moral conclusions, we will show that it can yield fresh lessons in such contemporary problems as climate change, government insolvency, the evolution of institutions, and the relationship of heterarchy and hierarchy in problem solving.

OVERVIEW OF THE ROMAN EMPIRE

For an agrarian empire activated by solar energy, the territory most efficiently administered would have been the Mediterranean Basin and fringing lands. This is because travel and transport by land was up to 56 times more costly than by sea (Jones 1964). As the Romans pushed into northwestern Europe and the interior of Anatolia, they conquered peoples who were less developed economically than those of the Mediterranean littoral, and at the same time incurred higher costs in administration. The cost of the Rhine garrisons, for example, may have equaled the tax revenues of Gaul north of Provence, leaving the central government with no net profit on the region (Drinkwater 1983, p. 65).

For an agrarian empire, the highest net returns are realized in the conquest phase, when the accumulated surpluses of the subject peoples are appropriated. These surpluses are the stored accumulation of past solar energy, transformed into the production of precious metals, works of art, and peasant populations. As have many empire-builders, Rome found her conquests initially to be highly profitable. In 167 B.C., for example, the Romans captured the Macedonian treasury, and promptly eliminated taxation of themselves. When Pergamon was annexed in 130 B.C. the state budget was doubled, from 25 million to 50 million denarii. (The denarius, discussed below, was a coin initially of very pure silver.) After he conquered Syria in 63 B.C., Pompey raised it to 85 million denarii. Julius Caesar relieved the Gauls of so much gold that its value in Rome fell 36 percent (Lévy 1967, pp. 62–65).

Once these accumulated surpluses are spent, the conqueror must assume responsibility to garrison, administer, and defend the province. These responsibilities may last centuries and are typically financed from yearly agricultural surpluses. The concentrated, high-quality resources available at conquest give way to resources derived from dispersed subsistence agriculture, which yields little surplus per capita (about 1/2 metric ton per hectare, and a yield of 3 to 4 times the seeding rate, in Roman and Medieval Europe [Smil 1994, pp. 66, 74]). Costs rise and benefits decline. When fresh problems arise, they must be met by taxing the populace, and if tax rates are insufficient they will likely be raised. Paying for continuity in such a system depends on establishing a bureaucracy to aggregate the small surpluses of individual producers. Empires can be supported in this way because, in an agrarian landscape, there are a lot of people to tax. Because of these constraints, imperial taxation systems tend to be elaborate and costly (Tainter et al. 2006).

Even the first emperor, Augustus (27 B.C.–A.D. 14), complained of fiscal shortfalls, and relieved the state budget from his own wealth. Facing war with Parthia and the cost of rebuilding Rome after the Great Fire, Nero (54–68) began in A.D. 64 a policy that later emperors found irresistible. He debased the primary silver coin, the denarius, reducing the alloy from 98 to 93 percent silver (Figure 5.1). It was the first step down a slope that resulted two centuries later in a worthless currency and an insolvent government.

Figure 5.1 depicts an extraordinary data set. It is the only glimpse we have into the year-to-year fiscal status of an ancient government. Since 90% of government revenue came from agricultural taxes (Jones 1964), it is clear both that the government ran on a very thin margin in good times, and that over the long-run such a complex imperial system could be sustained on solar energy only if no crises emerged that would require extraordinary expenditures.

Crises, of course, are normal and inevitable. In the half-century from 235 to 284 a concatenation of crises nearly brought the Empire to an end. There were foreign and civil wars almost without interruption. The period witnessed 26 legitimate emperors and perhaps 50 usurpers. Cities were sacked and frontier provinces devastated. The Empire shrank in the 260s to Italy, the Balkans, and North Africa. By prodigious effort the Empire survived the crisis, but it emerged at the turn of the fourth century A.D. as a very different organization.

In response to the crises, the emperors Diocletian and Constantine, in the late third and early fourth centuries, designed a government that was larger, more complex, and more highly organized. They doubled the size of the army. To pay for this the government taxed its citizens more heavily, conscripted their labor, and dictated their occupations. Villages were responsible for the taxes on their

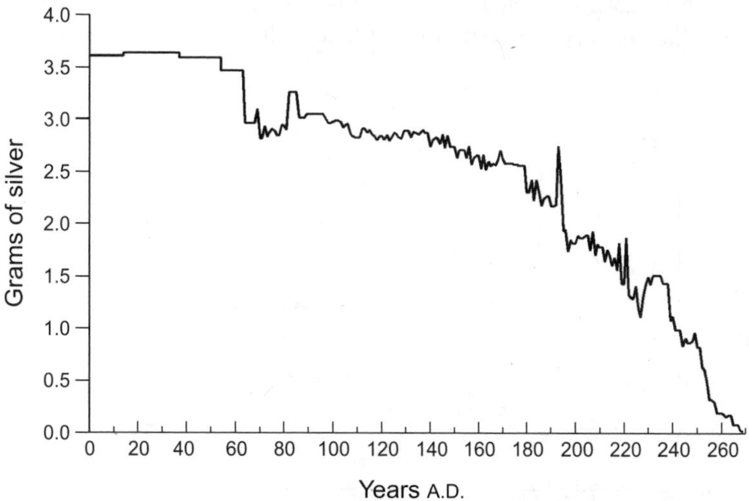

Figure 5.1 Debasement of the denarius to A.D. 269. Data from Cope (1969, 1974, and unpublished analyses on file in the British Museum); King (1982); LeGentilhomme (1962); Tyler (1975); and Walker (1976, 1977, 1978); see also Besly and Bland (1983, pp. 26–27) and Tainter (1994, p. 1217).

members, and one village could even be held liable for another. Despite several monetary reforms a stable currency could not be found (Figure 5.2). As masses of worthless coins were produced, prices rose higher and higher (Figure 5.3). Money-changers in the east would not convert imperial currency, and the government even refused to accept its own coins for payment of taxes.

With the rise in taxes, the population could not recover from plagues in the second and third centuries. There were chronic shortages of labor. Marginal lands went out of cultivation. Faced with taxes, peasants would abandon their lands and flee to the protection of a wealthy landowner. By A.D. 400 most of Gaul and Italy were owned by about 20 senatorial families.

From the late fourth century the peoples of central Europe could no longer be kept at bay. They forced their way into Roman lands in western Europe and North Africa. The government came to rely almost exclusively on troops from Germanic tribes. When finally they could not be paid, they overthrew the last emperor in Italy in 476 (Boak 1955; Russell 1958; Jones 1964, 1974; Hodgett 1972; MacMullen 1976; Wickham 1984; Williams 1985; Tainter 1988, 1994; Duncan-Jones 1990; Williams and Friell 1994; Harl 1996).

LOCAL EFFECTS: THE ROMAN EMPIRE IN GAUL

To bring this broad overview to a local level, we focus on Gaul, a part of the Roman Empire that both of us know well. Crumley directs a long-term project in Burgundy (e.g., Crumley and Marquardt 1987), while Tainter is preparing a synthesis of the Roman period in the lower Rhône Valley. Gaul figured prominently

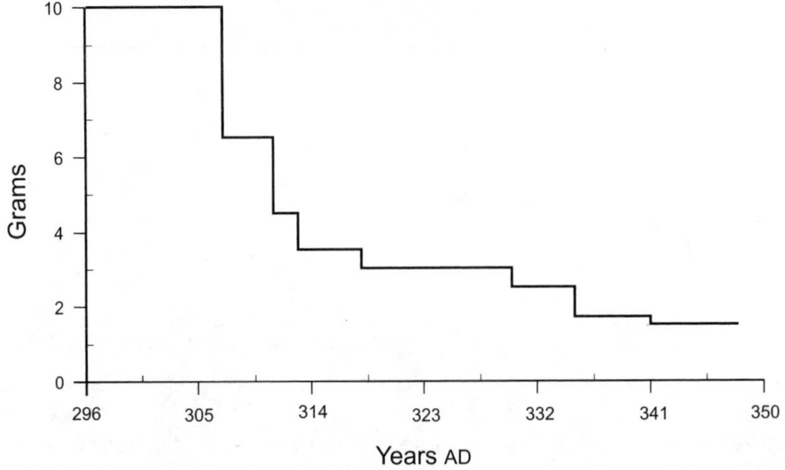

Figure 5.2 Reductions in the weight of the follis, A.D. 296 to 348. Data from Van Meter (1991, p. 47).

in the political, military, and economic history of the Empire, not least because of its proximity to the Rhine frontier.

Gaul was important to the Empire's economy and the government's finances. About 15% of the army's budget went to the Rhine garrisons (Drinkwater 1983, p. 65). While the Romans found Gaul productive, its agricultural output is variable. Over Burgundy, for example, three climatic regimes converge: the Atlantic (Greenland High), the Continental (Siberian High), and the Mediterranean (Azores High). These combine with terrain to produce a fourth climatic zone in central France, the Mountainous Zone over the Massif Central. The continental regime is characterized by summer dominant rainfall and cold, dry winters. The oceanic regime is cool and wet in the summer but mild and wet to dry in the autumn, when it receives most of its rain. The Mediterranean regime is hot and dry in the summer, receiving most of its rain in the winter (Crumley and Green 1987, pp. 28–32; Crumley 2003, pp. 141–142).

Between approximately 300 B.C. and A.D. 300, northwestern Europe experienced a prolonged period of warm and dry weather that is termed the Roman Warm Period or the Roman Climatic Optimum. The Azores high pressure system dominated much of western Europe during this period. The Mediterranean regime produced hot, dry summers and winter rains. The Romans expanded into southern Gaul in the second century B.C., and the remainder in the first. They brought with them Italian crops and production systems. Spatially diffuse Celtic systems of multiple-species agriculture and pastoralism gave way in many areas to intensive commercial production. It is in this period that we hear of grapes growing in southern Britain (Crumley 1993).

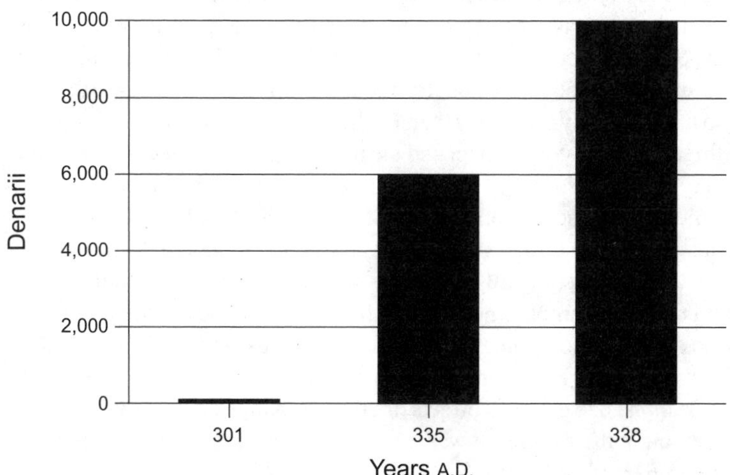

Figure 5.3 Denarii per modius of wheat in the early fourth century A.D. A modius was about nine liters. Data from Jones (1964, p. 119).

The end of this warm period is of interest (Denton and Karlén 1973; Gunn et al. 2004). Beginning in the second century A.D., ice rafting increased in the North Atlantic (Bond et al. 2001, p. 2131), signaling a transition in northwestern Europe toward conditions that were increasingly cool, moist, and variable. This change would have affected Italian-style agriculture practiced in northwestern Europe. As Crumley (2003, p. 142) has noted , "There would have been increasing instances of crop failure (due to late spring frosts and/or cool, damp summers characteristic of the temperate European pattern) and ruin at harvest (hailstorms) or upon storage (blight)."

The Romans had found in Gaul a mixed pattern of agriculture and husbandry that was tied to environmental variation. Diversity in production was linked to exchange systems and the institution of clientage. The system was flexible, appropriate to local conditions, and resilient. In many areas the Romans replaced native production systems with nonresilient, inflexible, cash cropping that was linked hierarchically to urban markets and the needs of the state. Outside of the Mediterranean zone for which it was developed, such a system lacked the productive flexibility that the changing climate required (Crumley 1987). As agricultural production became less certain, so did taxes.

In an area as large and diverse as the Roman Empire, many climatic regimes affected production, from the Greenland High to the rains of east-central Africa that feed the Nile. One might conclude that government revenues would hardly be affected by agricultural problems in only one region. Production elsewhere would compensate. But during the crisis of the third century A.D., the government's fiscal status experienced a pronounced downturn. The precious-metal content of the silver currency began its final plunge, reaching a nadir of 1.5% in 269 (Cope 1969). The debasements (Figure 5.1) produced such inflation (Figure 5.3) that the government could no longer fulfill its procurement needs with money. Soldiers were particularly affected. Apparently their pay in coinage no longer sufficed to buy supplies. By the time of Diocletian (284–305) the state was so unable to rely on money that it collected taxes in the form of supplies useable directly by the government and the military. Soldiers now received much of their pay in supplies, and would for the next century. With transport costly, it was desirable to have these supplies produced near military bases.

Thus, by the late third century, local production mattered. If local production was inadequate, not only might soldiers not be paid, they might not have enough to eat. Discontent among the Rhine garrisons was to be avoided. For most staples, no doubt the government had back-up sources, and Roman officials would not hesitate to seize grain supplies even if peasants went hungry. Thus we should not overemphasize the local effects of climate change on the government's ability to maintain the frontier. It was, rather, another of the multiple problems that the government faced at this time.

The crises of the third century affected Gaul profoundly. A raiding party of Franks crossed Gaul into Spain in 262, sacking Tarraco. There is a great increase in Gallic coin hoards from 259/260 onward. Significantly, many hoards were never recovered. Some fine silver treasures were also buried at this time, including temple treasures. These surely would have been recovered had the depositor been able to do so (King 1990, p. 174; Watson 1999, pp. 33–34). The rural population declined, either killed or captured by barbarians, starving, or deserting their fields to join bands of brigands. Town populations fell also, sometimes to the size of the Celtic villages that preceded the Empire (Boak 1955, pp. 19, 26, 38–39, 55–56, 113; MacMullen 1976, pp. 18, 183; Rostovtzeff 1926, p. 424). Vienne, for example, shrank from 200 to 20 hectares, Lyon from 160 to 20, and Autun from 200 to 10 (Randsborg 1991, p. 91). Paris contracted to the Île de la Cité (Williams 1985, p. 20).

Gaul north of Provence, along with Spain and Britain, broke away from the Roman Empire in 260 and remained independent until 274. Troops were withdrawn from the Rhine for the final battle with the Empire in 274, which left the frontier weakened (Watson 1999, pp. 93–94). An assortment of Vandals, Franks, Burgundians, Alamanni, and others broke through in the years 274–276, sacking Trier and many other towns (King 1990, p. 176; Watson 1999, pp. 95–98, 102). Cities across the Empire built new fortifications at this time, including Rome itself.

Part of Diocletian's reform was to ensure that the Empire had the fiscal resources to pay for a government and military that were larger and more complex. Diocletian developed an Imperial innovation: Rome's first government budget. Each year calculations were made of anticipated expenses, and a tax rate established to provide the revenue. Just to establish the tax system was an immense affair, requiring a complete census of people and land across the Empire. The tax rate was established from a master list of the Empire's resources, broken down province by province, city by city, household by household, field by field. Diocletian's successors revised the rates ever upward. Taxes apparently doubled between A.D. 324 and 364 (Williams 1985, pp. 118–125; Jones 1974, p. 82).

During the late Empire there was substantial abandonment of arable, and formerly cultivated, land. This problem first appeared in the late second century, perhaps due to plague, and was a subject of Imperial legislation from before Diocletian's time to that of Justinian (527–565). Aurelian (270–275) held city councils responsible for the taxes due on deserted lands. In some eastern provinces under Valens (364–378), from one-third to one-half of arable lands were deserted.

The tax system of the late Empire seems to have been to blame, for the rates were so high that peasant proprietors could accumulate no reserves. At Antaeopolis, Egypt, ca. 527, tax assessments in kind and money totaled one-fourth to one-third of average gross yields. At Ravenna, ca. 555, the situation was similar, with a tax:rent ratio of 57:43 (Wickham 1984, p. 11). If 50% of the yield went to seed and subsistence (Smil 1994, p. 74), then tax amounted to one-half to two-thirds of the surplus (or in bad years, all of it). If barbarians raided, or drought or locusts diminished the crop, farmers either borrowed or starved. Eventually their lands passed to creditors, to whom they became tenants.

Whatever crops were brought in had to be sold for taxes, even if it meant starvation for the farmer and his family. Farmers who could not pay their taxes were jailed, sold their children into slavery, or abandoned their homes and fields. In times of famine, farmers flocked to cities, where stores of grain were to be had. The state, moreover, always had a backup on taxes due, extending obligations to widows or orphans, even to dowries. It is no wonder that the peasant population failed to recover from plagues in the second and third centuries. Conditions did not favor the formation of large families.

Under these circumstances it became unprofitable to cultivate marginal land, as too often it would not yield enough for taxes and a surplus. And so lands came increasingly to be deserted. Faced with taxes, a small farmer might abandon his land to work for a wealthy neighbor, who in turn would be glad to have the extra labor.

The tax system of the late Empire could not accommodate the fluctuations in yield that would have afflicted Gaul in this period. The resulting fiduciary chaos reflects perennial tensions between social and political systems under local control and those at a distance, often pitting systems that value individual and group differences (hierarchies) against those that value more egalitarian and networked relations (heterarchies). Such differences in management structure and style have implications for the movement of information, for the diversity of solutions to problems, and for internal and external security.

Heterarchies are systems (or subsystems) in which each element possesses the potential of being unranked (relative to other elements) or ranked in a number of different ways (Crumley 1979, p. 144; Ehrenreich et al. 1995). For example, widely shared local knowledge in an agrarian economy allows individuals and communities more effectively to manage changes in environmental and economic conditions, while decisions made at a distance and with little detailed information, exacerbate local circumstances (Scott 1998). Similarly, strongly networked societies have incentives to cooperate and avoid the need for costly coercion. In general, heterarchical, networked societies, while doubtless with problems of their own, are more resilient to local environmental challenges than societies with marked social, political, and economic inequities (Crumley 2001, 2003, 2005). This understanding parallels the role of diversity in biotic communities (Holling et al. 2002, p. 21).

Heterarchy poses knowledge that is local (i.e., contextualized) and behavior that is consensual against hierarchies, whose knowledge is distant and decontextualized and whose approach to action is coercive. Diocletian's tax system was a massive exercise in hierarchy. Tax rates were set each year in anticipation of the government's needs. This was a distant, centralized process that ignored local knowledge, and could not take into account variable conditions. Taxes had to be paid regardless of fluctuations in yields. Delinquent taxes were remitted occasionally in the early and middle fourth century, but so frequently after 395 that it appears there was a general agricultural breakdown in the West (Boak 1955, p. 52). In Gaul, rebellious bands called the bagaudae persisted for decades at a time. In the mid fifth century a deputation of property owners and municipal authorities invited the Burgundians to occupy some of their lands (Isaac 1971, p. 127). "[B]y the 5th century," concludes Adams (1983, p. 47), "men were ready to abandon civilization itself in order to escape the fearful load of taxes."

On the last day of 406 an alliance of Vandals, Suevi, and Alans crossed the frozen Rhine and virtually overnight a major part of Gaul was lost. In time some of these people moved on to Spain or North Africa, while others remained in Gaul as nominal Roman allies. Rome was able eventually to reestablish a measure of control over Gaul and Spain. In 429, however, the Vandals crossed to North Africa and took Carthage in 439. Rome's North African food supply was gone forever.

The Western Empire was by this point in a downward spiral. Lost or devastated provinces meant lower government income and less military strength. Lower military strength in turn meant that more areas would be lost or ravaged. By 448 Rome had lost most of Spain (Barker 1924, pp. 413–514). In 458 the Emperor Majorian (457–461) remitted all taxes in arrears (Wickham 1984, p. 19). After the fall of Majorian in 461, Italy and Gaul had little connection. The Empire shrank to Italy, Raetia, and Noricum. The most important ruler in the West was no longer the Roman Emperor but the Vandal King, Gaiseric (Ferrill 1986, p. 154; Wickham 1981, p. 20).

IMPLICATIONS FOR THE IHOPE PROJECT

While the end of the Western Roman Empire presents many enduring lessons, as generations of scholars can attest, we wish to emphasize two in particular.The first lesson concerns the relationship of hierarchy to heterarchy, and global to local, in solving environmental and social problems. We see in the case of Roman taxation what happens when information about environmental capacity is decontextualized. Diocletian's distant, inflexible tax system exemplifies what Scott has called "environmental and social taxidermy" (1998, pp. 93, 228)—an attempt to order administratively that which is inherently flexible and changing. Administrative systems impose categories on the social and environmental realities of local life. Inevitably these categories—in the Roman case designed to accommodate all circumstances from Britain to Egypt—winnow the variability of local conditions. Once administrators establish such categories, they then try to force social reality to conform to them (Scott 1998). Hierarchical systems tend to commit themselves to specific structures and solutions, establishing brittleness where flexibility may be required. For peasants living on small margins, the consequences can be disastrous.

Local, contextualized environmental information is well integrated into heterarchical organization, which is better able to incorporate varieties of experience. Heterarchical systems operate either by consensus or by ad hoc consensual leadership. While the process of adjusting to new circumstances in heterarchy is unavoidably slow, it derives its legitimacy from consensus. Those who have participated in developing a heterarchical consensus are intrinsically committed to implementing it and will experience strong social pressures should they fail to do so (McIntosh et al. 2000, p. 31; Crumley 2003, pp. 138–139).

Today's approach to understanding environmental problems—on the part of both policy makers and scientists—has been largely hierarchical: authoritative, distant, and too often decontextualized. Our tendencies to develop abstract, aggregated models and to formulate international agreements exacerbate this problem (McIntosh et al. 2000; Crumley 2000; Tainter 2001). The upper levels in any hierarchical system act only on aggregated, filtered information and respond slowly to signals from below. The local information that is important in environmental conservation and productivity cannot be developed from a distance. There is wisdom in the exhortation of the environmental movement to think globally but act locally. Even well-meaning academic exercises may unintentionally harm the very people they are designed to assist, if they are pursued exclusively at an abstract, aggregated level. We are concerned that conventional, abstract modeling will no more be able to accommodate the flux of local circumstances than could Diocletian's tax system. While we certainly do not recommend that abstract, aggregated modeling not be pursued, it is important to proceed in awareness of what this approach overlooks. Just as all news is local and all politics are local, all environmental problems are local to the people whose sacrifices will be needed to effect solutions.

This is not to deny a role for distant, abstract analysis. Global change is, of course, global, and some aspects of it must be addressed in a decontextualized, aggregated manner. We recommend that within the IHOPE project, modeling will proceed simultaneously, and in concert with, locality-based studies of how people perceive their environments, transmit information, respond to external interventions, and recognize and accommodate change. IHOPE's approach should be neither hierarchical nor heterarchical exclusively, but a synthesis of both. This is challenging, but to attempt less is to fail in advance.

The second lesson concerns the evolution of complexity in problem-solving institutions, which the Roman Empire illustrates well. Complexity increases as systems differentiate in structure and increase in organization. Humans employ complexity as a response to problems. Much complexity and costliness in human societies emerges from resolving problems that range from mundane to vital. As the problems that institutions confront grow in size and complexity, problem solving grows more complex as well. Think of the attacks of September 11, 2001, and the growth of bureaucracy and regulation that followed. Much of the immediate response to the attacks was to increase the complexity of public institutions, by establishing new agencies, absorbing existing agencies into the federal government, and exerting control over behavior from which a threat might emerge. As seen in this example, we resort to complexity to solve problems because changes in organization can often be implemented quickly.

Complexity has great utility in problem solving, but it also costs. The evolution of complexity is a benefit–cost relation. The costs of complexity may be measured in energy, metabolic rates, labor, money, time, or any other unit of accounting. At the time a problem arises, increments to complexity may seem small and affordable. It is the continual accumulation of complexity and costs that becomes detrimental. As a benefit–cost function, complexity in problem solving can reach diminishing returns and become ineffective. In their complex phase, institutions may lack the fiscal reserves to address new challenges, whether the new challenges are hostile neighbors or environmental perturbations. A society that has adopted much costly complexity may lose resilience and become vulnerable to challenges that it could once have overcome, and even become more likely to collapse.

It is important not to think of complexity as inherently detrimental in human societies. In the early phases of problem solving, increasing complexity is typically effective, giving increasing returns and creating synergistic effects and positive feedbacks among such variables as population, agricultural production, political organization, fiscal strength, and military strength. In the Roman Empire of the late third and early fourth centuries A.D., increasing complexity allowed the government to resolve the multiple crises described above—at least for the short term. The problem with complexity comes when additional expenditures fail to produce proportionate benefits. A society entering this phase is weakened fiscally, in that resources must be allocated to activities that are needed just to maintain the status quo; it is also weakened in its legitimacy, because the support population is alienated by high taxes that produce few discernable benefits. Collapse becomes a matter of mathematical probability, as inevitably an insurmountable crisis will emerge (Tainter 1988, 2000).

Rising complexity in problem solving drives resource consumption. Problems occur in the present but environmental damage may be deferred. Thus the link between benefits and costs is often hidden, and contemporary decisions may have little connection to whether an effort fails or succeeds (Tainter 2000). Once environmental problems are evident, their resolution usually requires still more complexity and expenditure, the predicament in which we find ourselves today (Allen et al. 2003).

For humanity today, a number of major problems are clearly on the horizon and will manifest themselves increasingly over the next generation or two. In addition to questions of climate change and other environmental transformations, there are rising costs of energy, the growing costs of security, decaying infrastructure in many nations, aging populations and the problem of funding retirement pensions, increasing requirements in education driven by competition and technological changes, increasing reluctance of governments to tax, and so on. Addressing any of these problems would require societal expenditures that are large, but perhaps within the capacities of industrial nations. The challenge will be to address all of them simultaneously without reducing accustomed standards of living.

A major problem faced by the late Romans was that increasing complexity and expenditures were undertaken just to maintain the status quo. No new lands were conquered and no major new resources acquired. The benefit–cost ratio of Imperial rule declined, reducing its legitimacy. Toward the end the Empire sustained itself by consuming its capital resources: productive land and peasant population.

Many of the problems noted above fall into the same category of undertaking higher costs merely to maintain the status quo: energy costs, security, replacing infrastructure, funding retirement pensions, and paying for education. Much money will be spent restoring the environmental damage caused by previous economic activity and mitigating the effects of climate change. Given budgetary constraints in every nation, funding for much of this activity will be inadequate and some problems may not be addressed at all, unless there are major redirections of national and international priorities. If addressing the problems we foresee should cause the industrial standard of living to stagnate or fall, existing forms of government may lose legitimacy. Societies may polarize around "progressive" factions favoring an environmental restoration agenda, and "conservative" factions arguing that environmental conditions are irrelevant to prosperity. This bifurcation is, of course, already evident, and is exacerbated by special interests comprised of people who benefit from the status quo. A worthwhile undertaking for IHOPE's modeling efforts would be to assess the costs and benefits of addressing the problems that are foreseeable in our future. Will addressing these problems bring continuing growth in the complexity of industrial societies? How will the costs and benefits of addressing emerging problems intersect with the fiscal capacities of future economies, and with competing demands for public and private funds? Will addressing these problems require a reduction in accustomed standards of living or a reconceptualization of them? We all hope that our future will not be like that of the Roman Empire—so constrained by the costs of problem solving as to impoverish and alienate the very people on whom the future depends.

ACKNOWLEDGMENTS

We are pleased to thank our colleagues at the 96th Dahlem Workshop for their comments and suggestions on the earlier version of this chapter.

REFERENCES

Adams, R.M. 1983. Decadent Societies. San Francisco: North Point.

Allen, T.F.H., J.A. Tainter, and T.W. Hoekstra. 2003. Supply-Side Sustainability. New York: Columbia Univ. Press.

Barker, E. 1924. Italy and the West, 410–476. In: The Cambridge Medieval History, vol. 1, The Christian Roman Empire and the Foundation of the Teutonic Order (2d ed.), ed. H.M. Gwatkin and J.P. Whitney, pp. 392–431. Cambridge: Cambridge Univ. Press.

Besly, E., and R. Bland. 1983. The Cunetio Treasure: Roman Coinage of the Third Century AD. London: British Museum.

Boak, A.E.R. 1955. Manpower Shortage and the Fall of the Roman Empire in the West. Ann Arbor: Univ. of Michigan Press.

Bond, G., B. Kromer, J. Beer et al. 2001. Persistent solar influence on North Atlantic climate during the Holocene. *Science* **294**:2130–2136.

Cope, L.H. 1969. The nadir of the imperial Antoninianus in the reign of Claudius II Gothicus, A.D. 268–270. *Numis. Chron.* **7, IX**:145–161.

Cope, L.H. 1974. The Metallurgical Development of the Roman Imperial Coinage during the First Five Centuries A.D. Ph.D. diss., Liverpool Polytechnic.

Crumley, C.L. 1979. Three locational models: An epistemological assessment for anthropology and archaeology. In: Advances in Archaeological Method and Theory, vol. 2, ed. M.B. Schiffer, pp. 141–173. New York: Academic.

Crumley, C.L. 1987. Celtic settlement before the conquest: The dialectics of landscape and power. In: Regional Dynamics: Burgundian Landscapes in Historical Perspective, ed. C.L. Crumley and W.H. Marquardt, pp. 403–429. San Diego: Academic.

Crumley, C.L. 1993. Analyzing historic ecotonal shifts. *Ecol. Appl.* **3**:377–384.

Crumley, C.L. 2000. From garden to globe: Linking time and space with meaning and memory. In: The Way the Wind Blows: Climate, History, and Human Action, ed. R.J. McIntosh, J.A. Tainter, and S. Keech McIntosh, pp. 193–208. New York: Columbia Univ. Press.

Crumley, C.L. 2001. Communication, holism, and the evolution of sociopolitical complexity. In: Leaders to Rulers: The Development of Political Centralization, ed. J. Haas, pp. 19–33. New York: Plenum.

Crumley, C.L. 2003. Alternative forms of social order. In: Heterarchy, Political Economy, and the Ancient Maya: The Three Rivers Region of the East-Central Yucatán Peninsula, ed. V.L. Scarborough, F. Valdez, Jr., and N. Dunning, pp. 136–145. Tucson: Univ. of Arizona Press.

Crumley, C.L. 2005. Remember how to organize: Heterarchy across disciplines. In: Non-linear Models for Archaeology and Anthropology: Continuing the Revolution, ed. C.S. Beekman and W.W. Baden, pp. 35–50. Aldershot: Ashgate.

Crumley, C.L., and P.R. Green. 1987. Environmental setting. In: Regional Dynamics: Burgundian Landscapes in Historical Perspective, ed. C.L. Crumley and W.H. Marquardt, pp. 19–39. San Diego: Academic.

Crumley, C.L., and W.H. Marquardt, eds. 1987. Regional Dynamics: Burgundian Landscapes in Historical Perspective. San Diego: Academic.

Denton, G.H., and W. Karlén. 1973. Holocene climate variations: Their pattern and possible cause. *Quat. Res.* **3**:155–205.

Drinkwater, J.F. 1983. Roman Gaul: The Three Provinces, 58 B.C.–A.D. 260. Ithaca: Cornell Univ. Press.

Duncan-Jones, R. 1990. Structure and Scale in the Roman Economy. Cambridge: Cambridge Univ. Press.
Ehrenreich, R.M., C.L. Crumley, and J.E. Levy, eds. 1995. Heterarchy and the Analysis of Complex Societies. Archaeological Papers of the American Anthropological Assn. 6. Washington, D.C.: Am. Anthropological Assn.
Ferrill, A. 1986. The Fall of the Roman Empire: The Military Explanation. London: Thames and Hudson.
Gunn, J., C.L. Crumley, E. Jones, and B.K. Young. 2004. A landscape analysis of Western Europe during the early Middle Ages. In: The Archaeology of Global Change: The Impact of Humans on Their Environments, ed. C.L. Redman, S.R. James, P.R. Fish, and J.D. Rogers, pp. 165–185. Washington, D.C.: Smithsonian Institution Press.
Harl, K.W. 1996. Coinage in the Roman Economy, 300 B.C. to A.D. 700. Baltimore: Johns Hopkins Univ. Press.
Hodgett, G.A.J. 1972. A Social and Economic History of Medieval Europe. London: Methuen.
Holling, C.S., L.H. Gunderson, and D. Ludwig. 2002. In quest of a theory of adaptive change. In: Panarchy: Understanding Transformations in Human and Natural Systems, ed. L.H. Gunderson and C.S. Holling, pp. 3–22. Washington, D.C.: Island.
Isaac, J.P. 1971. Factors in the Ruin of Antiquity: A Criticism of Ancient Civilization. Toronto: Bryant.
Jones, A.H.M. 1964. The Later Roman Empire, 284–602: A Social, Economic and Administrative Survey. Norman: Univ. of Oklahoma Press.
Jones, A.H.M. 1974. The Roman Economy: Studies in Ancient Economic and Administrative History. Oxford: Blackwell.
King, C.E. 1982. Issues from the Rome mint during the sole reign of Gallienus. In: Actes du 9ème Congrès International de Numismatique, pp. 467–485. Louvain-la-Neuve: Assn. Intle. Numismates Professionels.
King, A. 1990. Roman Gaul and Germany. Berkeley: Univ. of California Press.
LeGentilhomme, P. 1962. Variations du titre de l'Antoninianus au IIIe siècle. *Revue Numismatique* **VI(IV)**:141–166.
Lévy, J.-P. 1967. The Economic Life of the Ancient World, trans. J.G. Biram. Chicago: Univ. of Chicago Press.
MacMullen, R. 1976. Roman Government's Response to Crisis, A.D. 235–337. New Haven, CT: Yale Univ. Press.
McIntosh, R.J., J.A. Tainter, and S. Keech McIntosh. 2000. Climate, history, and human action. In: The Way the Wind Blows: Climate, History, and Human Action, ed. R.J. McIntosh, J.A. Tainter, and S. Keech McIntosh, pp. 1–42. New York: Columbia Univ. Press.
Randsborg, K. 1991. The First Millennium A.D. in Europe and the Mediterranean: An Archaeological Essay. Cambridge: Cambridge Univ. Press.
Rostovtzeff, M. 1926. The Social and Economic History of the Roman Empire. Oxford: Oxford Univ. Press.
Russell, J.C. 1958. Late ancient and medieval population. *Trans. Am. Phil. Soc.* **48(3)**.
Scott, J. 1998. Seeing Like a State: How Certain Schemes to Improve the Human Condition Have Failed. New Haven, CT: Yale Univ. Press.
Smil, V. 1994. Energy in World History. Boulder, CO: Westview.
Tainter, J.A. 1988. The Collapse of Complex Societies. Cambridge: Cambridge Univ. Press.

Tainter, J.A. 1994. La Fine Dell'Amministrazione Centrale: Il Collaso Dell' Impero Romano in Occidente. In: Storia d'Europa, vol. Secondo: Preistoria e Antichità, ed. J. Guilaine and S. Settis, pp. 1207–1255. Turin: Einaudi.

Tainter, J.A. 2000. Problem solving: Complexity, history, sustainability. *Pop. Env.* **22**: 3–41.

Tainter, J.A. 2001. Sustainable rural communities: General principles and North American indicators. In: People Managing Forests: The Links between Human Well-Being and Sustainability, ed. C.J. Pierce Colfer and Y. Byron, pp. 347–361. Washington, D.C.: Resources for the Future and Bogor, Indonesia: Center for Intl. Forestry Research.

Tainter, J.A., T.F.H. Allen, and T.W. Hoekstra. 2006. Energy transformations and post-normal science. *Energy* **31**:44–58.

Tyler, P. 1975. The Persian wars of the 3rd century A.D. and Roman imperial monetary policy, A.D. 253–68. *Historia, Einzelschr.* **23**.

Van Meter, D. 1991. The Handbook of Roman Imperial Coins. Nashua, NH: Laurion Numismatics.

Walker, D.R. 1976. The Metrology of the Roman Silver Coinage. I. From Augustus to Domitian. British Archaeol. Reports, Suppl. Ser. 5. Oxford: Brit. Archaeol. Reports.

Walker, D.R. 1977. The Metrology of the Roman Silver Coinage. II. From Nerva to Commodus. Brit. Archaeol. Reports, Suppl. Ser. 22. Oxford: Brit. Archaeol. Reports.

Walker, D.R. 1978. The Metrology of the Roman Silver Coinage. III. From Pertinax to Uranius Antoninus. Brit. Archaeol. Reports, Suppl. Ser. 40. Oxford: Brit. Archaeol. Reports.

Watson, A. 1999. Aurelian and the Third Century. New York: Routledge.

Wickham, C. 1981. Early Medieval Italy: Central Power and Local Society 400–1000. London: Macmillan.

Wickham, C. 1984. The other transition: From the ancient world to feudalism. *Past and Present* **103**:3–36.

Williams, S. 1985. Diocletian and the Roman Recovery. New York: Methuen.

Williams, S., and G. Friell. 1994. Theodosius: The Empire at Bay. New Haven, CT: Yale Univ. Press.

6

Integration of Climatic, Archaeological, and Historical Data

A Case Study of the Khabur River Basin, Northeastern Syria

Frank Hole

Dept. of Anthropology, Yale University, New Haven, CT 06520–8277, U.S.A.

ABSTRACT

In the Khabur River Basin of northeastern Syria, rain-fed agriculture and livestock husbandry, with abundant natural reserves of arable land and pasture, were generally sustainable until the 20th century, but rapid population growth, expansion of settlements, and mechanized agriculture now strain the resilience of the natural and human systems. The 10,000-year-long history of human occupation of the Khabur River Basin is one of periods of successful agriculture and settlement, followed by long stretches when the region served primarily as grazing land for the flocks of mobile herders. Each successive episode was marked by distinct combinations of social organization, economy, and technology as people adapted to changing environmental and social circumstances. The modern era is yet another example of system change, as people create new social, economic, and technological means to deal with environmental impacts that far exceed in scale any in the past. The long history of the region verifies the essential fragility of the land to intensive use and to periodic shifts in the patterns of precipitation: fundamental changes in the natural and social systems are inevitable, without predicting when.

THE KHABUR RIVER BASIN

Northeastern Syria is part of a broad expanse of semiarid steppe (the Jazirah) that lies between the Euphrates and Tigris rivers (Figure 6.1). The northernmost part of the Jazirah is known as the Fertile Crescent, a narrow stretch of land suitable for rain-fed agriculture where annual precipitation reaches 350 mm. Southward, rainfall decreases markedly, and is <150 mm where the Khabur River

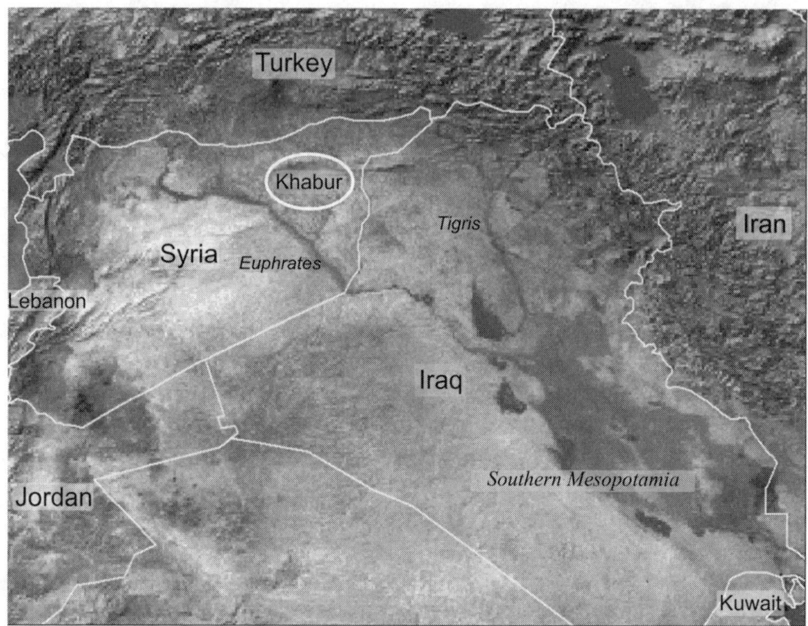

Figure 6.1 MODIS 8-day mosaic, 31/10–7/11/2000. The study area lies in the transitional zone between rain-fed agriculture and semiarid steppe. The image shows irrigated crops in green in late autumn at the beginning of the rainy season in the Khabur and Balikh valleys. Annual grasses and rain-fed crops have not yet begun their growth cycle.

enters the Euphrates. The normal limit for rain-fed agriculture today is 200–250 mm, a marginal zone that fluctuates spatially in response to interannual variability in precipitation that approaches 100 percent (Figure 6.2). The focus of this chapter is on the marginal zone where periodic multidecadal to centennial shifts in precipitation are reflected in human settlements (Table 6.1).

SOURCES AND NATURE OF DATA

There are three main sources of information about the past: archaeological and historical records tell of local human occupations and adaptations, whereas climate proxies inform on regional and global changes. The challenge to interpreting the past is to coordinate these disparate data chronologically and spatially so as to identify drivers of change in the human systems and to use this knowledge to forecast possible changes in the future.

Archaeological data are of two types: records of the presence, size, and ages of sites, based on surface indications; and excavations that provide precise dates and the nature of the settlements. Although there is no comprehensive survey of the entire region, there have been several surveys centered on excavated sites or

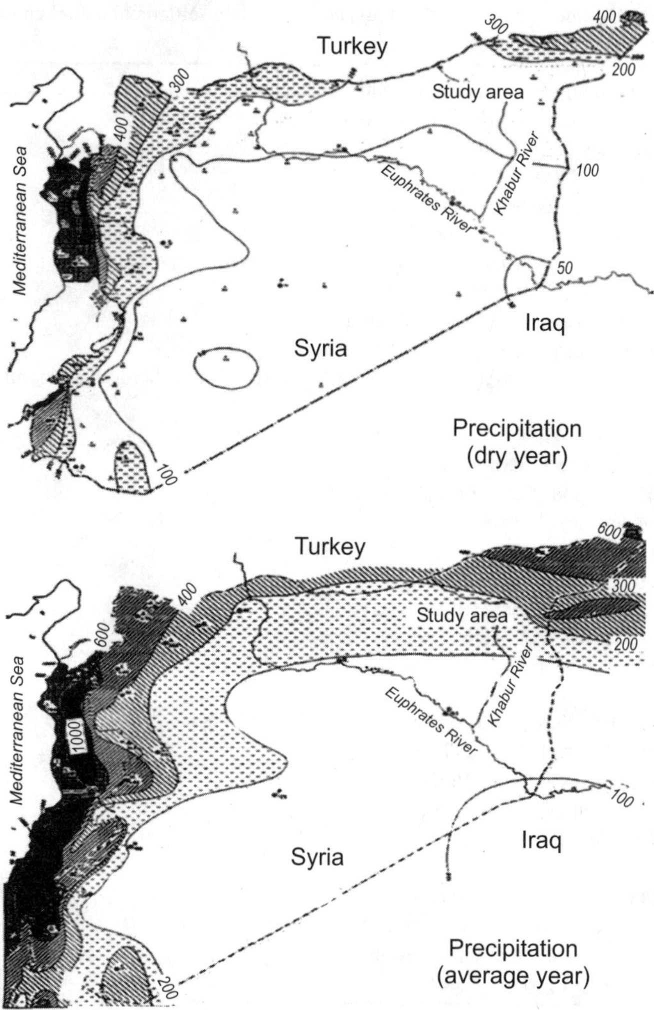

Figure 6.2 Both the Khabur and Balikh valleys lie in a part of the Near East where agriculture is especially sensitive to changes in precipitation. The study area lies in the transitional zone where rain-fed agriculture is problematic in most years. Precipitation maps from Wirth (1971, Map 4).

sub-regions of the Khabur, including the marginal zone (Hole 1997a, b; Bernbeck 1993; Kouchoukos 1998; Lyonnet 1996a, b; Wilkinson 2000; Ur 2002). Dating of the surveyed sites normally depends on recognizing the stylistic variability of the ceramics, or rarely on historical documents or the chance finding of coins. Often ceramics can be dated only to a span of 100–300 years.

Until the 20th century, historical sources rarely focused specifically on the Khabur but they do provide a regional political and economic context.

Table 6.1 Outline of significant climatic anomalies and settlement in the semiarid zone of the Khabur River drainage.

Holocene Climatic Optimum 9000–4000 B.C.
- First agricultural settlements, ca. 6500 B.C.
- Small sites until ca. 4000 B.C.

Arid Event, ca. 4200 B.C.
- Hiatus in settlement 4000–3000 B.C.

Increased Precipitation ca. 2600–2200 B.C.
- Early Bronze Age large towns, small sites, maximum population

Very Arid Event 2200–1900 B.C.
- Collapse of Akkadian and Old Kingdom
- Nomadic tribes in Khabur by 2000 B.C.
- Neo-Assyrian Iron Age, ca. 1000–700 B.C., small sites, herders; town on lower Khabur

Hiatus 700–200 B.C.
- Nomadic sheep herding tribes

Roman Wet Period 90 B.C. to A.D. 240
- Forts and small farmsteads

Arid Period
- Arab conquest from A.D. 640
- Small sites and some irrigation

Arid Period in 1200s?
- Mongol invasion A.D. 1260
- Abandonment of settlements
- Nomadic tribes

Little Ice Age A.D. 1400–1850
- Ottoman government 1600–1915
- Few settlements, nomadic tribes

French Mandate 1925–1946
- Settlement of nomads and refugees

Syrian State 1946 to the present
- Intensive agricultural development
- Degradation of steppe vegetation
- Depletion of ground water

Occasional travelers' descriptions provide specific dated observations of life in the region in the 19th to early 20th centuries (Layard 1853; Blunt 1879), and Ottoman tax records provide some information on the extent of settlements. Consistent and specific reference to the Khabur region follows the defeat of the Ottoman Empire and the dividing of the Near East between England and France after WWI. Both the League of Nations and French sources produced reports. Following the establishment of the Syrian State in the 1950s, official records and reports of international development organizations have produced a steady flow of printed information.

Instrumental climate data for the Khabur are available only for the last few decades, before which there are some regional climate records extending back as far as a century, and only information about isolated climate events before that. The climate fits within the Mediterranean climate system (Köeppen classification Csa) that pervades the Near East (wet winter, dry summer), but local effects are determinant for agriculture and are insecurely known. Various climate proxies have been developed, including pollen, isotope ratios, tree rings, speleothems, lake levels, flood events, and sedimentation rates (see references in Hole 1997a, b). Some of these are not specific to the Khabur and may not accurately reflect its climate. Furthermore, these data sets are of variable temporal precision.

Remotely sensed data include aerial photos dated to the 1930s that feature some of the prominent sites, the Corona series of spy satellite photos taken in the 1960s–1970s, and the various satellite platforms that came on line in the late 1970s and continue today with considerably enhanced features and frequency of coverage. These sources give a good view of the landscape before modern development had advanced very far and allow us to monitor the rate and direction of changes over the past half century.

Global climate models (GCMs) are based on grids of hundreds of kilometers, far too gross to account for the effects of local topography. Newer mesoscale models (such as RegCM2, with a 25-km grid) currently being developed (Evans et al. 2004) offer promise of more accurately modeling current weather patterns and ultimately of reconstructing past climates. Other GCMs include the Milankovitch cycles which are based on solar insolation (the 100 ky obliquity, the 41 ky tilt, and the 22 ky precession cycles). Cycles have also been recognized in ice cores, marine sediments, and glacier fluctuations, as well as in a wide range of local proxy data. In particular, a ~1500 year cycle "is well established" (Mayewski et al. 2004). Some six periods of global rapid climate change occurred at 9000–8000, 6000–5000, 4200–3800, 3500–2500, 1200–1000, and 600–150 B.P. In addition there are shorter-lived and perhaps more localized episodes such as El Niño, which fluctuates on annual to several year cycles. While aridity at the end of the third millennium B.C. has been convincingly recognized locally and globally, other climate anomalies in the Khabur are also seen in changes in hydrology and sedimentation (Blackburn 1995; Blackburn and Fortin 1994; Courty 1994; Deckers and Riehl 2004; Engel 1996; Ergenzinger et al. 1988; Ergenzinger and Kühne 1991; Frey and Kürschner 1991; Gremmen and Bottema 1991; Rösner and Schäbitz 1991).

KEY FINDINGS

It is no longer in doubt that precipitation has fluctuated periodically and, through agriculture, impacted humans throughout the Near East, although usually not with the devastating effects of the late third millennium B.C. Climate proxies from the region clearly show this 4.2 K Event to have been particularly strong and long-lived, with major historical consequences throughout the Near

East and beyond (Weiss 1997), but periodic changes have continued to recent times (Roberts et al. 1994; Issar and Zohar et al. 2004; Bar-Matthews 1999). Resilience of the cultures has varied with sociopolitical conditions as well as with the balance between agriculture and herding.

Families mediate between natural and larger social systems, and can respond to external changes or prospect of changes quickly if alternatives are available. The larger social systems are usually slower to react, despite what their individual components do. Changes in the surrounding natural systems—weather and its effects—occur episodically (e.g., floods or droughts) and may be of variable duration, from hours to years and even decades. Resilience depends on information and alternatives. For example, during periods of drought or warfare, agricultural families could migrate to better regions or adopt a migratory life as herders until agriculture was feasible once again. Historically there has been a regular shift between these two modes, but these avenues to resilience are no longer available so that historical solutions cannot be used as a guide to the future: unused land is not available and national borders prevent migration. Thus, while we can be fairly certain that adverse climate episodes will occur, new solutions to sustainability will have to be devised. We can also be fairly certain that unprecedented environmental impacts will be created in the Syrian Khabur both through current practices of intensive cultivation, and through capture of water from the Khabur catchment area by Turkish irrigation projects (Beaumont 1981, 1996). Farming families at the bottom of the water distribution pyramid are presented with an uncertain future and few alternatives other than leaving the land and seeking wage employment when conditions become untenable (Rida et al. 2004).

In the 1950s when Syria was a newly formed state and modeled its development on Soviet-style Five Year Plans, there was much optimism because the land was underdeveloped. Accordingly, plans to dam the major rivers for irrigation and domestic use, open rangeland to cultivation, and settle Bedouin tribes were widely applauded. Within a few decades it became clear that some of these policies were detrimental to the land but in the meantime the population grew at some 3% a year. By the time the large irrigation projects came on line, the assumptions on which they had been planned had been rendered moot by the opening of the economy to entrepreneurial development. In a belated attempt to mitigate destruction of rangeland, plowing of the steppe (<200 mm of precipitation) was banned; recently summer irrigation has also been curtailed to prevent further salinization. In short, the production targets, already not being met, have been lowered, with unknown cost to the farming families who are now facing a "drought" of human rather than natural origin.

The social and natural systems in northeastern Syria today are not well synchronized; each has its own pace and mode of reaction to changes in the other. The pace and extent of changes today are new, for over previous millennia this landscape had seen only seasonal grazing and limited cultivation—a social and natural system in approximate equilibrium, where interannual variability could be readily accommodated with only minor adjustments. Even when climatic or political events disrupted agricultural settlement, the rangeland itself, free from deep plowing and destruction of native plants, retained its viability.

On a millennial timescale, settlement of the marginal rainfall zone (<250 mm) fluctuated, with periods of occupation on the order of 1–300 years; abandonments lasted a thousand or more years. During periods when agricultural villages were not present, there may have been pastoralists in the area seasonally, but the steppe was largely in a "natural" state. Even when villages were present, the vast majority of land was only lightly exploited through grazing and fuel gathering. Shifts in settlement may have occurred fairly frequently over spans as short as a few years when drought temporarily overwhelmed the resilience of the families. When these adverse conditions lasted a century or more, there would have been no social memory of better times so that recolonization may have taken far longer than abandonment. In recent years changes occur on an annual to decadal scale as the pace of changes in the economy, government policies, and the land itself have accelerated. Today there is little "regeneration," except perhaps in the recent abandonment of summer cultivation, which at best, may retard further degradation, but does little to improve the soil. Where in the past, responses to change occurred at the family level and could be implemented in a moment, today changes in agricultural policies and even mass resettlement of people occur at the government level, but while they are being implemented families are left with few adaptive strategies other than abandoning their lifestyle (Rida et al. 2004).

Syrian agriculture is dominated by national needs and is supported directly through production quotas, provision of seed and fertilizer, inexpensive fuel and water, and agricultural loans for the purchase of equipment. Long-range planning and construction have generally been based on optimistic projections rather than sound assessment of sustainability. The growth of human populations and their increasing wealth have contributed to the problem through rising demand for goods that can be supplied only through further intensification of pressure on the land. While this is a national problem, each region contributes in different ways. National-scale plans are ill-suited to respond to unanticipated local needs and consequently often require preventative reactions to ameliorate harm. In ancient history, planning and implementation took place on a human scale where the actors could comprehend the system and make rapid and appropriate adjustments based on the limited tools at their disposal.

CAN THE PAST INFORM THE FUTURE?

There are indicators from past experience that can help us to anticipate future problems. One of these concerns long-term changes; the other relatively short-term ones, either of which could potentially reduce or destroy the sustainability of agriculture in the sensitive zones. The first indicator derives from climatic history. During the past century droughts of 2–3 years' duration occurred, but none approaching the severity of decades-long droughts of the past. Although we do not have specific information from the Khabur, there are enough examples of very severe droughts or other climatic anomalies in other parts of the world during the past two millennia to imply that similar events impacted the Khabur (Woodhouse and Overpeck 1998). Further, the millennia-scale events recorded globally can be expected to recur. The archaeological record as well as regional climate proxies indicate that these were arid periods. Even without global warming it is fair to say, "if it happened in the past, it can happen again." When scientists recover more specific local evidence of climatic fluctuations and modelers determine their causes, it may be possible to produce plausible forecasts of future events, whether on millennial, centennial, or decadal scales.

Indicators in archaeological data of potential problems reside primarily in the spread of agriculture into zones that today have too little precipitation. All such settlements have failed in the past and one can predict that they will also fail in the future. Today's unprecedented settlement of the semiarid steppe is surely a predictor of potential dire consequences despite the increasingly elaborate means of sustaining agriculture that are in use today. For instance, it is well known that in hot, arid lands, all irrigation in poorly drained soil eventually succumbs to salinization and this process is already advanced in Syria (Oweis 1999). It remains to be seen whether genetically modified crops can be developed to surmount this problem; even if they can, the depletion of groundwater at a rate in excess of regeneration would likely result eventually in the failure of both crops and settlements. Although these impacts will be most severe in the marginal zones, history shows that large-scale emigration to the more favorable zones may be unsupportable and lead to eventual system failure.

Predictions made twenty years ago would have proven wrong because of the continual influx of new conditions. Predictions today will no doubt have similar value, but information will be the key to successful anticipation of problems and evaluation of solutions. One way to do this is to monitor both the spatial and temporal rates as well as the extent and impact of changes on the land. Although this can be done on the ground, owing to national borders and the spatial scale on which natural processes operate, its success, particularly over the long term, is problematic. An obvious solution is to monitor the landscape via satellites, which can provide both quantitative and qualitative information on repeat cycles ranging from daily to weekly to seasonally or over decades (Kouchoukos et al. 1998; Hole and Smith 2004; Hole and Zaitchik, unpublished). Timely and accurate information can allow social systems to anticipate and react to change, and the tools to achieve this on a scale large enough to impact decision making in regions and even countries are available. Unfortunately there is no local or historical experience by policy makers in the use of such means to assess the viability and sustainability of the present mix of social, economic, political, and natural factors. Under such circumstances we may expect policy to react to adversity rather than to attempt proactively to forestall it.

The present situation is contingent upon the past, but in the Khabur the relevant past for understanding the present is relatively short because technology in the form of the use of fossil fuels during the mid-20th century transformed the ability of humans to impact their environment. Over this same short span of time, society reorganized from one based on tribal custom and land use, partly under French control, to a national government whose model for development was the Soviet Union with a strong emphasis on central planning. Despite these changes, the plans and policies that could be taken by the Syrian government were contingent upon the fact that the land had lain relatively undeveloped and were thought to promise extraordinary returns on investment. In the absence of convincing documentary evidence of climate anomalies in the past, it could be assumed that present-day conditions were stable and that technology could create the means to buffer minor, predictable episodes of drought. On the other hand, a few years of greater than average precipitation encouraged the opportunistic exploitation of marginal land at relatively little cost, but with great consequence for the land itself. These dual impacts—expansion of irrigation and degradation of the steppe—will have a lasting effect for generations to come and further limit the options for sustainability (Hole and Smith 2004).

ACKNOWLEDGMENTS

Archaeological research was funded by NSF–SBR–9510543, NSF–SBR–9515394, and NSF–BNS–9012337. Nicholas Kouchoukos dated the survey ceramics and settlements. Satellite analysis, funded by NASA NAG5–9316, was done in the Center for Earth Observation at Yale University. A digital atlas of photographs of the region as well as a poster showing land-use changes in the Khabur can be seen at http://www.yale.edu/ceo/Projects/swap.html.

REFERENCES

Bar-Matthews, M., A. Ayalon, A. Kaufman, and G. Wasserburg. 1999. The Eastern Mediterranean paleoclimate as a reflection of regional events: Soreq Cave, Israel. *Earth Planet. Sci. Lett.* **166**:85–95.

Beaumont, P. 1981. Water resources and their management in the Middle East. In: Change and Developments in the Middle East: Essays in Honor of W. B. Fisher, ed. J.I. Clarke and H. Bowen-Jones, pp. 40–72. London: Methuen.

Beaumont, P. 1996. Agricultural and environmental changes in the Upper Euphrates catchment of Turkey and Syria and their political and economic implications. *Appl. Geogr.* **16**:137–157.

Bernbeck, R. 1993. Steppe als Kulturlandschaft. Berliner Beiträge zum Vorderen Orient-Ausgrabungen, vol 1. Berlin: Reimer.

Blackburn, M. 1995. Environnement géomorphologique du centre de la moyenne vallée du Khabour, Syrie. *Bulletin, Can. Soc. Mesopotamian Stud.* **29**:5–20

Blackburn, M., and M. Fortin. 1994. Geomorphology of Tell'Atij, northern Syria. *Geoarchaeology* **9**:57–74

Blunt, L.A. 1879. Bedouin Tribes of the Euphrates. New York: Harper.

Courty, M.-A. 1994. Le cadre paléogéographique des occupations humaines dan le bassin du Haut-Khabour, Syrie du Nord-Est. Premiers résultats. *Paléorient* **20**:21–59

Deckers, K. and S. Riehl. 2004. The development of economy and environment from the Bronze Age to the Early Iron Age in northern Syria and the Levant: A case-study from the upper Khabur region. *Antiquity* **78**:http://antiquity.ac.uk/projgall/deckers/index.html

Engel, R. 1996. Archaeobotanical analysis of timber and firewood used in third millennium houses at Tall Bderi/Northeast-Syria. In: Houses and Households on Ancient Mesopotamia, vol. 78, 40e Rencontre Assyriologique en Leiden 1993, pp. 105–115. Leiden/Istanbul: Nederlands Historisch-Archaeologisch Institut te Istanbul.

Ergenzinger, P.J., W. Frey, H. Kühne, and H. Kürschner. 1988. The reconstruction of environment, irrigation and development of settlement on the Habur in north-east Syria. In: Conceptual Issues in Environmental Archaeology, ed. J.L. Bintliff, D.A. Davidson, and E.G. Grant, pp. 108–128. Edinburgh: Edinburgh Univ. Press.

Ergenzinger, P.J., and H. Kühne. 1991. Ein regionales Bewässerungssystem am Habur. In: Die Rezente Umwelt von Tall Seh Hamad und Daten zur Umweltrekonstruktion der Assyrischen Stadt Dur-Katlimu, ed. by H. Kühne, pp. 163–190. Berlin: Reimer.

Evans, J.P., R. Smith, and R.J. Oglesby. 2004. Middle East climate simulation and dominant precipitation processes. *Intl. J. Climatol*. **24**:1671–1694.

Frey, W., and H. Kürschner. 1991. Die Aktuelle und potentielle natürliche Vegetation im Bereich des Unteren Habur. In: Die Rezente Umwelt von Tall Seh Hamad und Daten zur Umweltrekonstruktion der Assyrischen Stadt Dur-Katlimu, ed. H. Kühne, pp. 87–103. Berlin: Reimer.

Gremmen, W.H.E., and S. Bottema. 1991. Palynological investigations in the Syrian Gazira. In: Die Rezente Umwelt von Tall Seh Hamad und Daten zur Umweltrekonstruktion der Assyrischen Stadt Dur-Katlimu, ed. by H. Kühne, pp. 105–116. Berlin: Reimer.

Hole, F. 1997a. Evidence for mid-Holocene environmental change in the western Habur drainage, northeastern Syria. In: Third Millennium B.C. Climate Change and Old World Collapse, ed. H.N. Dalfes, G. Kukla, and H. Weiss, pp. 39–66. NATO ASI 149. Berlin: Springer.

Hole, F. 1997b. Paleoenvironment and human society in the Jezireh of northern Mesopotamia 20,000–6,000 B.P. *Paléorient* **23**:39–49.

Hole, F., and R. Smith. 2004. Arid land agriculture in northeastern Syria: Will this be a tragedy of the commons? In: Land Change Science: Observing, Monitoring, and Understanding Trajectories of Change on the Earth's Surface, vol. 6, Remote Sensing and Digital Image Processing, ed. G. Gutman, A.C. Janetos, C.O. Justice et al., pp. 209–222. Dordrecht: Kluwer.

Issar, A. S., and M. Zohar. 2004. Climate Change: Environment and Civilization in the Middle East. New York: Springer.

Kouchoukos, N. 1998. Landscape and Social Change in Late Prehistoric Mesopotamia. Ph.D. diss. Yale Univ., New Haven, CT.

Kouchoukos, N., R. Smith, A. Gleason et al. 1998. Monitoring the distribution, use, and regeneration of natural resources in semi-arid Southwest Asia. In: Transformations of

Middle Eastern Natural Environments: Legacies and Lessons, ed. J. Albert, M. Bernardsson, and R. Kenna, , pp. 467–491. Bulletin 103, Yale School of Forestry and Environmental Studies. New Haven, CT: Yale Univ. Press.

Layard, A.H. 1853. Discoveries in the Ruins of Nineveh and Babylon. London: Murray.

Lyonnet, B. 1996. La prospection archéologique de la partie occidentale du haut-Khabur (Syrie du Nord-Est): Méthodes, resultats et questions autour de l'occupation aux IIIe et IIe millénaires av. N. É. *Amurru* **1**:363–376.

Lyonnet, B. 1996. Settlement pattern in the uper Khabur (N.E. Syria), from the Achaemenids to the Abbasid period: Methods and preliminary results from a survey. In: Continuity and Change in Northern Mesopotamia from the Hellenistic to the Early Islamic Period, ed. K. Bartl and S.R. Hauser, pp. 349–361. Berliner Beiträge zum Vorderen Orient-Ausgrabungen. Berlin: Reimer.

Mayewski, P.A., E.E. Rohling, J.C. Stager et al. 2004. Holocene climate variability. *Quat. Res.* **62**:243–255.

Oweis, T. 1999. A little goes a long way. *ICARDA Caravan* **11**:14–15

Rida, F., A. Aw-Hassan, and A. Bruggeman. 2004. Sustainable use of groundwater in Syria. *ICARDA Caravan* **20/21**:22–23.

Roberts, N., H. Lamb, N. El Hamouti, and P. Barker. 1994. Abrupt Holocene hydro-climatic events: Palaeolimnological evidence from northwest Africa. In: Change in Drylands, ed. A.C. Millington and K. Pye, pp. 163–175. New York: Wiley.

Rösner, U., and F. Schäbitz. 1991. Palynological and sedimentological evidence for the historic environment of Khatouniye, eastern Syrian Djezire. *Paléorient* **17**:77-87.

Ur, J.A. 2002. Settlement and landscape in Northern Mesopotamia: The Tell Hamoukar survey 2000–2001 *Akkadica* **123**:57–88.

Weiss, H. 1997. Late third millennium abrupt climate change and social collapse in West Asia and Egypt. In: Third Millennium BC Climate Change and Old World Collapse, ed. H.N. Dalfes, G. Kukla, and H. Weiss, pp. 711–723. NATO ASI 149. Berlin: Springer.

Wilkinson, T.J. 2000. Archaeological survey of the Tell Beydar region Syria, 1997: A preliminary report. In: Tell Beydar: Environmental and Technical Studies, ed. K. Van Leberghe and G. Voet, pp. 1–37. Turnhout: Brepols.

Wirth, E. 1971. Syrien: Eine geographische Landeskunde. Darmstadt: Wissenschaftliche Buchgesellschaft.

Woodhouse, C.A., and J.T. Overpeck. 1998. 2000 Years of drought variability in the central United States. *Bull. Am. Meterol. Soc.* **79**:2693–2714.

7

The Trajectory of Human Evolution in Australia

10,000 B.P. to the Present

Timothy F. Flannery

South Australian Museum, North Terrace, Adelaide, S.A. 5000, Australia

ABSTRACT

This chapter summarizes human–ecological interactions in Australia. By reviewing the literature, two periods of rapid, human-mediated environmental change are identified: at first contact 46,000 years ago, and between 4,000 and 1,000 years ago. Both of these periods remain poorly understood from the perspective of human–ecosystem interactions. One striking aspect of the last 46,000 years is the great period of relative stability between 45,000 and 5,000 years ago. This period of stability is marked by a lack of vertebrate extinctions, which is truly remarkable since it spans the rapid climate shifts of the Last Glacial Maximum and early Holocene. In addition, there is little evidence of human cultural change over this period.

INTRODUCTION

It was only in the early 21st century, with the development of optically stimulated luminescence (OSL) and electron spin resonance (ESR) dating, along with advances in ^{14}C dating, that researchers developed a plausible chronology for human arrival in Australia and for megafaunal extinction. This new knowledge has dramatically altered our understanding of human–environment interactions in Australia. The emerging consensus is that humans arrived in Australia between 45,000–55,000 B.P. (Bowler et al. 2003) and that the continent underwent a dramatic ecological transformation at around the same time. This transformation included the extinction of many large vertebrates (Roberts et al. 2001) and a narrowing of diet for surviving species, such as the emu (Miller et al. 2005). It suggests a collapse of ecosystem productivity and occurs concomitantly with evidence of a widespread change in fire regime, vegetation, and possibly hydrology.

The best evidence of vegetation change at this time comes from pollen analysis of sediments preserved in Lynch's Carter, a lake in the rainforest-covered Atherton Tablelands of northeast Queensland, which preserves a pollen record stretching back at least 100,000 years. Around 45,000 years ago fire-sensitive dry rainforest, dominated by araucarias, gave way to fire-tolerant eucalypts (Turney et al. 2001; Flannery 1994). Other changes, which are less securely dated, have been observed in cores from ocean sediments. These indicate widespread replacement of auracarian dry rainforest with eucalyptus and a rapid growth in mangrove habitat (possibly due to sediment accumulation in estuaries resulting from catchment degradation). In the inland salt lakes, changes in rainfall, possibly due to altered transpiration resulting from the replacement of rainforest with eucalypts, have also been detected (Miller and Magee 1992).

I would like to give you my best estimate of precisely when these momentous events occurred, over what timescale, and in what order. This is highly contentious, but I think that enough data now exist to make a case. As of 1998, no remains of Australian megafauna had been convincingly dated. Today (2006) around 30 sites (published and unpublished) have been dated using OSL and ESR, and all are older than 46,000 B.P. Over 100 archaeological sites have been dated to older than 10,000 B.P., but only two (both cave deposits in Arnhem Land) have widely accepted dates older than 46,000 B.P. No site yet discovered in Australia has been convincingly demonstrated to preserve evidence of the co-occurrence of humans and megafauna. Although the precise chronology of events is still hotly debated, these data suggest to me that the overlap between humans and megafauna was brief, that it occurred around 46,000 B.P., and that the event irreversibly and dramatically altered Australia's terrestrial ecosystems.

Several hypotheses—none of which can be adequately tested at present—have been put forward to explain these observations. Miller et al. (1999) argue that human burning of fire-sensitive vegetation deprived megafauna of food, precipitating them into a fatal decline; I have argued (Flannery 1994) that rapid overkill of the large herbivores led to the accumulation of plant biomass, which in turn led to a changed fire regime. Natural ignition sources are abundant enough in Australia to make this credible.

In summary, it is now clear that Australia underwent a profound transition in ecosystem function around the time that people arrived on the continent. Precise timing and cause and effect are still being vigorously debated, but there is now little doubt that humans were key factors in this change. As a result of the shift, the biological productivity of Australian ecosystems plummeted, and its climate was possibly changed.

The archaeological record, from this point on, however, is one of stability. For 40,000 years there is little evidence of cultural change and no evidence whatsoever of faunal extinction, which is remarkable given that this period covers the Last Glacial Maximum (LGM: 25–15,000 years ago) during which ecosystems were profoundly reorganized worldwide. In Australia, there is evidence of a marked cooling (around 9°C), increased wind, and lowered water tables at the time of the LGM. As a result, patterns of vegetation distribution alter markedly, as do those (to a lesser extent) of vertebrates. Yet no extinctions are seen, nor other perturbations on the scale observed in the Americas and Europe. Human prey items do, however, shift to reflect climate as, presumably, did human distribution.

Following the LGM, the early Holocene is marked worldwide by a remarkable, independent shift in human ecology at several regional centers, involving the domestication of animals and crops as well as the developments of villages and other settlements. In Australia, however, this phase passes without major shifts, at least as far as the archaeological record has revealed thus far. This is equally as remarkable as the stability evidenced through the LGM.

It is only over the last 5,000 years that further environmental perturbations, as evinced by the extinctions of the thylacine (*Thylacinus cynocephalus*) and Tasmanian devil (*Sarcophilus harrisi*) on the Australian mainland, are seen. This event coincides in time with the introduction of the dingo (around 4,000 B.P.), and it is thought that competition with the new predator may have caused the extinction of the large marsupial carnivores. It is also at this time that Aboriginal culture underwent a number of changes, the most striking of which is the spread of the Pama-Nyungan language family across most of the continent (Evans and Jones 1997). Linguists postulate this language family's origin lies in a small region of northeast Arnhem Land, from which, beginning around 5,000 years ago (using linguistic dating), it spread over 80–90% of the mainland. By 1788, the only regions where non-Pama-Nyungan languages survived were parts of the Kimberley, Arnhem Land, and Tasmania.

Changes in Aboriginal technology are also evident over the past 5 millennia. The introduction of the "small tool" stone technology dates to round 4,000 years ago. It appears earlier in Southeast Asia and may have arrived as part of the same cultural package that included the dingo. At this time, the Austronesian people, who had well-developed maritime technologies, began to expand into the region from their base in Taiwan. Knowledge of how to treat the toxic cycad fruit in order to make it edible also appears in Australia around this period. As this technology is widespread in the Pacific and Southeast Asia, it may also have been an introduction. The ability to treat cycad fruit to make it edible may have had a considerable impact on Aborigines living where cycads grew, for the seeds represent a large, temporarily storable source of carbohydrate which may have enabled large gatherings to take place. This was certainly the case by the historic period.

By 1,000 years ago Aboriginal people had begun to grind grass seed to form storable seedcakes. Archaeologist Bruno David (pers. comm.) argues that, in central Australia at least, this led to a marked division of labor among the sexes, as well as to the co-option of female labor to serve male ends. This occurred through the production by women of large volumes of seedcake that were used to provision participants in the ceremonies that are so central to the lives of the central Australian Aborigines. There is also evidence that long-distance trade (e.g., in axheads, which were only quarried at a few locations) accelerates at this time, perhaps as a result of the development of a transportable, durable, and high-energy food source.

How can we account for these mid- to late Holocene changes in Australian ecosystems and Aboriginal economy? Archaeologists recognize a period of "intensification" that is coincident with these changes, and which they argue marks a phase of increasing human population density in Australia. However, as yet the relationship between these disparate events and population density is unclear. As a hypothesis, I suggest the following scenario to explain these changes, which may or may not have been accompanied by changes in population.

The extinction of the large marsupial carnivores suggests that the arrival of the dingo had a profound effect on Australian ecosystems. The medium-sized and larger mammals, which were the mainstay of the Aboriginal economy until that point, were relatively naive to canid predators, and it can be anticipated that their numbers declined as dingoes multiplied. A similar situation was observed in the early 20th century, when foxes devastated the population of smaller marsupials. Some Aboriginal people are an exception among human societies in tropical and temperate regions in being heavily reliant on animal protein. Before the arrival of the dingo, this may have been even more the case.

For those Aboriginal clans that first received dingoes, and who learned how to use them as companion animals, dingoes may have represented a huge advantage, for they would have delivered a bonus yield of meat by unsustainably harvesting marsupials. As they spread into the landscape, dingoes may have affected the dingo-less Aboriginal clans they encountered. Presumably these people would have treated dingoes as wild animals and thus the dogs did not share meat with them. Instead, dingoes would have rapidly depleted the stock of larger marsupials, denying a vital resource to the clans. In this situation, it is easy to envisage the spread of a dingo-owning clan into the lands of the dingo-less people. This may account for the spread of the Pama-Nyungan language throughout most of Australia, but not into most of Arnhem Land and the Kimberley, where Aborigines may have independently acquired dingoes.

Once the initial bounty supplied by dingoes was exhausted, a new equilibrium would have been reached, in which the numbers of large marsupials was much reduced. This indeed was the situation encountered by early European observers in the late 18th and early 19th centuries. It was only with the decline of the dingo, and of Aborigines, that kangaroo numbers expanded (e.g., Stewart 1910). Furthermore, it seems possible that the origin of grass-seed grinding in Australia is tied in with ecological changes initiated by the dingo. With herbivore numbers suppressed, grasses may have had the opportunity to switch from predominantly vegetative reproduction to sexual reproduction. Previously, sexual reproduction may have been rare because of predation on seed-heads by marsupials. With more seed, the opportunity existed for Aborigines to move down the trophic pyramid from meat to seed.

Other innovative life-ways developed in the region during the Holocene. Between 50,000 and 10,000 B.P. New Guinea and Australia were joined; even today the genetic and cultural relationships between Aborigines and New Guinean highlanders are close. By the mid-Holocene, New Guineans had begun to develop agricultural systems in wet valley bottoms, presumably to grow taro, sugar cane, and *Australimusa* bananas, all of which originated in New Guinea (Denham et al. 2003). The development of agriculture in New Guinea represents one of just a half-dozen or so independent plant domestication events known, and it resulted in the development of the world's most valuable crop—sugar cane. This stands in stark contrast to the situation in Australia, and it probably reflects New Guinea's good soils as well as rainfall reliability and abundance.

Australia's Aborigines never pursued agricultural life-ways, but there are nevertheless extraordinary incidents of landscape modification, of which eel farming in Western Victoria is perhaps the best-documented example. The origins of eel farming on Victoria's Western District basalt plains are obscure, but it is clear that by A.D. 1800 extensive weirs and ditches, which were manipulated to maintain optimum conditions for eels, covered at least 75 square kilometers of the basalt plains. Harvests from this system were used in long-distance trade, and human population densities seem to have been unusually high. In the southwest sector of Western Australia, intense use of yam-daisies may have approached agriculture in that the fields seem to have been dug over with an intensity that would amount to plant management. Whether these practices pre-date the mid-Holocene is not known, but it can be surmised that Aboriginal culture as we know it today, with its emphasis (at least in the drier regions) on secret male ceremonies and long-distance trade, results from mid-Holocene shifts in human ecology. Far too little study has, however, been done on the ecology of the past 5,000 years to be certain about the details of this shift.

SUMMARY

Human interactions with the environment in Australia have followed a very different trajectory from that seen elsewhere. Impacts were early (46,000 B.P.) and profound, resulting in the extinction of 95% of vertebrate genera exceeding 40 kg in weight, and permanently altering ecosystem productivity and possibly the continent's climate. The LGM and early Holocene, which elsewhere is the scene of widespread vertebrate extinctions and the development of agriculture and settled human communities, pass in Australia with very little change. The mid-late Holocene, in contrast, is a period of rapid ecological and cultural re-organization. I would argue that a dingo-driven revolution had wide impacts both on ecosystems and human cultures, the end result of which was a further impoverishment of the environment and a shift in human economy from animal protein to grass seeds.

Although Australia offers a striking contrast with human–environment experiences elsewhere, some aspects of that interaction are depressingly familiar: with each major perturbation, the Australian ecosystem has become less diverse and productive. The long period of evident stability between 46,000 B.P. and 5,000 B.P., which occurred despite dramatic climate change, does offer some hope that people can strike a stable balance, long-term, with their environment. Unfortunately, so little archaeological research has been done on the nature of this balance that we are as yet unable to grasp its key components.

REFERENCES

Bowler, J., H. Johnston, J.M. Olley et al. 2003. New ages for human occupation and climate change at Lake Mungo, Australia. *Nature* **421**:837–840.

Denham, T.P., S.G. Haberle, C. Lentfer et al. 2003. Origins of agriculture at Kuk Swamp in the highlands of New Guinea. *Science* **301**:189–194.

Evans, N., and R. Jones. 1997. The cradle of the Pama-Nyungans: Archaeological and linguistic speculations. In: Archaeology and Linguistics: Aboriginal Australia in Global Perspective, ed. P. McConvell and N. Evans, pp. 385–418. Melbourne: Oxford Univ. Press.

Flannery, T.F. 1994. The Future Eaters: An Ecological History of the Australasian Lands and People. Chatswood, NSW: Reed Books.

Miller, G.S., and J. Magee. 1992. Drought in the Australian outback: Anthropogenic impact on regional climate. Abstracts American Geophysical Union 104. Washington, D.C.: Am. Geophys. Union.

Miller, G.H., J.W. Magee, M.L. Fogel et al. 2005. Dietary reconstructions from eggshells of *Dromaius* and *Genyornis* suggest a human-induced ecosystem shift caused extinction of the Australian megafauna. Abstract. 10th CAVEPS symposium, Narracourte, S.A. April 2005.

Miller, G.H., J.W. Magee, B.J. Johnson et al. 1999. Pleistocene extinction of *Genyornis newtoni*: Human impact on Australian megafauna. *Science* **283**:205–208.

Roberts, R., T. Flannery, L. Ayliffe et al. 2001. New Ages for the Last Australian Megafauna: Continent-wide extinction about 46,000 years ago. *Science* **292**: 1888–1892.

Stewart, D. 1910. Notes on the Buandik Aborigines of the South East of South Australia. Adelaide: Archives of the South Australian Museum.

Turney, C.S.M., A.P. Kershaw, L.K. Fifield et al. 2001. Redating the onset of burning at Lynch's Crater (North Queensland): Implications for human settlement in Australia. *J. Quat. Res.* **16**:767–771.

8

Toward a Comparative Study of Hegemonic Decline in Global Systems

The Complexity of Crisis and the Paradoxes of Differentiated Experience

Jonathan Friedman

EHESS, 75006 Paris, France and Department of Social Anthropology, Lund University, 21100 Lund, Sweden

ABSTRACT

This chapter addresses the nature of hegemonic decline as a global systemic process. It argues first for a complex model of expansion and contraction that is broadly applicable and where the dynamic of growth is based on a logic that has never been adapted to the environmental conditions of social reproduction. This accounts for the ubiquitousness of social crisis and collapse. The few historical ethnographic examples of apparent adaptation are understood as historical products of long-term devolutionary processes. While the structures of global systems are quite different than those preceding the former, there is a common accumulative dynamic that leads to numerous internal incompatibilities among the various logics that constitute such systems. The model is of a long-term process of cyclical expansion/contraction and geographical shift in the centers of accumulation with periodic larger declines and "dark ages." The state of nature at the thresholds of such major declines is clearly one of depletion, although it will be argued that it is not the case that the depletion itself "causes" the decline. The final section of the chapter explores the opacity of social logics with respect to their conditions of existence. As the goals of social strategies do not include the adaptation to external limits of reproduction, the usual tendency is that such limits can easily be transgressed.

INTRODUCTION

In this chapter I review some of our own research on the above subject and a thought experiment of sorts—one meant to be suggestive only. It consists primarily of a series of propositions concerning the issues of decline and a model of the dynamics involved therein, especially of hegemonic decline in global systems. Although the discussion begins with expansion and decline in "tribal" social orders, the primary focus is with phenomena that have been taken up in a recent publication (Friedman and Chase-Dunn 2004), which focuses on what might be called comparative hegemonic decline. The discussion is also very much a history of a particular approach to research on global[1] processes. While this might strike some as overly personal, it is maintained here because there is a certain developmental logic linking the different areas that we have researched.

Social and cultural anthropology has been occupied for a hundred years with the issues of human variation, cultural creativity, technological innovation, and what is often referred to as social and cultural evolution. In interaction, often quite intense, with archaeologists and ancient historians, there has been a vast and varied production of models of the social history of our species. I have myself been very much involved in this broad attempt to discover the nature of long-term processes, a perspective which has become much diminished in the field of anthropology since the 1980s, not least under the aegis of social and cultural anthropologists themselves. During this period there was a focus on ethnographic research, but there was also an increasing dominance of a strongly present-oriented ethnography that eschewed both historical and comparative research.

In the late 1960s and 1970s at Columbia University[2], which was, during this period, the major center of cultural materialism and neo-evolutionism, a number of very important contributions to the field were produced in a relatively short period of time, associated with major figures in the field such as Marshall Sahlins, Morton Fired, Andrew Vayda, Roy Rappaport, and Marvin Harris. Cultural materialism and cultural ecology were highly functionalist and what might be called adaptationist in their approach. The latter emerged in the late 1950s as an attempt to demonstrate the rationality of exotic institutions and practices often assumed to be irrational or at least a-rational manifestations of primitive and even not so primitive cultural schemes or mentalities. The rationale of this new approach was one in which apparently strange phenomena (e.g., pig feasts, the potlatch, and sacred cows) were accounted for in terms of their practical functions. Rather than functionally integrating the social order, the rationality of such institutions concerned the relation between society and the natural environment. There was little place for crises, for systems that did not work, and explanation was mostly a question of ecological adaptation itself. A reaction

1 Global process refers here to processes that define a field of social reproduction including a number of political units. Such processes can be quite limited geographically as many civilizations and even smaller regional systems discussed below. The word *global* should not be conflated with the notion of the global as in the whole world.

2 I was a doctoral student at Columbia from 1968 to 1972 when I finished my thesis, discussed in part below.

developed to this approach, which had become quite dominant during the 1960s in the United States and even, albeit to a lesser extent, in the Marxist anthropology emerging in Europe at the time (mid- to late-1960s). The approach that was proposed was one in which social systems are not conceived as adaptive machines, but as systems of social reproductive processes whose properties are incompatible with one another and which in their dynamics often lead to crisis, even collapse, and historical transformation (Friedman 1974; Sahlins 1969; Murphy 1970).

This approach was applied quite early to a study of the Kachin and neighboring societies of Highland Burma, Assam, and Yunnan (Friedman 1979, 1998), and made substantial use of historical ethnography, ancient history, and archaeology to try to understand the way in which different social forms could be said to be historically related to one another. The model contained a number of hypothetical frameworks. The first was that the dominant social strategies of social reproduction are not self-monitoring for major functional incompatibilities, especially those that might generate systemic crisis. They are not organized around negative feedback functions. They are not, then, preorganized for adaptation but rather for accumulation of power and control. This entails that they can easily outstrip their limit conditions of reproduction.

In the analysis of tribal social reproduction among the Kachin of Highland Burma it was found that logics of reproduction tended toward the hierarchization of lineages and the development of increasingly larger chiefly polities. The reproductive process is structured through two kinds of circulation: the alliance structure and the ritual feasting structure. Prestige is gained via competitive feasting at particular ritual occasions in which a great deal of food and beer is distributed and cattle are sacrificed. The ability to give large-scale feasts implies that one is closer to the ancestor-gods who provide fertility, which means that one is closer in kinship terms to these gods or that one's immediate ancestors are closer or even identical to such gods. This is a process of transformation of prestige into descent from high ranked ancestor-deities. Prestige is then transformed into rank via a system of "generalized exchange" in which women descend from higher- to lower-ranked groups. The global effects of this logic are the accumulation of sacred rank at the top of the social order and the production of indebted slaves at the bottom. This leads to demographic expansion of the chiefly domain as increasing numbers of lineages are connected, via the extension of exchange links, at the same time as those of lower rank become indebted as the cost of alliance increases. This is sometimes referred to in terms of the inflation of bridewealth, one led by chiefly competition. The chiefly lineages are able to maintain the competition because of their increasing access to the labor input of a larger portion of the population, as well as the import of labor via warfare.

The entire process occurs on the basis of swidden agriculture in sensitive conditions where a fallow rate of 1/14 is the limit for the maintenance of soil fertility. This ratio deteriorates, however, with intensification, and productivity declines as a result, which in turn means that increasing amounts of land become necessary to maintain the demand for total production. Chiefs can carry this out but commoners cannot since they neither have access to more land nor to slaves. Thus, commoners drop out, becoming slaves themselves or at least living under conditions of increasingly difficult indebtedness. Ultimately there develops a situation of revolt. The chief may be deposed or even killed and people re-disperse into the forest areas living in small egalitarian communities. But the social logic remains unchanged, and the process is repeated. This is what I have referred to as the short cycle of expansion and contraction.

There is a longer cycle as well. Historically the cycles of expansion lead to long-term degradation of larger areas and in their turn to the transformation of the social order. Declining levels of productivity and higher population densities prevent the rapid emergence of hierarchy for a complex of reasons that are detailed in other publications (Friedman 1979, 1998). There is an emergence of more egalitarian but still competitive social forms, a major increase in warfare, and the transformation of the nature of power. The generous chief is historically transformed into a warring big-man, an "anti-chief," and finally a headhunter.

The emergence of anti-chiefs along this historical trajectory reveals a great deal about the transformation of power. These chiefs, often chosen from among youth, against their will, are not allowed any role in fertility rites, to give feasts or gifts, nor to engage in sex with their wives for a specified number of years. Fertility rites themselves concern increasingly the maintaining or recapturing of previous standards than with increasing wealth. Although the expansionist chiefdoms call their acquired slaves "grandchildren," the imploded end-version of this cycle is one where ancestral power is a scarcity, so that potential slaves are killed and transformed into ancestral fertility and the heads of victims are placed in sacred groves where they too are sources of ancestral force.

This model delineates a particular logic of accumulation, one that transformed production into prestige, then into rank, and finally into sacred position defined within an already extant cosmology in which rank is inscribed into a cosmological hierarchy of ancestral efficacity in the providing of increasing fertility. The short cycle is related to the expansion and collapse of chiefly hierarchies, and the long cycle (Figure 8.1) is linked to a more permanent transformation of the social order itself. This is because the latter cycle is also part of a dialectical relation between the particular logic of accumulation and the conditions in which it occurs, producing long-term ecological degradation. The latter can be represented as in Figure 8.1. This figure is a condensed summary of a series of other graphs that detail the transformation of "tribal" systems and which include the development of states in some conditions and the devolution of the social order in others. This particular graph refers only to the long-term devolutionary process. It might seem paradoxical that swidden agriculture is associated with expansion and irrigation with decline. This is because terracing

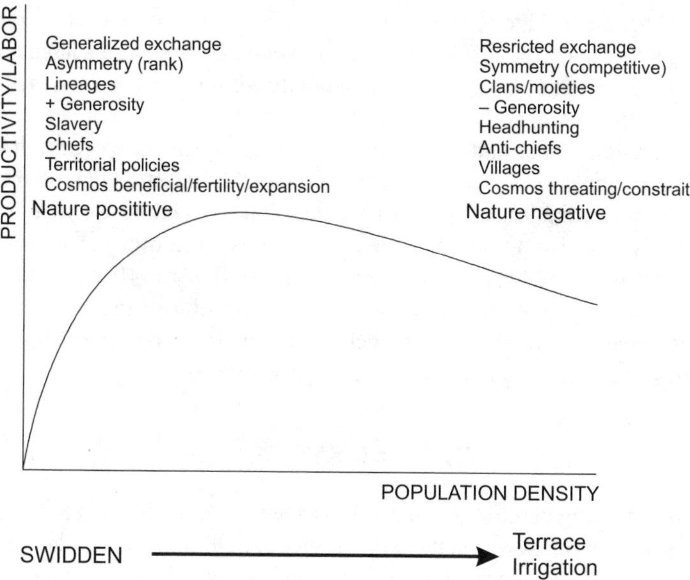

Figure 8.1 The long cycle of intensification.

here occurs in a rainfall zone in which swiddening has become far too unproductive to support the social order. Terracing allows the maintenance of total production quotas but is very difficult to expand without major inputs of labor. It is also far less productive in terms of labor inputs, thus not really adequate to an expansionist and competitive economy. The conditions of hydraulic irrigation in riverine plains are quite different with respect to both productivity and growth capacity.

Numerous works have tended to reduce the relation between the social order and nature to a quasi-Darwinian relation in which societies "chose to fail or survive," primarily by destroying or improving their ecological base (Diamond 2005). Diamond is fully aware of the complexity of such phenomena, but his analysis of "collapse" is one that simply lists different factors such as foreign invasion, trade, climate change, etc., that interact in the process of decline in which ecological factors are primary. The approach here, while focused on issues of "crisis" and contraction, suggests that such phenomena—although related to ecological conditions—are better accounted for in terms of the internal incompatibilities of social systems themselves. Where Diamond's assumptions imply a rather direct relation between whole societies and environmental limits, our own model conceives of environmental limits as the outermost but not necessarily the determinant cause of decline. Diamond's suggestions might be expected from a natural science perspective—one that underestimates the dynamic and quite specific character of the social systems that organize the natural within their own strategies while being subject to their constraints. This accounts for the need to seek culprits, the particular factor, or even a combination of factors that accounts for decline. However, if such factors have no independent existence but are linked systematically within larger processes of social reproduction, then the different weights accorded to one or another, to climate versus invasion, can be understood as parts of the same overall process. It is not a question of either-or, nor even both-and, rather it is a question of systemic relations between variables or aspects of a larger process. It is also worth noting that Diamond's work is reminiscent of the discussions concerning social evolution that occurred in the 1960s (i.e., Harris 1977, 1979; Boserup 1965; Harner 1970) referred to at the start of this discussion. The added feature is the assumption that societies make choices with respect to survival, a species-metaphor that conflates strategic actors and strategies with whole societies.

GLOBAL SYSTEMS

What was missing in the above analysis was the contextualization of this particular pulsating process within the larger regional field of imperial states and major trade systems. In subsequent work, following several initial essays by Kajsa Ekholm Friedman (1976, 1980; also Ekholm and Friedman 1979, 1980; Friedman 1976), we developed what became known as global systemic anthropology. Crucial here is the point of departure for the analysis, the processes of social reproduction of populations, processes that can transgress the borders of any particular polity and which define the larger arena within which political units are dependent for their survival.

The global systemic approach was applied on a "small" scale to the macrohistory of the Pacific, a history that has been the subject of some of the classical evolutionary writings in anthropology and which is exemplified in most major textbooks. Here the logic of accumulation was not based on feasting, even if feasting continued to have a role in the social order of these societies. Instead control over valuable trade goods, or prestige goods, which for any particular population either came from the outside or were monopolized by a central political node, implied control over an entire population's social conditions of reproduction. The specific character of such goods is that they are necessary for marriage and other life-cycle exchanges as well as being a sign of social status. Such systems are widespread in the ethnographic and even the archaeological literature (see Hedeager 1978, 1992; Kristiansen 1998).

While evolutionists had envisioned the Pacific in terms of a movement from egalitarian and big-man-based Melanesia to chiefly Polynesia, it is striking that it was Melanesia that had the most densely populated territories and were clearly the oldest in terms of settlement. The results of research led to a model in which a proto-typical social order based on monopoly over trade goods/prestige goods associated with the Lapita period of Oceanian history was transformed along different trajectories in the eastern and western parts of the Pacific. Here the classical interpretation was reversed. The Lapita societies were conceived as stratified societies based on the monopoly over prestige goods trade (long distance), which in Melanesia became increasingly fragmented due to loss of monopolies, and which in eastern Polynesia were transformed into a system based on warfare and theocratic feudal power as a result of the collapse of long-distance trade.

In western Polynesia and southern Melanesia (Tonga-Fiji-New Caledonia); the original prestige goods model was more or less maintained for a longer period due to the stability of monopoly trade systems in the area. Here expansion/contraction and transformation followed a different pattern than that found in highland Southeast Asia discussed above. It was loss of monopoly over external exchange that was the cause of political fragmentation, even if this could imply increasing production intensities at the local level. It is only where strategies of accumulation implied intensification of land use that ecological limits became significant. In the history of the Pacific, this occurs periodically in Melanesia, especially in more recent periods where big-man systems replaced earlier regional prestige good systems leading to local over-intensification. Similarly in central and eastern Polynesia, where agricultural intensification also became crucial in the accumulation processes with the decline of prestige goods based economies, we find cases of extreme depletion and political decline (e.g., in the Marquesas, Mangareva, and Easter Island).

COMPARING HEGEMONIES IN WORLD SYSTEM HISTORY

In the two examples discussed thus far, the focus is on social orders organized on the basis of kinship. More recent work has taken us in the direction of so-called civilizations, including our own. Here the issue of hegmony in world systemic terms is fundamental. One project has concentrated on ancient world systems, including the Middle and Late Bronze Age and the Hellenistic era (Ekholm Friedman 2005 and in prep.; Friedman 2005; Friedman and Chase-Dunn 2004). These periods display developed and complex economies with relatively clearly defined center/periphery structures. The competition among a number of centers, shifts in hegemony, and long-term intensification are characteristics of such systems. The issue of the existence of capitalism in these periods is one of our principal interests. It might be suggested that competition and accumulation in such systems can be understood in terms of the Weberian model of capital based on his concept of abstract wealth or even Marx's notion of capital as "real abstraction."

The Hellenistic imperial project grew out of the decline of the Athenian hegemony in Greece, the export of capital to both the Middle East and to Macedonia, and the military expansion of Alexander, following the plans of his father Phillip. The goal of this expansion was the conquest of the known world,

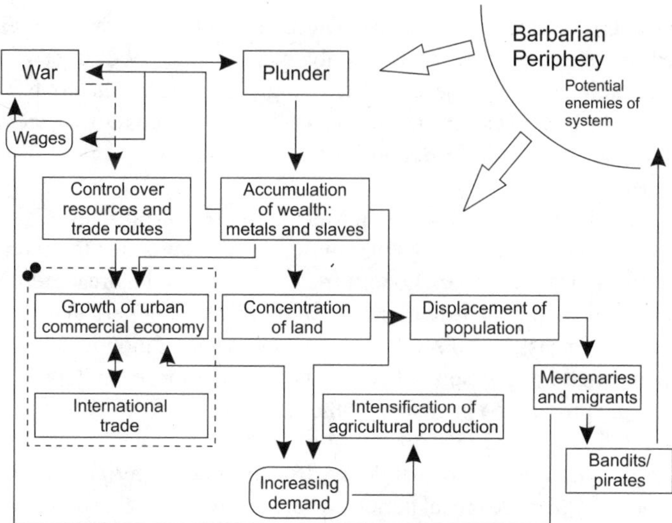

Figure 8.2 Hellenistic state economies: a military–industrial complex.

including the Mediterranean and Egypt as well as the Middle East as far as India, the "playing field" previously established in the Bronze Age. One might suggest that the shifting hegemonies in the Hellenistic period, in which Ptolemaic Egypt became the major economic center, ended with the expansion of Rome, which can be understood, in such terms, as the last Hellenistic state. The entire period, only a couple of centuries in length, was organized in terms of a military-based economy which, extending the Roman model (Hopkins 1978), can be represented as in Figure 8.2. Here the capitalist economy develops within the envelope of the military machine, which is also the principal wage labor employer.

ENDPOINT

The Roman consolidation of the Hellenistic field of conquest represents the final stage of this dynamic period. There are excellent descriptions of economic and political processes throughout this era, and there is a wealth of literature and archival material, including letters, that provide important insights into the social life of the entire period. Here we can speak of a truly complex economy in which the accumulation of abstract wealth or money plays an important role, a capitalist system I would maintain that dates back to earlier epochs in the Middle East. The typical trajectories of such systems include the geographical shift of centers of accumulation over time. The reason for the latter is related to the increasing costs of maintenance in centers of accumulation, which become relatively expensive in relation to the surrounding hinterlands and potential competitors. This leads to a period of decentralization of accumulation as "capital" is exported to other areas and the center eventually becomes dependent on the import of products from its own exported capital. It is important to situate the specific historical context of the Hellenistic states in order to articulate properly what might seem specific to the more general aspects of the dynamics of the period. This was an era of increased and increasing competition, one that might be understood as an intensification in relation to the earlier Classical Greece/Persian Empire period.

The principal aspects of the Hellenistic decline can be summarized as follows:

1. States confront increasing financial difficulties.
2. Polities begin to fragment into smaller entities that become independent and even attempt to conquer state power.
3. Cultural politics increase within the larger state and imperial orders. This is exemplified by the combined rise of Jewish nationalism in Israel and the diasporization of Jewish identity in Egypt.
4. There are invasions by "barbarians" from central Europe and Asia. It appears that unstable relations exist throughout the period, but there is evidence that the northerners, whether Celts/Gauls or Scythians, are engaged as mercenaries or specialist producers, as peripheral functions of the larger states in periods of hegemonic expansion. It is in the decline of such states, under situations where they cannot afford to pay mercenaries or purchase peripheral goods to the same degree, that invasions become more aggressive and often successful.
5. Increasing internal conflicts take the following forms:
    a. class conflict including slave revolts,
    b. ethnic conflict in the eastern states,
    c. intra-dynastic conflicts over succession,
    d. increasing warfare between states and within former states now fragmented into smaller entities.
6. Increase of oriental cults that fill the gap following the decline of the Olympian gods and the disintegration of "modernism" (Walbank 1993, p. 220).
7. It is interesting that of the philosophies of the period, stoicism was adopted by expanding Rome, representing as it does a kind of universalist modernism whereas cynicism and epicureanism are closer to a postmodernist relativism. Diogenes, for example, although a cosmopolitan, was closer to the cultural version of that position in his relativistic openness to all differences. As an historical comparison, today's neo-Confucianism in East Asia has embraced figures such as Habermas (Tran Van Doan 1981) in its search for a common modernism, while the West has imploded in postmodern tendencies. This is not to say that China is today's Rome, but simply that there is an interesting distribution of tendencies that needs to be understood in systemic terms.

HEGEMONIC DECLINE

The issue of decline depicted above is one that can be applied to many so-called "civilizations." It is the common endgame of a series of waves of expansion/contraction of centers that replace one another as hegemonic in particular periods. This is also the case with the history of Western development since the end of the Middle Ages, a period which, in terms of the global field, coincided with the simultaneous decline of the Arab empires in which Europe had previously been a supply zone periphery. The general model, much oversimplified here, is depicted in Figure 8.3: the envelope curve might be said to represent the limits of expansion of the system as a whole based on technological capabilities and limits of extraction of "natural resources" defined as a function of a particular socially organized technology. This is a complex issue, of course, and cannot be addressed here in depth. It might be understood in terms of the limits of a particular energy resource, such as wood, or oil in today's world, which has set off a scramble for alternative forms of energy. However, it might also be a resource related to the maintenance of exchange, such as tin in the Bronze Age. It is the demands of the social order that specify which resources are strategic, rather than the converse.

There are also more general limits of functioning of a social system or world system that are directly linked to degradation of the environment. Such degradation must be understood as the impairment of the conditions of functioning of the social order. The fact that Highland New Guinea, as much of Northern China, displays high levels of ecological degradation does not necessarily correspond to social decline. It all depends on the specific needs or demands of the social system itself. In our own research, the envelope is rarely the direct cause of

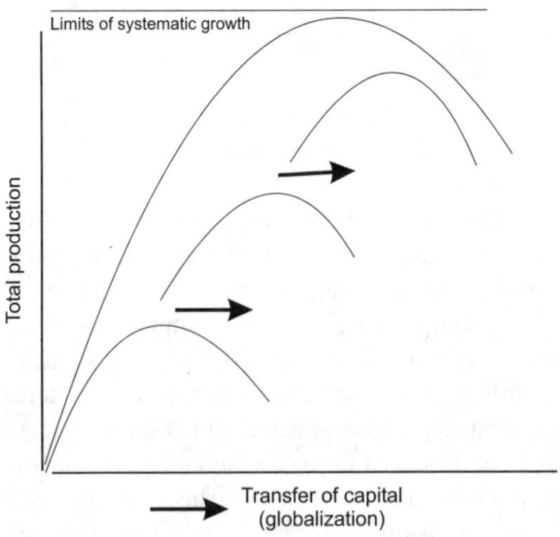

Figure 8.3 Hegemonic shifts and system cycles.

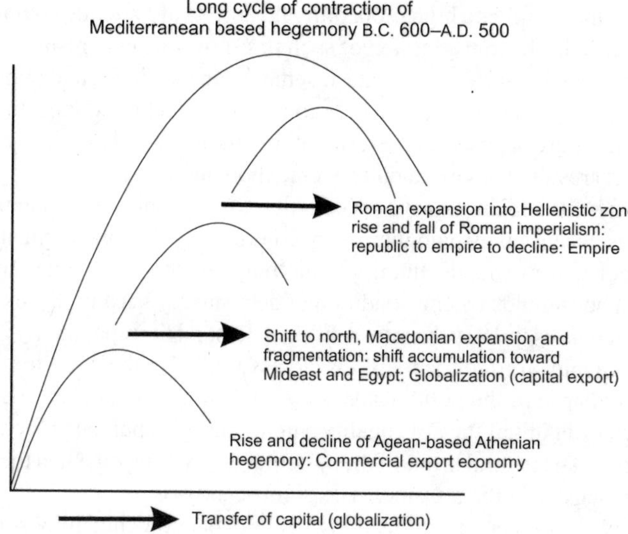

Figure 8.4 Mediterranean hegemonic cycles.

decline. The ecological basis of an expansive system might well be put under a great deal of strain, but there are intervening mechanisms that tend to play a more immediate role in decline even if there is clearly an interaction between levels of exploitation of nature and dominant strategies and relations within the social order. Applied to the particular historical period stretching from the decline of Athenian hegemony to the Roman Empire, we might suggest the trajectory depicted in Figure 8.4.

The limits of systemic growth can be understood as the threshold of exhaustion of the system. This limit, as suggested above, is related to both techno-economic factors and to contradictory tendencies within the system as a whole. It is rarely, as suggested, reducible to ecological limits as such, even if rates of depletion can be extreme. The structure of human social orders in general can be understood as consisting of processes of reproduction containing numerous strategic logics that are autonomous in relation to one another in terms of their internal properties yet joined in the larger reproductive process. The autonomy of internal properties implies the possibility of incompatibility among the various logics of the reproductive process.

Almost all human social systems are accumulative and competitive,[3] and the dynamics of such systems can easily be described in terms of positive feedback

3 Why this should be the case is beyond the confines of this chapter, but it is becoming increasingly clear from archaeological research that accumulative dynamics play an essential role in all of human history. Even the notion that hunting and gathering societies represent stable equilibrium has been challenged for some time. It is based on the conflation of ethnographic examples drawn from the marginal zones of larger social systems with the archaeological past.

processes, which implies a high probability that limits of systemic compatibility can be exceeded. The transgression of such thresholds usually results in crises, and the latter can be more or less catastrophic for the societies and populations involved. The examples referred to here contain such contradictions. In the first example, chiefs are deposed as the result of revolts and chiefly polities collapse, not as a direct result of environmental degradation but as a result of the way such degradation is filtered through the social order. Chiefs continue their expansionist activities because they gain increasing control over the total population, access to labor that offsets declining productivity. At the same time increasing numbers of households enter a condition of debt slavery because the expansion of chiefly activities inflates the total debt (i.e., demand for repayment in the system). The Kachin word for debt, *hka*, is closely related to the word for revolt or feud. The collapse of the polity leads, as we indicated, to population dispersal and a return to an initial state of equality among lineages, but the process begins all over again. The natural limits of the system are only reached when the general population density of the area increases significantly.

Similarly in commercial civilizations, declines occur within the absolute limits of the system because the primary contradictory tendencies of accumulation of abstract wealth lead to periodic breakdowns and hegemonic shifts, for reasons that are not directly related to environmental or resource thresholds. The well-known example of over-accumulation of capital in our own world system is one in which further productive investment declines and fictitious accumulation of "paper" wealth becomes dominant, leading eventually to liquidity crises. Italy, Holland, and England did not decline as hegemonies for environmental reasons. However, as the system as a whole continues to expand, such pulsations can eventually strain the system to its resource limits. This accounts for the fact that, with hindsight, it often appears that collapsed civilizations, are in fact accompanied by major ecological exhaustion as well.

In the models of civilization that we have used in our research, the crises are related to growth itself and to the nature of such growth in terms of the articulations of local/global relations. Emergent hegemony is based on a combination of military conquest, the transfer of wealth, and the development of export production.[4] A hegemon is a dominant actor within a larger group of polities of which some are allied, some competing, with a peripheral zone of lesser developed or underdeveloped suppliers of raw materials and manpower. Although a number of authors have stressed the importance of systemic limits in understanding decline (e.g., Yoffee and Cogwill 1988; Tainter 1988), their analyses have not been set within a global systemic framework.

4 Population growth is sometimes assumed as an independent variable in discussions of growth and decline. In the approach suggested here, population growth and population density are outcomes of other strategies of accumulation that lead to increasing rates of fertility and/or import of labor.

HEGEMONIC SHIFTS

Much of the cyclical argument detailed here revolves around the reasons for and effects of hegemonic exhaustion and accompanying shifts in accumulation within global arenas for both the decline of some regions and the rise of others. While decline and accompanying shifts of hegemony are slow processes (relatively), they also include quite rapid changes. The shock of emergent awareness is an important part of this. Since the 1970s, there has been a rather rapid shift of capital accumulation from the West to East and, to a lesser degree, South Asia. However, recent media coverage has reinforced cognizance that China has become a world center of production, producing innumerable panicky discussions about where the world is headed. Chinese production as a percentage of world production for several commodities in 2001 was:

- tractors: 83%
- watches/clocks: 75%
- toys: 70%
- penicillin: 60%
- cameras: 55%
- vitamin C: 50%
- laptop computers: 50%
- telephones: 50%
- air conditioners: 30%
- televisions: 29%
- washing machines: 24%
- refrigerators: 16%
- furniture: 16%
- steel: 15%.

The Chinese diaspora numbers 30–50 million and account, in this same period, for 75% of all foreign investment in China. They controlled also, for example, 60–70% of the gross domestic product of Indonesia.

China can be said to be rapidly occupying a position that was once that of the United States and before it Britain. This particular shift of the locus of production took place over a number of years, but thresholds of awareness occur quite suddenly. Reports in the media indicate the nature of such rapid change, as evidenced by headlines during a time frame of just a few weeks in April, 2005:

- Chad: Young Street Children Recruited into Islamist Groups
- Shipments of Children (primarily Philippines) Major Industry
- Micro-credits, a Form of Microfinancing in the Third World Now Being Employed Extensively in Europe
- New York City for Sale
- Burnout Phenomenon Linked to Fear of the Social Stress of the Work Place: Fear of Conflict

- Euro-scepticism Has Been on the Rise All over Europe
- Increasing Level of Violence against Women in "Progressive" Sweden Reported on in France as a Scandal
- Chinese Are Making a Major Effort to Repatriate Chinese Researchers in Other Parts of the World with Offers of High Salaries and Good Working Conditions
- China and India Meet to Establish Alliance in which Taking Increasing Global Responsibility is Central Issue.

What do such headlines reveal: increasing disorder, violence, and dissonance in the West and in the weak links of declining Western hegemony at the same time as there is consolidation in East and South Asia. The surprise that is often indicated in such reports is one that is very much the product of a lack of information concerning the context within which such events and transformations occur. Now it is of course true that a major crisis in the West would lead in turn to a world crisis that would cripple Asia as well, since the latter is very much dependent on Western markets.[5] But the trend is quite clear; it consists of a shift in global hegemony, no matter how this may eventually strain the system to its absolute limit. In terms of our model of nested cycles, such an absolute limit might indeed be on the horizon, which would imply that without a major technological revolution in basic energy resources, our entire world system could be facing a terminal crisis. The point, of course, is that this is not the first time humanity has had to face such a situation.

THE OPACITY OF STRATEGIC SOCIAL OPERATORS

The examples here seem to indicate that major social change is cyclical in nature, even if the technological trend is toward increasingly larger systems that have now approached the geographic limits of the Earth in their scope. At various levels of social organization we confront periodic expansions and contractions. In global systems this cycle logically implies the decline of older centers and the shift of wealth accumulation to new centers. We have argued that there is a high degree of determinism in the dynamics of global and even of smaller systems of social reproduction. One question that always arises is the degree to which such determinism actually exists. My argument is that this seems empirically to be the case and that it is so for reasons that have not always been researched in the social sciences. These reasons, though, are well captured in certain literary works, most powerfully in the work of Kafka, on the way in which

5 The recent East Asian economic crisis was very much a product of a situation in which this region, a major attractor of industrial capital, outstripped in output the effective consumer demand, largely in the West, on which it depended. However, the crisis was certainly not the end of the game as can be seen today.

actors are imprisoned within the social universes and cultural orders which are instrumental in the production of personal knowledge of the world. The immediacy of what is taken for granted lies in the fact that it presents itself as an existential/phenomenological reality. This is also a primary reason for the difficulty of overcoming the disastrous situations that humankind has often found itself confronting.

Survival is a short-term operation for most social actors and the premises of survival are constituted within larger sets of assumptions about reality that are not often subject to real reflexive analysis about systemic conditions of material reproduction and survival. Bourdieu (1977) suggested the importance of this phenomenon in discussing the notion of *doxa*, a set of assumptions that form the rules of the game and the shape of the playing field. Within the field of *doxa* there may be a multitude of positions and counterpositions, but they are all logically related to the invariant structures of the field itself. The same notion was also touched upon by Marx in his discussion of the fetishism of commodities. The immediate appearance of the commodity can be said to conceal its "inner" or essential properties (i.e., the social relations under which it is produced). This can be extended to all dominant social relations in all social systems.

In the tribal societies referred to earlier, the sacred chief is sacred because he is able to produce (i.e., his lineage) more wealth than others, which implies that he has a closer relation to those who supply fertility (i.e., the ancestors of the entire group). Under the logic of this particular social order, closeness can only be expressed as kinship proximity, so that the chief is by definition the nearest descendent of the ancestors/gods and is thus himself sacred. It would make no sense to claim that this is an inversion of the real situation of control over other people's labor. In the same sense, accumulation of abstract wealth—capital—in many ancient and modern forms of capitalism is a process that cannot be said to reflect underlying technological or natural realities. Speculation, as other forms of accumulation, creates a redistribution of access to the "real" resources of the world. It also generates inflation and may even slow down the growth process by increasing the ratio of fictitious value to real costs of reproduction. If the financial and/or nonproductive component of the process of reproduction becomes an increasing proportion of the latter, it appears in the "incorrect" form of declining productivity and "diminishing returns."

Differential rates of accumulation also generate a differential gradient of costs of production so that a particular area becomes expensive compared to other less-developed places (e.g., the West's recent relation to China and India), that is, areas with lower overall costs of social reproduction. Capital, as accumulated wealth, does not contain within itself information concerning the conditions of productivity in technological (or energetic) terms nor about the conditions of its own reproduction over time. It is a fetish that fundamentally misrepresents significant properties of the reality that it organizes, not by intention, of course, but as a mere aspect of the way social life is constituted.

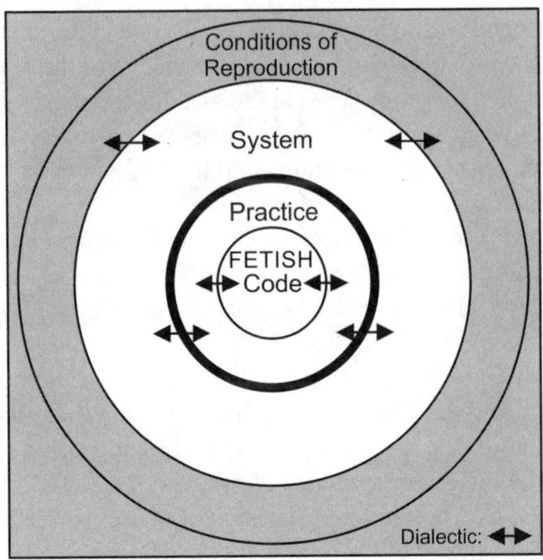

Figure 8.5 Dialectic: fetish/practice/system.

Figure 8.5 attempts to capture the degree to which social orders do not learn, or at least the limits within which such learning might be said to occur. I would suggest that this framework accounts for the ultimate ubiquity of decline and for the logical tendency toward cyclical patterns in history as well as to the experienced suddenness of the former. The notion suggested by Diamond (2005) that societies make choices, metaphorically or literally, concerning their own survival is grossly reductionist in the terms set out here. One must clearly identify the actors that are able to make such choices. There is no evidence that societies, small or large, are like subjects. On the contrary, one finds that there are dominant actors who are themselves caught in the webs of reproductive logics that escape their own intentionality and which, in fact, are rarely correctly cognized. This perspective on the history of humanity might account for phenomena such as the *Economist*'s reports on various national and regional economies. Thus when Asia was growing rapidly the *Economist* reported that Asia was doing something very right—company flags, discipline, state intervention, and the like—which we in the West ought to imitate. However, when the crisis came, the journal turned around and blamed Asia for precisely too much state intervention, for corruption and waste, and called for reforms closer to the Western model.

CONCLUSION: ALTERNATIVE SCENARIOS/ ALTERNATIVE PERSPECTIVES

The recent catastrophe in New Orleans is an excellent example of the complex process of accumulated contradictions and the decline of centrality in the center of a world system. With some hindsight and without access to the actual activities involved in the collapse of the dykes and the flooding of this large city as well as its abandonment, it might easily be attributed to a more general phenomenon such as climate change, whether global warming due to human activity or to natural cyclical changes. On closer examination, however, we see a global hegemony under stress. The hurricane that struck the city did not cause the greatest damage in itself. In fact, first reports were that New Orleans had escaped relatively unscathed. But then the levees collapsed in strategic places and the city was quickly flooded. The results were catastrophic, with hundreds of deaths (some predicted 10,000), and the question of why and whose fault immediately became media hype.

If we were to look at the abandonment of the city, and its destruction, as archaeologists, we might connect climate changes to the heating up of the Caribbean and the production of more intense hurricanes. Either human-induced global warming or natural climatic cycles could then be causal links to the decline of this urban area. If this were correlated with the declining power of the U.S.A., then climatic changes could be argued to be the cause of the decline of this major historical hegemony. Thus the Thera Syndrome would be perfectly replicated. The volcanic eruption of Thera—whose dating is disputed but most specialists now believe that it was closer to 1640 B.C. and not 1450 B.C.—is believed by some to be the cause of the decline of Crete, since the combined result of a tidal wave that struck the North Coast and the spread of volanic ash in the region would have caused serious agricultural problems (although for only a short period of time). There is no clear evidence, however, and the correlation with a possible spread of 200 years is a very weak argument for the effects of the supposed natural catastrophe, especially in the light of other more socially based accounts of the decline of Minoan civilization and its conquest by the Mycenians, one in which the changing balance of power between the Greek mainland and Crete is a historical process that can be accounted for in terms of social systemic analysis without any necessary natural catastrophic intervention.

Assuming that U.S. hegemony is on the decline, and has been for some time at least in economic terms, then the argument for climatic or natural catastrophic causes ought to seem misleading. It is of course true that nature is not reducible to the social even if it is social strategies that largely determine the properties of nature that are relevant for the survival of human populations. It is also true that there is a significant articulation between social processes and natural processes on which the social processes ultimately depend. This suggests, however, that catastrophes entailing the destruction of cities and the large-scale loss of lives is more of an expression of a social situation than a question of natural impact. It is increasingly clear that the reason for the flooding in New Orleans was related to a complex of factors that lead back to a decline on the part of the state's capacity to address such problems. The factors involved were:

1. The lack of coordination among agencies, from city and state to federal government.

2. The lack of troops available due to deployment in the Iraq conflict.
3. The lack of maintenance of the levees although it was well known that this was a crucial problem.
4. Conflicts among different social actors over the years, for example, between ecologists and government agencies concerning the ecological impact of the levee complex on the "natural" environment.

These are social factors that led to what looked like a natural disaster, and they are all related to a decreasing capacity to finance the activities necessary to prevent catastrophe. The zero-sum aspect of the lack of personnel, the lack of maintenance of levees, and conflicts in the chain of command are phenomena typical of situations of decline. Now it is of course true that there was a major storm related in an indirect way to the warming of the Caribbean, but the consequences of the storm in this particular case can only be understood in terms of the failure of state structures to respond to this natural phenomenon in a situation where there was in fact adequate knowledge of the consequences available.

There are a number of issues involved here. It is clear that the expansion of accumulative systems leads to increased stress on environmental systems. Global warming would seem to be a generally accepted result of human activity although it might be difficult to establish the actual causality involved. However, the accumulation process also creates strains on the social system which weaken its capacity to respond to the human-induced natural catastrophes that might have been under other circumstances adequately addressed. Of course, the story has not by any means reached its end. The city could be rebuilt—an age-old scenario—but one must nevertheless enquire into the capacity for reconstruction in the long term.

Needless to say, a typical remnant of declining civilization is the decline and abandonment of urban areas. To grapple with the mechanisms involved in the kinds of declines discussed in this chapter, it is necessary to achieve a level of precision that clearly identifies social actors, strategies, and logics that are directly involved in the relevant processes, the results of which we often interpret without the benefit of such a perspective.

REFERENCES

Boserup, E. 1965 The Conditions of Agricultural Growth: The Economics of Agrarian Change under Population Pressure. London: Allen and Unwin.

Bourdieu, P. 1977. Outline of a Theory of Practice. Cambridge: Cambridge Univ. Press.

Diamond, J.M. 2005. Collapse: How Societies Choose to Fail or Succeed. New York: Viking.

Ekholm, K., and J. Friedman. 1979. "Capital" imperialism and exploitation in Ancient World Systems. In: Power and Propaganda: A Symposium on Ancient Empires, ed. M.T. Larsen, pp. 44–58. Copenhagen: Akademisk.

Ekholm, K., and J. Friedman. 1980. Towards a global anthropology. In: History and Underdevelopment, ed. L. Blussé, H. Wesseling, and C. Winius, pp. 61–76. Leiden and Paris: Center for the Study of European Expansion.

Ekholm Friedman, K. 1976. Om studiet av det globala systemets dynamik. *Antropologiska Studier* **14**:15–23.

Ekholm Friedman, K. 1980. On the limitations of civilization: The structure and dynamics of global systems. *Dialectical Anthrop.* **5**:155–166.

Ekholm Friedman, K. 2005. Structure, dynamics and the final collapse of Bronze Age civilizations in the second millenium B.C. In: Declining Hegemonies: Present and Past, ed. J. Friedman and C. Chase-Dunn, pp. 51–87. Boulder: Paradigm.

Friedman, J. 1974. Marxism, structuralism and vulgar materialism. *Man* **9**:444–469.

Friedman, J. 1976. Marxist theory and systems of total reproduction. *Crit. Anthrop.* **7**:3–16.

Friedman, J. 1979. System, Structure and Contradiction in the Evolution of "Asiatic" Social Formations. Copenhagen: Natl. Museum of Copenhagen.

Friedman, J. 1998. System Structure and Contradiction in the Evolution of "Asiatic" Social Formations. Repub. of 1979 book with a new introd. and a new appendix. Walnut Creek, CA: AltaMira.

Friedman, J. 2004. Plus ça change: On not learning from history. In: Hegemonic Decline: Past and Present, ed. J. Friedman and C. Chase-Dunn. Boulder: Paradigm.

Friedman, J., and C. Chase-Dunn, eds. 2004. Hegemonic Decline: Past and Present. Boulder: Paradigm.

Harner, M. 1970. Population pressure and the social evolution of agriculturalists. *Southw. J. Anthrop.* **26**:67–86.

Harris, M. 1977. Cannibals and Kings: The Origins of Cultures. New York: Random House.

Harris, M. 1979. Cultural Materialism: The Struggle for a Science of Culture. New York: Random House.

Hedeager, L. 1978. A quantitative analysis of Roman imports in Europe north of the limes (A.D. 0–400), and the question of Roman-Germanic exchange. Copenhagen: Natl. Museum of Denmark.

Hedeager, L. 1992. Iron-age Societies: From Tribe to State in Northern Europe, 500 BC to AD 700. Oxford: Blackwell.

Hopkins, A. 1978. Economic growth and towns in classical antiquity. In: Towns in Societies, ed. P. Abrams and E.A. Wrigley, pp. 35–77. Cambridge: Cambridge Univ. Press.

Kristiansen, K. 1998. Europe before History. Cambridge: Cambridge Univ. Press.

Murphy, R. 1970. Basin ethnography and ecological theory. In: Languages and Cultures of Western North America, ed. E. Swanson, pp. 152–171. Pocatello: Idaho State Univ. Press.

Sahlins, M. 1969. Economic anthropology and anthropological economics. In: *Soc. Sci. Info.* **8**:13–33.

Tainter, J.A. 1988. The Collapse of Complex Societies. Cambridge: Cambridge Univ. Press.

Tran Van Doan 1981. Confucianism and the modern world: A critical evaluation. Proc. 1st Intl. Conf. on Sinology, vol. 1, pp. 1687–1693. Taipei: Natl. Tapei Univ.

Walbank, F.W. 1993. The Hellenistic World. Rev. ed. Cambridge, MA: Harvard Univ. Press.

Yoffee, N., and G. Cogwill, eds. 1988. The Collapse of Ancient States and Civilizations. Tucson: Univ. of Arizona Press.

Left to right: João Morais, Carole Crumley, Vern Scarborough, Yoshinori Yasuda, Joe Tainter, Fekri Hassan, Frank Riedel, Chuck Redman, and Peter Turchin

9

Group Report: Millennial Perspectives on the Dynamic Interaction of Climate, People, and Resources

Charles L. Redman, Rapporteur

Carole L. Crumley, Fekri A. Hassan, Frank Hole, João Morais, Frank Riedel, Vernon L. Scarborough, Joseph A. Tainter, Peter Turchin, and Yoshinori Yasuda

INTRODUCTION

The dynamic interactions between human societies and their environments are best understood from a perspective that accounts for long-term patterns and processes and addresses questions from an integrated, often interdisciplinary, perspective on human societies and biophysical environments. As proposed by the Earth System Science Partnership/International Council for Science to the World Summit on Sustainable Development in Johannesburg 2002, the development of a science for sustainability will require the development of a long-term perspective. "Archives from the past—e.g., ice cores, coral cores, tree rings, archaeological and historical records—must be studied more vigorously to provide paths of change, baseline conditions, insights into past societal resilience or fragility and perspectives on projections of future change" (ICSU 2002, p. 2).

In this chapter, we go beyond simply comparing data on environmental change and human activities to address more fundamentally their interaction and the characteristics that make systems more or less resilient in face of perturbations. In other words, what contributing conditions or circumstances led to effective or ineffective response to climatic change? Who benefited from these responses and who suffered? How enduring were these solutions, or did they lead to new, unforeseen challenges? What are the critical thresholds beyond which societies and environments find it difficult to recover? We are aware that different cultural elements (social and political structure, traditional practices, and beliefs, to name a few) enable or constrain specific responses. We are also aware that even global-scale events do not affect all regions at the same time or with equal intensity. Thus, to understand the past better and develop tools to meet the challenges of the future, we are constructing a framework to evaluate how societal characteristics and environmental conditions affected the abilities of past societies to cope with climatic change.

We have chosen two periods in climate history for which researchers agree that evidence for global climate change is present: the *4.2 K Event*, a drier and cooler period that occurred between 2200 and 1800 B.C. and the *7th- to 10th-Century Episode*, a drier and warmer period.

For both cases, we have chosen societies with detailed archaeological and historical information, including China, India, Egypt, the Near East, Europe, and the Americas. Some of these societies prospered in changed circumstances, some struggled and survived, and some collapsed. By analyzing both the impact on the environment/natural resources and on society, we can begin to discern what structures, practices, and attitudes cope best with environmental change.

Why expend our intellectual resources examining processes on a millennial timescale, well beyond the duration of a single life, and focus efforts on the distant past where some would argue the data are imprecise and incomplete? First, contemporary and short-duration studies usually have to content themselves with investigating an historical cycle that is truncated, that is, not having completed what future observers will identify as a coherent episode of history such as a dynasty or what resilience theorists define as an adaptive cycle (Holling et al. 2002). Archaeological and other long-term historical case studies can provide not only completed cycles, but multiple completed cycles. A continuum of studies allows us to better understand the dynamics of phases of a single cycle, linked cycles, and how they might change as systems reorganize. It also permits more in-depth monitoring of the slow processes and low-frequency events that are likely key to ultimate system resilience (Scheffer et al. 2001; Gunderson and Folke 2003). Although ecologists know that ecosystem structure and function may take decades or centuries to respond fully to disturbance, ecological studies almost exclusively examine ecosystem dynamics over intervals of a few days to a few years. Rare decade- to century-scale studies suggest that some human impacts are enduring, yet only a few integrative ecological studies of human land use cover timescales longer than a century (Roberts 1998; Foster 2000; Mann 2002; Heckenberger et al. 2003; Vitousek et al. 2004). Based on this growing data base, these ecologists strongly assert that the condition of many landscapes and the dynamics that govern them cannot be understood without close attention to the effects of historic land use (Foster et al. 2003).

Second, the "deep-time" perspective can reveal the ultimate, as well as proximate, causes of the collapse of socioecological systems. That is to say, human social, political, economic, and religious institutions can be elaborately stratified (hierarchies) or networked (meshworks or heterarchies); both are considered complex. In particular, the long-term history of human–environment interactions contained in the archaeological record reveals that many human responses and strategies, while beneficial in increasing production in the short term (even over a few generations), nonetheless led to serious erosion of resilience in the long term, resulting in the collapse of both social and environmental systems (McGovern et al. 1988; van Andel et al. 1990; Crumley 1994, 2003; Redman 1999; Redman et al. 2004; McIntosh et al. 2000; Kirch et al. 1992; Kohler 1992; Rollefson and Köhler-Rollefson 1992). Only with the deep-time perspective can we identify which of the many, seemingly beneficial, near-term actions truly contribute to long-term resilience and identify the ways in which seemingly rational choices can lead to undesirable outcomes. The converse of this scenario is social adaptations or cultural traditions that appear inefficient or "illogical" when viewed in the short term, but serve to reduce risk and increase resilience in the long term (Butzer 1996).

Third, the fields of archaeology and history, when supplemented with other social-science perspectives, allow a rich understanding of the linked dynamics of human behavior, social dynamics, and ecological systems across broad scales of organization—from individual households to hamlets, villages, cities, and civilizations. Few other social sciences encompass such a broad spectrum, preferring to concentrate more narrowly on "bottom-up" (household, village) or "top-down" (nation, state) levels of organization. If a system's operation is predicated on linked dynamics across scales—particularly the interaction of "fast" and "slow" variables or the "mismatch" of scales at which social and ecological variables interact—then it is crucial to examine these linkages from both social and ecological perspectives (Folke et al. 2002).

RESILIENCE AND VULNERABILITY TO ENVIRONMENTAL CHALLENGES

There is a danger in historical research of resorting too readily to explaining change by simple environmental causation. It is easy to correlate factors in the environment—such as climate—with changes in human societies. Climate and environment always vary, and extreme events are inevitable. Similarly, cultures change continuously and sometimes precipitously. If one seeks to link cultural change to environmental change, one can always find supporting evidence. A stronger framework would be to adopt a Popperian approach and attempt to falsify such interpretations. To link environmental change to cultural change is so attractive that this falsification approach seems never to be attempted.

Part of the reason we should hesitate to correlate culture and environment in historical research is that episodes of environmental challenge seem to produce quite variable cultural responses. Desiccation in the Near East at the site of Tell Leilan, for example, has been interpreted as leading to such diverse cultural responses as increasing complexity on the one hand, and collapse and abandonment on the other (Weiss et al. 1993). In both the Hohokam area of the U.S.

Southwest for the mid-second millennium A.D. (Doyel 1981) and in the north of China in the 6th to 4th millennium B.C. (Yasuda et al. 2004), similar interpretations have been advanced. To fulfill the goals of the IHOPE project, we need to understand how environmental fluctuations can be linked to such variable cultural processes. Is there any reason to hope that we may find regularities in the environment–society relationship?

There is reason for hope, but the relationship is often subtle, indirect, and variable. Part of the challenge is that human societies experience internal transformations in complexity, resilience, and vulnerability. Often these transformations respond to factors other than the biophysical environment. When societies respond to changes in the biophysical environment, those changes have themselves been mediated by social, political, and economic institutions. There are circumstances when a society will be resilient to all but the most extreme perturbations. There are also circumstances when a society will be so vulnerable that it may be unable to cope with problems of comparatively small magnitude (Tainter 1988). It is easy to see why human societies are viewed as variable and fascinating to social scientists, yet frustrating to biophysical scientists.

Some definitions are in order. To explore this perplexing relationship, we must consider complexity and resilience. Although several conceptions of complexity exist in the sciences, the term is used here to mean differentiation in social, political, and/or economic structure combined with organization that integrates diverse structural parts into a whole (Tainter 2000b). This conception is consistent with how anthropologists and archaeologists have approached variation and evolution in human organization (e.g., Service 1962; Plog 1974; Tainter 1988; Crumley 2005). Subsumed in our use of the term is a concern with the nature of diversification versus integration operating in either the horizontal or vertical dimensions within a society. Resilience, in turn, is termed the capacity of an institution to adjust to perturbations. An institution, such as a society, that lacks this capacity, or has lost much of it, may be vulnerable to challenges.

We do not intend to imply the engineering definition of resilience that relates to stability around a single state, but rather the possibility of multiple, stable states that maintain the primary functional relationships of the socioecological system. This definition is taken from a group of ecologists and economists who use Resilience Theory to understand the source and role of stability and change in systems—particularly the kinds of change that are transforming and adaptive (Gunderson et al. 1995; Levin 1999; Holling et al. 2002). Four key features of ecosystems yield the underlying assumptions of Resilience Theory (Holling and Gunderson 2002, pp. 25–27; Redman and Kinzig 2003; Redman 2005); we believe these features and assumptions are worthy of evaluating for coupled socioecological systems and from a long-term perspective. First, change is neither continuous and gradual, nor consistently chaotic. Rather, it is episodic, with periods of slow accumulation of "natural capital," punctuated by sudden releases and reorganizations of those legacies. Episodic behavior is documented in our case studies as the interactions between fast variables, such as a particular flood or series of low-flow years, and slow variables, such as increasing vulnerability of agricultural systems due to declining soil fertility from overcropping. Second, spatial and temporal attributes are neither uniform nor scale invariant; rather, patterns and processes are patchy and discontinuous at all scales. Therefore, scaling up from small to large cannot be a process of simple aggregation. Third, ecosystems do not have a single equilibrium with homeostatic controls; rather, multiple equilibriums commonly define functionally different states. Destabilizing forces are important in maintaining diversity, flexibility, and opportunity, while stabilizing forces are important in maintaining productivity, fixed capital, and social memory. Fourth, and finally, policies and management that apply fixed rules for achieving constant yields, independent of scale and changing context, lead to systems that increasingly lose resilience—that is, to systems that suddenly break down in the face of disturbances that previously could be absorbed.

To operationalize a resilience and vulnerability approach in anything but a general, theoretical arena, one must define "resilience (and vulnerability) of what, to what." For an archaeologist or historian, the unit of interest in answering the "of what" question traditionally has been the continuing existence of a historically identifiable society. For an ecologist, it may mean a landscape or region with a particular set of biotic communities and biophysical operating processes. To a climatologist, it may mean the capacity to absorb greenhouse gases and still maintain today's climate parameters. Because our goal in this volume is to investigate the dynamic relationships among these domains, we must be concerned with units that acknowledge this interdependence. Hence, the entity of concern is the integrated socioecological system that includes the full range of biotic communities, human organizational features, as well as landscape and climatic characteristics. Although understanding these as integrated entities is an objective of this volume, the authors of this chapter immediately acknowledge that these socioecological systems have many components and respond to forces in diverse ways and at different spatio-temporal scales. It is, in fact, the complexity of these components and interactions that leads to both the difficulty and promise of this inquiry. Essential to a resilience perspective is that many different stable states may exist for any particular socioecological system and that these different states may reflect substantial changes in some key aspect of the system, such as the agricultural system, but relative stability in other aspects, such as the social organization. This scenario is reflected in the response of some settled agriculturalists in northern Mesopotamia to the 4.2 K Event, when they abandoned their aggregated settlements and aspects of their productive economy for a more mobile, dispersed existence.

We also need to recognize that there are many different levels of resilience reflecting degrees of vulnerability to change. Some systems will be so effectively organized and generate sufficient resources that they are extremely resilient in the face of even major perturbations, while most others are at an intermediate level of resilience where changes in one or another subsystem or at one scale will allow the system to continue in the face of most perturbations. Other systems have worked themselves into situations of low resilience where even a modest perturbation will set the system on a trajectory toward relatively complete collapse. Although these changes should be monitored according to many system attributes, for the sake of this chapter we will focus on changing values of three measurable and socially significant characteristics: population, productive system, and sociopolitical complexity.

When we specify "resilience to what," we must be concerned with both external perturbations and internal reorganization, which is often a response to perturbation. From an empirical perspective, it would be good to be able to monitor a socioecological system's resilience in the face of a change of, for instance, 1°C, 10-cm drop in mean annual precipitation or some significant change in the pattern of variability of these and other biophysical characteristics. Although these parameters may be available in some cases, the significance of their impact will vary with conditions of each case study as well as its organization. We seek to develop an approach that monitors challenges that may emanate from the biophysical world, but translate into changes in drivers that humans recognize and to which they respond. Hence, we seek to monitor climate changes that modify the carrying capacity of a landscape, alter the human quality of life, occur over various spatial scales, or are of varying duration or of a repetitive nature.

EFFECTIVENESS IN PROBLEM SOLVING

One dynamic internal to societies that is particularly relevant to understanding responses to environmental challenges is the evolution of problem-solving strategies and institutions. Problem solving tends to evolve from solutions that are simple and inexpensive to solutions more complex and costly. Much complexity and costliness in human societies emerge from developing solutions to problems that range from mundane to vital. A good illustration of this dynamic is the response of the U.S. government to the terrorist attacks of September 11, 2001. Much of the immediate response was to increase the complexity of public institutions, by establishing new agencies, absorbing existing agencies into the federal government, and exerting new forms of control over behavior that might represent a threat. As seen in this example, we resort to complexity to solve problems because organizational changes can often be implemented quickly. At the time a problem arises, the solution may seem small, incremental, and affordable. It is the continual accumulation of complexity and costs that most often does the damage. Rarely do societies eliminate institutions or willingly simplify their structure. In a condition of high complexity and costliness, institutions may lack the fiscal reserves to address new challenges, whether these challenges come in the form of environmental perturbations, hostile neighbors, or something else. A society that has adopted costly problem solving in major sectors of public investment may lose resilience and become vulnerable to challenges that it once could have overcome (Tainter 1988, 2000a).

The environment, including climate, may be one source of problems to which a society adapts. In ancient Egypt, Mesopotamia, and China, for example, many bureaucratic and religious organizations emerged to meet the challenge of delivering water to farmlands. Similarly, among the ancient Maya and Hohokam, there seems to have been a strong connection between political organization and landscape transformation as a mechanism for water management. Changes in climate, in such productive systems or in others, can stimulate further developments in problem solving. These further efforts, and their attendant costs, may promote the productivity of the agricultural system, or they may burden it. Within a given society and agricultural regime, both outcomes may emerge, but at different times (Tainter 2000a, b).

It is important not to think of complexity as inherently detrimental in human societies. In the early phases of problem solving, complexity is typically effective, giving positive returns and creating positive feedbacks among variables such as population, agricultural production, fiscal strength, and military strength. The difficulty comes when problem solving produces diminishing returns, so that additional expenditures fail to produce proportionate benefits. The net cost to society of maintaining costly problem solving must be absorbed somewhere in the system. These costs are often displaced to subjugated populations, other regions or economic sectors, or delayed. A society experiencing diminishing returns to problem solving in major public sectors will be weakened fiscally. Resources must be allocated just to maintain the status quo, and the legitimacy of leadership suffers because high taxes that yield few discernable benefits alienate the population. The Roman Empire encountered this problem in its late phase (Tainter and Crumley, this volume). The Roman government increased its size, complexity, and costliness just to maintain the status quo. This increase consumed resources so that the empire lost its resiliency. Collapse became a matter of mathematical probability, as an insurmountable challenge was bound to arise.

This perspective suggests one way to examine the relationship of society to climate and environment over time. Why do climatic extremes, such as those at 4.2 K and between A.D. 700 and 1000, seem to overwhelm some societies and strengthen others, while other societies seem completely unperturbed? We will understand such variable responses better when we investigate both environmental perturbations and the internal trajectory of a society's capacity to cope with challenges (Tainter 2000a). The evolution of complexity in problem solving is one such internal trajectory. There are others that should be considered as well. To undertake such investigation makes the research more challenging, but also more reliable.

As a society's resource needs fail to be met, whether from a change in climate or other factors, a common response is to intensify production. Intensification involves increasing inputs to the productive system to obtain higher outputs. Complexification, as discussed above, is closely related to intensification. Because complexity costs, the expense of inputs increases in more complex systems. In agriculture, inputs include such factors as labor, water, and fertilizers. Increasing these may lead to higher outputs, but this may be short-lived. Sometimes the productive system becomes degraded and productivity falls. As discussed by Tainter and Crumley (this volume), for example, the late Roman government attempted to increase production (and thus tax yields) by requiring that taxes be paid on each parcel of land as if it were under cultivation. Outputs initially increased, but overall production declined as peasants abandoned marginal lands. Degradation is not, however, an inevitable outcome of intensification. A system of subsistence agriculture using night soil as a fertilizer, for example, may be sustainable for a long time if the ratio of population to land can be held relatively constant. Also in the case study of rice-paddy agriculture in the Yangtze River region of China, long-term resilience was achieved through substantial investment in terraced fields and water systems. Intensified production in these and other cases may be sustainable if the inputs continue to be affordable.

HOW CASES WERE CHOSEN

At least six global or semiglobal periods during which the amplitude of change (precipitation, temperature, wind) was somewhat higher than before characterize the Holocene. During historical times, the signals of the "Little Ice Age" (A.D. ~1250–1880 with major effects between A.D. 1550 and 1850) and the "Medieval Warm Epoch" (A.D. ~750 to 1200) with the "Little Optimum" (A.D. ~800 to 1000) can be widely detected, but mainly in parts of the Northern Hemisphere. However, there is increasing evidence that global-scale teleconnections existed. The prehistorical Holocene periods of significant climate change were between 9 and 8 ky, 6 and 5 ky, 4.2 and 3.8 ky and 3.5 and 2.5 ky. Of these the 4.2 to 3.8 ky period is of special attraction because, on the one hand, there is evidence that changes took place at the global scale, with a trend to drier and cooler conditions in most of the lower latitudes (but partly warming in higher latitudes) and, on the other hand, records exist which propose that there was a particularly rapid change around 4.2 ky ("4.2 K Event") thus giving the opportunity of integrating chronologies of other societal events related to our case studies (see Figure 9.1). To contrast the prehistorical "4.2 K Event," we selected the historical "Medieval Warm Epoch" of which we know that at least during the "Little Optimum" there were some extreme dry events on both sides of the Atlantic.

The 2200 to 2000 B.C. climatic event (we refer to this as the 4.2 K Event, although it probably represents a change that extended from roughly 4200 to 3800 B.P.) was most likely due to severe cold spells more characteristic of an Ice Age rather than the warmer conditions that prevailed before. The weak circulation over the Atlantic in the ice cores from Greenland confirms this supposition, as

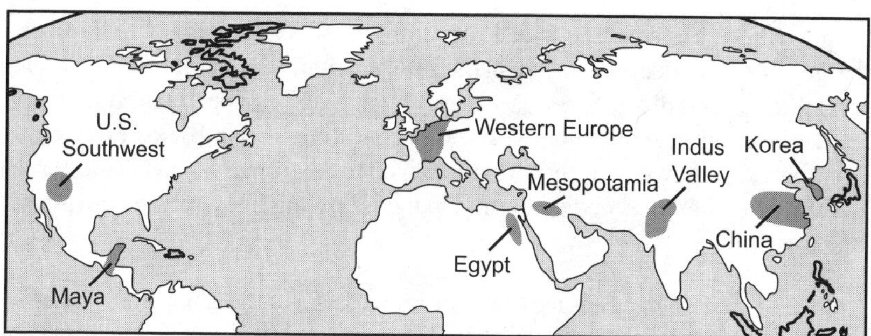

Figure 9.1 World maps with locations of the case studies.

does the transition from birch and grassland vegetation to arctic conditions in Iceland at 2150 B.C. (Rose et al. 1997). The global impact of the cold phase was felt beyond Europe and the North Atlantic in areas affected by the Intertropical Convergence Zone, such as is documented in Western Tibet, Gulf of Oman, Atlas Mountains, and the Andes (Gasse and Van Campo 1994).

Societies and the socioecological systems of which they are a part regularly experience alternating episodes of stability and change. In selecting our case studies, we attempted to focus on cases well understood by the archaeological and/or historical community that have experienced a relatively dramatic transformation of their demographic, social, political, or economic conditions in concert with independently documented, but equally dramatic, climatic changes. These changes may have contributed to expansion of the complexity and economy of the society, as in the 4.2 K Harrapan case or the 10th-century European case, to transformation in the character of the system as in the 10th-century Southwest U.S. case, or to contraction or outright collapse of the society as in the 4.2 K Egyptian, Mesopotamian, and Chinese cases and the 10th-century Mayan and Korean cases. Of substantial interest are case studies of socioecological systems that weathered the same climatic perturbations without undergoing major transformations as well as how the cases we describe remained relatively stable during different episodes of climatic change.

The 4.2 K Event and the Fall of Old Kingdom Egypt

Nothing prepared Egypt for the collapse of the Egyptian Old Kingdom in 2180 B.C., made manifest in the eclipse of royal power and widespread poverty and famine following the death of Pepy II, who ruled for more than 90 years. What factor weakened the monarchy and allowed provincial governors to assume royal power over their regions? One possibility is an invasion by Asiatics, Bedouins from Levant. However, there is little evidence that Asiatics invaded Egypt at that time and, even if they did, what had weakened the state to the point where invaders would be successful? More convincingly, the fall of the Old

Kingdom seems to coincide with a catastrophic reduction in the Nile flood discharge over a 2–3 decade episode in the 2180 B.C. period (Hassan 1997). The ensuing agricultural disruption was so severe that famine gripped the country and undermined political institutions. The dire conditions are aptly expressed by an Egyptian Sage Ipuwer whose work is believed to have been written in the period just after the First Intermediate Period in the following lines excerpted from his long treatise:

> Lo, the desert claims. The land towns are ravaged, Upper Egypt became a wasteland. Lo, everyone's hair (has fallen out). Lo, great and small say, 'I wish I were dead.' Lo, children of nobles are dashed against walls. Infants are put on high ground. Food is lacking. Wearers of fine linen are beaten with [sticks]. Ladies suffer like Maidservants. Lo, those who were entombed are cast on high grounds. Men stir up strife unopposed. Groaning is throughout the land, mingled with laments. See now the land deprived of kingship. What the pyramid hid is empty. Lo, people are diminished (Gardiner 1909, pp. 95–110; Lichtheim 1975, pp. 149–163).

The repeated failure of Nile floods led to a calamitous reduction in agricultural production. This decline, in turn, caused widespread famine and depopulation. Those who remained alive ransacked palaces, temples, and royal granaries. Local governors fled or were perhaps killed. Gangs roamed the country pillaging and searching for food and killing those who opposed them. The king was besieged. The administrative hierarchy of tax collectors was disrupted. The king was bereft of revenues and his authority was undermined. The sequence of events could be schematically recognized as follows:

- Abrupt decadal climate variability ca. 2180 B.C. led to significantly reduced rainfall in the Ethiopian and equatorial sources of the Nile, causing severe shortage in Nile flood discharge.
- Repetition of low Nile flood discharge was associated with a change in sediment/load ratio and texture of sediments, which led to: (a) repeated loss of cultivable land; (b) frequent shifts of the course of the Nile, rate of siltation, and adjustment of flood plain morphology; and (c) shallowing or drying up of the Nile channel.
- Reduction of cultivable land because most of the floodplain could not be irrigated, resulted in dramatically reduced agricultural productivity.
- Altered floodplain morphology led to property disputes and a loss of cultivable land for many.
- Loss of cultivable land and reduced agricultural productivity led to severe famines.
- Repeated famine led to morbidity, high mortality rates, emigration, and attacks on palace, temples, and royal granaries.
- Accumulation of corpses and morbidity led to worsening hygienic conditions and epidemics.

- Depopulation as a result of famine led to a dramatic reduction in food producers.
- Urban dwellers, craftspeople, artisans, scribes, and others who depended on barter with farmers or gifts from the king for their livelihood, perished.
- Attacks on nobility in the course of ransacking food from granaries led to violent confrontations ending with mass revolts and the overthrow of provincial nobility.
- Demise of provincial nobility led to destroyed administrative hierarchy and paralyzed state bureaucracy.
- Disruption and collapse of state administration coupled with famine, strife, and the drying up and shallowing of the river channel drastically reduced revenues to the king from the provinces.
- Loss of revenues to the king dramatically cut back on financial support for the royal household, high state officials, and nobility at the nation's capital. The king and high state officials rapidly descended into poverty.
- Reduced functional power of the king and the state prevented serious remedy to the problems in the provinces, thus aggravating local famines and allowing strife to go unabated.
- More raiders roamed the country in search for food, causing havoc everywhere they went.
- Local droughts concomitant with low Nile flood discharge in the Eastern Mediterranean prompted "invasions" of the East Delta by Bedouin marauders.
- Failure of the king's power to mobilize troops led to the advance of Bedouin marauders unhindered into East Delta and even into Middle Egypt. Kings rose and fell in rapid succession as a result of the serious economic and political problems.
- Since the king "failed" to meet his responsibilities for maintaining order and prosperity as a divine intermediary with the gods, people lost faith, undermining the ideological underpinnings of kingship. Royal pyramids, tombs, and other royal edifices were ransacked, and the secrets of kingship were debauched.

Why did this series of low floods lead to dynastic collapse where other low-flood periods did not? One answer may be that the scope of the shortfall and the repetition of low-flood conditions may have been more severe than in the past (finer-resolution data are needed to support this conclusion). Whether or not this episode was more severe, the expanding costs of the Pharaonic hierarchy had weakened the sociopolitical situation (i.e., the cost of complexification discussed above). The cumulative costs of extraordinary monument construction, bureaucracy, priesthood, and elaborate rituals pushed the system to the brink. Shortfalls in production were destined to undermine the top of the system that had no productive lands under its control, but relied upon taxes from provincial rulers.

The 4.2 K Event in Rural and Urban Mesopotamia

The Mesopotamian case study offers us both a long duration of civilizations and substantial environmental and historical diversity. In focusing upon the 4.2 K Event, we examine two geographically different regions of Mesopotamia: the rain-fed steppe of northern Mesopotamia and the highly urbanized, irrigated floodplain of the south. From the fourth millennium B.C. onward, interactions between the two zones became increasingly important. In the north, small towns and cites emerged in a dispersed, rural landscape, while urban centers, connected to outlying villages through extensive canal systems, dominated the south. We should recognize, however, that despite the contrast between northern rain-fed and southern irrigation agriculture, each region was internally diverse and offered differential access to surrounding areas, factors important to the historical dynamics.

The north lies within the Fertile Crescent, an arc of favorable precipitation and soils that support productive rain-fed agriculture from the eastern Mediterranean to the flanks of the Zagros Mountains. Nevertheless, this arc is narrow and rapidly transitions to semiarid and then arid steppe, as precipitation drops from north to south. Vegetation responds quickly to fluctuations in precipitation in this region, and even small annual changes may imperil subsistence agriculture, especially in the drier zones. Northeastern Syria, in the Khabur River drainage, enjoys some of the highest rainfall (>450 mm) as well as some of the lowest (<150 mm). At about the 250 mm precipitation isohyet, agriculture becomes problematic and below 200 mm is normally not possible. This "marginal" zone is particularly sensitive to annual variability and longer-term climatic fluctuations, and agricultural settlement through the millennia reflects such changes. Archaeological survey of this marginal zone in the western Khabur drainage has revealed periodic human settlement that responded to the 4.2 K Event (Hole 1997).

During the late 5th to early 4th millennium B.C., the wadis held water perennially, indicating a more favorable water balance than today (Courty 1994). Sometime during the fourth millennium, owing to flooding that may have resulted from loss of vegetation due to aridity, much of the upper Khabur, including remnants of older settlement, became covered with two meters of sand and gravel. By the early third millennium B.C., the land surface had stabilized and archaeological evidence suggests that, in the mid-third millennium B.C., precipitation was especially favorable. Sites then moved away from the river and wadis to which they had been confined. Many sites, many with encircling walls, and as large as 30 ha, were established on landscapes that have rarely seen settlement since, indicating excellent conditions for agriculture and sheepherding. The first true urbanization of the region also occurred in the wettest zones: Tell Leilan at 90 ha, was walled by 2600 B.C. (http://www.research.yale.edu/leilan). By the mid-third millennium B.C., settlements in the marginal zone began to be abandoned and, within a few centuries, most of the towns and cities even in the northernmost wetter zone were abandoned or much reduced in size. This transformation occurred at the time of the 4.2 K climatic event, just a few generations after the Akkadian conquest of the region, and may have contributed to the collapse of that empire and others in the Old World (Weiss 1997a, 2000; deMenocal 2001; Weiss and Bradley 2001). After abandonment, the wetter zone was reoccupied within three centuries, but the marginal zone was not again resettled until the first millennium B.C. Subsequent, but less intense, occupation of the marginal zone occurred during the Iron Age, Roman/Byzantine, and Islamic periods. The pattern became short periods of settlement separated by 1000 to 1500 years of abandonment (Hole 1997). After Medieval times and, until recently, nomadic tribes of sheepherders occupied the region primarily.

Changes in precipitation patterns during the third millennium B.C. probably account for the changes in human settlement, although depletion of fragile steppe environments was a contributing factor. Advancing aridity probably led to depopulation of the marginal zone where even a modest decrease in precipitation would have seriously affected subsistence agriculture. This scenario may have resulted in the dislocation of large numbers of people, some perhaps through forced resettlement (Weiss 1997b, p. 343), who converged on cities and settlements in the more favorable areas. The greater density of population would have added to stresses that were becoming more severe throughout the region. There are, of course factors other than climate and demographic change. The growing demand for wool to supply cities outside the region may have sparked the establishment of settlements across the arid Khabur. Furthermore, grain may also have been exported from the outlying settlements to the rapidly growing cities in the Khabur, such as Tell Leilan, Mozan, and Brak, to support their infrastructure (Wilkinson 2000). Failure of agricultural production in the outlying settlements would have intensified production in the wetter zones through construction of canals to distribute water as well as through reduction of fallow. A predictable result would have been declining yields. Akkadian imperial demands exacerbated the situation: one cuneiform tablet documents the export of 29 metric tons of grain from the Khabur (http://www.research.yale.edu/leilan), but it is not clear whether this was an annual tax or another kind of transaction. In view of the dual pressures of an overstressed agricultural system and the need to ship grain out of the region, combined with the onset of aridity at 4.2 K, the system was too fragile to withstand the shock.

We see two situations in the Khabur. First, economies dependent on subsistence agriculture began to fail; then city-based economies dependent on these outlying settlements, but still required to export grain, lacked the resilience to buffer these losses. The collapse happened quickly: Weiss maintains that construction of a palace stopped abruptly, literally in midcourse (http://www.research. yale.edu/leilan) and that there was a general "migration of people to riverine environments with sustainable agriculture" (Weiss 1997a, p. 714), leading to the construction of the Repeller-of-the-Amorites Wall in the south

(Gasche 1990; Weiss et al. 1993). With the collapse of the Akkadian Empire, "the entire Near East reverted to a system of independent states" (van De Mieroop 2004, p. 67), which reconsolidated soon after during the Third Dynasty of Ur. The immigration from the north may have doubled the population in the post-Akkaian Ur III period, most of which settled in villages in southern Mesopotamia (Adams 1981, pp. 142–143). It has been suggested that, as a result of excessive expectations for agricultural production during this dynasty, much of the irrigated lands in southern Mesopotamia suffered from some degree of salinization (Redman 1992). This salinization led to diminished productivity from 2000–7100 B.C. and a breakup of the centralized political administration. The timing of this process is late compared to the 4.2 K Event and may be due more to regional socioenvironmental factors than to global climate change.

The contrast of the rain-fed north with the irrigated south could scarcely be more pronounced. Nearly all settlements in the south were on the major canals and the entire countryside had a very urban look in the late third millennium. "90% of the population lived in settlements larger than 30 ha" (Nissen 1988, p. 131). In Mesopotamia, changes in precipitation affected the flow of the Euphrates River which, in turn, provided water for irrigation. Serious fluctuations in its flow, as in the case of the Nile, reduced the amount of cultivable land. The reduction of Euphrates flow, with a consequent tendency for the channels to meander, was countered by further linearization of irrigation canals (Adams 1981, p. 164). There was another problem, however: the courses of the Euphrates and Tigris were not well established and shifts in their beds occasionally isolated cities. Throughout the third millennium B.C., the canal systems were expanded to serve the cities and their fields, as well as provide avenues for transport of goods. The land that lay outside the canals' reach was used for grazing livestock whose forage was sensitive to precipitation, but canal drainage buffered such fluctuations. The availability of water through extensive human management mitigated natural changes so that the south did not respond with wholesale regional depopulation and long periods of abandonment that regions on the margins of the north experienced. Rather, there was a fluctuation between city-based and more dispersed settlements. After the Akkadian collapse, the south saw increasing hostilities among city-states competing for water and land (Nissen 1988, p. 142) and control of irrigation water was one reason for the repeated consolidation of city-states into regional states (Nissen 1988, p. 145; van De Mieroop 2004, pp. 66–67). This consolidation was first manifest in the Akkadian "empire," a grouping of cities into a single political entity across Mesopotamia.

In summary, let us think in terms of integrated systems. We begin in the fifth to fourth millennia B.C. with semiautonomous towns and villages, supported by subsistence agriculture dependent upon rain. The number of settlements across the Khabur gradually increase and some grow to large size. During this time, stresses on the land could have been overcome through movement of a settlement. A severe and prolonged aridification felt across the Near East at the end of the fifth millennium may have precipitated a population crash. This event or process can be seen on the ground in heavy erosion and deposition.

By the end of the fourth millennium, populations had begun to rebound, but were curtailed by another episode of aridity at 5.2 K. Again, settlement was disrupted and many regions were effectively abandoned. The third millennium saw all regions rebound, grow to unprecedented levels of population and urban density, and then collapse at 4.2 K. It is interesting that in both the late fifth and late fourth millennia, religious activity as seen in the building of temples, intensified. This activity occurred in both the north and the south, possibly as an attempt to deal with unpredictable and often violent forces of nature. Urbanism followed in the third millennium, with the rising importance of secular institutions and the formation of the first city-states. These large polities began to acquire goods from outside their regions, such as wool, and in the process introduced a commodification of that product. Groups previously on the periphery emerged ready to supply the growing need for wool. Much of the expansion of settlements onto the Khabur steppe can be seen in this light, although these settlements could not have sustained themselves without dependable agriculture. Hence, favorable climate made large-scale production of wool feasible. The growing city centers required grain to feed burgeoning urban populations and to support military activities. Eventually, especially during the Akkadian period, extractive taxation may have become extremely burdensome. Perhaps in an attempt to solve this problem, labor was enlisted from the outlying settlements to construct canals and cultivate fields that could generate the surpluses for local consumption and export. Intensification of agriculture through canal building was somewhat risky and labor intensive, as the repeated filling of the canals with sediment shows (Weiss 1997b, p. 344). Moreover, another strategy, the shortening of the fallow cycle, would have led to reduced yields, thus requiring further attempts at intensification and extraction of more produce from outside. The system was probably stretched to its theoretical limits as none of the largest cities and towns controlled enough land in their immediate vicinity to feed the resident populations (Wilkinson 2000). These factors alone could, as Wilkinson has suggested, engender collapse. One could also point to disaffected populations who were called upon to supply the cities, as another source of systemic problems (van De Mieroop 2004, pp. 66–67). It is likely that a combination of social and environmental stresses had built up just at the time of the 4.2 K Event. A conjuncture of natural, physical, and social processes thus stressed the system beyond the point of resilience.

The 4.2 K Event in China

There is good evidence that the 4.2 K Event was felt in East Asia where the monsoon belts were depressed to the south, yielding a drier and cooler episode for much of China for several centuries. Archaeological evidence suggests that

Chinese civilizations responded in ways distinct to this climatic change. Of special interest is the variation of the response given the nature of the landscapes and productive systems of two major regions. Among the dry-field farming and stockbreeding people of Yellow River region of north China, this and other similar climatic perturbations seriously impacted agricultural productivity and pastoral economies. In turn, the already semisedentary groups of the region migrated from the region, disrupting those in more established settlements and, under extreme conditions, challenging the occupants of the early cities in the region. The 4.2 K Event negatively affected the productivity of rain-fed cereal farmers and reduced the amount of pasture land available for stockbreeders (Figure 9.2). A drop in total population and increasing mobility of the inhabitants of the region ensued.

The 4.2 K Event appears to have impacted the wet-rice cultivating and fishing people of the Yangtze River region of central China differently (Yasuda et al. 2004). Due to the major investment in rice-paddy construction and water systems, the effect of the drying and cooling episode was serious but significantly

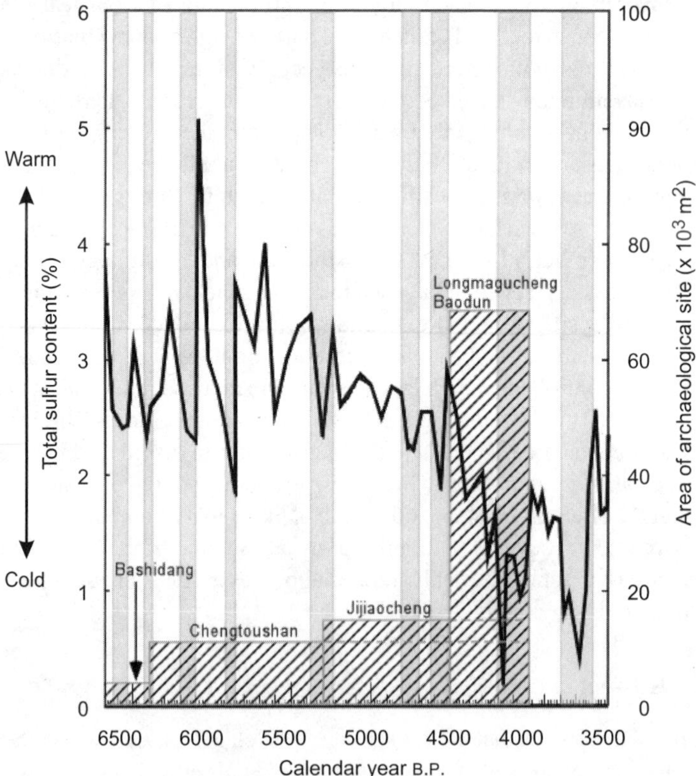

Figure 9.2 Climate and civilization in ancient China.

buffered. These more intensive agriculturalists did not experience the dramatic productivity and population declines among the dry-field farmers. Although still vulnerable to climatic perturbations, investments in water-control infrastructure, which ensured minimal water delivery even under drought conditions, substantially diminished the impact. In addition, because of the difference in the landscapes and productive technologies, the Yangtze region did not experience major population migrations, nor the ensuing pressure that would be otherwise exerted on the sedentary populations. Hence, at least for the prehistoric period, variations in landscapes and productive technologies appear to have engendered different responses in two regions of China. The Yangtze River region appears more resilient to climatic perturbations while the Yellow River region appears far more vulnerable.

The 4.2 K Event among the Harappans of the Indus Valley

Unlike the three Old World civilizations described above, the Harappans of the Indus Valley and adjacent regions approached a population peak approximately coincident with the 4.2 K Event. Although precise dating is debated, most agree that the initial precipitous rise of socioeconomic and sociopolitical complexity occurred by 2550 B.C. and continued until 1900 B.C. (Possehl 1990, p. 275). One of the two "type sites," Mohenjo-daro, had a population of 40,000 with a density of 16,000/km^2 (Wheeler 1968; Fairservis 1971; Scarborough 2003, p. 145). Kenoyer (pers. comm.) suggests a population of 80,000 over an area of 1.5–2 km^2, signifying a density of twice or even thrice this figure, more than an order of magnitude denser than the Maya described below. Cities were highly regimented, with standardized grid plans of streets and alleys in the case of Mohenjo-daro and even a set of proportional brick sizes. Although population density dropped rapidly outside the main urban nodes, the civilization was dispersed widely and understood to have the broadest geographical reach of any Old World "primary" state—though frequently tethered in one way or another to the greater five distributaries of the Indus as well as the recently surveyed "dry" Ghaggar-Hakra river channel (Mughal 1997; cf. Possehl 2000). Given the homogeneity of styles in the material record, knowledge and information exchange was constant, redundant, and predictable during this time interval. A form of writing developed, though it remains undecipherable. Aspects of the sociopolitical system appear highly centralized, but "power architecture" (such as pyramids or ziggurats) or grand depictions of rulers are absent. The Harappans were a Bronze Age culture except that most functional tools were likely made from ceramic, stone, and wood. Nevertheless, good evidence exists for the plow and bullock cart (wheel), as well as maritime and riverine boat traffic. Riverine systems fed a conjectured canal-irrigation system (Scarborough 2003, p. 142), but 12 m of Himalayan sediment at Mohenjo-daro (Wheeler 1968, p. 38) and elsewhere buried much of the present floodplain and prevent an accurate assessment of agricultural intensification. Given aspects of the flow dynamics of the Indus and the sophistication of known waterworks, the engineered landscape was probably altered not unlike Pharanoic Egyptian bund field systems along the lower Nile. Natural resources were abundant given the waterway distributaries of the greater Indus River. Wheat and barley (together with millet, rice, cattle, goats, and sheep) were staples, but we know little of the means, let alone modes, of production. Nevertheless, the economic system seems akin to other early states of the Old World with a wide exchange of highly portable objects, though luxury items to set elites apart were few. In fact, indigenous investment in the production of "status tokens" was significantly limited, a condition adding to the "faceless" character of the society (diminutive statuary and figurines exist, but grand portraiture monuments as identifiable in other primary states at comparable periods of development do not).

Another aspect of Harappan society was its exceptional health, even antiseptic by comparison to all other primary states (Miller 1985; Scarborough 2003, pp. 144–145). Although conflict appears to be a facet of all societies, little evidence for violent warfare is documented until perhaps 1900 B.C. or later. The picture that unfolds is that of a highly repetitive, controlled, and all-pervasive ideology with strong prohibitions against ostentatious expressions of wealth and power, and a deeply embedded ethos of equality among members of society. Several of the largest cities have elevated mounds with "baths" and fire altars (Allchin and Allchin 1982; Rao 1979, p. 77; Wheeler 1968). Standardized house sizes and shapes as described for Mohenjo-daro (Sarcina 1979) coupled with carefully constructed water and sewage systems for most members of the largest cities further suggest highly representational governance like few others known.

The resilience of Harappan society associated with a stable foundational ideology was an outgrowth of the equitable organizational strategies that evolved and were adopted (Possehl 2002; Kenoyer 1998). The role of environmental diversity also contributed to economic resilience as the various occupied ecological settings permitted the harvesting of a variety of natural and domesticated resources that, in turn, were rapidly exchanged by way of extensive riverine arterials and connected overland routes, to say nothing of the maritime linkages. Unlike the highly linear and narrow floodplain associated with the environmentally circumscribed Nile Valley or even the relatively limited watersheds of the Tigris–Euphrates or the Yangtze, the highly splayed distributary system of the Indus, abutting the highest and most rugged watershed known, allowed for the use of innumerable microenvironments and a natural region interconnected by waterways.

The previous Old World examples present a good case for the consequences of an acute drying trend associated with the 4.2 K Event. The resilience of the Harappans during this period is likely threefold. First, the society was structured differently than its neighboring peers, allowing a degree of within- and between-community commonality less seen in other ancient experiments in statecraft. Secondly, the extensive distribution of Harappan towns and cities within a myriad of environmental settings frequently connected by waterways provided a degree of resource risk-sharing management unlike other early states. Third, the massive glacial ice packs of the Himalayan Mountains and neighboring ranges may have buffered against water deficits from a prolonged drought. Unlike the other early Old World states along major rivers, the Harappans were less affected by dropping water levels or increasingly severe sedimentation loads. The impact of the 4.2 K Event manifest elsewhere may have been neutralized by the recharging glacial melt waters into the Indus Plain in concert with the other buffering conditions—settlement dispersion and "community" ideology.

We may never fully understand the nuanced interplay between social structure and the role of the biophysical setting. We do know that to dismiss either will surely prevent progress in understanding the complexity of human behavior.

The 7th- to 10th-Century Episode: The Origin of the High Middle Ages in Western Europe

European climate during the period A.D. 700–1000 becomes increasingly consistent (despite several notable excursions in the first two centuries, such as the freezing of the Nile in A.D. 829). Solar emissions increase and sea level rises to above that of the present day, both indicators of global warming (Landscheidt 1987; Eddy 1994). Widespread population movements, invasions of northern continental Europe by the Norse (a strong invasion beginning in A.D. 886), of southern Europe by the Saracens (A.D. 890s), and of western Europe by the Magyars in the early A.D. 900s reflect the early, less consistent character of the larger warming trend (Gunn 1991, 1994).

By A.D. 1000, the warming trend is well established as new populations settled, converted to Christianity, and became a bulwark against eastern invasions. The synthesis of southern grain-complex culture with northern vegetable and cattle adaptations through the three-field rotation system increased agricultural versatility, and the experimental and inventive approaches to gardening by monastic establishments increased agricultural versatility (Gunn et al. 2004).

We focus on the 7th- to 10th-Century Episode because its climatic characteristics differed from the 4.2 K Event and because the New World examples will contrast with the Old World. The event itself was of a different sort, leading to drier and warmer episodes and offering greater regional variability. Nevertheless, it is exciting to investigate teleconnections of this episode across the Atlantic region and compare its impact among far different cultural and technological regimes.

Western Europe's dominant theme in this period was a political cycle of centralization and decentralization. During the 8th century, the Carolingian Empire briefly unified most of western and central Europe, achieving the largest political unification (in terms of territory) between the Roman Empire and today's

European Union. After Charlemagne's death in 814, the Carolingian Empire was repeatedly partitioned and, by the end of the century, was seriously fragmented. The 10th century saw another centralizing trend under the Saxon (as well as Ottonian) and Salian emperors. This medieval German Empire began fragmenting around 1100. The formerly frontier territories, such as France, Castile, England, and Denmark, took the next round of centralization.

From 700 to 1300, the population of France more than tripled, from <6 million to 18–20 million. However, the population increase was not linear and roughly paralleled the centralization–decentralization trend. Population numbers stagnated or even declined during the 9th century. Especially hard hit were frontier areas that suffered from constant raiding from the Vikings, Saracens, Magyars, and Polabian Slavs. Resumption of population growth varied according to region, but after the mid-12th century, a sustained population rise appeared to occur all over western Europe. The bulk of population increase from the Roman level to the High Middle Ages was achieved in one century, 1150 to 1250.

Building activity and trade followed the centralization cycle. Church building accelerated during the so-called Carolingian renaissance. Trade increased until the 830s and then declined due to the collapse of central authority and increased Viking and Saracen raids. The 10th century was the period of great monasteries (such as the Cluniacs). Monastic expansion went hand-in-hand with the Ottonian centralization (the Ottonian emperors, in fact, used the monasteries as support nodes for their power network). The post-1000 period saw another increase in church building (the Romanesque and Gothic periods) and the rise of trade (Champaigne fairs, Italians). Interestingly, the run-up to this episode saw major dislocations and disruptive movements of people across Europe, but through their assimilation into the religious system and the emergence of a productive economic network, what had been a period of regional decline was transformed into an era of growth and innovation in the technological and construction realms.

The 7th- to 10th-Century Episode in East Asia: The Fall of Bal He Kuk, Korea

Establishing a systematic method of varved sediment analysis has enabled researchers to reconstruct signals from the past on a yearly basis and made it possible to investigate the relationship between environmental history and the history of human civilization at a high level of precision. The Japanese archipelago is one of the few places in the world with exceptionally well-preserved and continuous varved sediment sequences. Specific conditions are required for well-preserved sediments, and several lakes in Japan seem to have provided the perfect environment.

Figure 9.3 illustrates an example of varved sediment cores from the lake bottom in Megata, Akita prefecture, Japan, and the results of analysis. Previous studies have attributed the sudden fall of the Bal He Kuk Kingdom in A.D. 926 to

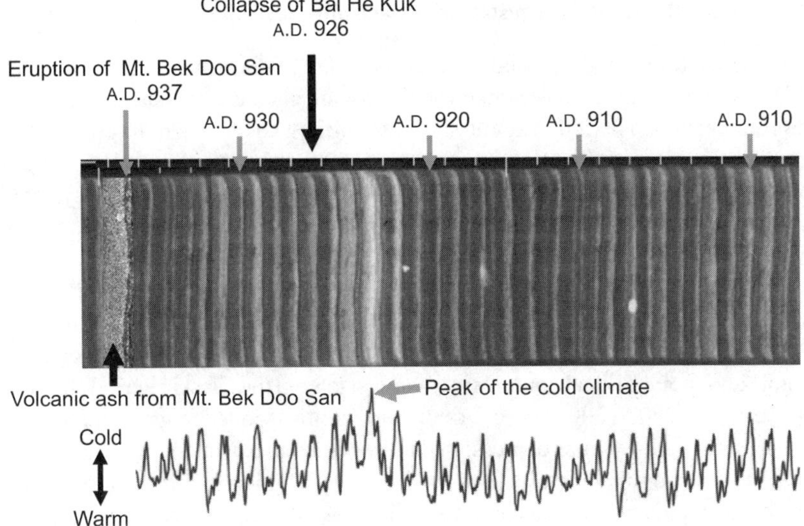

Figure 9.3 Correlation between the historical events and environmental changes (Yasuda 2005, after Fukusawa).

a large volcanic eruption of Mount Bek Doo San. Some of the archeological sites of the Bal He Kuk Kingdom were covered with volcanic ash flow from Mt. Bek Doo San, which had been dated at 920±20 by ^{14}C dating methods, resulting in a misconception that "the Bal He Kuk Kingdom is the Pompeii of East Asia." However, the results of varved sediment analysis of Megata samples reveal that the eruption of Mount Bek Doo San occurred in A.D. 937, 11 years after the Kingdom's fall. Thus, it may be concluded with certainty that the Mt. Bek Doo San eruption was irrelevant to the fall of the Bal He Kuk Kingdom. Furthermore, results of the various analyses on varved sediments indicate the presence of an extremely cold period around A.D. 920, immediately preceding the fall, which suggests that a short episode of climatic cooling may have been a contributing factor that led to the fall. This short period of climatic cooling during the Medieval warm period would have severely impacted grain production of the Kingdom, which is at the northern limit of grain cultivation, especially rice. The weakened Kingdom was then swiftly destroyed by an invasion by the nomadic Khitan. As discussed below, Maya civilization had met a similar fate. Just as the short period of climatic cooling around A.D. 920 destroyed the Bal He Kuk, Maya civilization also collapsed at roughly the same time, with especially dry conditions triggering drought.

Varved sediment analyses have revealed a startling fact that only a few years may be required for a civilization to fall after a significant climate perturbation, as in the cases of the Bal He Kuk Kingdom, Old Kingdom Egypt, and the Maya. Future studies will investigate whether the same holds true for other civilizations.

The 7th- to 10th-Century Episode and the Collapse of the Lowland Maya

The ancient Maya suggest another case study of resilience and vulnerability, but this time, one of apparent sustained socioecological resilience during a more or less uninterrupted period of about 1500 years followed by one of the classic collapses of ancient civilization. As early as 1500 B.C., settled populations dispersed throughout the Yucatán Peninsula following the few riverine systems into the interior. Settlements away from the limited number of surface drainages were made possible when the vast system of internally draining, seasonal swamps (bajo) of today functioned as a series of shallow ancient lakes in antiquity. During the first experiments in statecraft about 300 B.C. to A.D. 1, or the Late Preclassic Period, populations developed most of the trappings of socioeconomic and sociopolitical complexity manifest later in the Classic Period fluorescence of A.D. 250 to 800. These former communities tended to gravitate toward the sizable shallow water depressions or "concave microwatersheds." The success of these populations was their environmental undoing in that short-fallow slash-and-burn agricultural activities excited lake margin deforestation and soil erosion, in part displacing the water source that attracted them to shallow lake settings initially (Scarborough, this volume). When coupled with a warming trend throughout the Yucatán and groundwater-recharge interruptions caused by the new sedimentation loads eroding into the basin catchments, water access became untenable (Dunning et al. 2002).

The transition from the Late Preclassic Period (300 B.C. to A.D. 1) into a new period of social complexity occurred in the context of a relatively fixed set of ideological and economic parameters. These parameters were established during the ecologically adaptive, though passive, water and land-use system associated with the "concave microwatershed." Although clearly wedded to the past, the ancient Maya were nevertheless forced to redesign their engineered landscape. This transition was marked by an initial threshold of development resulting in subsequent growth and landscape alteration, but an overall transition characterized by a slow and incremental tempo commensurate with the fragility of a semitropical forest harvested by humans (Scarborough 2003).

The Classic Period (A.D. 200–800) represented the movement of settlements away from the immediate margins of desiccated and displaced lake settings to adjacent hillocks and ridges—a "convex microwatershed." By carefully and systematically altering the summits of these hills, large communities arose through the construction of huge paved and canted plaza surfaces on which pyramids, palaces, and ball courts were erected. In the wake of these most visible construction projects were the sizable quarry scars, depressions that were rapidly converted into tanks and reservoirs. What the Classic Period Maya produced was a built environment in which seasonal precipitation was directed into large elevated tanks that in turn were "opened" through a series of dams or overarching causeways. These mechanisms allowed the release of water onto a hill-flanking residential zone during the four months of drought associated with the wet-dry tropical forest of the region. Water was further recycled as gray water through a system of bajo-margin tanks at the bottom of the occupied hillock for final release into the sediment-infilled margins of the earlier Late Preclassic shallow lake zone (Scarborough, this volume), these latter settings now reclaimed as field plots.

This socioecological adaptation was very successful but required a high degree of cooperation and coordination—an ideological construct likely established by the earliest Preclassic sedentist in the Maya lowlands (Scarborough 1998). Challenging the notion of a "tragedy of the commons," the social community cohesion of living in proximity to a common and most visible primary resource like the shallow lakes of the Preclassic Period set the stage for subsequent interdependencies as groups increased in population size and expanded into less desirable ecological zones. From the onset, the Maya were economically organized across the landscape in a hierarchical manner by way of specializing at the village and town level. These "resource-specialized communities" focused on one or two products for exchange with other communities in an evolving network of information and material-goods flow. Although each village could maintain its own independent economic base, to minimize risks in the event of access to crucial resources from a neighboring village were disrupted, each community cultivated local alliances and access to prevailing regional economic and political conditions by participating in highly flexible and resilient exchange networks. Communication and information flow was rapid and verifiable. With time, a few truly sizable centers or nodes developed to coordinate the movement and distribution of goods, services, and information. With the advent of the "convex microwatershed center," nodal economic expression converged with a new way of living on the landscape. The wedding of the long-established ideological constructs for what it meant to be Maya with the economic and ecological drivers resulted in a highly sustainable and resilient social structure and built environment. By the middle of the Classic Period, the southern Maya Lowlands were a hugely successful socioecological adaptation (Scarborough and Valdez 2003; Scarborough, this volume).

Nevertheless, they collapsed. Two unfortunate vectors may have overridden other explanatory variables. The population densities in the Maya area likely accelerated the consumption of finite resource availability, even given their careful stewardship of the landscape. Environmental stewardship by the ancient Maya was likely hard won by many early mistakes involving attempts to rapidly exploit microhabitats or patches when access to a specific species for food or fiber was manifest in limited numbers—a simple ecological principle of semitropical settings and a slow process of escalating vulnerability in resilience theory terminology. The other vector was the role of drought—three quick but acute blows over a 100-year period (Peterson and Haug 2005), or a fast variable in a resilience perspective. These two ecological variables appear to have enervated the network of economic and political organization. Even the resilience of the reinvented environment in which the Maya lived could not adjust to the frequency of these drying climatic spikes.

There was, however, another element that may have undermined the system and illustrates complexification. By the 7th century, the power of some of the largest centers seems to have grown beyond any earlier examples of control in the Maya area. Coupled with the consequences of the above vectors, elite hubris and attempts to graft hierarchal, even hegemonic, controls from these centers onto the well-defined networks of flexible information exchange divorced the highly influential hinterlands from the centers and disrupted the fundamental linkages that sustained longevity in a fragile environmental setting (Scarborough, this volume).

The 10th-Century Event among the Hohokam in the U.S. Southwest: An Era of Growth and Transformation

The term Hohokam denotes a distinctive constellation of archaeological materials found in the Sonoran Desert of central and southern Arizona in the Southwest U.S. (Haury 1976; Wilcox 1979). A long period of dependence on wild resources and riparian agriculture was transformed around A.D. 700, as multi-village cooperatives constructed and operated irrigation canals that watered thousands of hectares on the terraces above the Salt and Gila rivers. In fact, this irrigation system was the largest in the New World north of the Andes, accomplished without significant use of domestic animals, metal, or the wheel. A network of communal ball courts was established that eventually spanned a territory of some 80,000 km^2 and represented early steps toward complexification (Doyel 1991). This network became the basis for the Hohokam regional system, a true regional economy involving large-scale specialized production and exchange of bulk goods such as pottery, shell, ground stone, and agricultural crops (Abbott 2003). Yet, around A.D. 1070, this network, and possibly the regional economy as a whole, began to face rapid collapse.

The fluorescence of the Preclassic Period Hohokam appears to correlate nicely with the 7th- to 10th-century climatic interval, but a closer examination of regional patterns reveals a more complex picture. Based largely on observed patterns of aggradation–degradation of floodplains and dendroclimatic variability, it is possible to partition this time period into relevant episodes of climatic–social interactions (Masse 1991; Dean 2000, 2004). The record indicates high levels of aggradation and effective moisture with little dendroclimatic variability in the period A.D. 600 to 750 that would have favored the development of irrigation networks as built by the Hohokam at that time. One would expect that the following period of 750 to 900, characterized by floodplain degradation, low effective moisture, and high interannual variability, would have been detrimental to agriculture; however, the Hohokam may have had a relative advantage over their neighbors due to the buffering effects of their established irrigation system. Two centuries of relatively high effective moisture and region-wide conditions favorable for agriculture followed. The period A.D. 900 to 1100 coincided with a major population increase across the entire Southwest, including in the heartland of the Hohokam.

Interestingly, it is on the heels of this long period of regional growth that the regional economy of the Hohokam was undermined, perhaps by the growing independence of the neighboring societies. Beginning around A.D. 1150, a new balkanized social order developed, characterized by localized political hierarchies likely symbolized by platform-mound construction with restricted access. This period, known as the Hohokam Classic, was witness to social fragmentation and local self sufficiency, heavy reliance on irrigation agriculture, population pressure and, eventually, environmental degradation and poor human health. The system eventually collapsed, leading to regional depopulation and the abandonment of major portions of the Hohokam heartland that had been occupied for more than a millennium. The Hohokam sequence encompasses two complete adaptive cycles (as defined by resilience theorists) that involve a variety of capitalization processes in both social and infrastructural realms, apparently leading to major vulnerabilities and resulting in major transformations with heavy human costs. Although the 7th- to 10th-century climatic interval and its aftermath exerted powerful influences on this marginal farming environment, the responses and interactions were far more diverse than a one-to-one relationship. Subinterval variations and local climatic impacts led to intraregional differences in agricultural success that allowed the Hohokam to flourish throughout a long period bracketing the interval, yet led to a fundamental transformation of the society and its environmental relations. One mechanism of change was the regional economy that was not seen in other Southwestern cases, and it might have increased the system vulnerability to local perturbations, such as social conflict on the one hand and variable streamflow on the other. Canal irrigation could have produced large surpluses—possibly fueling political competition—but it would also have increased vulnerability to streamflow fluctuations and flooding. As part of this process the Classic Period reorganization was clearly at a higher level of complexification with its attendant costs and pressures to increase agricultural productivity. This complexification, in turn, engendered a long-term decline in soil fertility, possibly increasing the threat of salinization and eventual collapse in the face of climatic shocks that otherwise could have been absorbed.

SYNTHESIS OF CASE STUDIES

Case studies are essential to comparative historical research. The cases presented in this chapter show diversity in societal responses to climate change. The diversity makes generalization difficult, but also conveys an important lesson. The 4.2 K Event profoundly affected both Old Kingdom Egypt and

Mesopotamia; it contributed to collapse in Egypt and to collapse and regional abandonment in the Khabur River region of northern Mesopotamia. In contrast, southern Mesopotamian polities managed canal systems to avoid the need to abandon regions and also fluctuated between urban and dispersed settlement. Contests for water and land led to the formation of small regional states. In China, the 4.2 K Event had serious negative effects on rainfall, cereal farmers, and stockbreeders of the Yellow River region of North China, while rice-paddy farmers of the Yangtze River region buffered climatic problems through water control. The Harappans of the lower Indus Valley may have survived the 4.2 K Event through connections among communities, by dispersion of settlements across diverse environmental settings, and by the subsidy of meltwater from the Himalayan glaciers.

The 7th- to 10th-Century Medieval Warm Period produced equally variable responses. Northern Europe saw periods of imperial expansion and contraction, which covaried with building activity and trade. This period also witnessed the start and expansion of Viking raids. In East Asia, extremely cold periods around A.D. 920 would have affected rice production, weakening the Bal He Kuk Kingdom. Among the Maya, drought seems to have coincided with escalating vulnerability, political complexification, and alienation of hinterland populations.

If there is commonality to these cases, it may be in the observation that societies least altered by climatic events seem to have had recourse to options in settlement (southern Mesopotamia) and to buffering through interaction (Harappa), natural subsidies (Harappa), and/or water management (southern Mesopotamia, central China, Hohokam). The societies most profoundly affected—Old Kingdom Egypt, northern Mesopotamia, north China, the Bal He Kuk Kingdom, and the Maya—seem, for reasons still not fully understood, to have had fewer options to address the climatic challenge. The options available to the first group appear to have conveyed possibilities for resilience that the second group lacked.

ISSUES IN DEVELOPING A NEW RESEARCH PARADIGM

To develop a new research paradigm for the interaction of climate, people, and resources, we suggest a twofold approach that focuses upon: (a) the dynamics of human adaptive systems; and (b) issues of data quality and recovery.

There are unresolved questions about the conditions when societies are vulnerable to pronounced environmental perturbations, and when they are not. These questions revolve as much around characteristics of the societies as around characteristics of the environment. If we are to understand human–environment interactions beyond noting superficial correlations of events, major progress in this former area is essential. Earlier in this chapter, we suggested that the problem-solving abilities of individual human societies change over time. The complexity of problem solving has much to do with this trajectory (Tainter

1988, 2000a, b). As problem solving grows complex and costly, it saps fiscal resources and leaves societies vulnerable to stresses that otherwise they might have overcome. Fiscal distress and declining capacity in problem solving alienate support populations and reduce the legitimacy of rulers and regimes. These internal dynamics combine with environmental variation to make societies vulnerable to catastrophic state changes such as collapses. Societies may be most vulnerable when a major environmental perturbation coincides with a state of fiscal weakness and reduced legitimacy brought on by complex and ineffective problem solving. An approach to human–environment interaction that incorporates problem solving would focus, in part, on understanding how problem-solving institutions change over time.

Another approach has been offered by Holling (e.g., 2001). Termed "panarchy," it is a part of Resilience Theory, as described above. Within this approach, change is in part a function of internal system dynamics. Living systems, including human societies, are proposed to cycle through phases of growth and accumulation, followed by release and reorganization. The first phase maximizes production and accumulation. The second maximizes invention and reassortment (Holling and Gunderson 2002, p. 47). The whole loop is termed the *adaptive cycle*. Societies in these phases vary in their flexibility, inventiveness, and capacity to adopt noncatastrophic change (e.g., Redman and Kinzig 2003). We suspect that societies in these phases also vary in their ability to adjust to major environmental perturbations. Much work needs to be done to understand such variations in adaptive capacity.

Approaches focusing on problem solving and adaptive cycles are not mutually exclusive. It is likely that they could be fruitfully combined. The evolution of problem-solving institutions from simple to complex, and the fiscal constraints that this imposes, may help to clarify why societies appear to vary in their flexibility and inventiveness. As suggested above, discontinuous changes (such as collapses) may emerge from the intersection of fast and slow variables (Scarborough 2003). In human societies, slow variables might include such factors as capital accumulation, heterarchical decision making, transformations in cosmology, and growth in the complexity of problem solving. Fast variables would include the rapid onset of adverse environmental conditions. Understanding the relationship of fast and slow variables adds another level of detail that must be mastered in historical cases, but also another level of richness to explanations of change. We recommend continued research along these lines.

At the beginning of this chapter, we affirmed the value of an interdisciplinary, long-term perspective on the dynamics of human and natural systems. By reviewing our knowledge of even a few of the better-known case studies, a variety of data-quality issues become apparent and must be addressed to move to the next level of understanding. We follow recommendations for advances in environmental data collecting with suggestions for analogous improvements in archaeological and historical data recovery.

- *Data Acquisition*: We must acquire data in the microregional context. For example, lakes of a certain size create a microclimate, the signals of which may not be detected from environmental archives located at some distance. Similarly, topographical relief features such as a mountain range between target area and archive produce climatic differences. A significant amount of climate change data is from marine drilling sites (i.e., the Ocean Drilling Program), which is not where people have lived. Just as environmental data must move to cover microregional contexts, archaeological and historical information must be sought from more continuous and geographically comprehensive units. When paleoclimatic data are available from only limited locations, historical data must be sought in related locations.
- *Data Selection*: Which archives are used? We need the entire available spectrum, from all kinds of sediments, geomorphologic features, ice shields, trees, corals, and speleothems. We need vegetation and other biological associations, tree rings, growth rhythms of corals and mollusks, isotope ratios, all kinds of chemical and physical sediment properties (e.g., erosion processes, mineralization processes). The multiproxy approach is widely accepted but there are substantial qualitative differences in corresponding records. We must ensure that we design archaeological and historical inquiry to support the kinds of measurements and monitoring suggested at the beginning of this chapter. Population, productive technology, and social organization are basic parameters, but other characteristics related to responses to environmental perturbations must be sought. Detailed analysis of early historic records and texts is a promising line of evidence that is seriously understudied, and the condition of some of these records is seriously threatened. Disciplinary differences have led to major inequalities in funding sources and employment possibilities. The specter of a time when no scholars can read the various ancient languages and reveal the insights of thousands of texts is not too far-fetched.
- *Interpretation*: One would hope that processes or conditions are learned from modern systems and then applied to paleosystems. However, training sets, which would allow for statistically based interpretations, are sorely lacking. Training sets for biological transfer functions, for example, have to be regional. What are the controlling factors of proxies? In one example, primary production in a lake can be induced by local input (sediments from erosion or agricultural fertilizers, both as a consequence of land use) or by eolian input coming from a remote source. In another example, a lake-level drop need not mirror salinity changes as is often assumed. A lake with a low level can still yield potable water. Yet another example is suggested when water temperature is interpreted from biological associations but may not reflect the climatic setting because of the ecological setting (e.g., hydrothermal vents). For the same reason, isotopic signals may not match the climatic setting. Trivial as it sounds, we have to understand the whole

system and our research design has to be adjusted correspondingly. Of course, archaeological interpretations rely heavily on the arguments linking proxy measure to variables of substantive interest, particularly the parameters of interest here.

- *Resolution*: Low-resolution climate data show only general trends, while high resolution is necessary to pinpoint rapid or abrupt climate changes. Centennial and decadal resolution has been termed "high" in the literature. Annual and seasonal resolution are needed, but such high-resolution records are extremely scarce. For example, due to insufficient resolution, Hoddell et al. (1995) suggest an extremely long drought from A.D. 800 to 1000 hitting Maya civilization. High-resolution data from the Ocean Drilling Program, however, suggest extreme drying events around A.D. 810, 860, and 910. The resolution of information for some aspects of society (subsistence, settlement, technology) may have a high geographic resolution, but it is almost always constrained by temporal resolution no finer than a human generation and more likely a century or two. This constraint seriously inhibits our interpretation of the kinds of fast processes and rapidly changing patterns that archaeologists can monitor. Of enormous concern is the correct placement of events and responses in their correct chronological order. This concern will continue to be a high priority among archaeologists.
- *Dating*: One of the main problems is that ^{14}C dating includes an error which is often "smoothed" by similar events from other archives. In using this approach, events become isochronous that may be time-transgressive or not related at all, or events appear to have prevailed much longer than they actually did. Global signals may be produced by this too. Varved (rhythmic) sediments and tree rings represent adequate calendars which, however, cannot always be applied. Given the urgency of our need to understand long-term human–environment interaction, it is difficult to wait for technical innovation in both environmental and archaeological studies.

PARTING OBSERVATIONS FROM TAKING A CASE-STUDY PERSPECTIVE

Even a quick perusal of the selected case studies reveals that any assertion of a simple, deterministic relationship between climate and social change is not merely oversimplistic, but wrong. It is clear that even with relatively uniform global climatic events, societies respond in both parallel and diverse ways. Moreover, the same society at one point in time may respond in a positive direction to the input and at another point respond negatively. As suspected, there are organizational, technological, and perceptual mechanisms that mediate and redirect the response to particular climatic inputs and those institutions. Furthermore, climatic inputs change over time and often manifest themselves differently over space. Beyond the contextualized response of a society to any particular climate signal, there is both a time-series sequence to many climate signals and social responses as well as lag in the impact, perception, recognition, decision, and response to any particular signal.

Recognizing the diverse factors that determine responses to climatic perturbations and the exploratory nature of the inquiry in this chapter, we nevertheless are confident that regularities, or at least parallels, emerge in a case-study approach. These regularities will appear not as simple correlations, nor as direct cause and effect, but as subtly changing relationships between the challenges that societies experience and the abilities of those societies to adapt.

REFERENCES

Abbott, D.R. 2003. Centuries of Decline during the Hohokam Classic Period at Pueblo Grande. Tucson: Univ. of Arizona Press.

Adams, R.M. 1981. Heartland of Cities. Chicago: Univ. of Chicago Press.

Allchin, B., and R. Allchin. 1982. The Rise of Civilization in India and Pakistan. Cambridge: Cambridge Univ. Press.

Butzer, K.W. 1996. Ecology in the long view: Settlement histories, agrosystemic strategies, and ecological performance. *J. Field Archaeol.* **23**:141–150.

Courty, M.-A. 1994. Le cadre paléogéographique des occupations humaines dans le bassin du Haut-Khabour, Syrie du Nord-Est: Premiers résultats. *Paléorient* **20**: 21–59.

Crumley, C.L., ed. 1994. Historical Ecology: Cultural Knowledge and Changing Landscapes. Santa Fe: School of American Research Press.

Crumley, C.L. 2003. Alternative forms of societal order. In: Heterarchy, Political Economy, and the Ancient Maya, ed. V.L. Scarborough, F. Valdez, Jr., and N. Dunning, pp. 136–145. Tucson: Univ. of Arizona Press.

Crumley, C.L. 2005. Remember how to organize: Heterarchy across disciplines. In: Nonlinear Models for Archaeology and Anthropology: Continuing the Revolution, ed. C.S. Beekman and W. Baden, pp. 35–50. London: Ashgate.

Dean, J.S. 2000. Complexity theory and sociocultural change in the American Southwest. In: The Way the Wind Blows: Climate, History, and Human Action, ed. R.J. McIntosh, J.A. Tainter, and S.K. McIntosh, pp. 89–118. New York: Columbia Univ. Press.

Dean, J.S. 2004. Anthropogenic environmental change in the Southwest as viewed from the Colorado Plateau. In: The Archaeology of Global Change: The Impact of Humans on the Their Environment, ed. C.L Redman, S.R. James, P.R. Fish, and J.D. Rogers, pp. 191–207. Washington, D.C.: Smithsonian Books.

deMenocal, P.B. 2001. Cultural responses to climate change during the late Holocene. *Science* **292**:667–673.

Doyel, D.E. 1981. Late Hohokam Prehistory in Southern Arizona. Gila Press Contributions to Archaeology 2. Globe, AZ: Gila Press.

Doyel, D.E. 1991. Hohokam culture evolution in the Phoenix Basin. In: Exploring the Hohokam: Prehistoric Desert Peoples of the American Southwest, ed. G.J. Gumerman, pp. 231–278. Amerind Foundation, Dragoon, AZ, and Univ. of New Mexico Press, Albuquerque.

Dunning, N., S. Luzzadder-Beach, T. Beach et al. 2002. Arising from the Bajos: Anthropogenic change of wetlands and the rise of Maya civilization. *Ann. Assn. Am. Geogr.* **92**:267–283.

Eddy, J.A. 1994. Solar history and human affairs. *Hum. Ecol.* **22**:23–36.

Fairservis, W.A. 1971. The Roots of Ancient India. 2d ed. Chicago: Univ. of Chicago Press.

Folke, C., S. Carpenter, T. Elmqvist et al. 2002. Resilience and Sustainable Development: Building Adaptive Capacity in a World of Transformations. Scientific Background Paper on Resilience for the World Summit on Sustainable Development. Report for the Swedish Environmental Advisory Council. Stockholm: Ministry of the Environment. http://www.sou.gov.se/mvb/pdf/resiliens.pdf.

Foster, D.R. 2000. Conservation lessons and challenges from ecological history. *Forest Hist. Today* **Fall 2000**:2–11.

Foster, D., F. Swanson, J. Aber et al. 2003. The importance of land-use legacies to ecology and conservation. *BioScience* **53**:77–88.

Gardiner, A.H. 1909. The Admonitions of an Ancient Egyptian Sage, from a Hieratic Papyrus in Leiden. Leipzig: Hinrich.

Gasche, H.H. 1990. Mauere. In: Reallexikon der Assyriologie und vorderasiatischen Archaeologie, ed. E. Ebeling, D.O. Edzard, and E.F. Weidner, vol. 7, pp. 591–595. Berlin: de Gruyter.

Gasse, F., and E. Van Campo. 1994. Abrupt post-glacial climate events in West Asia and North Africa monsoon domains. *Earth Planet. Sci. Lett.* **126**:435–456.

Gunderson, L.H., and C. Folke. 2003. Toward a "science of the long view." *Conserv. Ecol.* **7**:15.

Gunderson, L.H., C.S. Holling, and S.S. Light. 1995. Barriers and Bridges to the Renewal of Ecosystems and Institutions. New York: Columbia Univ. Press.

Gunn, J. 1991. Influences of various forcing variables on global energy balance during the period of intensive instrumental observation (1958–1987) and their implications for paleoclimate. *Clim. Change* **19**:393–420.

Gunn, J. 1994. Introduction: A perspective from the humanities–science boundary. Global climate–human life: Physical contexts of historical ecology. *Hum. Ecol.* **22**:1–23.

Gunn, J., C.L. Crumley, E. Jones, and B.K. Young. 2004. A landscape analysis of Western Europe during the Early Middle Ages. In: The Archaeology of Global Change: The Impact of Humans on their Environment, ed. C.L. Redman, S.R. James, P.R. Fish, and J.D. Rogers, pp. 165–185. Washington, D.C.: Smithsonian Books.

Hassan, F.A. 1997. Nile floods and political disorder in early Egypt. In: Third Millennium B.C. Abrupt Climate Change and the Old World Collapse, ed. H.N Dalfes, G. Kukula, and H. Weiss, pp. 1–23. NATO ASI 149. Berlin: Springer.

Haury, E.W. 1976. The Hohokam. Tucson: Univ. of Arizona Press.

Heckenberger, M.J., A. Kuikuro, U.T. Kuikuro et al. 2003. Amazonia 1492: Pristine forest or cultural parkland? *Science* **301**:1710–1714.

Hoddell, D.A., J.H. Curtis, G.A. Jones et al. 1995. Possible role of climate change in the collapse of classic Maya civilization. *Nature* **375**:391–394.

Hole, F. 1997. Evidence for mid-Holocene environmental change in the western Habur drainage, northeastern Syria. In: Third Millennium B.C. Climate Change and Old World Collapse, ed. H.N. Dalfes, G. Kukla, and H. Weiss, pp. 39–66. NATO ASI 149. Berlin: Springer.

Holling, C.S. 2001. Understanding the complexity of economic, ecological, and social systems. *Ecosystems* **4**:390–405.

Holling, C.S., and L.H. Gunderson. 2002. Resilience and adaptive cycles. In: Panarchy: Understanding Transformations in Human and Natural Systems, ed. L.H. Gunderson and C.S. Holling, pp. 25–62. Washington, D.C.: Island.

Holling, C.S., L.H. Gunderson, and D. Ludwig. 2002. In quest of a theory of adaptive change. In: Panarchy: Understanding Transformations in Human and Natural Systems, ed. L.H. Gunderson and C.S. Holling, pp. 3–24. Washington, D.C.: Island.

ICSU (International Council for Science). 2002. Science as a Foundation for Sustainable Development: A Report by the Global Environmental Change Programmes, Feb. 4–6, 2002. Paris: ICSU, http://www.igbp.kva.se/congress/downloads/Paris_workshop_summary.pdf

Kenoyer, J.M. 1998. Ancient Cities of the Indus Valley Civilization. New York: Oxford Univ. Press.

Kirch, P.V., J.R. Flenley, D. Steadman, F. Lamont, and S. Dawson. 1992. Ancient environmental degradation. *Natl. Geogr. Res. Expl.* **8**:166–179.

Kohler, T. 1992. Historic human impact on the environment in upland North American Southwest. *Pop. Env.* **13**:255–268.

Landscheidt, T. 1987. Long-range forecasts of solar cycles and climate change. In: Climate History, Periodicity, and Predictability, ed. M. Rampino, pp. 421–445. New York: Van Nostrand Reinhold.

Levin, S.A. 1999. Fragile Dominion: Complexity and the Commons. Reading, MA: Perseus.

Lichtheim, M. 1975. Ancient Egyptian Literature, vol. 1. The Old and Middle Kingdom. Berkeley: Univ. of California Press.

Mann, C. 2002 The real dirt on rainforest fertility. *Science* **297**:920–923.

Masse, W.B. 1991. The quest for subsistence sufficiency and civilization in the Sonoran Desert. In: Chaco and Hohokam: Prehistoric Regional Systems in the American Southwest, ed. P.L. Crown and W.J. Judge, pp. 195–223. Santa Fe: School of American Research Press.

McGovern, T.H., G. Bigelow, T. Amorosi, and D. Russell. 1988. Northern islands, human error, and environmental degradation: A view of social and ecological change in the medieval North Atlantic. *Hum. Ecol.* **16**:225–269.

McIntosh, R.J., J.A. Tainter, and S.K. McIntosh, eds. 2000. The Way the Wind Blows: Climate, History, and Human Action. New York: Columbia Univ. Press.

Miller, D. 1985. Ideology and the Harappan civilization. *J. Anthrop. Archaeol.* **4**:34–71.

Mughal, M.R. 1997. Ancient Cholistan: Archaeology and Architecture. Rawalpindi: Ferozons.

Nissen, H.J. 1988. The Early History of the Ancient Near East 9000–2000 B.C. Chicago: Univ. of Chicago Press.

Peterson, L.C., and G.H. Haug. 2005. Climate and the collapse of Maya civilization. *Am. Sci.* **93**:322–327.

Plog, F.T. 1974. The Study of Prehistoric Change. New York: Academic.

Possehl, G.L. 1990. Revolution in the urban revolution: The emergence of Indus urbanization. *Ann. Rev. Anthrop.* **19**:261–282.

Possehl, G.L. 2000. The drying up of the Sarasvati: Environmental disruption in South Asian prehistory. In: Environmental Disaster and the Archaeology of Human Response, ed. G. Bawden and R. Reycraft, pp. 63–74. Albuquerque: Maxwell Museum of Anthropology and Univ. of New Mexico Press.

Possehl, G.L. 2002. The Indus Civilization: A Contemporary Perspective. Walnut Creek, CA: AltaMira.
Rao, S.R. 1979. Lothal: A Harappan Port Town (1955–62), vol. 1. Archaeological Survey of India, Memoir 78. New Delhi: Archaeological Survey of India.
Redman, C.L. 1992. The impact of food production: Short-term strategies and long-term consequences. In: Human Impact on the Environment: Ancient Roots, Current Challenges, ed. J.E. Jacobsen and F. Firor, pp. 35–49. Boulder, CO: Westview.
Redman, C.L. 1999. Human Impacts on the Ancient Environment. Tucson: Univ. of Arizona Press.
Redman, C.L. 2005. Resilience in archaeology. *Am. Anthropol.* **107**:70–77.
Redman, C.L., S.R. James, P.R. Fish, and J.D. Rogers, eds. 2004. The Archaeology of Global Change: The Impact of Humans on their Environment. Washington, D.C.: Smithsonian Books.
Redman, C.L., and A.P. Kinzig. 2003. Resilience of past landscapes: Resilience theory, society, and the longue durée. *Conserv. Ecol.* **7**:14.
Roberts, N. 1998. The Holocene: An Environmental History. 2d ed. Oxford: Blackwell.
Rollefson, G.O., and I. Köhler-Rollefson. 1992. Early Neolithic exploitation patterns in the Levant: Cultural impact on the environment. *Pop. Env.* **13**:243–254.
Rose, J., C.A. Whiteman, J. Lee et al. 1997. Mid- and Late-Holocene vegetation, surface weathering and glaciation, Fjalljokull Southeast Iceland. *Holocene* **7**:457–471.
Sarcina, A. 1979. A statistical assessment of house patterns at Mohenjo-daro. *Mesopotamia* **13–14**:155–199.
Scarborough, V.L. 1998. The ecology of ritual: Water management and the Maya. *Latin Am. Antiq.* **9**:135–159.
Scarborough, V.L. 2003. The Flow of Power: Ancient Water Systems and Landscapes. Santa Fe: School of American Research Press.
Scarborough, V.L., and F. Valdez, Jr. 2003. The engineered environment and politcal economy of the Three Rivers region. In: Heterarchy, Political Economy and the Ancient Maya: The Three Rivers Region of East-Central Yucatán, ed. V.L. Scarborough, F. Valdez, Jr., and N. Dunning, pp. 3–13. Tucson: Univ. of Arizona Press.
Scheffer, M., S. Carpenter, J.A. Foley, C. Folke, and B. Walker. 2001. Catastrophic shifts in ecosystems. *Nature* **413**:591–596.
Service, E.R. 1962. Primitive Social Organization: An Evolutionary Perspective. New York: Random House.
Tainter, J.A. 1988. The Collapse of Complex Societies. Cambridge: Cambridge Univ. Press.
Tainter, J.A. 2000a. Global change, history, and sustainability. In: The Way the Wind Blows: Climate, History, and Human Action, ed. R.J. McIntosh, J.A. Tainter, and S.K. McIntosh, pp. 331–356. New York: Columbia Univ. Press.
Tainter, J.A. 2000b. Problem solving: Complexity, history, sustainability. *Pop. Env.* **22**:3–41.
van Andel, T.H., E. Zangger, and A. Demitrack. 1990. Land use and soil erosion in prehistoric and historical Greece. *J. Field Archaeol.* **17**:379–396.
van De Mieroop, M. 2004. A History of the Ancient Near East ca. 3000–323 B.C.. Malden, MA: Blackwell.
Vitousek, P.M., T.N. Ladefoged, P.V. Kirch et al. 2004. Soils, agriculture, and society in precontact Hawai'i. *Science* **304**:1665–1669.
Weiss, H. 1997a. Late Third Millennium abrupt climate change and social collapse in West Asia and Egypt. In: Third Millennium B.C. Climate Change and Old World

Collapse, ed. H.N. Dalfes, G. Kukla, and H. Weiss, pp. 711–723. NATO ASI 149. Berlin: Springer.

Weiss, H. 1997b. Leilan, Tell. In: The Oxford Encyclopedia of Archaeology in the Near East, ed. E.M. Meyers, vol. 3, pp. 341–347. New York: Oxford Univ. Press.

Weiss, H. 2000. Beyond the Younger Dryas: Collapse as an adaptation to abrupt climate change in ancient West Asia and the Eastern Mediterranean. In: Confronting Natural Disaster: Engaging the Past to Understand the Future, ed. G. Bawden and R. Reycraft, pp. 75–98. Albuquerque: Univ. of New Mexico Press.

Weiss, H., and R.S. Bradley. 2001. What drives societal collapse? *Science* **291**:609–610.

Weiss, H., M.-A. Courty, W. Wetterstrom et al. 1993. The genesis and collapse of third millennium north Mesopotamian civilization. *Science* **261**:995–1004.

Wheeler, M. 1968. The Indus Civilization. 3d ed. Cambridge: Cambridge Univ. Press.

Wilcox, D.R. 1979. The Hohokam regional system. In: An Archaeological Test of Sites in the Gila Butte-Santan Region, South-Central Arizona, ed. G.E. Rice, D. Wilcox, K. Rafferty, and J. Schoenwetter, pp. 77–116. Anthropological Research Papers 18. Tempe: Arizona State Univ.

Wilkinson, T.J. 2000. Settlement and land use in the zone of uncertainty in Upper Mesopotamia. In: Proc. Third MOS Symp. (Leiden 1999), vol. 3, ed. R.M. Jas, pp. 3–35. Istanbul: Nederlands Historisch-Archaeologisch Instituut te Istanbul.

Yasuda, Y., ed. 2005. Catastrophic Disaster. Tokyo: Wedge.

Yasuda, Y., T. Fujiki, H. Nasu et al. 2004. Environmental archaeology at Chengtoushan site, Hunan Province, China, and implications for environmental change and the rise and fall of the Yangtze River civilization. *Quat. Intl.* **123**:149–158.

The Centennial Timescale: Up to 1000 Years Ago

10

Revolutionary Weather

The Climatic and Economic Crisis of 1788–1795 and the Discovery of El Niño

Richard H. Grove

Resource Management in the Asia Pacific Programme, Research School of Pacific and Asian Studies, Australian National University, Canberra, ACT 0200, Australia

ABSTRACT

Archival evidence and historical analyses strongly suggest that the 1788–1795 El Niño event was remarkable for its intensity and prolonged global impact. This event may have been among the most severe in the available written record. Climate anomalies associated with the event played a role in key historical events. Severe, prolonged drought starting in 1788 accompanied the European settlement of New South Wales, Australia and challenged the agricultural strategies of the new settlers. The event was manifest as a failure of the Indian monsoon, starting about 1780 and resulting in the famine and associated high mortality throughout most of India. Unusual harvest failures were reported in the Caribbean. In Africa, the event is associated with record low levels of the Nile from 1790 to 1797. Highly abnormal weather patterns in western Europe starting in 1787 led to severe crop failures, which, in turn, exacerbated social pressures that culminated in the French Revolution and the Catalonian Revolt. In North America, during this period unusually hot weather and mild winters were associated with disease outbreaks. The global scope of this severe El Niño event sheds light on the interactions of globally scaled climate events and regional historical events.

INTRODUCTION

Australia has above all other places a claim on the epithet, "the El Niño continent." It is regularly beset by this phenomenon, which links climate anomalies across the globe. During an ENSO event, a spatial pattern of climate fluctuations develops: heavy rains fall on the Pacific coast of South America, while the lands to the west of the Pacific Ocean—from Australia and Indonesia through to southern and northeast Africa, and including South Asia—suffer severe droughts. The distinct feature about the phenomenon is that fluctuations appear in many locations almost simultaneously. Climatologists call the relations between fluctuations "teleconnections" (Nicholls 1992; Grove 1997).

The remarkable correlations and connections between the strength of the El Niño current and Southern Oscillation (ENSO) have been a particular focus in climatology and oceanography since about 1982, but research into historical records can extend data backward to approximately A.D. 1500 with some reliability. From this long data series garnered from the scrutiny of the archival record, some ENSO events stand out as particularly severe. None is worse that the 1788–1795 crisis, which coincided with the first European settlement of New South Wales in Botany Bay, right at the heart of what we now know to be the part of Australia most affected by ENSO. The main perennial drinking water supply, the Tank Stream, dried up. The same event produced very prolonged droughts in South Asia. The severity of the situation warranted comment in contemporary sources in these and other places. Thus some of the earliest documentation of this global climate phenomenon was achieved.

Despite the regularity of claims that the "latest" El Niño (whichever it may be) "is the worst in history" (e.g., by U.S. Vice-President Al Gore in 1997–1998[1]) and the concomitant fears that such events may be linked to global climate change, the historical as well as the prehistoric record suggests that worse events have occurred in earlier eras (Grove 1998). Documentary evidence suggests that, even in the last thousand years, very much stronger and longer El Niño events than those experienced since 1982 have been experienced globally. The strongest of them took place during the Little Ice Age between about 1250 and 1900, and especially in the closing years of the 18th century, when an El Niño event of exceptional strength and length appears to have affected the entire global environment. This last event is of particular interest in understanding the economic and social history of its time. It also suggests the intrinsic value of reconstructing the characteristics and impact of pre-instrumental El Niño events.

EL NIÑO, AUSTRALIA, AND THE REST OF THE WORLD

The extent to which Australia's history has been significantly shaped by El Niño events has been underexplored. This chapter considers the severe environmental events that shaped the world at the time of the first seven years of European settlement in New South Wales. Conditions were remarkably contrary to supporting the imperial agricultural vision for the newly annexed land.

On November 5, 1791, Governor Arthur Phillip reported that the normally perennial "Tank Stream" flowing into Sydney Harbor had been dry for "some months" (Governor Phillip's diary in McCormick 1987). It did not flow again

1 See statement by U.S. Vice-President Al Gore and NOAA scientists, The White House, June 8, 1998, claiming the 1997–1998 El Niño to be the most significant climatic event of the century and suggesting (without any evidence being produced), that this implied an acceleration of global warming. Other examples include Trenberth and Hoar (1996, pp. 57–60).

until 1794. The drought had begun, Phillip records, in July 1790 and no rain had fallen at all by August 1791. In a letter to W. W. Grenville on March 4, 1791, he noted that "from June (1790) until the present time so little rain has fallen that most of the runs of water in the different parts of the harbour have been dried up for several months, and the run which still supplies this settlement is greatly reduced, but still sufficient for all culinary purposes…I do not think it probable that so dry a season often occurs. Our crops of corn have suffered greatly from the dry weather" (Nicholls 1988, pp. 4–7).

The settlement of New South Wales was a struggle because of environmental conditions, yet the knowledge base and networks supporting the agricultural venture were growing. As early as 1791, Phillip was chancing a comparison: such a dry a season was rare. There are many problems involved in reconstructing the conditions and severity of El Niño events that took place before about 1870, when the period of modern instrumental observation began. However, Phillip's intuitive hunch (perhaps informed by Aboriginal insights) that this drought was exceptional can, in fact, be confirmed with hindsight.

Corroborating evidence for the severity of the 1788–1795 El Niño event comes from contemporaneous reports of drought in southern Africa, South Asia, the Atlantic, the Caribbean, and Mexico, where severe famines were reported in some regions. From the early 18th century, and even more so after about 1780, the richness of the available historical record, especially in the South Asian tropics, allows us to reconstruct the global footprint of a major El Niño event with very reasonable accuracy and resolution. This is due partly to the growing contemporary detail of Indian weather and population data gathered by the British East India Company. Useful contemporary global weather data from voyages and new settlements at the time are also available, particularly from the Southern Hemisphere, not least from parts of Australia and Oceania from which weather data prior to the 1780s is largely nonexistent.

The drought years of 1788–1794 in India were first recognized as having had a global impact by Alexander Beatson, Governor of St. Helena. Beatson suggested in 1816 that drought events of 1791, which occurred simultaneously in different parts of the world (he referred particularly to India, St. Helena, and Montserrat), had been part of the same connected phenomenon (Beatson 1816, p. 10). Beatson witnessed drought-stricken parts of South India before arriving in St. Helena and also experienced the effects of the 1791 drought in Mysore.

INDIA AND THE DROUGHTS OF 1788–1795

The El Niño episode of 1788–1795 was actually the culmination of a succession of unusual weather episodes that had begun in about 1780, and which were characterized by extreme events in both temperate and tropical latitudes in Europe and Asia (Kington 1988). These years were especially serious throughout South Asia (though of course there is no formal data for Australia before 1788). One year, 1783, which brought famine to almost all India, was memorialized in popular culture throughout India under the name of the *chalisa*. The word itself, which emphatically associates the Hindi number "forty" with a particular variety of famine, may suggest a characteristic return interval of 40–50 years for severe droughts, an interval which is, very roughly, borne out in reality during the Little Ice Age. The social disruption caused by this particular event was long term, since nearly 4% of all villages in the Tanjore district of the Madras Presidency were entirely depopulated in the early 1780s and over 17% in the Sirkali region (see Census of Tanjore). Up to eleven million people may have died in South Asia as a direct result of the passage of this event.

The earliest indications of the event are contained in the manuscript records of meteorological observations made for the East India Company by William Roxburgh, a Company surgeon, at Samulcottah in the northern Circars of the Madras Presidency (modern day Samalkot) (Roxburgh 1793a). Roxburgh accumulated a fourteen-year set of temperature and pressure data from the early 1770s and was thus able to recognize the exceptional nature of the droughts that began about 1789 (Roxburgh 1778). These droughts had previously been approached in intensity, he reported to the Company, only by those of a century earlier, in 1685–1687 (Roxburgh 1793b). These latter years are also now believed to have been characterized by "very severe" El Niño conditions in the eastern Pacific, but a year later, in 1687–1688. Roxburgh's rainfall figures record the consecutive failure of the South Asian monsoon between 1789 and 1792, with the most severe failure being experienced in 1790 (see Table 10.1).

Of particular note is the indication that the first major rainfall deficit associated with the event was experienced in 1789 in southern India, more than a year *before* similar deficits were experienced, toward the end of 1790, in Australia, Mexico, the Atlantic islands, and southern Africa. The possibility that the Indian monsoon is an active rather than a passive feature of tropical circulation and that monsoon failure may be efficient in *foreshadowing* El Niño (rather than being predicted by it) has been suggested (Krishnamurthy and Goswami 2000). It should be noted that some El Niño events, such as that of 1997, appear to articulate only with a failure of the Southeast Asian monsoon rather than with a South Asian failure. In the case of the 1789–1793 El Niño event (and the 1685–1688 El Niño event), a failure of the monsoon in both regions appears to have occurred. By November 1792, over 600,000 deaths were being attributed directly to the prolonged droughts in the 167 districts of the northern Circars of the Madras Presidency alone; up to half the population there died in 1792 (Table 10.2).[2]

The long drought periods were interspersed by very short periods of intense and highly destructive rainfall. In three days at Madras, in late October 1791, 25.5 inches of rain fell, "more than...has been known within the memory of

2 Manuscript reports on remissions of Company land revenues due to prolonged droughts and famine in Northern Circars districts of the Madras Presidency, East India Company Boards Collections, ref. no. F/4/12–12–14, sections 735–743, British Library IOLR, London.

Table 10.1 Monthly rainfall at Samulcottah, Andhra Pradesh, India May–November 1788–1792 (in inches and twelfths of an inch) as measured by Roxburgh (1794).

	1788	1789	1790	1791	1792
May	15.4	1.0		4.0	3.6
June	7.2	6.0	1.8	4.1	5.0
July	22.3	6.1	4.9	5.6	6.4
August	12.2	21.1	3.8		1.8
September	8.9	1.4	4.8	3.9	7.5
October	5.9	10.1	1.5	3.3	13.11
November	6.0	1.3	1.2	6.4	
TOTAL	77.5	43.1	17.4	26.11	37.10

man" (*Madras Courier* 1791). Throughout India, the famines of 1788–1794 resulted in very high mortality. In limited areas, such as the Northern Circars, East India Company officials attempted to estimate total mortality statistics. In other regions a much rougher but still useful guide is provided by the figures for deserted village sites. In the Gorakhpur district of Bihar, for example, the 19,600 villages extant in 1760 had fallen to 6,700 by 1801, with a mere third of the district falling under cultivation. Not all of these desertions can be attributed to to famine mortality, but a high proportion of them probably were (figures quoted in Commander 1989, p. 50). A similar pattern of mortality obtained in southern India during the period, where a pronounced pattern of village desertion can be established, and up to 30% of villages were deserted, for example, in some parts of

Table 10.2 Deaths from famine in the Madras Presidency in 1792.

Muglatore	141,682
Havelly 1	53,956
Havelly 2	4,874
Peddapore	184,923
Pittapore	82,937
Nandeganah	11,376
Sullapelly	9,018
Poolavam	16,204
Goulatah	12,639
Cotapilly	4,851
Corcoudah	9,035
Ramachandrapuram	7,430
Cottah	7,800
Somapah villages	2,306
Noozeed	96,210
Char Mahar	16,245

Salem district (for an overview, see Murton 1984; see also Lardinois 1989). Extrapolating from these kinds of figures we may attribute a total famine mortality during 1788–1794 of perhaps eleven million to the extended El Niño conditions of the period. However, although the human cost of the episode was particularly high in the subcontinent, severe consequences were also felt elsewhere, especially further east.

EL NIÑO PATTERNS EAST OF AUSTRALIA TO THE CARRIBEAN

The rainfall deficiency of South Asia spread toward the East, with unseasonably severe droughts being experienced in Java (Quinn et al. 1975). In the Pacific region, temperatures were unusually low. There is some limited evidence of cold and severe drought in the western Pacific from the journals of D'Entrecasteaux, who visited New Caledonia in 1793 (Thierry Correge, Orstom, Noumea, New Caledonia, pers. comm.). Another source of data from the late 1780s is the logbook of *HMS Bounty*. Temperature readings were made every four hours while at sea, and the lowest and highest temperatures were read while in port. In the December 6, 1788, entry of his logbook, Bligh wrote while his ship sheltered at Matavai Bay, Tahiti, that:

> I experienced a scene of today of Wind and Weather which I never supposed could have been met with in this place. The wind varied from ESE to the NW and the Therm. stood between 78 and 81.5 degrees. By sunset a very high breaking sea ran across Dolphin Bank, and before seven o'clock it made such a way into the Bay that we rode with much difficulty and hazard. midnight it increased still more and we rode until eight in the morning in the midst of a heavy broken sea which frequently came over us. The Wind at times dying away was a great evil to us for the Ship from the tremendous Sea that broke over the reefs to the eastward of Point Venus, producing such an outset thwarting us against the surge.

Bligh's observations of atypical weather conditions for the time of year suggest that by December 1788, the El Niño episode was already well under way. A few months later the El Niño might have actually saved the life of Bligh and the men who were cast adrift by the mutineers. Instead of the very hot conditions that would normally have confronted the 23-foot open boat between Tofua in the South Pacific and Timor during the periof of April 29 to June 14, Bligh and his men encountered cold conditions throughout the voyage. Furthermore, instead of a dry heat that would probably have been deadly, the rainfall they experienced was so cold that Bligh instructed his men to soak their clothes in warm seawater and then wear the wet clothes to keep warm! The next month, Bligh again recorded on June 18, 1789, heavy rain, "which enables us to keep our stock of water up." The crew all complained of rheumatic pains and cold. Furthermore, the supply of rainwater allowed Bligh to avoid making landings on hostile Pacific islands for water, as they would have in a normal year. So the El Niño that began in 1788 probably meant that the history of Captain Bligh and the *Mutiny* returned to Britain, instead of an uncertain note about the "mysterious disappearance of both the *Bounty* and Captain Bligh."

In Mexico the prolonged aridity that developed during 1791 was recorded in the steady fall in the level of Lake Patzcuaro between 1791 and 1793, giving rise to disputes over the ownership of the new land that unexpectedly emerged (Endfield and O'Hara 1997). As in Europe, these events were preceded by summer crop failures. On August 27, 1785, a hard night frost and the ensuing crop failure precipitated the great famine of 1785–1786 (Ouweneel 1996, p. 92). In the 1790s in Mexico, not one annual maize crop yielded an abundant harvest. This was entirely due to droughts every year, primarily in June and July. The severest droughts of the 1788–1795 El Niño event did not strike Mexico until 1793, so that the onset of full El Niño conditions did not affect rainfall for more than two years after the same event had caused monsoon failure in India. But wholesale failures of the wheat and maize crops took place in 1793 and 1797 (Ouweneel 1996, pp. 75–91). In 1794 the maize crop was again very poor due to almost complete drought. In 1795 the crop returned to near normal, although one might note that drought conditions persisted in that year in many Caribbean islands.[3]

Along the Peruvian coast, the great strength of the El Niño current itself during 1791, with its resulting degradational effects on agriculture and fisheries, was also documented by contemporary observers. Flooding of normally arid areas was especially widespread during that year. The first indications of the onset of unusual drought in the Caribbean were felt in the most drought-prone islands of the eastern Antilles, especially on Antigua and Barbuda. Antigua had already suffered from a long drought in 1779 and 1780 in an earlier El Niño episode. In 1789 the drought occurred again, but with "redoubled severity." Even as late as 1837 this year was still referred to by Antiguans as "the year of the drought." As a chronicler notes: "What miseries the Antiguans then suffered I am of course from experience unable to say; but if they exceeded those endured in that eventful year 1837 [a later severe El Niño] they must have been terrible indeed" (Oliver 1844, pp. 189–191).

By August 1791 the desiccating effects on the islands of the Antilles were already the severest recorded since the late 17th century. On the islands of St. Vincent and Montserrat no measurable rainfall had been recorded by the middle of the month, according to the colonial archival documentation. Landowners made formal requests for tax relief due to harvest failure.[4] The drought continued on

---

3 Further ENSO-caused crop failures also took place in the summers of 1808–1811, bringing about a wholesale restructuring of the economy of Central Mexico.

4 Petition dated Aug. 13, 1791 by William McKealy on behalf of the Council of Montserrat, Leeward Islands, British West Indies; Montserrat Legislative Assembly Proceedings, Government Archives, Plymouth, Colony of Montserrat.

Montserrat until at least November 1792.[5] The timing varied slightly from place to place. For example, the extended and abnormal periods of drought (1791–1794) on St. Helena in the South Atlantic were later than those in the Caribbean. On St. Vincent and on St. Helena the droughts drove calls by government naturalists for the formal gazetting of forest reservations to encourage rainfall.[6] The great El Niño event continued longest for the eastern Caribbean and Atlantic, with the Times Index indicating that in 1795 there was still an unrelieved drought in Antigua (Michael Chenoweth, pers. comm. [1998], who referred to the Times Index of 1796).

AFRICAN DROUGHTS AND EL NIÑO

The recorded incidence of the El Niño current, as documented by Quinn and Neal (1987), may not be, by itself, an adequate guide to the true dimensions, impact, and longevity of the 1788–1795 event. Turning to Africa, record low levels of the Nile from 1790 to 1797 suggest its extraordinary reach and true severity, in much reducing monsoonal rainfall on the Ethiopian highlands. In Egypt three successive years of exceptionally low floods led to famine and soaring wheat prices. This was followed in 1789 by the plague (called *Ta'oun Ismail Bey*), which lasted for five months. In 1791 and again in 1792, a slight drop below the long-term mean and only two cubits or about one meter from the optimal level led to a severe famine. According to Antoune Zakry (1926), people were forced to eat dead horses and donkeys and even children. Another series of low floods in 1794, 1795, and 1796 led to a peasant revolt. This experience was comparable only with that of 1877, when the flood was two meters below average, leaving 62% of Qena Province and 75% of Girga Province unirrigated (Willocks and Craig 1913).

Evidence from much of the rest of Africa is scanty. However, prolonged droughts in Natal and Zululand between 1789 and 1799 resulted in the *Mahlatule* famine (Webb and Wright 1976). This was the most severe drought known in the written record to have affected southern Africa prior to an El Niño event of 1862.[7] The low rainfall shows up very clearly in dendrochronological records (Hall 1976, p. 702). This ten-year drought began with an extraordinary record of cold sea conditions off the Cape. On December 24, 1788, while

---

5 As stated in letter of March 6, 1792, from the Commissioner of the Council of Trade and Plantations to the Council of Monserrat, Government Archives, Plymouth, Montserrat.

6 Letter from the Directors of the East India Company to the Governor of St. Helena, dated March 7, 1794, St. Helena Records, Government Archives, St. Helena, South Atlantic, reproduced in Janisch (1908); see also article by Grove (1992).

7 The 1862 event caused the worst droughts in southern Africa in the 19th century. Dendrochronological records point to the early 1860s and the 1790s as long dry spells.

carrying vital supplies to New South Wales, the *Guardian* foundered on an iceberg right up near the Cape of Good Hope.[8]

THE "GREAT EL NIÑO" AND THE FRENCH REVOLUTION

In the normally temperate regions of western Europe, highly abnormal weather patterns were making themselves felt as early as 1787. There was an unusually cold winter in western Europe in 1787–1788, followed by a late and wet spring, and then a summer drought, which resulted in severe crop failures that helped to stimulate the critically explosive social pressures that culminated in the French Revolution (Neumann 1977; Neumann and Dettwiller 1980).

The extreme summer droughts and hailstorms of 1788 were decisive in their effects, as recorded by a peasant winegrower from near Meaux (Desbordes 1961):

> In the year 1788, there was no winter, the spring was not favorable to crops, it was cold in the spring, the rye was not good, the wheat was quite good but the too great heat shrivelled the kernels so that the grain harvest was so small, hardly a sheaf or a peck, so that it was put off, but the wine harvest was very good and very good wines, gathered at the end of September, the wine was worth 25 livres after the harvest and the wheat 24 livres after the harvest, on July 13 there was a cloud of hail which began the other side of Paris and crossed all of France as far as Picardy, it did great damage, the hail weighed 8 livres, it cut down wheat and trees in its path, its course was two leagues wide by fifty long, some horses were killed (Le Roy Ladurie 1972, p. 75).

This hailstorm burst over a great part of central France from Rouen in Normandy as far as Toulouse in the south. Thomas Blaikie, who witnessed it, wrote of stones so monstrous that they killed hares and partridge and ripped branches from elm trees. The hailstorm wiped out budding vines in Alsace, Burgundy, and the Loire, and laid waste to wheat fields in much of central France. Ripening fruit was damaged on the trees in the Midi and the Calvados regions. In the western province of the Beauce, cereal crops that had already survived one hailstorm on May 29 succumbed to the second blow in July. Farmers south of Paris reported that, after July, the countryside had been reduced to an arid desert. These kinds of conditions led in late summer 1788 to serious rural unrest. Small-scale rural revolts took place in the areas worst affected by the summer droughts: Provence, Hainault, Cambresis, Picardy, the area to the south of Paris, eastward in Franche-Comté, around Lyon and Languedoc, and westwards in Poitou and Brittany (Lefebvre 1932, pp. 47–53).

---

8 The *Australian*, Dec. 24, 1998. (Note that similar unusually heavy ice conditions caused the wreck of Shackleton's ship *Endurance* in Oct. 1914.)

To make matters worse, the prolonged drought was followed by the severest winter since 1709.[9] Rivers froze throughout the country and wolves were said to descend from the Alps down into Languedoc. In the Tarn and the Ardeche, men were reduced to boiling tree bark to make gruel. Birds froze on the perches or fell from the sky. Watermills froze in their rivers and thus prevented the grinding of wheat for desperately needed flour. Snow lay on the ground as far as Toulouse until late April. In January, Mirabeau visted Provence and wrote:

> Every scourge has been unloosed. Everywhere I have found men dead of cold and hunger, and that in the midst of wheat for lack of flour, all the mills being frozen (Monahan 1993).

Occasional thaws made the situation worse: the Loire burst its banks and flooded onto the streets of Blois and Tours.

All of these winter disasters came on top of food shortages brought on by the droughts of the 1787 summer and the appalling harvests of summer 1788. Warm, dry spring–summers are usually favorable to grain in northern France and northwestern Europe, but at certain times, they can be disastrous. For example, a spell of dry heat during the growth period, when the grain is still soft and moist and not yet hardened, can wither all hope of harvests in a few days. This is what happened in 1788, which had a good summer, early wine harvests, and bad grain harvests. The wheat shriveled, thus paving the way for the food crisis, the "great fear," and the unrest of the hungry, when the time of the *soudure* or bridging of the gap between harvest came in the spring of 1789. No one expressed this fear better than the poor woman with whom Arthur Young walked up a hill in Champagne on a July day in 1789:

> Her husband had a morsel of land, one cowe, and a poor litte horse, yet they had 42 lbs. of wheat and three chickens to pay as a quit-rent to one seigneur, and 168 lbs of oats, one chicken and one sou, to pay another, besides very heavy tailles and other taxes. She had 7 children, and the cow's milk helped to make the soup. It was said at present that something was to be done by some great folks for such poor ones, but she did not know who or how, but God send us better, *car les tailles et les droits nous écrasent.*[10]

The price of bread doubled between 1787 and October 1788. By midwinter 1788–1789, clergy estimated that one-fifth of the population of Paris had become dependent on charitable relief of some sort. In the countryside landless laborers were especially badly affected. Exploitation of the dearth by grain traders and hoarders made the situation steadily worse. It was in this context that the French King requested communities throughout France to draw up *cahiers* of complaints and grievances to be presented in Paris. During February to April

9 In 1709, the red Bordeaux was said to have frozen in Louis XIV's goblet (see Monahan 1993).

10 "For the cuts and taxes will ruin us" (Young 1953, p. 173).

1789 over 25,000 *cahiers* were written. From these we can assess not only the accumulation of long-term grievances but some idea of the intense dislocation of normal economic life that the extreme weather conditions had brought.

The excessive cold and food shortages of early 1789 drove increased poaching and stealing. There were regular attacks on grain transports both on road and river. Bakeries and granaries were also robbed. Rabbits, deer, and other game were slaughtered, irrespective of ownership or regulation. Gamekeepers and other symbols of authority who opposed such actions were killed. The populace became accustomed to a level of resistance that soon developed into broader reaction and violent protest. Anger at the price of grain and bread in Paris soon found suitable targets for rioting and violence, particularly where the large population of quayside laborers were out of work because the Seine was still frozen in April. A number of pamphlets printed at this time proclaimed that the supply of bread should be the first object of the planned Estates-General: the very first duty of all true citizens was to "tear from the jaws of death your co-citizens who groan at the very doors of your assemblies" (quoted in Schama 1989, p. 331).

In the summer of 1789 much of France rose in revolt; in cities, urban crowds rioted. How far the resulting course of revolution had its roots in the anomalous climatic situation of the period is open to debate. What is certain is that the part played by extreme weather events in bringing about social disturbance during the French Revolution simply cannot be neglected. It may be, as de Tocqueville put it, that had these responses to anomalous climatic events not occurred, "the old social edifice would have none the less fallen everywhere, at one place sooner, at another later; only it would have fallen piece by piece, instead of collapsing in a single crash" (de Tocqueville 1952, p. 96).

THE CATALONIAN REVOLT AND THE REST OF EUROPE

Connections between an accumulation of unusual and extreme weather events and popular rebellion were by no means confined to France. In northern Mediterranean Spain, the cold winter of 1788–1789 was even more unusual than in France. Here, too, persistent summer droughts were followed by a winter of intense cold and heavy snowfall (Barriendos et al. 2000). One observer wrote:

> Autumn this year was colder than normal...and no one alive has ever experienced the weather so cold in El Prat. It was extraordinary, both what was observed and the effects it caused....on the 30th and 31st December the wash of the waves on the beach froze which has also never been seen or heard of before. Likewise it was observed that the water froze in the washbasins in the cells where the nuns slept at the Religious Order of Compassion....the river channels froze and the carriages passed over the ice without breaking it (Salva 1788).

Between August 1788 and February 1789, cereal prices in Barcelona rose by 50% (Fontana 1988, quoted in Barriendos et al. 2000). Between February and

March, 1789, there was a revolt in the city of Rebomboris de Pa, when the population set fire to the municipal stores and ovens. The authorities attempted to pacify the angry crowds by handing out provisions and offering supplies at reasonable prices. The privileged classes also provided money and contributions in kind to pacify the underprivileged, while the military and police authorities stepped back to allow events to run their course.

The authorities then took refuge in the two fortresses that controlled the city and put up powerful defenses in case events got out of control. Despite these measures, chaotic riots ensued and in the aftermath six people were executed (Mercader 1986; Riera i Tuebols 1985; Torras i Ribe 1978, all quoted in Barriendos et al. 2000). Similar riots took place on other parts of Catalonia when the poor outlook for the 1789 harvest became clear and profiteers and hoarders made their appearance. Revolts and emergency actions by municipal authorities took place both in the coast and inland with documentary reports being made in cities such as Vic, Mataro, and Tortosa (Barriendos et al. 2000). The fact that these social responses to cold and crop failure did not lead to the same degree of social turmoil and rebellion as in France should not disguise the fact that they were highly unusual.

Even where there were no revolutions, extraordinary weather patterns were reported. In the English winter of 1790, for example, Parson Woodforde's diary tells of unusually high temperatures and summer-like weather in January and February of 1790 (Woodforde 1985).[11]

One of the advantages in trying to understand European history in terms of the succession of prior climatic stresses is that it provides an environmental context for events of the major revolutions, rather than analyzing them as purely social phenomena. As de Tocqueville commented: "The French Revolution will only be the darkness of night to those who see it in isolation; only the times which preceded it will give the light to illuminate it" (de Tocqueville 1952, p. 249).[12]

The whole social edifice of *Ancien Regime* France collapsed at a single blow, and the fact that this was in the midst of one of the worst El Niño episodes of the millennium is something that should be taken into account when examining this history.

THE GREAT EL NIÑO IN NORTH AMERICA

High winter temperatures were also experienced in North America in the early 1790s. Contemporary observers commented that horse herds expanded greatly in numbers. This facilitated expansion and migrations by the Cree, Assiniboine, Blackfoot, and Gros Ventre in parts of Washington, Montana, and Wyoming.

---

11 Temperatures were comparable to the high temperatures of January, 1998.

12 For a more developed discussion of the agrarian background to the French Revolution, see Davies (1964).

The first three years of the 1790s were very warm and dry on the northern Great Plains. Fur traders in the region repeatedly remarked about how warm and snowless those winters were. High temperatures were accompanied by high rainfall events. On January 13, 1791, the first of a series of very heavy thunderstorms in the region was recorded in Saskatchewan. This produced hardships for bison hunters, but horse herds multiplied. Hostilities among native groups were rare in those years. That warm episode ended abruptly in 1793–1794. At the end of the El Niño event, the return of cold winters provoked wars between the Indian tribes as conditions deteriorated for them and their horses. Horse herds were decimated, and warfare reached a climax in the next year.[13]

Further south, conditions also gave rise to heavy rainfall and high temperatures in the United States, which brought an inexorable rise in the mosquito population. As a result, by 1793 conditions were ideal for the spread of mosquito-borne diseases. On August 19, 1793, Dr. Benjamin Rush, a doctor in the relatively northerly city of Philadelphia, noted his first cases of yellow fever (Foster et al. 1998). This is a disease normally spread in tropical America by *Aedes aegypti*, a mosquito with a pronounced tropical range. By October 1793 the epidemic had killed over 5000 people in Philadelphia alone, and the epidemic was only ended by a severe frost in November, which killed the mosquitoes. It seems that the epidemic had spread from the French colony of Saint Domingue (now Haiti), where a slave rebellion had been aggravated by prolonged bad weather and crop failure in Europe. Refugees from the rebellion carried the yellow fever with them to the East Coast ports of the U.S., where the aberrantly high mosquito population allowed the disease to flourish. The political consequences of this disease event were far-reaching and resulted, not least, in the French losing Haiti and relinquishing most of their American colonies.

Other diseases also flourished in North America during the period of the Great El Niño. This was particularly the case with influenza, a disease whose epidemics had previously rarely affected the North American mainland. An epidemic spread from Georgia in the southern U.S. to Nova Scotia, Canada, between September and December 1789, and again in spring 1790. Its most famous victim was George Washington. Dr. Warren recorded that "at New York, as far as I can learn, its appearance was somewhat later than here, and our beloved President Washington is but now on the recovery from a very severe and dangerous attack of it in that city" (Pettigrew, quoted in Thompson 1852, p. 234). Thomas Pettigrew observed that "the summer preceding the fall disease, was remarkably hot…the last winter was uncommonly mild and rainy. The diseases of that season numerous, [including] both synocha and typhus" (Pettigrew, quoted in Thompson 1852, pp. 199–202). Certainly the very hot summers and mild winters which characterize El Niño conditions in much of

---

13 I am indebted for archival details of the weather in the Great Plains in 1790–1793 to Ted Binnema. See also Binnema (2001).

North America appear to have encouraged the spread of epidemics in several different diseases, and not least in 1788–1794.

CONCLUSIONS

Although further archival research is needed to characterize the 1788–1795 event more fully, the evidence of an intense and prolonged global impact already suggests that it may have been among the most severe in the available written record. The data from this well-documented period can throw light on exceptional global weather patterns. The correlation between a weak phase in the North Atlantic Oscillation (NAO) and an extreme El Niño episode elsewhere suggests a pattern tying NAO and ENSO together via the South Asian monsoon. There were already well-established strong correlations between interannual variability in the NAO and Indian summer monsoon rainfall, specifically between summer monsoon rainfall and the NAO of the preceding year in January.

The early stages of the Great El Niño event were observable in southern India *more than a year before* the warm current was recorded along the Peruvian coast, the usual key indicator of an El Niño phenomenon. This was also the case for the major El Niño event of the 17th century in 1685–1688. Moreover, it appears that in both episodes, a presumed weak phase of the NAO led to a very cold winter, high pressure over Europe, a cold wet spring and a summer drought preceding later monsoonal blocking and drought in South and Southeast Asia.

We are only now realizing that this sequence of events and phasing of activity seems to form a global pattern. The mechanisms of the NAO, the Asian monsoon, and the Pacific ENSO appear to be closely articulated in the case of the severest El Niño episodes.

The connection ("teleconnection") between the failure of Indian monsoons and Australian droughts was noted in Australia by Charles Todd as early as 1888 (Grove 1998). In 1896, the New South Wales Government Observer, H. C. Russell, carefully documented the coincidences between India and Australia over the period from 1789 to 1886 and showed the correlations to be strongest in the case of the worst episodes (Table 10.3).

The developmental sequence of the 1788–1795 El Niño may provide a model by which we can compare other subsequent, very severe El Niño events that have occurred since that time: 1895–1902, 1982–1983, and 1997–1998. It may also serve as a template for understanding much earlier events in the historical record, where data are sketchier. Such an understanding would give retrospective comfort to the observant Arthur Phillip, that his hunch was right. The events do not directly correlate, but rather form a pattern that may take several years to work sequentially through the global weather and ocean system (Grove 1998). "Persistent" or extended El Niños, such as that of 1788–1795, suggest a sequence for these patterns.

Table 10.3 Coincidence of Australian and Indian droughts (after Russell's account; see Grove 1997).

Australian droughts	Indian droughts
1789–1791	1790–1792
1793	
1797	
1798–1800	
1802–1804	1802–1804
1808–1815	1808–1813
1818–1821	
1824	1824–1825
1827–1829	1828
1833	1832–1833
1837–1839	1837–1839
1842–1843	
1846–1847	
1849–1852	
1855	
1857–1859	1856–1858
1861–1862	
1865–1869	1865–1866
1872	
1875–1877	1875–1877
1880–1881	
1884–1886	1884

REFERENCES

Barriendos, M., J.C. Pena, J. Martin-Vide, P. Jonsson, and G. Demaree. 2006. The winter of 1788–1789 in the Iberian Peninsula from meteorological readings, observations and proxy-data records. In: Giuseppe Toaldo e il Suo Tempo, 1719–1797: Scienza e Lumi tra Veneto ed Europa, pp. 921–942. Padova: Centro per la Storia dell Università di Padova.

Beatson, A. 1816. Tracts on the Island of St. Helena. London: Bulmer.

Binnema, T. 2001. Common and Contested Ground: A Human and Environmental History of the Northwestern Plains. Norman: Univ. of Oklahoma Press.

Bligh, W. 1788/1789. *Log of Captain William Bligh*. HM Admiralty Records PRO London. Facsimile portion of log publ. by Pagemaster Press, Guildford, England, 1981.

Census of Tanjore. Papers of the Walkers of Bowland, MS 13615 B. Edinburgh: National Library of Scotland.

Commander, S. 1989. The mechanics of demographic and economic growth in Uttar Pradesh: 1800–1900. In: India's Historical Demography: Studies in Famine, Disease and Society, ed. T. Dyson. London: Curzon.

Davies, A. 1964. The origins of the French revolution of 1789. *History* **49**:24–41.

Desbordes, J.M. 1961. La chronique villageoise de Varreddes (Seine et Marne):Un document sur la vie rurale des XVIIe et XVIIIe siècles. Paris: Edition de l'Ecole.

de Tocqueville, A. 1952. L'Ancien Regime et la Revolution, ed. J.P. Mayer. Paris: Gallimard.

Endfield, G.H., and S.L. O'Hara. 1997. Conflicts over water in the "Little Drought Age" in Central Mexico. *Environ. Hist.* **3**:255–272.

Fontana, J. 1988. Historia de Catalunya: La fie de l'Antic regim i la industrialitzacio, 1787–1868. Vol. 5. Barcelona: Edicions 62.

Foster, K.R., M.F. Jenkins, and A.C. Toogood. 1998. The Philadelphia Yellow Fever epidemic of 1793. *Sci. Am.* **279**:68–74.

Grove, R.H. 1992. The origins of western environmentalism. *Sci. Am.* **267**:42–47.

Grove, R.H. 1997. The East India Company, the Australians and the El Niño. In: Ecology, Climate and Empire: Colonialism and Global Environmental History 1400–1940, by R.H. Grove, pp. 124–46. Cambridge: White Horse.

Grove, R.H. 1998. The global impact of the 1789–1793 El Niño. *Nature* **393**:318–319.

Hall, M. 1976. Dendrochronology, rainfall and human adaptation in the Later Iron Age of Natal and Zululand. Pietermaritzburg, South Africa: Annals of the Natal Museum.

Janisch, H.R., ed. 1908. Extracts of the St. Helena Chronicles and Records of Cape Commanders. Jamestown, St Helena.

Kington, J. 1988. The Weather of the 1780s over Europe. Cambridge: Cambridge Univ. Press.

Krishnamurthy, V., and B.N. Goswami. 2000, Indian monsoon-ENSO relationship on interdecadal time scale. *J. Clim.* **3**:579–595.

Lardinois, R. 1989. Deserted villages and depopulation in rural Tamil Nadu, c. 1780–1830. In: India's Historical Demography: Studies in Famine, Disease and Society, ed. T. Dyson, pp. 16–48. London: Curzon.

Lefebvre, G. 1932. La Grand Peur de 1789. Paris: Societé d'Edition d'Enseignement Superieur.

Le Roy Ladurie, E. 1972. Times of Feast, Times of Famine. London: Allen and Unwin.

Madras Courier. 1791. Report on Nov. 3, 1791.

McCormick, T. 1987. First Views of Australia, 1788–1825: A History of Early Sydney. Sydney: David Ell Press.

Mercader, J. 1986. Els capitans generals. Barcelona: Viviens–Vives.

Monahan, W.G. 1993. Year of Sorrows: The Great Famine of 1709 in Lyon. Columbus: Ohio State Univ. Press.

Murton, B. 1984. Spatial and temporal patterns of famine in southern India before the famine codes. In: Famine as a Geographical Phenomenon, ed. B. Currey and G. Hugo, pp. 71–89. Dordrecht: Reidel.

Neumann, J. 1977. Great historical events that were significantly affected by the weather: Part 2, The year leading to the Revolution of 1789 in France. Bull. *Am. Meteorol. Soc.* **58**:163–168.

Neumann, J., and J. Dettwiller. 1980. Great historical events that were significantly affected by the weather: Part 9, The year leading to the revolution of 1789 in France (II). *Bull. Am. Meteorol. Soc.* **71**:33–41.

Nicholls, N. 1988. More on early ENSOs: Evidence from Australian documentary sources. *Bull. Am. Meteorol. Soc.* **69**:4–7.

Nicholls, N. 1992. Historical El Niño/Southern Oscillation variability in the Australasian region. In: El Niño: Historical and Paleoclimatic Aspects of the

Southern Oscillation, ed. H.F. Diaz and V. Markgref, pp. 151–173. Cambridge: Cambridge Univ. Press.

Oliver, V.L. 1844/1894–99.The History of the Island of Antigua…from the first settlement in 1635 to the present time. 3 vols. London: Mitchell and Hughes.

Ouweneel, A. 1996. Shadows over Anahuac: An Ecological Interpretation of Crisis and Development in Central Mexico, 1730–1810. Albuquerque: Univ. of New Mexico Press.

Pettigrew, T.J. 1817. Memoirs of the Life and Writings of J. Coakley Lettsom, vol. 3. London: Nichols and Bentley.

Quinn, W.H., et al. 1975. Historical trends and statistics of the Southern Oscillation, El Niño and Indonesian droughts. *Fishery Bull.* **76**:663–678.

Quinn, W.H. and V.T. Neal. 1987. El Niño occurrences over the past four and a half centuries. *J. Geophys. Res.* **92**:14,449–14,461.

Riera i Tuebols, S. 1985. Ciencia i tecnic a la illistracio; Francesc Salva i Campillo (1751–1828). Barcelona: La Magrana.

Roxburgh, W. 1778/1790. A meteorological diary kept at Fort St. George in the East Indies. *Phil. Trans. Roy. Soc. Lond.* **68**:180–190; **80**:112–114.

Roxburgh, W. 1793a. Letter to Sir Charles Oakley, dated Jan. 23, 1793, East India Company Boards Collections. Ref no. F/4/99, p. 29. London: British Library, India Office Library and Records [IOLR].

Roxburgh, W. 1793b. Report to the President's Council, Privy Council Letters, vol. clxxxl, dated Feb. 8. Tamil Nadu State Archives.

Salva, F. 1788. Archive of the Royal Academy of Medicine, Barcelona. *Tablas Meteorologicas* **1.**

Schama, S. 1989. Citizens: A Chronicle of the French Revolution. London: Penguin.

Thompson, T. 1852. Annals of Influenza or Epidemic Catarrhal Fever from 1510 to 1837. London: Sydenham.

Torras i Ribe, J.M. 1978. La Catalunya borbonica: Evolucio i reaccions contra el nou regim. In: Historia de Catalunya, vol. 4, pp. 178–203. Barcelona: Salvat.

Trenberth, K., and T.J. Hoar. 1996. The 1990–1995 El Niño-Southern Oscillation event: Longest on record? *Geophys. Res. Lett.* **23**:57–60.

Webb, C., and J.B. Wright. 1976. The James Stuart Archive of Recorded Oral Evidence relating to the History of the Zulu and Neighbouring Peoples, vol. 1. Pietermaritzburg, South Africa: Univ. of Natal Press.

Willocks, W., and J.I. Craig. 1913. Egyptian Irrigation. 3d ed., vols. 1 and 2. London: E and F.N. Spon.

Woodforde, J. 1985. A Country Parson: James Woodforde's Diary 1758–1802. Oxford: Oxford Univ. Press.

Young, A. 1953. Travels in France, ed. C. Maxwell. Cambridge: Cambridge Univ. Press.

Zakry, A. 1926. The Nile in the Times of the Pharoahs and the Arabs [in Arabic]. Cairo: Maktabet Al–Ma'arif (reprinted 1995 by Maktabet Madbouli).

11

The Lie of History

Nation-States and the Contradictions of Complex Societies[1]

Fekri A. Hassan

Institute of Archaeology, University College London, London WC1H 0PY, U.K.

ABSTRACT

Interpreting the relationship between people and their environment for developing future scenarios of people on Earth requires, as a first step, an understanding of human history beyond current accounts, which are mostly informed by the political ideologies of the "modern" nation-state. This chapter contends that the nation-state, which still shapes the current political arena, is beset by inherent contradictions that are characteristic of complex societies, and that the depletion of natural resources, pollution, explosive population growth, and urbanization are historical outcomes of tendencies to increase productivity and sustain economic growth, often at the expense of masses of people. The emergence of multinational corporations and hegemonic international political institutions (political and financial) represent a menace to the future of people on Earth because the legitimization of the sovereignty of the nation-state is being eroded and masses of poor, disadvantaged individuals in many different settings have become cognizant of their predicament and are disenchanted with current hegemonic political regimes and ideologies; these individuals are opting for alternative models of identity and orientations.

With glaring differences in lifestyles exposed by the media, rising expectations, limited growth potential (both due to disparities in wealth appropriation), and with a burgeoning world population, coupled with a massive assault on almost all world biomes everywhere in the world, the framework for modeling the future cannot be limited to the interplay of environmental and economic variables. Before the world is exhausted from environmental fatigue, it is more likely to descend into a nightmare of civil unrest, violence, and despair as a result of the failure of current political systems to address social grievances, the problematic of identity (within the current context of "multiculturalism"), and the loss of hope in resolving outstanding disparities. The situation is now exacerbated by a disgruntled younger generation well-tuned to the ideologies and technologies of violent protest that have characterized the recent history of European nations. In

1 I dedicate this contribution to my daughter, Mona Hassan, who has now the power to make and write history.

this milieu, attention to climatic extreme events beyond the scale of the last century may be minimized, placing the planet at great risk.

The historical account presented in this chapter aims at clarifying the present as both a result of a historical continuity and an unprecedented state caused by the progressive expansion of population size, number of nonfood producers, intensity of production, spatial extensification of production areas, and use and demand for critical resources (other than food) fostered by psychological tendencies grounded in power differential, disparities in status, and the importance of material goods in determining self-worth. From this historical perspective, modeling the future requires an integration of psychological, social, and economic dimensions. It also demands an integration of micro-models of regional landscapes and individual agents with macro-models of a global dimension. Special attention must be paid to the role of the Internet in creating networks of instantaneous information transfer among individuals, which may lead to unexpected (nonlinear) events.

SYNOPSIS

Nations play a major force in making history. In a world undergoing radical transformations, we may benefit from a cautious approach to apologetic historic writings as well as from a deeper, multifaceted understanding of how individuals shape historical events. Interpreting human history has not only been influenced by royalist, religious, and, more recently, nationalist agendas, but also by a lack of integration of insights from psychology, sociology, and anthropology. A fuller understanding of the historical process requires an explanation of how the thoughts, communications, and actions by individuals in a social matrix lead to norms and modalities that endure at different timescales. Examination of historical modalities from this perspective reveals inherent contradictions in large, complex societies since their emergence more than 5,000 years ago in the wake of the shift to agriculture. Such societies, down to modern nation-states, depend upon a managerial elite, who legitimize their role and privileges through a variety of ideological technologies and, as needed, coercion.

There are various forms by which complex societies are managed, but they invariably consist of a top ruling elite in alliance with an ideological establishment and prominent individuals/groups (e.g., landlords, provincial leaders, or industrial magnates). In practice, throughout history, individuals in these hegemonic subgroups usually vie for power and work at aggrandizing their privileges, often at the expense of other segments in society. This has led to a turbulent history of social control, on one hand, and to territorial expansion, often by force or the threat of force, to enlarge revenues on the other. This, in turn, leads to turbulent intersocietal dynamics.

Today, world history is being shaped by individuals who operate within what has come to be known as the modern nation-state in which the notion of "nation" is buffed up as a monolithic collective whole. The governing of such nations takes place through this nationalist construction in an *inter*-national arena. This construction masks, however, the heterogeneity and contradictions inherent in the "nation-state" as well as the *trans*-"national" links that pre-date and run parallel to national affiliations.

In recent decades, the impact of multinational corporations and information communication technologies have progressively undermined the monopolies of governments on information and weakened their ability to control their nationals. By promising economic welfare, prosperity, and security, the modern ruling elite have unleashed an unprecedented wave of rising expectations. Such expectations can hardly be met even with continued economic growth because those who have still want more, while those who do not have try by any means to gain what is, in the ideology of the modern state, their unalienable right. The consequences of this historical development over the last 100 years have been far reaching. Military colonial expansion became unworkable and indefensible; education to supply professional manpower (and markets) for industry and commerce has fueled rising expectations for equality, justice, prosperity, and peace; escalation of scientific discoveries and technological innovations to increase financial gains from trade has, among other things, had a deleterious impact on the Earth's life-support natural resources; the increase in the size of rural families to minimize poverty in poor countries has led to a global demographic explosion.

Poverty and harsh inequalities in a world growing more transparent through information technologies and the spread of education pose a tough challenge to the commercial–financial–industrial elite. Such elites, forming different geographical/historical (transnational) blocks, are competing for global dominance and, in the process, are marshalling a ragbag of ideologies (backed with military supremacy) to rally "their" own nations in the struggle for economic power against "Others," while trying to win the "hearts and minds" of those who are targets for domination and subjugation.

The realities of oppression (using modern technologies to quell opposition and dissent); dire, intolerable, absolute, and relative poverty (aggravating poor health and ignorance); and rising expectations for a better life are at present an unstoppable historical tide. It would be unwise to cast this situation in terms of "international relations," to portray it as a "clash of civilizations," to reduce it to a conflict of political ideologies, to blame the situation on "terrorists" who can be captured and annihilated, or to propagate the slogans of a certain ideology as a cure to all ills. If history is construed as the continued monopoly and social control by an elite to their own advantage with disregard to the suffering of others who are entrusted to their care, then, perhaps history is coming to an end.

INTRODUCTORY NOTE

In this chapter, I present a very brief overview of an integrated theory of history, combining insights from psychology, sociology, archaeology, and cultural anthropology. This theory shares many threads with other contemporary theories, especially those that highlight agency and structure. On the basis of this theory, I offer a synoptic, transcultural view of the salient, structural lineaments of state societies: from the earliest agrarian societies to the modern nation-state. This leads to an exposé of the inherent contradictions of state societies that are now exacerbated by the growing influence of multinational corporations and information communication technologies. This is mainly an interpretative essay with occasional reference to examples from Ancient Egypt and the Near East, as well as from the historical and contemporary dynamics of western Europe. To provide more substantive historical data would have lengthened the text and reduced its clarity. I hope that the limited selection of references is helpful in directing the reader to additional sources of data. Although a great deal of my published work has focused on paleoenvironmental issues, I have opted to redress the balance inherent in paleoenvironmental investigations by focusing on a theory of history that can serve as a means for addressing both the cultural as well as the historical interactions between human societies and their environment. I have given a closer look at the latter subject in substantive contributions, especially in connection with the emergence and sustainability of Egyptian civilization and the impact of abrupt climatic changes on the cultural developments in Africa over the last 10,000 years (Hassan 1993, 1997, 2000, 2002). A readable, pertinent prognosis of whether civilization is sustainable and its implications for the future, published in 1995, predicted that "without an ethos of global sharing and amity, a rise in violence, terrorism, and hatred will turn industrial nations into beleaguered fortresses under siege" (Hassan 1995, p. 30). With bombs exploding at the door steps of my institute in London as I write this contribution (July 2005), as a follow up to other incidences over the last decade in many parts of the world and by many different groups, I hope that we will begin to redirect the course of our destiny.

HISTORY AS A HEURISTIC CONSTRUCT

History is an imaginary construct of moments of action (events) by individuals situated and inexorably linked to other individuals in conceptual aggregates often referred to as "societies." As constructs, history as well as "culture" are nevertheless active ingredients in "reality" inasmuch as they contribute to the cognitive apparatus by which individuals act in the world.

The making of history is a social process consisting of active or passive differential retention and organization of the memory of "past" events. There are various histories in any society that vary in scale, purpose, and certainly content and organization. There are, for example, individual and family histories, histories of the "church," the histories of "ethnic" groups, and often the master narrative of history promoted by the state.

Norms, Modalities, and Cognitive Schemata

Events persist as traces in memory. With its inherent capabilities, the human mind is conditioned by the way we are socialized. We are trained to perceive, interpret, and organize sensory data. We process information in sets, sometimes related, of cognitive schemata that allow us to act in the world, to interact with others, and to make sense of the world as a means by which we can take decisions and follow certain life projects. Throughout the process, we develop "profiles" of ourselves and our "personalities" and become active agents in a world of others with a sense of "self" and "being." We may or may not question notions about the world, history, or ourselves that are instilled in us in childhood: notions that may be continually reinforced in all the stages of our lives by family, friends, governments, religious establishments, universities, and media. Faced with competing or alternate views, we are likely to resist, dismiss, mollify, or ridicule those notions that can undermine our "being." Certain notions are fundamental to our sense of being: integrity, *raison d'être*, and "life projects." Such axial notions are bastions of resistance and tend to contribute to the salient lineaments of our perceived "essence" and sense of "security." As long as many individuals continue to hold fundamental views of themselves and the world, there is reason to claim that there is a "society" and a "social structure."

We, as individuals, may adopt new ideas or circumnavigate fundamental, axial notions if we perceive or convince ourselves that the new ideas are beneficial to us and are not likely to compromise our strongly held beliefs. We are often faced with such "cognitive dissonance" and are likely to overcome such dissonance by dismissing new notions or adopting them as an isolated package with no attempt to integrate them in our axial sets of beliefs. However, it is hardly likely in a population of thousands or millions of individuals to find those who will be swayed to adopt radical ideas either because of the temptation of perceived gains (psychological or economic), dissatisfaction of current situations, or/and the lack of a strong belief in traditional norms. Moreover, an aggregate of thousands or millions of individuals are not a homogeneous mass but rather a heterogeneous aggregate of "groups" self-identified by descent, residence, religion, and socioeconomic status. There is also now a growing emphasis on other groups defined by gender, age, and occupation.

In a complex society, certain groups (families, sodalities, clans, interest groups) manage to achieve greater "power" than others by virtue of their special ecological setting, (historical) background, appropriate decisions, or just as a result of chance. Following an initial kickoff point, individuals in these groups will guard their own interests and aim to aggrandize, legitimize, and perpetuate their power. Such groups are often referred to as "elite" or "status groups" (see discussion in Barnes 1995, pp.130–150).

It would be impossible to provide any adequate understanding of social developments and of the course of civilization following the rise of state societies without reference to such groups that play a fundamental role in the structuring of such complex societies. Those who succeed in assuming managerial positions, which allow them to gain revenues and taxes from the masses, and who succeed in developing viable administrative and legitimizing ideological formulations, are most likely to form a dominant, hegemonic group exercising far greater power in "structuring" society.

Social Structure

Although the concept of "structure" (Archer 1982, 1995; Giddens 1984) has been highlighted in recent discussions in various disciplines, such a concept is still underdeveloped and requires further elaboration. The term is often used to refer to prominent, persistent, resilient, transgenerational, regular sets of social relationships that are responsible for endowing a society, a culture, or a civilization with its particular diagnostic characteristics (cf. Dark 2000, pp. 113–117, esp. p. 115, footnote 31). Although social structure is an interpretative, imaginary construct, it may serve as a basis for action and is the result of adherence to certain modes of practice.

It is thus possible to develop models of "social structure" by examining, analyzing, and interpreting historical and archaeological data representing tangible traces of individual actions of various sectors of society through medium-range (200–300 years) or long-term (over millennia) time spans. It is only possible to construct a "structure" if certain actions are repeated over and over again, reproduced from one generation to the next over multiple generations, and involving a sizable sector of the population. The regular repetition of certain actions is often referred to as *habits*, *customs*, or *traditions* (hereafter *modalities*). Modalities are not the result of mindless, mechanical behavior but are embedded and reflexive with *norms*, *values*, and *beliefs* (hereafter "*norms*"). The persistence of the set of relations that allows an observer to construct a structure is a function of the strength of the norms, the social context of control, and the situations that may lead to attenuation, modification (adjustment, amendment, alteration), or rejection of a norm.

Since structures refer to social (collective) modalities and norms, they are often erroneously thought to be detached from the actions of individuals. Also, since structures are trans-subjective, they are also erroneously believed to be "objective." Moreover, because norms and structures are "abstract," persistent, and transgenerational features, they are mistaken for static formulations. However, norms do not exist as independent abstract notions, even when they are encoded in books. Norms are active and "real" only inasmuch as they are a part of the cognitive makeup of individuals. In addition, norms are not static templates for specific actions; they are elements of "cognitive schemata" (D'Andrade 1981, 1995).[2]

Cognitive schemata provide the basis for the persistence of norms in a stable formation of ideas, which allows individuals to reproduce preexisting

---

2 A schema (pl. schemata) is a generic formula that allows an individual to relate various elemental norms together in a cohesive set of relations, and to translate general axial norms into specific situations.

modalities as well as to create new modalities. In agreement with Goffman (1959, 1967, 1970), norms (the schemata in my interpretation) are actualized through human encounters in specific situations. This is concordant with the insights by Becker (1970) on how individuals adjust to new situations (a process of situational adjustment). The malleability of norms, however, should not be overestimated. The issue here concerns the "weight" of a certain norm, which depends on its level in the hierarchy and axiality of the cognitive schemata.

Norms rarely exist as rarefied abstract notions. They are intertwined with "encounters" and "interactions" among individuals. Encounters shape norms and schemata by the way they are expressed through "performances" or "appearances." Encounters are charged emotional interactions grounded in the primacy of the body and its functions, as well as its embeddedness in "nature" through biological, physiological, and psychological connections.

Starting from an encounter between individuals (as sentient, feeling, and thinking bodies in a place at a certain time) to the selection of interindividual modalities for action and intersubjective norms and ending with the transgenerational transmission of norms and nested schemata for proper action, an aggregate of people become a society with a distinct culture. Deviations from norms are not only a result of chance encounters, they are also due to the changing nature and circumstances of encounters, the re-assessment of the efficacy or relevance of norms, and the perceived impact of modalities on the acceptance of self and others.

ORIGINS OF ELITE POWER AND THE DISCONTENT OF CIVILIZATIONS

Ten thousand years ago, after millions of years of peripatetic existence as hunters and gatherers roaming the Earth in small bands (15–25) affiliated to loosely organized hordes of 500–1000 persons, our ancestors, who did not number more than 10 million, began in certain parts of the world to develop an agrarian mode of life. Agriculture encouraged the aggregation of large, sedentary groups and made a large family advantageous because of the greater overall yield per person, and the greater security gained when a large group is involved. This incentive led to a slight relaxation of family (birth) controls and brought about a rapid proliferation of hamlets and villages. An increase in risk and uncertainty was characteristic of early agrarian economies that became vulnerable because of their sedentary mode of life, the reduction in the diversity of food resources, and their negative impact on their ecological setting (Hassan 1992).

Below I will suggest that reduction of risk and uncertainty was achieved through religious ideologies and managerial strategies that linked communities together in a safety network. Religion/ideology and governance have since been the hallmark of complex societies that emerged from the early agrarian (Neolithic) experience. By 5000 B.C., the development of hierarchical organizational formats culminated in the emergence of the first state societies headed by kings, legitimized by a religious ideology, and allied to formal religious establishments. Early states varied in the degree to which they were allied with the chiefs of administrative districts and clan leaders in their realm. The disparity in wealth and power between those who were affiliated with the seat of power and the people led to persuasive and coercive control strategies and created a source for disjunction and discontent between the rich and the poor. This remains one of the main ills of our own times. The differential in power and wealth also led to envy and mimesis (coveting the possession or attainment of others, and desiring to be the other through imitation of their appearance, gestures, actions, and possessions, respectively). Individuals in the royal household, high priests, provincial rulers, top administrators, and military generals were tempted by the prospects of gaining the topmost power position. Maneuvering to achieve this position has been one of the main sources of social dynamics in societies.

The ruling elite, past and present, develop or adopt power technologies to set them apart and above others whom they come to dominate and exploit in the name of a divine mandate or serving the people and ensuring their prosperity and well-being. The maintenance of such power technologies depends on the deployment of substances or practices that are beyond the means of ordinary individuals or rival groups. Possession of rare materials such as gold, silver and diamonds, precious metals, stones, and minerals, and goods that require exceptional skills by master artisans and crafts people are often a key feature in the arsenal of power technology. Even ordinary, common items such as denim jeans can be made into power items by the use of brand labels!

I submit that one of the axial elements in the structure of complex state societies is the existence of a power differential between ruling elite and the masses, which are the main source of income for the elite. From this and the deployment of costly power paraphernalia it follows that the elite must convince or persuade the masses to continue to supply them with revenues to sustain their power. This is a structural link that characterized such societies and is tackled in one or a combination of strategies. One of the common strategies is that the elite are critical for the good floods, good weather, bountiful harvests, and for preventing natural catastrophes, upheavals, and turbulence. Another strategy is that they are indispensable for social order and harmony. Yet another is that they are essential to protect the country from outside enemies. These strategies are often integral elements of the royal or ruling ideology, rhetoric, discourse, rituals, and monuments. In addition to such persuasive measures that become a part of the ingrained schema of the masses through bodily encounters, performances, and manipulation, the elite may and do resort to violence or the threat of violence (on Earth as well as in the afterlife). Certain individuals and groups can be used to set an example to others. Exclusion, monopoly, and siphoning of revenues from the masses cannot be sustained without some form of control and ideological notions of order. Accordingly, religion, "national" ideologies, and police forces are likely to characterize complex societies (for a survey of the formative histories of some of the major civilizations and early state societies, see Maisels 1999 and Feinman and Marcus 1998).

PATHWAYS TO INEQUALITY

The initial emergence of a governing elite may be a result of consensual agreement in the expectation that there is a payoff from supporting a person or persons placed in charge of resolution of internal community disputes or in coordinating exchanges among members of the community and between communities to enhance food security, to coordinate group defense or task forces, to ensure the flow of needed goods from outside the community, or to serve as a ritual mediator between the gods and people. Complex societies (i.e., large societies with differentiated roles) appeared around 7000 years ago in the wake of the advent of an agrarian mode of life, which encouraged the emergence of large groups of people who resided permanently close to the fields. The erratic and uncertain productivity from agricultural activities provided the incentive to increase the domain of sharing, cooperation, and exchanges among neighboring communities, who in the initial stages were bonded by kinship. The recurrence of regional food shortages and attacks by marauders to rob the community of its crops and cattle provided opportunities to expand the circle of cooperation and management beyond the spatial domain of kin-related communities.

The key roles of the "managers," who were either self-promoting or chosen by the communities concerned, included (a) enhancing food security through food exchanges and procurement of materials needed for better production, (b) defense, and (c) resolution of disputes. Ritual was an element of all such activities, in addition to its utility in facilitating group cohesion (which is advantageous for agrarian communities) and the provision of psychological assurances of a good harvest in years of food shortage. Collective support for community managers thus depended on perceptions of certain payoffs that warranted allocating a portion of one's productivity to support those who are entrusted with vital services to the state elite.

Through time "societies" with large populations spread over hundreds or thousands of kilometers emerged as a result of the expansion of managerial tasks to coordinate relations among numerous regional groups. This, in turn, facilitated in the long term (hundreds and up to a couple of thousand years) increasing cultural similarities between groups as a result of long-term social interactions, especially those that concern the religious ideology which appear to provide a payoff at no economic cost, at least initially (Renfrew 1986). The process of expanding societies beyond the domain of kin or immediately neighboring communities is not possible without a general expectation among diverse ethnic, regional, or other groups that there is a positive payoff from joining together with other peer polities, and in the mean time, the establishment of a monopolistic, exclusive, network of prominent elite from the different communities in the emerging confederation. The appropriation of secret knowledge (including writing), ritual, and ideological devices by the trans-community elite was an essential ingredient in early state societies. It paved the way for the development of a hierarchical structure of dominance that secured the flow of revenues from the communities to a king who, in turn, was expected to reward the elite of various communities with luxury and ritual goods by which they are set apart, respected, and feared by their own communities.

THE PERSISTENCE OF THE STATE AND ITS INHERENT CONTRADICTIONS

The elementary structure of power in early complex societies set the stage for some of the key sets of relations in state societies in general; a system that has lasted for more than 5,000 years. Its success as a model of governance is attested in many regions and at different times. Even when state societies break down, they are restarted by other "dynasties." Their structure is also imitated by splinter regional groups or ethnic and religious factions.

The continuity and persistence of the state model reside in part in the continued perception of the utility of a governing sector, but it has also, in the past, been a function of the persistence of the belief in the divine right of kings, a belief that has been successfully assailed in some countries only over the last 200 years. However, the role of religion in supporting monarchies and even "secular" governments is still visible and has become a key ingredient in the recent policies of the United States of America. This is primarily because religious leaders and followers in many parts of the world still cling to the idea that God is the fount of state governance. The persistence of the state as an organizational structure is also a result of its historical legitimacy and of its benefits (material, social, and psychological) to those who manage to head the state, as well as to those who are its beneficiaries.

States, however, carry their own contradictions and inner conflicts. One of the fundamental contradictions lies in its role as a means of ensuring food security and peace. In this context, the elite, who are responsible for such worthy goals, are voracious consumers of goods as key elements in their technology of power. The increase in the number of "managerial" elite and the unchecked growth in their expenditure and enhancement of living conditions lead to progressively more demands from the productive masses. This reduces the food security of the masses as they are forced, under traditional food-producing technologies, to yield a higher percentage of their income or work harder to meet the growing demands of the elite. This is likely to engender more conflict, which has to be addressed by coercion or greater doses of religious ideology involving costly expenditures to be paid from higher rates of taxation. Any increase in agrarian output is likely to be a result of prolonging the number of workdays, the length of a working day, or the use of child labor, a situation that enhances misery, reduces satisfaction, and encourages revolts or desertion. In general, early societies were not structured on the basis of a profit-economy aimed at expanding productivity by the deliberate deployment of technology and industry. The main focus was on ritual and bureaucratic technologies that ensured the perpetuation of kingship. Trade and industry were primarily geared to provide the goods and material correlates of elite ideologies and rituals rather than as a source of "profit." Royal monopolies of trade and key luxury and ritual industries ensured that no one else could achieve their status, and made them the sole controllers and providers of emblems of power and status which they could confer on their favorite subjects (those who were obedient and who distinguished themselves in the service of the head of state).

It is for this reason that one of the main revolutionary developments in the history of state societies was the curtailment of the power of kings and church in trade and industry and the emergence of a nonroyal, powerful elite. Undoubtedly, the rise of commercial–financial–industrial elite in Europe since the Renaissance with an ideology predicated upon wealth, profit, and ostentatious consumption has created a cultural wave that still collides with other forms of governance steeped in historically antecedent traditions and their offshoots.

STATE DYNAMICS

Envy and Mimesis

The cost of the managerial elite in early state societies escalated as more regional groups and distant units were included in the union (such groups would have been attracted by the promise of security and peace promoted by the state). The escalating costs of administration and management was a function of increasing distance of control, transport, establishment of religious institutions and buildings, policing, and the manufacture of rare goods to be used as gifts to pacify, mollify, or coopt community leaders. The emergence of differences not only in material prosperity but also in one's religious standing (manifest in provision for elaborate tombs and rituals) led to envy and mimesis (Gebauer and Wulf 1995). The religious clergy, who buttress kingship, and the provincial elite were in a position to make bids for assuming the ultimate seat of power whenever the time was ripe. By contrast, it was in the interest of all those on the top of the social ladder to keep the masses under control and to instill a sense of obedience to God and King as well as to instill fear of severe punishment here and in the afterlife. At times, King and God were offered reconciliatory rewards for pious [most obedient] subjects. The mutual dependencies between the King (and his court), the high priests, and the country on the one hand, and the coercive/conciliatory relationship between those governing and the masses provided "fault" lines that were often activated to reshape power relations depending on perceived opportunities, historical exigencies, ecological crises, or invasions. Rivalry, even within the royal household, was an additional source of inherent instability.

Empires and Militarism

The state and consequent modes of extravagant elite consumption and display of power created a pervasive sense of inequality within and between societies. It also engendered a clash between neighboring expanding state societies, including secondary states created in imitation of primary states. The expansive character of state societies is predicated upon the need to secure foreign imports of luxury and ritual goods necessary for the perpetuation of state religions and privileges, and by the necessity, as local land and labor resources fall short, of supplying the revenues required to sustain the operations of the state. For this reason many state societies began to annex or conquer their neighbors. A prosperous state society is also a prize to be sought by pastoral nomads who live in the surrounding territories exploiting any opportunity to take advantage of the settled agrarian communities. Such enemies of the state become symptomatic of the "dangers" that legitimize the role of the state as a defender of the people against uncivilized barbarians who symbolized terror and anarchy.

The rise of military elite, as a consequence of armed conflict between state societies and rival polities, added to the management cost of the state and engendered a new segment of the society endowed with power and privileges, creating yet another principal node of power along with that of high priests, provincial rulers, and the royal court. Military conquests secured more land and labor, but pacification and control of vast territories with a great diversity of ethnic, religious, and language groups was at the cost of garrisons, forts, repeated military expeditions to quell revolts and uprisings, religious conversion, and gifts to local rulers. Eventually the benefit from adding more land was outweighed by the cost of extending control to distant lands; consequently, over time, control of an expansive "empire" ceased to be feasible as returns diminish relative to the rising cost of subduing [pacifying] the colonies, especially as the elite in the colonized territories become more attuned to statecraft and aware of the weaknesses and vulnerabilities of the imperial power. In addition, the rising demands by the military, internal competition for power within the center, or/and discontent, revolts and desertion in the colonies often undermined the ability of the empire to control its distant territories.

The history of empires in the Mediterranean region started about 1500 B.C. and was characterized by military empires beginning with the Hittites and the Egyptians, followed by the Assyrian and Persian empires. The stage of early empires culminated in the rise of the Roman Empire which encompassed 60 million people, one-fifth of the world population at the time. By A.D. 180, the Empire was too large to remain stable. Diocletian (A.D. 284–305) enlarged the army from 300,000 to 500,000, but the cost of such an army, the growing cost of bureaucracy, increased taxation, and inflation contributed to the demise of the Empire.

Militaristic regimes linked with imperial expansion often lead to disruption of farming activities, industry, depopulation, and emigration as a result of battles. They can also cause major disruptions of life-support systems as a result of their impact on neglect of soil conservation, irrigation of water works, aggregation of refuges in marginal habitats, and intensive utilization of wild food resources. The cost of rehabilitating traumatized communities and disrupted life-support systems may also prove to be too costly.

It is hard not to concede that the rise of militaristic elite and military forces over the last 3,500 years has been a key ingredient in the sociopolitics of nation-states. Today, military power is an important element in international relations. Policy makers and state elite do not question the use of war, military interventions, and clandestine actions as a legitimate option in foreign policy. To achieve their goals abroad, governments do not hesitate to sponsor strikes and riots, mount *coups d'état*, or organize, train, and arm insurgents. Such interventions have become common since the 19th century, first as a means to stamp out independence movements but later as a means of achieving political objectives.

RELIGION FOR THE MASSES

Subversion of Royal Ideologies and Resistance

One of the principal developments in the Mediterranean region and under the Romans was the emergence of a religious ideology (Christianity) that was independent and even in opposition to the state. This represented a radical shift in the role that religion played in the past (i.e., as the handmaid of the state). In Christianity, the masses found a religion of charity and compassion independent of a terrestrial state. Unlike Judaism, this religion was not linked to an ethnic group, a clan, or a tribal unit. Christian brethren formed their own nation independent of ethnicity, language, and racial background. Islam promotes the same ideology. Both religions thus have the potential to be mobilized against state societies. However, both religions were coopted by state organizations to serve as the legitimizing ideology of Byzantium, the Islamic caliphates, and the Ottoman Empire. Nevertheless, both religions still maintain their mass appeal and power and have been mobilized in recent history against state organizations.

The Mercantile Turn: Merchants, Bankers, and Industrialists

The remaking of Europe since the High Middle Ages was characterized by the emergence of entrepreneurs who sought to aggrandize their wealth and social status through long-distance commerce. A new ethic of profit fostered improvements in manufacture and distribution of goods with a dramatic impact on the development of sciences, technology, and industry. Prudent financial management and investment in supporting technologies, including those of warfare, gave European countries by the 18th century a clear edge over the last of the traditional empires, the Ottoman Empire. In the process, European imperialism consisting of rival mini-empires succeeded in subjugating the majority of the world countries. Eventually, as in previous empires, the cost of controlling colonies by military force was no longer affordable, and a new world order was established. Many of the ex-colonies are still in the process of reconfiguring their political system. Russia and China, burdened with rural economies, have struggled to cope and keep up with the industrialized western European nations. Russian political economy has undergone a major shake-up following internal breakdown caused by excessive military expenditure, rigid bureaucracy, and failure to capture world markets. China is still developing its economic strength and improving its trade exports. By contrast, Japan, with a much smaller population, was capable, like Europe, of transforming its agrarian economy to produce higher yields, and to foster commercial activities. The activities of the merchants were safeguarded by powerful families, religious institutions, and the military. Again as in Europe, the mercantile transition led to industrial developments that placed Japan as one of the main players in the current political scene.

Modern States: Modern Histories

The emergence of the modern state, a form of social organization that has, over the last 200 years, become the most dominant form of polity, confounds the notion of the "state" with that of the idea of a "nation."

Appearing first in Spain, France, and England, the modern nation-state fostered an ideology of national identity that surpassed and subsumed all other identities. The state separates the nation from others by well-guarded borders. Nation-states claim sovereignty and develop national institutions to manage the nation as a unified whole. Nation-states have their own versions of their national history and that of the world.

The rise of modern nation-states in Europe has evidently overshadowed the search and construction of a "universal" history and has pushed aside theological versions of history. It has also obliterated, underrated, or coopted ethnic histories. However, the modern state created a political system that eventually recognized the role of common individuals in government and economy, following the demise of divine kingship, the attack by "common" entrepreneurs (merchants, bankers, proto-industrialists) on the economic monopolies by King and Church.

As long as the state was operated and dominated by economic elite capable of propagating versions of history that legitimized their actions, the master [state] narrative of history was rarely disputed by those to whom the state provided a sense of security, pride, prosperity, and hope.

The integrity of the modern state and its historical narrative has been assailed by the wars among European "nations" and by repeated economic failures.

During the second half of the 20th century, the hegemony of European countries over their colonies receded in the wake of the escalation of independence movements. As the modern state had responded in the past to demands for better wages and living conditions of its labor masses, first by violence and then by reconciliation and novel methods of control, European countries since the 1960s have begun to reshape their international relations with what has become known as the "Third World" or "Developing nations." This turn in the international scene and the emergence of the "United Nations" as a response to the dilapidating effect of inter-European wars, which led to the rise of the U.S. as a principal international player, was coupled with the aggrandizement of the "services" market and the "electronic" revolution.

New Agents: Commoners and Consumers

In Europe, the advent of modern industry, mass production, the steam engine, and subsequent technological advances in industrial production, innovations, and transport was matched with the greater opportunities for social mobility, which significantly increased the number of potential consumers. This represents a major turn in the human condition and state organization. Ordinary individuals became recognized as agents in society. Their purchase power and their votes are coveted by the new elite (commercial–industrial–financial) who became a ghost state utilizing the government as their front. The transformation in Europe and in Japan entailed a switch from rural to urban with attenuation with the links with nature. The impact of the new sociopolitical changes and attendant technological–industrial developments has been unprecedented. As more food is produced and health conditions are ameliorating, the poor have tried to improve their conditions by having larger families, which has led to a demographic explosion and a perpetuation of poverty.

Beyond the Nation-State: Rewriting Histories

In recent decades, the world order has been reshaped from one dominated almost entirely by European "modern" industrial, nation-states (having succeeded in undermining the power of the Ottoman Empire) and marginalizing "older" European powers (such as Spain and Portugal), to a global network of power relations influenced by giant multinational corporations run by a cadre of business elite operating from "world" capitals of superpower nation-states with gigantic military apparatus even in peacetime.[3]

---

3 It is not possible to discuss the vast number of topics concerning nation-states and imperialism, which led to a global rift that sets the stage for our current predicament. The reader is referred to Stavrianos (1981) and Abernethy (2000). For a cogent discussion of corporations and world affairs, see Korten (1995).

Discontent engendered by the economic failures and wars, the weakening of the intellectual hegemony of the state as a consequence of the information revolution, and the erosion of the economic and political integrity of nation-states, either by multinational corporations or international monetary organizations, have emboldened those that at one time were either silenced or duped by the "integrative" ideology of the modern state. The result has been a resurgence of notions of "ethnic" and "religious" identities, and indeed the valorization of notions of "Identity" and "Diversity." European modern nation-states were party to a global wave of imperialism that spread, among other things, Western models of "civilization." Schopflin (2002, p. 141) concludes that empires fail when the imperial culture begins to lose its attraction. This happens when individuals of subjugated cultures conclude that they have less and less to gain from adherence to the imperial models of culture. Such individuals begin to search for anti-imperialist alternatives.

The shake-up of the integrity of the nation-state and the search for extra-national identities has led both to the revival and construction of ethnic identities, which are in some cases counternationalistic. Religion also became a viable ideology for those disenfranchised with the nation-state and a world order that leaves them impoverished, humiliated, "insignificant," or lost. Paradoxically, current counternationalist movements use the monolithic and dogmatic ideology of the modern state as their model (cf. Abernethy 2000, p. 330). The need to provide natives with a modicum of schooling to use them in colonial ventures has the side-effect of increasing the capacity of individuals to organize political movement (Abernethy 2000, p. 335), initially at the scale of their own colonies but later on a global scale. The movements were initially led by political leaders who eventually emerged as national heroes and heads of state. However, in the aftermath, as they struggled to deal with devastated economies, ruined political infrastructure, and great expectations from the masses, while, at the same time, becoming pawns in the postcolonial hidden or overt struggle for world dominance among the world "great" powers, the leaders of the ex-colonies opted, on occasion, to resort to repugnant oppressive strategies or were forced to relinquish power to war lords, religious zealots, or endemic civic disorder. It did not help that many of the leaders were from a military background (again created by or in imitation of Western powers) or paramilitary organizations mobilized to resist military occupations. Moreover, it did not help that those who were colonizers at the beginning of the 20th century chose to exploit the political turmoil in the "developing" countries to their advantages, and have not acted vigorously to rehabilitate the colonies so as to alleviate their economic situation. People of many nations that were once colonized have thus become disenchanted not only with Western hegemonic stances, but also with their own regimes.

It is also evident that the wars of liberation have inspired various political movements inside Western societies (including the U.S.), ranging from pacifist activities to violent confrontations (e.g., Brecher and Costello 1976; Hayden et al. 1966). The civil rights movement in the U.S. and later the "feminist" movements exposed some of the internal contradictions of Western societies. Many nations and disadvantaged peoples had high expectations by the end of WWI and following the Universal Declaration of Human Rights in 1948. Wallerstein (1997, p. 197) concluded that there was a world revolution in 1968 that centered on the theme of the false hopes generated by global liberalism and the "nefarious motives behind its program of rational reformism." He adds that "the self-contradictions of liberal ideology are total. If all humans have equal rights, and all peoples have equal rights, then we cannot maintain the kind of inegalitarian system that the capitalist world economy has been and always will be. But if this is openly admitted then the capitalist world economy will have no legitimacy in the eyes of the dangerous (i.e., the disposessed) classes. It will not survive."

Although I am not particularly keen on political rhetoric and the narrow focus on "capitalism," it is abundantly clear that the issue of "human rights" and rights-based movements have a global resonance because they transcend nationalist agendas and because they have been advocated by Western powers. Ignatieff (2000, p. 139) thinks that the modern legitimization of public institutions was based on a gamble consisting of being attentive to difference while treating *all* as equal. He considers this as a new gamble, conceived in the 17th century by the founding fathers of liberal political philosophy, like John Locke. Their original ideas were confined exclusively to white, proprieted males. However, once this ideal became an element in the propaganda of the state, the die was cast. Women, nonwhites, and all others who were once excluded from political life could no longer by excluded by right.

The problem today is that the postmodern relativism has eroded the basis for transcultural human rights. Many groups assume that it is their right to pursue their own agendas and ideologies regardless of how this influences the rights of individuals or other groups. The solution is not in the assertion of the "righteousness" of the ideologies of hegemonic nation-states, but in bolstering a common transnational ethical declaration that acknowledges common core human values of different cultures and religions.

The situation at present is highly volatile. The Internet encourages and promotes simultaneous behavior (including, surprisingly, the assembly of hundreds of nude individuals in chilly weather for art photography!). This is a new neural fabric that can engender unexpected nonlinear events, which begin with individuals and can build into explosions (Farrell 2000, p. 206). The behavior may be simply based on "mimesis" rather than any "real" causes. The historical transformation of individual behavior to social modalities, and subsequently normative values and structures in theory discussed above is now in peril. Freak nonlinear events can emerge unexpectedly within and through existing normative structures. These events may be totally based on rumors or imagined scenarios and can hasten the demise and collapse of organization or even perhaps the world order.

HISTORY AND THE PROSPECTS OF HUMAN–ENVIRONMENT SYSTEMS

The theory of history that I propose here provides a basis for considering the relationship between societies and their environments from an integrated perspective. Environmental events are perceived and processed through preexisting social norms, canons, and ideologies subject to the interests of elite who provide a master narrative of aesthetics, work, and attitudes to nature. In pre-state societies, interactions with nature are direct and immediate. In agrarian state societies, farmers respond to fluctuations in rainfall or floods on the basis of previous experiences within the span of living memory. However, some extreme events may be coded in oral traditions that may persist for millennia. When farming begins to depend on an urban population for technology, energy, cultigens, stock, and markets, the direct link with the local environment becomes attenuated. As a state society becomes increasingly complex, the link between environmental decision makers and environmental cues are further attenuated leading to policies that may not be in harmony with natural rhythms and ecological stability, and perhaps even not in conformity to traditional, resilient food-getting strategies. The industrial state is notorious in this regard. Not only is industry often in conflict with farming (e.g., conflict over water resources), it may also have a deleterious effect on the environment and agricultural productivity through the promotion of cash crops (i.e., crops produced strictly for profit rather than sustainability), wage labor, plantations, monocropping, use of fertilizers, and pesticides. Industrial pursuits may thus lead to water contamination, deforestation, and dramatic ecological effects that have long-term effects on the integrity of landscapes, biodiversity, and the resilience of the ecosystem.

Our current environmental crisis, and all those in the human past, are primarily a function of how people place themselves at the mercy of their environment. The advent of agriculture entailed a major transformation of how people changed their mode of life to one that entailed greater risk and vulnerability. Large, sedentary communities subsisting on a few staple crops became prone to long-term fluctuations in rainfall and floods. They also began to impact surrounding areas through intensive utilization of wildlife, wood for building purposes, tools, and fuels. The emergence of top-level "management" elite began to siphon a progressively increasing portion of agricultural production to procure luxury goods, patronize artisans, build monuments, and support functionaries who controlled local communities and ensured the flow of goods to the state center. This led to waves of agrarian intensification, extensification, and expansion of the labor force (thus beginning the slow, but eventually catastrophic, increase in world population). Agriculture, in many places in the world, thus fostered a state of society that sowed the seeds of progressively increasing population size, large localized communities, higher population density, higher rates of consumption, higher ratio of nonfood producers to food producers, expansion into marginal, fragile areas, and intensive impact on natural resources (through high rates of consumption and technological innovations from weeding to tractors). Three other developments attributable to agriculture are organized religions, urbanization, and systematic military warfare. Warfare as a means of appropriating more land and labor eventually reaches a limit where there is no more land to conquer within striking distance, and not only has a direct ecological cost, but also favors wanton exploitation of resources in fragile regions (e.g., Roman exploitation of desert resources) within a complex system of trade and exchanges). In addition, the accelerating demands by military personnel for payments, land and prestige items as well as the demands for armaments (from the chariot to bombers and from the iron spears to the nuclear bomb) propel the system toward a high state of vulnerability. Imperial extensification eventually leads to intensification using agrarian technologies and experimenting with new crops (again as in the Roman experience in Egypt and elsewhere with the introduction of new crops and a high-maintenance water technology). This creates a system that depends on imperial links, which is likely to collapse with the fall of the empire. Urbanization as the aggregation of relatively large numbers of political, religious, and craft personnel in administrative and power spatial modes eventually led to the expansion of urban centers beyond the local capabilities of their habitats (both in terms of water and food). The advent of state religions was perhaps primarily to legitimize the appropriation by certain individuals of extraordinary powers and to give reasons and explanations of the social order, thus fostering social cohesion. Some religions also provided an ideology that justified conquest and mastery of nature, elite dominance, and military conquests. Several religions were incubators for religio-ethnic identities and sectarian conflicts. Thus urban centers became linked with religious establishments (represented by cathedrals, mosques, and temples), administrative institutions (palaces), and military installations (forts). Our current environmental situation is now bonded to such urban agglomerations that were unimaginable in the early days of the state. The cancerous expansion of urbanization and explosive (and implosive) increase in cities has placed humanity today at unprecedented levels of risk. The recent aggrandizement of cities and their unstoppable spread was enabled by higher rates of agricultural production: mass, fast, transport and distribution of food products and other resources, and food processing. Although this system is potentially capable of serving humanity through judicious integration of food resources and distribution to ensure equity and harmony with ecological resilience, it has led instead to socially disruptive and ecologically disastrous consequences as certain urban centers in rich, industrial states use the world (mainly outside their borders) for ruinous short-term agricultural ventures, cash-cropping, deforestation, and intensive use of machinery, fossil fuel, pesticides, and fertilizers. This attack on the planet exceeds by far any previous historical instances.

From a scientific point of view, climate change is now a fact. How climate change is perceived, manipulated, and dealt with has, however, very little to do with science but rather much more to do with the ideologies and interests of powerful nation-states (which have fashioned a new world order through political and financial organizations and are now represented by self-appointed alliances). Scientists are in general sponsored through state funds, intergovernmental, international foundations and industrial ventures. They constitute, ideally, a segment of society concerned since the Industrial Revolution with the discovery of how nature and societies work for the service of industrial development, trade, and economic growth. Following the droughts in the 1970s and the realization that climate change can influence international security and prosperity, national and international agendas began to increase attention to climate change. First, attention was paid to verifying and assessing ongoing climatic changes. More recently focus has shifted to integrating climate change scenarios with social models of human responses to environmental change. Although this is a welcome addition to our overall understanding of climate change and its implications for our human future, well-reflected in the objectives of IHOPE, the political, ideological, and economic matrix of current climatological investigations requires critical reflection on the social acceptance and implementation of scientific conclusions and interpretations. The colossal economic implications of climatic change (e.g., in terms of costs related to shortages in agricultural production, industrial failures, urban plight, coastal submergence, and disruption of transport) make data and results of climatic change an extremely valuable political card. The potential impact of climate change to the livelihood of individuals in urban and rural settings all over the whole spectrum of different occupations (through the effect of climate change on drinking water, agricultural fields, pastures, and even tourism) affecting both human survival and lifestyles makes climate change a matter of personal concern. Climate change thus became a media subject because of its sensational potential. Mediascience, which links the scientific establishment and the public, can thus be easily manipulated by various social bodies depending on their particular concerns.

Mitigating factors against the manipulation of scientific data and interpretations by nationalist, industrial, sectarian, or partisan profiteers reside in the ideology of scientific inquiry and conduct. Scientific ideology emerged in contradistinction to magic, theology and ethnic narratives and epistemic traditions to valorize transcultural and transsubjective documentation of observable phenomena and modes of interpretation and verification (assessment of the provisional adequacy) with the proviso that all observations and conclusions are tentative and subject to revision and reevaluation. Professional scientists thus have emerged as the first cadre of individuals who have an allegiance to transcultural "truth" (no matter how tentative, uncertain, and nondogmatic). Greater reliance on scientists in industrial societies has given them a certain credibility which became the basis for the emergence of "experts" as a national and international phenomenon (Brint 1994). In a sense, the scientist and the expert (the latter coopted beyond the realm of science) have appropriated one of the primary functions of religious elite who in the past provided explanations of the natural and cultural world. In the past, this was both linked to state operations and hence to ideological objectives (note, e.g., the changes in the interpretations of Christianity by successive generations of theologians in tandem with social and economic changes in world over the last 2,000 years). Scientists, however, have given the "subjective" nature of ideology from linking their scientific activities with ideological objectives. This became a problematic issue when the explanations of how atoms work could be manipulated for mass destruction of humanity. Today the ethics of scientific investigations is a main domain of concern to many scientists. Climate change,with its enormous implications for international conflict and economic gain, requires critical reflections by scientists on how their results are manipulated and packaged by politicians, multinational corporations, and business communities. These groups often seek to aggrandize their own benefits with traditional blindness to either long-term consequences or negative implications for other parts of the planet (i.e., outside where they live). This lack of attention to scale, both temporal and spatial, should be avoided when ecological issues are addressed. Various temporal scales, as reflected in the organization of this Dahlem Workshop—from decadal to millennial—are essential for understanding both social and environmental issues informing the present by reference to microscale events that become woven from one generation to another in a historical tapestry that "evolves" or "emerges" through previous normative structures through social modalities under changing conditions and innovations to constitute trends and transgenerational structures. Long-term normative structures may be also compared and correlated with transregional scales that transcend spatial peculiarities at the local scale. The scientific strategy for an incisive understanding of the link between environment and societies would thus benefit not only from an understanding of how a managing elite make decisions on environmental issues, but also from a multi-scalar approach that views the world in terms of interrelated and interactive cultural landscapes with human agents who are differentiated by geography and occupation but unified by the recurrent need to cooperate for mutual benefits. As the domain of cooperative units expands to a global scale, new forms of organization and ideology are needed to ensure the adherence of the transglobal communities to common codes of behavior that ensures manifest benefits to all participants. An ideology of environmental and social ethics that valorizes trans-temporal resilience and "recharge" of natural resources (using groundwater aquifers for analogy) while concurrenty ensuring opportunities for ameliorating the life of those who have been disadvantaged in the recent past is, in my opinion, a primary prerequisite for any future human survival with or without the complications of external climatic oscillations. The elements of this ideology are now emerging, and they will be met with resistance and will not be readily accepted by those who are short-sighted or too selfish to realize that their current economic pursuits, steeped in previous historical normative structures, are ruinous for everyone in the long run. There will be no life boats in a sea of misery, inhumane behavior, and horrendous suicidal acts. Indeed, now, as in the past, there are high costs involved in the policing and control of the disenfranchised who fall prey to fanatical ideologies (again steeped in parochial nationalistic and religious zealotry) by those who seek to establish their self-worth through desperate acts of imagined glory and those who are in historical collision with secular regimes that undermine their sense of identity and support systems and who had no recourse or opportunities to join in benefiting of "modernity." In the struggle to maintain the integrity, resilience, and productivity of our global environment and given the limited resources available in the world, social problems that drain financial resources and demoralize people must be addressed. It would be wise, in fact, to mobilize those who feel outside the realm of hope to join in ameliorating the environment through simple means that can be rewarding both morally and financially.

The emergence of a global economy, privatization, and the valorization of individual ethos at the expense of social cohesion, consumerism coupled with the surge of urban expansion, the weakening of the nation-state, and the tide of religious zealotry are significant developments in a world reaching toward a new phase beyond nationalism as well as colonial and post-colonial hegemonies. Neither our knowledge of how climate changes nor how we can cope with the future of humankind under changing climatic conditions can be advanced if the social threats to our existence are not balanced by a new global ideology of equity and a commitment to environmental ethics. New technologies of communication, information processing, networking, and long-distance transport promise to be tools for managing our planet within an appropriate vision of the primary lesson from our human past. No society has ever managed to last under conditions of inequity and injustice. The historical game consisted of maintaining societies through unjust exploitation of others. As long as "others" remained faceless, ill-informed, and too poor or dispossessed to sustain resistance, it was possible for certain societies to persist for 200–300 years at the expense of a mass of "others." Since the liberation wars after WWII and the advent of mass literacy and middle-class segments in poor societies, coupled with the spread of transpiration and communication technologies, "others" are no longer ill-informed or too poor to mount effective resistance. We have passed a critical threshold. The threat of climatic catastrophes is only made worse by the prospect of social strife and the breakdown of the rules and norms that made violence the exclusive pejorative of the state and those that restricted human treatment to their own peoples with no regard for the humanity of others.

Human Prospects in a Changing World

Commenting on human prospects in 1980, Robert L. Heilbroner (1980) realized that the current tendencies in global consumption, pollution, population increase, and urbanization will create serious problems to industrial nations with massive urban complexes and giant corporations. He predicted that dangerous military despots will be tempted to create centralized hegemonic regimes blending "religious" orientation with "military" discipline.

The arms industry, with no loyalties except for profit, has enabled subnational and supranational groups to amass the means for clandestine and military actions. Any attempt to simply counteract such movement by military retaliation is counterproductive. In addition, the adoption of surveillance measures and suspension of the civil liberties of suspects represents a major encroachment on the rights of citizens granted under the charters of the modern nation-state.

This can only backfire, increasing the stronghold of religious fundamentalists in industrial nations on science, technology, and social institutions in a backward slide to the "Dark Ages." Moreover, such fundamentalism is likely to engender and fuel counterreligious movements within the state and abroad, thus preparing the ground for more violent confrontations.

This grim prospect is today almost a reality: military action appears as a swift and decisive means to promote certain political agendas, but it is no longer an option because the "enemy" is no more a geographically definable, accountable, nation-state. In addition, most ex-colonial nation-states in Africa, Latin America, and Asia are weakened by their own colonial inheritance as well as by the pressures placed on them by superpowers to be capable of developing satisfactory solutions to their internal problems. Although collective intervention by the United Nations in the internal affairs of member states has become a legitimate practice for coping with domestic conflicts, any attempt by the "super-powers" to undertake unilateral actions is likely to undermine the credibility of the United Nations and isolate the superpower as a rogue state, undermining its authority and the legitimacy of its action.

The ideologues of this new version of world order are keen on "re-writing" a new version of history based on the notion of the clash of civilizations and the creation of a menacing religious "Other." By adopting this construction of world history, the "superpowers" lay claim to "moral" right and, employing a strategy once used in the Soviet Union, paint the world in terms of an "Evil Other." By selecting and framing a new world terror, which in the process it nurtures and aggrandizes, the world is already teetering on the verge of chaos and anarchy.

Redirecting History

The current world situation is not sustainable. Poverty and inequality breed resistance and resentment. The powerful elite, as in the past, have to recourse to the use of an ideology of order, peace, and prosperity or/and to the use of military force. Both are costly endeavors and without a real payoff in the long run; the rhetoric of order, peace, and prosperity will not withstand the reality of poverty, hunger, and disease which are the real "terrors" that we must fight. Military occupation or control can never be a long-term solution because of its cost and deleterious ecological effects. Cruttwell (1995, p. 199) in his *History Out of Control* concludes that "a new world order will reject the Economic of Wonderland" which he believes to be the "cause of environmental destruction, resource depletion, and, at the same time, the main cause-and-effect of the population explosion."

From an archaeological–historical perspective, the ills of civilization and complex societies can only be mitigated when the elite curb their voracious appetite for consumption and the levying of higher and higher share of revenues. Cutting down the cost of governments, reducing military expenditure, subsidizing improvements in health, education, and environmental protection, and curbing sharp and destabilizing inequalities are issues for consideration. Creating a culture of hope instead of despair and an environment of intercultural dialogue and understanding rather than an emphasis on "identity" and "diversity" may also be more conducive to a more stable world. In the final analysis, perhaps nothing can save humanity but a return to the virtue of charity and an ideology of cosmic, all inclusive justice, order, and goodness.

From a pragmatic point of view, a change in the course of history depends on the concerted action of individuals who are willing to diagnose and predict the probable outcome of current policies, and those who are willing by their words and actions to mobilize collective responses to ruinous individual, social, national, multinational, and international actions. Those who live in "democratic" states have greater responsibilities than others who are burdened by authoritarian regimes and should be called upon to examine the outcome of their own actions and those of their governments. The professionals who are the link between policy makers and the people have yet a greater responsibility because of the potential impact of their work. To recall Cruttwell's (1995, p. 206) words again: "Totally new political machinery is now needed if the inherent tensions and incompatibilities between (social) democracy and the free private market economy are to be resolved so that a human culture can survive and prosper in the third millennium."

A policy that ignores the call of the weak and downtrodden, that is blind to misery and disease, and which lives on borrowed time must sooner or later suffer from internal moral callousness and duplicity as well as from desperate resistance by those who have no hope for a better future and whose lives are no longer worth living. It may be that the dignity they gain in desperate actions is perhaps their only salvation in a world that has turned a deaf ear to their suffering, or those with whom they identify.

Redirecting history will not be achieved without addressing the anxieties and psychological proclivities of individuals who are experiencing bewildering, agonizing dilemmas. The valorization of "Identity" in the West since the 1960s, and elsewhere thereafter, betrays the impact of changing sociopolitical conditions on perceptions of self. On the one hand, modern nation-states have emphasized identity as the fundamental organizing principle of self; on the other, this identification has been compromised by conflicts with a malleable, flexible schema of self where different affiliations (geographical, occupational, ethnic, linguistic, socioeconomic, religious, and ideological) provided a network of vectors. The self, as Ahdaf Soueif (2004) remarked, consisted in a country like Egypt in the 1960s of a common ground of different identifications. In addition, the sense of national identity has been more recently undermined by the weakening of the nation-state as the source of self-fulfilling schema as a result of increasing global connectedness and transcultural flow of information. The modern state has also undermined the self-assurance gained from community and religion without providing a viable sense of financial, social, or emotional security. Moreover, the commercial, industrial ethos of competitiveness has led to a breakdown of conventional norms of cooperation, mutual support, and caring, especially within the principal life-support social units, namely family and friends. This was excerpted by increasing spatial mobility, life in stressful urban settings, and juxtaposition and jostling of different lifestyles and values belonging to various segments of the population.

Diversity has now become a pretext for cultural insulations and a barrier against harmonizing dialogue across the boundaries of "ethnic" and "religious" "communities." It has fostered a new social fiction that knows no loyalty to the greater civic society. It is a fiction fed by half-truths and lies that has the capacity to spread not because of any rational context, but because of its own "emotive" appeal.

The insightful analysis by Rollo May (1967) is revealing in this regard and presents the often underestimated link between history and psychology. May's perspective on the historical roots of modern anxiety in the 20th century may be extended to the anxieties of the global era. He notes that if a society is in a state of confusion and traumatic change, the individual has no solid ground on which to meet the situations which confront him (cf. p. 70). Anxieties are aggravated if the situations are perceived to be threatening. By buffing up mortal threats from an enemy, modern state societies only add to the panic and acute anxiety of individuals. Individuals are thus likely to cling to simplifying dogmas (rigid schemata) and unquestionable "truths" to resolve their anxieties. This explains chauvinistic nationalism, neo-Nazi racialism, evangelical and other forms of religious fundamentalism. The latter is becoming particularly virulent because of the decline of nationalism. Religious fundamentalism is not a revival of the ethical guidelines of religious ideologies, but instead a new form of identity, following that of the modern nation-state, complete with all the aspects of the state technology of power from monuments (the building or restoration of mosques and churches), dress codes, rituals, historical discourse, and the use of violence. In fact, the intrusion of violence in the fundamentalist movements of otherwise pacifist religions is a significant indicator of the radical rather than "fundamental" aspects of such postmodern religious movements. The situation is worsened by the historical experience of the colonial encounters during the 19th and 20th centuries, when European colonial powers used (Christian) missionaries and guns to colonize countries with different religious persuasions. Regardless of the ideology of the secular modern-state, the European West, presented itself to the world in a religious garb, and elicited responses based on mobilization of religious sentiments, thus dragging religion into the nationalist discourse of conflicting, monolithic identities. Even after independence, the legacy of religious leaders who fought colonialism and the movements (often militant) they created still presents a force that has repeatedly contributed to agitation against their own nation-state or opposition to the global forces that marginalize them economically and culturally; in so doing they have undermined their sense of identity and inner security.

The perpetuation and consolidation of dominant and reactive schemata in the name of "the only true religion," "civilization" (vs. Barbarism), consumerism, or Utopian political ideologies is a sure recipe for more violence, bloodshed, economic disasters, and ecological catastrophes. I believe that such schemata are the medium and the outcome of modalities shaped by deep anxieties resulting from problems in defining "self" and accepting a self that does not measure up to the demands of ostentatious consumption, status, and (socioeconomic) achievement. Other schemata may deal with this psychological situation by withdrawal from society, annihilation of self, adoption of a "false" self, or indulgence in overwhelmingly self-oriented, solipsistic activities. This abdication of social engagements, except on the most superficial level, is likely to lead to the making of history by those who actively engage perpetuating, resisting, or redirecting current social modalities, schemata, and norms.

I believe that a new course of history will only be possible by (a) promoting practices and norms of psychological fulfillment through interpersonal, transcultural, and transnational affiliations and mutual belonging; (b) by cultivating the means by which we can gain satisfaction from the unbound gifts of nature (which we must conserve); and (c) by encouraging ourselves and the young to explore creativity, each in his/her own way, and to cherish the thought that together as human beings we have the power to remake history.

REFERENCES

Abernethy, D. 2000. The Dynamics of Global Dominance: European Overseas Empires 1415–1980. New Haven, CT: Yale Univ. Press.

Archer, M.S. 1982. Morphogenesis versus structuration: On combining structure and action. *Brit. J. Sociol.* **33**:455–483.

Archer, M.S. 1995. Realistic Social Theory: The Morphogenetic Approach. Cambridge: Cambridge Univ. Press.

Barnes, B. 1995. The Elements of Social Theory. Princeton, NJ: Princeton Univ. Press.

Becker, H.S. 1970. Sociological Work: Method and Substance. London: Allen Lane.

Brecher, J., and T. Costello. 1976. Common Sense for Hard Times. Boston: South End Press.

Brint, S. 1994. In an Age of Experts: The Challenging Role of Professionals in Politics and Public Life. Princeton, NJ: Princeton Univ. Press.

Cruttwell, P. 1995. History out of Control: Confronting Global Anarchy. Devon: Resurgence.

D'Andrade, R.G. 1981. The cultural part of cognition. *Cog. Sci.* **5**:179–195.

D'Andrade, R.G. 1995. The Development of Cognitive Anthropology. Cambridge: Cambridge Univ. Press.

Dark, K.R. 2000. The Waves of Time: Long-term Change and International Relations. London: Continuum.

Farrell, W. 2000. How Hits Happen: Forcasting Predictability in a Chaotic Market. New York: Textere.

Feinman, G.M., and J. Marcus, eds. 1998. Archaic States. Santa Fe: School of American Research Press.

Gebauer, G., and C. Wulf. 1995. Mimesis: Art-Culture-Society. Berkeley: Univ. of California Press.

Giddens, A. 1984. The Constitution of Society: Outline of the Theory of Structuration. Berkeley: Univ. of California Press.

Goffman, E. 1959. The Presentation of Self in Everyday Life. Garden City, NY: Doubleday.

Goffman, E. 1967. Interaction Ritual. Garden City, NY: Anchor.

Goffman, E. 1970. Strategic Interaction. Oxford: Blackwell.

Hassan, F.A. 1992. The ecological consequences of evolutionary cultural transformations: The case of Egypt and reflection on global issues. In: Nature and Humankind in the Age of Environmental Crisis, ed. S. Ito and Y. Yasuda, pp. 29–44. Intl. Symposium 6. Kyoto: Intl. Research Center for Japanese Studies.

Hassan, F.A. 1993. Population ecology and civilization in ancient Egypt. In: Historical Ecology, ed. C.L. Crumley, pp. 155–181. Santa Fe: School of American Research Press.

Hassan, F.A. 1995. Is civilization sustainable? An anthropological perspective illuminates its fragility. *Universe* **8**:28–31.

Hassan, F.A. 1997. Climate, famine and chaos: Nile floods and political disorder in early Egypt. In: Third Millennium BC Climate Change and Old World Collapse, ed. H. Nüzhet Dalfes, G. Kukla, and H. Weiss, pp.1–23. NATO ASI 149. Berlin: Springer.

Hassan, F.A. 2000. Environmental perception and human responses in history and prehistory. In: The Way the Wind Blows: Climate Change, History, and Human Action, ed. R.J. McIntosh, J.A. Tainter, R. McIntosh, and S. Keech McIntosh, pp. 121–140. New York: Columbia Univ. Press.

Hassan, F.A., ed. 2002. Droughts, Food and Culture: Ecological Change and Food Security in Africa's Later Prehistory. New York: Kluwer/Plenum.

Hayden, T., R. Coles, S. Carmichael et al. 1966. Thoughts of Young Radicals. New York: Pitman.

Heilbroner, R.L. 1980. An Inquiry into the Human Prospect. 2d ed. New York: Norton.

Ignatieff, M. 2000. The Rights Revolution. Toronto: Anansi.

Korten, D.C. 1995. When Corporations Rule the World. London: Earthscan.

Maisels, C.K. 1999. Early Civilizations of the Old World. London: Routledge.

May, R. 1967. Psychology and the Human Dilemma. New York: Van Norstrand.

Renfrew, C. 1986. Introduction: Peer polity interaction and socio-political change. In: Peer Polity Interaction and Socio-Political Change, ed. C. Renfrew and J.F. Cherry, pp. 1–18. Cambridge: Cambridge Univ. Press.

Schopflin, G. 2002. Nations, Identity, Power: The New Politics of Europe. London: Hurst.

Soueif, A. 2004. Mezzaterra: Fragments from the Common Ground. London: Bloomsbury.

Stavrianos, L.S. 1981. Global Rift: The Third World Comes of Age. New York: Marrow.

Wallerstein, I. 1997. Insurmountable contradictions of liberalism: Human rights and the rights of peoples in the geoculture of the modern world-system. In: Nations, Identities, Cultures, ed. V.Y. Mudimbe, pp. 181–198. Durham, NC: Duke Univ. Press.

12

Little Ice Age-type Impacts and the Mitigation of Social Vulnerability to Climate in the Swiss Canton of Bern prior to 1800

Christian Pfister

Section of Economic, Social, and Environmental History, Institute of History, University of Bern, 3000 Bern 9, Switzerland

ABSTRACT

This chapter relates large-scale patterns of climate in central Europe to regional food vulnerability in the Swiss canton of Bern, taking into consideration the adaptive strategies of peasant farmers and the buffering strategies of the authorities. First, a short survey of the main properties of climate during the Little Ice Age is provided. From the 14th to the late 19th centuries, cold winters were more severe and more frequent; some years passed without a summer season as a consequence of enormous volcanic eruptions in the tropics. It is argued that changes in average temperature and precipitation are not adequate measures to assess impacts of climate on human societies.

Next an alternative approach is presented in which an attempt is made to set up impact models from the known properties of agro-ecosystems to assess biophysical impacts. Such a model was worked out for the area of the northern Alpine foothills and compared to a long series of grain prices for Munich in southern Germany. Results showed that the coincidence of cold springs with rainy and cool mid-summers was simultaneously related to peaks of grain prices and major advances of glaciers. Extensive measures of poor relief were taken by the cantonal authorities in Bern during the 18th century. From the example of the subsistence crisis in 1770/1771 it is demonstrated that a great number of people became permanently dependent on welfare. However, demographic effects of the crisis were virtually absent. More case studies of this kind are needed to explore the plurality of human responses in mitigating social vulnerability to climate variability.

PROPERTIES OF LITTLE ICE AGE CLIMATES

The Little Ice Age (LIA) is the most recent period when glaciers maintained an expanded position on most parts of the globe, as their fronts oscillated in advanced positions (Grove 2001). The LIA was a simultaneous, worldwide phenomenon; however, there was considerable regional and local variation.

In the Alps, three phases of maximum extension can be distinguished: the first occurred around 1385, the second in the mid-17th century, and the third around 1860 (Holzhauser 2002). Heinz Wanner (2000) coined the term of "Little Ice Age-type events" (LIATE) to designate the three far-reaching advances known from the last millennium. Each of the three LIATE was the outcome of a specific combination of seasonal patterns of temperature and precipitation (Luterbacher 2000). No single, long-term climatic trend supports the advanced position of glaciers during the LIA; rather, a multitude of interacting seasonal patterns of temperature and precipitation affected—either positively or negatively—the mass balance of glaciers. Extended cold spells during the winter half-year (October through March or April) characterized the climate throughout the LIA. Winter conditions were more frequent and severe—both in terms of duration and temperature—during the LIA compared to the Medieval Warm Period and the 20th century. However, the cold and dryness of winters did not significantly affect the mass balance of glaciers.

Far-reaching glacial advances occurred when very cold springs and autumns coincided with chilly and wet mid-summers. The last "year without a summer" occurred in 1816, but several others are documented to have occurred during the last millennium. They were the crucial elements underlying the LIATE. Most, if not all, were triggered by volcanic eruptions in the tropics, which generated a global veil of volcanic dust (Harington 1992). The spatial dimension of years without a summer was usually limited to mainland Europe north of the Alps, stretching from the Parisian Basin in the West to the Russian border in the East. Conditions in western France, Ireland, Iceland, and Russia were usually better, whereas those in the Mediterranean were fundamentally different (Luterbacher et al. 2002, 2004). The effect of years without a summer were counterbalanced periodically by clusters of warm and dry summers (e.g., in the 1720s) which caused a melt-back onto the glaciers.

Most historians became acquainted with the history of climate through the groundbreaking work of the French historian Emmanuel Le Roy Ladurie (1972). A student of Fernand Braudel (1902–1985), Le Roy Ladurie wrote his *History of Climate* according to the Braudelian scheme of historical temporalities. It is well known that Braudel defined three levels of history (*res gestae*), which were both chronological and operational. The superficial level is one of short-term historical events and individuals; the middle level comprises conjunctures (cyclical phenomena) that occur over medium-length timescales; the basal level consists of long-lasting structures. The short-term, rapidly changing levels of historical events, chance occurrences, and individual men and women comprise what Braudel viewed as the traditional approach to history, and it was against this that he reacted. He played down its importance, viewing events and individuals as the "ephemera" or "trivia" of the past. Among the temporalities of very long duration, Braudel (1972/1995) mentioned changes in climate.

Le Roy Ladurie's classical historiographical concept of the LIA fits into the Braudelian scheme of "long duration." From evidence based on glaciers and from vine harvest dates, Le Roy Ladurie concluded that all seasons more or less underwent a synchronous cooling at the end of the 16th century. Likewise he assumed that the warming from the late 19th century was more or less synchronous in all seasons (Le Roy Ladurie 1972). His book leaves the impression that there was a distinct LIA climate, which was predominantly cold and rainy. Consequently, Le Roy Ladurie was looking for human impacts of the hypothesized changes in long-term average climate. He concluded that "*in the long term* (my emphasis) the human consequences of climate seem to be *slight, perhaps negligible, and certainly difficult to detect*" (Le Roy Ladurie 1972, p. 119). For several decades this claim, made by an influential pioneer, served as a key argument to shun attempts to assess the human significance of past climate change.

Although the macrohistory of climate aims at reconstructing temperature and precipitation for the period prior to the creation of national meteorological networks, both in terms of time series and spatial patterns, a history of climate—tailored to the needs of the historian—should highlight *changes in the frequency and severity* of those extreme events that are known to have *affected everyday life* in the Early Modern period. Such extreme events include climate anomalies (e.g., droughts, long spells of rain, cold waves, untimely snowfalls) and natural disasters (mostly floods). They are usually coined in terms of weather histories, which address processes that have a duration from hours to seasons. In particular, weather is an overarching component in peasant memory because of its fundamental role in daily life (Münch 1992).

These two kinds of accounts on climate are difficult to reconcile. Reports on the microlevel focus on single climate anomalies and are close to the sources. They reveal the ways in which extreme events affected humans and their decision making. However, such episodes are too fragmented to be integrated into narratives of climatic change over long periods of time.

Histories of climate, however, supply impressive overviews of climate change without providing conclusive links to human history. Over several decades, differences in average temperature and precipitation are not convincing in this respect. They encounter the argument that in such situations people may adapt their way of living to a changing climate. Innovations may become accepted that are better suited to the new situation, whereas older outdated practices may tacitly disappear (Wigley et al. 1985). Nonetheless, climate history on the macrolevel can offer an interpretative framework in which significance may be attached to individual climate anomalies. It is important for human perception and interpretation—as well as for the measures being taken—to understand whether such episodes occur surprisingly after a pause of several decades, whether they remain isolated outliers, or whether they occur repeatedly.

According to Le Roy Ladurie, a conclusive investigation of climatic impacts should involve two steps. First, climate in the pre-instrumental period should be studied for its own sake, that is, separately from its impacts on the human world.

Second, the climatic evidence obtained should be used to explore the impacts of climatic variations on crops, food prices, demographic growth, and social disarray. Le Roy Ladurie suggested that a picture of climate without humankind in the pre-instrumental period might be reconstructed from data describing the meteorological nature of certain years, seasons, months, and days (i.e., from long series of documentary proxy data). The ultimate goal of such a reconstruction should be to set up a series of continuous, quantitative, and homogeneous climate indicators (Le Roy Ladurie 1972).

Historical climatologists acted upon Le Roy Ladurie's suggestion inasmuch as an approach to quantify qualitative observations in a more or less standardized way, was developed from the late 1960s. This consists of deducing continuous, quantitative, and *quasi*-homogeneous time series of intensity indices for temperature and precipitation from documentary data used as substitutes for instrumental measurements. In most cases, a reconstruction involves different kinds of documentary data supplemented by high-resolution natural proxy data. Thus far, such series have been set up for Germany, Switzerland, the Low Countries, the Czech Republic, Hungary, Andalusia, Portugal, and Greece (Brazdil et al. 2005).

A CASCADE OF EFFECTS

When series of continuous, quantitative, and quasi-homogeneous climatic indicators are set up for the pre-instrumental period, such series may be used to construct models that enable the exploration of the impacts of climatic variations upon economies and societies. The effects of climatic fluctuations on the "course of history," for instance, are difficult to demonstrate because most of the factors include many internal mechanisms that compensate for adverse climatic effects (Berger 2002).

It is frequently overlooked that both "climate" and "history" are blanket terms, situated on such a high level of abstraction, such that relationships between them cannot be investigated in a meaningful way according to accepted rules of scientific methodology. On a very general level, it could be said that beneficial climatic effects tend to enlarge the scope of human action, whereas climatic shocks tend to restrict it. Which sequences of climatic situations really matter depends upon the impacted unit and the environmental, cultural, and historical context (Pfister 2001). However, this statement needs to be restricted in the sense that the term "climatic shock" itself is ambiguous, as it is well known that some of the people and groups involved always take advantage of situations of general distress, both economically and politically.

To become more meaningful, "climate and history," as a collective issue, need to be broken down to lower scales of analysis, with a specific focus, for example, on food, health, or energy systems or on specific activities such as transportation, communications, and military or naval operations. Particular emphasis must also be given to short- and medium-term events. Moreover, concepts need to be worked out to disentangle the severity of climate impacts and the efficiency of measures for coping with them. The closer details are investigated, the higher the probability will be of finding significant coherences (Roy 1982).

Regardless of which event or impact is to be studied, an impacted group, activity, or area exposed to these events must be selected. In general, the focus is on individuals, populations, or activities in the form of livelihoods or regional ecotypes. Most examples provided in this chapter refer to the Swiss canton of Bern, which in 1760 comprised 8591 km^2 and had about 350,000 inhabitants. As of 1764, its statistical coverage is excellent (Pfister and Egli 1998). The canton may be roughly divided into three different ecozones. In the Lowlands (by Swiss standards), grain grown within the three-field system was the dominant crop. In the hilly zone, grain cultivation and dairy production had about an equal share. In the Alpine zone, dairy production was dominant; grain cultivation played a marginal role on a microscale.

The most difficult choices of study elements are those of impacts and consequences. Biophysical impact studies may help identify how climate anomalies affected crops, domestic animals, and disease vectors through a study of their climatic sensitivity. Social impact assessment studies can then examine how biophysical impacts (i.e., effects of climate anomalies upon biota) are propagated into the social and political systems.

Robert W. Kates (1985) suggested that such studies, in a first run, might be arranged in the order of propagation to events, although this arrangement may be arbitrary in the sense that the real-time process actually takes place simultaneously or that the sequence is unknown to climatic processes. Figure 12.1 depicts a simplified version of this approach, showing a cascade of effects for preindustrial societies.

Biophysical impacts focus on the production of food (e.g., yields per hectare, relation of seeded to harvested grains) and its availability for human production (including loss through storage). Economic impacts consider consequences on prices of food, animal feed, and firewood. Grain prices were by far the most important parameters for business activity; they also constitute the only economic data for which continuous series are widely available in Europe. Demographic and social impacts highlight consequences of subsistence crises such as malnutrition, social disruption, and food migration. A subsistence crisis is an integrated process in which nature and society interact. Its severity, however measured, depends on the magnitude of the biophysical impact as well as on the preparedness of the people involved and on the efficiency of the measures and strategies that are taken to address the crisis. The significance of human intervention in the process increased from top to bottom at the expense of climate impacts. There were few options available to dampen biophysical impacts, whereas economic measures and social assistance could considerably reduce social disruption. Of course, interactive models, including the societal responses to biophysical and economic impacts in terms of positive or negative

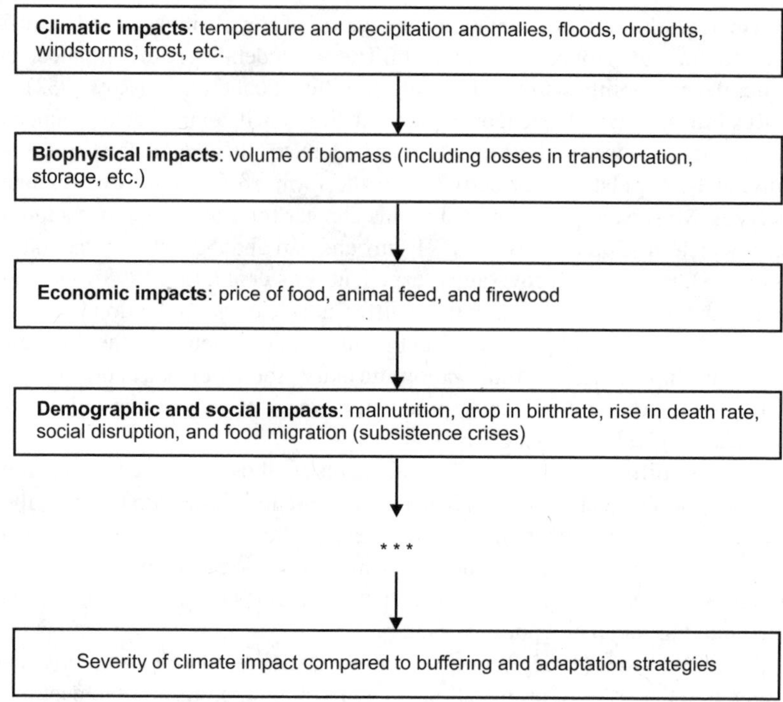

Figure 12.1 A basic model of climate impacts on society (modified after Kates 1985).

feedback, would be more realistic than linear models. Kates (1985, p. 14), however, correctly observes that it is easier to draw schematics than to describe what actually occurs.

ASSESSING AND MODELING LITTLE ICE AGE-TYPE IMPACTS

In western and central Europe, two kinds of impacts were detrimental for agriculture: cold springs and wet summers, where long wet spells during the harvest period prompted the most devastating impact. Continuous rains lowered the flour content of the grains and made them vulnerable to mold infections and attacks of grain weevil (*Sitophilus granarius*) (Kaplan 1976). During winter storage, huge losses were caused by insects and fungi, which increased grain prices in the subsequent spring. These effects can hardly be assessed today, let alone under the conditions of an Early Modern economy. In addition to long spells of rain in midsummer, cold springs harmed grain crops. From present-day agro-meteorological analyses it is known that grain yields depend on sufficient warmth and moisture in April (Hanus and Aimiller 1978). This implies that crops suffered from dry and cold springs, which were frequent during the LIA.

Extended snow cover was particularly harmful. When snow cover lasted for several months, until March or April, winter grains were attacked by the fungus *fusarium nivale*. Peasant farmers often ploughed the choked plants under and seeded spring grains to compensate for the lost crop.

To face the growing frequency of wet summers in the late 16th century, peasants changed their crop mix by growing more spelt instead of rye, since spelt ears bend and thus permit rainwater to drain off easily. Spelt is also moisture-protected by a sheath. In addition, in the Alpine area, production of hard (as opposed to soft) cheese spread rapidly during the late 16th century. This was perhaps an adaptive strategy, since hard cheeses can be stored for several years; thus they were better suited to bridge multiple bad summers in the 16th century, when cheese production was marginal (Pfister 1984). Whether these novel practices were indeed related to the changing climate and not to market incentives, however, remains to be confirmed.

In terms of the vulnerability of the main sources of food, based on present and historical knowledge, it appears that a given set of specific weather sequences over the agricultural year was likely to affect simultaneously all sources of food, leaving little margin for substitution. Table 12.1 depicts the properties of a worse-case crop failure and, inversely, of a year of plenty (Pfister 1984, 2005). It also summarizes the impact of adverse temperature and precipitation patterns on grain, dairy forage, and vine production during critical periods of the year under the following weather patterns: Cold periods in March and April lowered the volumes of the grain harvest and dairy forage production. Wet mid-summers (July–August) affected all sources of food production. Throughout September and October, cold spells lowered the sugar content of wine, and wet spells reduced the amount of grain sown, and lowered the nitrogen content of the soil. Most importantly, the simultaneous occurrence of rainy autumns with cold springs and wet midsummers in subsequent years had a cumulative impact on agricultural production. This same combination of seasonal patterns largely contributed to trigger far-reaching advances of glaciers. The economically adverse combination of climatic patterns is labeled Little Ice Age-type impacts

Table 12.1 Weather-related impacts affecting the agricultural production of traditional temperate–climate agriculture in Europe. *Italicized* text indicates weather conditions that affected the volume of harvests or animal production. **Boldface** text denotes weather conditions that affected the quality (i.e., the content in nutrients or sugar) of crops.

Critical months	Agricultural products		
	Grain	Dairy Forage	Vine
September–October	*Wet*	*Cold*	**Cold and wet**
March–April	*Cold*	*Cold*	*(Late frost)*
July–August	**Wet**	**Wet**	*Cold and Wet*

(LIATIMP) (Pfister 2005). Such biophysical impacts need to be understood as ideal types in the sense of Max Weber. They are heuristic tools against which a given body of evidence can be compared (Kalberg 1994).

The concept of LIATIMP provides a way to measure the severity of biophysical impacts. Compared to other parameters (e.g., grain prices and demographic data), it allows a direct assessment of the vulnerability of affected groups or societies to climate. The term of *vulnerability* represents a multitude of factors such as social stratification, the availability of substitute foods, the efficiency of provisioning buffers (e.g., private and public grain stores) as well as measures of poor relief.

Let us now consider the properties of a model used to simulate LIATIMP (Figure 12.2). The model is solely based on monthly temperature and precipitation indices. Biophysical climate impact factors (BCIF) were defined from the effect of weather on crops known from both contemporary reports and present-day scientific knowledge (Table 12.1). Grain prices were used as an indicator for socioeconomic impacts (i.e., subsistence crises). Since long-term multi-secular grain price series are not available for Switzerland, a series of rye prices from neighboring Bavaria were applied as a substitute. The weighing of terms in the equation was empirically done through fitting the BCIFs to the curve of grain prices (see Appendix 12.1).

Considering the graphical representation of BCIF from 1550 (not shown), it turned out that several breaks (1566/1567, 1629/1630, 1817/1818, 1843/1844) were statistically significant (see Appendix 12.1); multidecadal periods of low and high biophysical climate impact levels can be distinguished. This result challenges the view widely held by economic historians: Wilhelm Abel (1972, p. 35) assumed that climatic impacts on the economy need to be understood as a series of random shocks. Likewise, Karl-Gunnar Persson (1999, p. 98) does not envisage possible changes in climate over time. "That the price fluctuations… were triggered by output shocks is too obvious to dispute," he declared. Nobel laureate Robert Fogel (1992) even tried to disprove the existence of any relationship between climatic extremes and famine (i.e., between agriculture and climate) through questionable statistical manipulations. He simply claimed: "Famines were caused not by natural disasters but by dramatic redistributions of entitlements to grain" (see Landsteiner 2005, p. 99). James Jarraud, Secretary-General of the World Meteorological Organization (WMO), came to a somewhat different assessment. With a view to the present and the future, he wrote: "Climate variability affects all economic sectors, but agriculture and forestry sectors are perhaps the most vulnerable and sensitive activities to such fluctuations" (Jarraud 2005, p. 5). In addition, for historians who focus on people's perceptions and the established facts found in the sources instead of on equilibrium models and numerical data, climate variability mattered for the preindustrial economies. Its significance, however, cannot be simply implied. It needs to be established from case to case.

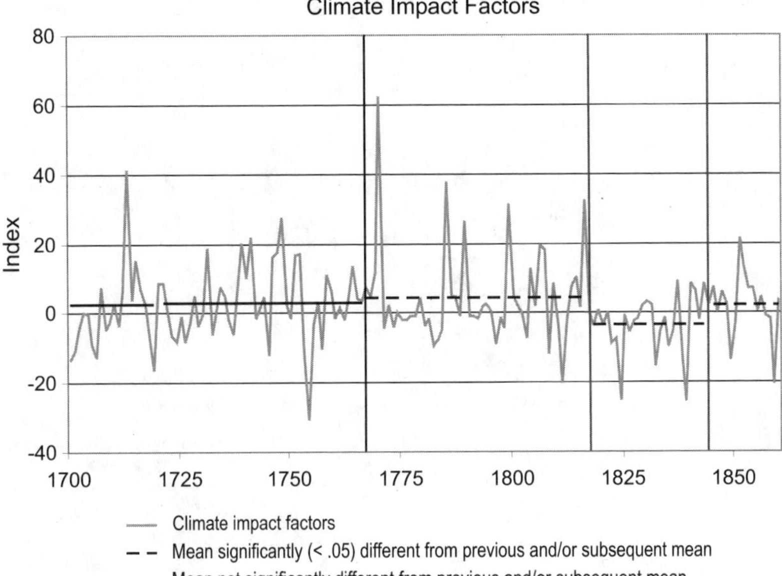

Figure 12.2 Biophysical climate impact factors (BCIF) in central Europe from 1700–1900. The well-known peaks of grain prices (around 1714, 1740, 1770, 1817, 1853–1855) clearly emerge.

The spatial dimension of a crisis is another important aspect (see below). Which regions in Europe were affected by LIATIMP, and which ones escaped?

Figure 12.3 compares the lowest and highest mean annual grain prices between 1760 and 1774 for 29 towns in Europe. The highest prices within this period are given as a percentage of the lowest, which are set to 100. The following characteristics are worth mentioning:

- North of the Alps, a clear West–East gradient stands out from the coast of the English Channel and the North Sea to the foothills of the Alps and the Carpathians.
- Whereas grain prices in Antwerp rose by 60%, in London by 70%, and in Paris by 110%, they more than tripled in Vienna and in Lwow (Poland).
- Price levels were generally lower south of the Alps with minima in Rome and Naples, where prices throughout the period remained almost at the same level (Abel 1972).

The zone of very high prices roughly coincides with the spatial dimension of the low pressure areas in the summers 1769 to 1771. People in areas close to the North Sea were less affected by the crisis because grain could be shipped there at low cost from surplus areas in the Baltic. The Mediterranean area was not affected by this kind of adverse weather; subsistence crises there were usually

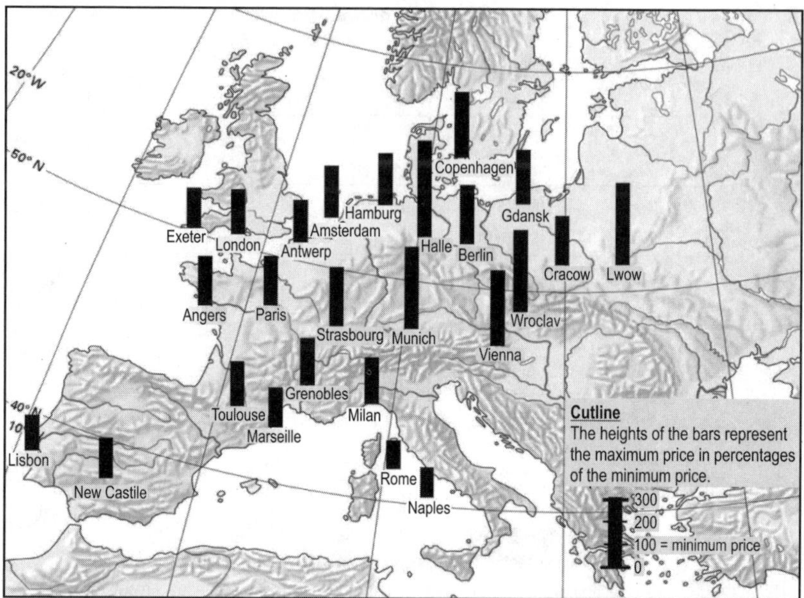

Figure 12.3 The rise of mean annual grain prices in Europe from 1760 to 1774. The highest prices within this period are given as a percentage of the lowest, which are set to 100. North of the Alps, a clear West–East gradient stands out from the coast of the English Channel and the North Sea to the foothills of the Alps and the Carpathians. Based on Abel (1972, p. 47).

caused by extended drought in the winter half-year (Xoplaki et al. 2001). Thus, the spatial dimension of climatic vulnerability depended to a considerable extent on the possibility of transporting imported grain at low cost on waterways. Mountainous areas in the interior of continents were particularly disadvantaged in this respect. On a local scale, price surges could still be more pronounced. In some areas of the Erzgebirge (Ore Mountains), which is the hilly borderland between the Czech Republic and Saxony, the price of a bushel of rye rose tenfold between spring 1770 and early summer 1772. To conclude from the report of a local parson, this led to outright starvation (Abel 1972).

MEASURES OF THE AUTHORITIES TO REDUCE VULNERABILITY IN THE CANTON OF BERN

Let us now turn to the crucial issue of social vulnerability. During subsistence crises, the ownership or disposal of agricultural land determined the availability of food resources. Wealthy peasant farmers and major landowners made disproportionately large profits, whereas the landless, cottagers, or peasants with only medium-sized holdings (i.e., about 80% of the population in the canton of Bern) spent a much larger proportion of their budget on food.

Outside of wars, the management of subsistence crises was among the most serious challenge faced by Early Modern authorities. Grain harvest shortfalls led to higher food prices, mounting unemployment rates, and an increase in the scale of begging, vagrancy, crime, and social disorder. Such conditions inevitably resulted in a welfare crisis of varying magnitude.

Environmental stress, economic hardship, and social disarray fostered overcrowding and other changes in normal community spacing arrangements. These conditions often appeared in the form of mortality peaks (Post 1990). Authorities took an interest in combating the effects of crises, particularly following the Thirty Years War, as they sought to increase the number of soldiers as well as tax income, both of which were dependent on the number of productive workers.

On a macroscale, the extended areas of high grain prices mask enormous differences of people's vulnerability on a microscale. On the local level, other factors need to be considered, such as the kind of crops grown, the degree of social inequality, and the efficiency of relief measures taken by the authorities. Generalizations are thus not possible.

Short-term measures usually included the symbolic persecution of hoarders and speculators, who were made responsible for the crisis, as well as the distribution of grain to the needy people in the capital. On a longer term, as of the late 17th century, a regional network of grain stores was implemented in the canton of Bern. Brandenberger (2004) concludes that the Bernese authorities shifted the focus of crisis management toward complementing traditional short-term measures with sustained efforts to promote agricultural productivity by facilitating the legal conditions for subdividing and privatizing communal pastures.

Poor relief was another strategy of reducing vulnerability: Bernese authorities oscillated between spending on welfare and taking economy measures. During the 18th century they established an area-wide relief system for the poor in which all communities had to participate. Erika Flückiger-Strebel (2002) demonstrated that in the wake of the crisis of 1770/1771, a great number of people became permanently dependent on welfare. Obviously, many working poor who were at the margin of impoverishment were not able to buffer the shock from the fall in real income during the crisis. They needed to sell most of their belongings for food and were henceforth dependent on continual assistance. The cost of social security rose much faster than any other entry of the budget, and communities had to bear a rising share of the burden.

How successful was this paternalist social policy? The severity of the crisis in demographic terms was assessed from the aggregate number of baptisms and burials, which is available for every parish since 1730 (Pfister 1995). From this data it can be concluded that both indications of excess mortality and a pronounced deficit of baptisms—which would point to some nutritional problem—were nearly absent. This suggests that vulnerability was substantially reduced. However, to put this argument on a firm basis, the situation in the canton of Bern should be compared with conditions in neighboring areas, where it is hypothesized that the crisis management was less efficient.

CONCLUSION

Environmental history addresses the role and place of nature in human life. It studies the interactions that past societies have had with the nonhuman world. Most scholars primarily understand this paraphrasing of the field by Donald Worster (1993) as being an account of humans transforming their natural surroundings and encroaching on them. Few, however, look at the other side of the issue and ask about nature's role as an independent agent in history. One reason may be that such a position runs the danger of being stigmatized as determinist. Another could be the current low standing of quantitative and ecological approaches within mainstream history, which unilateraly favors microstories and anthropological approaches. Indeed, results from many early quantitative studies were rather trivial, but this was mostly due to insufficient data.

In this chapter I have attempted to view socioenvironmental interactions at different temporal and spatial scales to demonstrate their complexity and to explore a new modeling approach: macroscale reconstructions of monthly air pressure and temperature for the entire European continent provide the starting point. These were obtained from a variety of high-resolution documentary and natural data. During the LIA, climate impacts never affected Europe as a whole; they were limited to certain climatic zones of the continent. Impacts in western and central Europe, for example, did not extend to the Mediterranean. Furthermore, spatial differentiations of vulnerability within the impacted areas need to be made according to ecozones and for the period prior to the construction of the rail network and access to cheap maritime transport. During a crisis, grain prices increased two to three times higher in the landlocked areas of the continent than in regions close to the sea. In terms of the temporal dimension, the modeling of biophysical impacts suggests that the frequency and severity of climate impacts changed over time, thereby leading to multidecadal periods of favorable and adverse climate.

The story of human vulnerability to climate told along a chain of causation, which runs from natural forcing to economics to the level of political and social decision making, requires a change from the macroscale of generalizations to the microscale of case studies. From the case of the canton of Bern during the crisis in the early 1770s, we can learn about the motivations and strategies of an enlightened elite to mitigate the stress of subsistence crises for the lower strata in the final decades of the Ancient Regime. Drawing on the available wealth of demographic data for this canton, we may even speculate about the efficiency of this early social policy in terms of avoided demographic losses. However, at this level it is unthinkable to arrive at more general conclusions for Europe or even parts of it. Even if many more case studies could be made, a comparative approach might well come to the conclusion that the differences between case studies are larger than their common characteristics. More studies of socioenvironmental interactions should be encouraged, but not with the intent of arriving at a universal picture of social vulnerability to climate impacts, as has been repeatedly attempted for global climate change over the last millennium. Quite the contrary, it would be worthwhile to illustrate the plurality of human responses and solutions in mitigating social vulnerability to climate variability.

ACKNOWLEDGMENTS

This work was supported by the Swiss NCCR Climate Program. Paul-Anthon Nielson is acknowledged for carefully reading the manuscript. Jonas Steinmann performed the computing work.

REFERENCES

Abel, W. 1972. Massenarmut und Hungerkrisen im vorindustriellen Deutschland. Göttingen: Vandenhoeck und Ruprecht.

Berger, W.H. 2002. Climate history and the great geophysical experiment. In: Climate Development and History of the North Atlantic Realm, ed. G. Wefer, W.H. Berger, K.-E. Behre, and E. Jansen, pp. 1–16. Berlin: Springer.

Brandenberger, A. 2004. Ausbruch aus der "Malthusianischen Falle": Versorgungslage und Wirtschaftsentwicklung im Staate Bern 1755–1797. Bern: Peter Lang.

Braudel, F. 1972/1995. The Mediterranean and the Mediterranean World in the Age of Philippe II, transl. S. Reynolds (1972). London: Fontana.

Brazdil, R., C. Pfister, H. Wanner, H. von Storch, and J. Luterbacher. 2005. Historical climatology in Europe: The state of the art. *Clim. Change* **70**:363–430.

Flückiger-Strebel, E. 2002. Zwischen Wohlfahrt und Staatsökonomie. Bern: Chronos.

Grove, J.M. 2001. The initiation of the "Little Ice Age" in regions round the North Atlantic. *Clim. Change* **49**:53–82.

Hanus, J., and O. Aimiller. 1978. Ertragsvorhersage aus Witterungsdaten. *Z. Acker- und Pflanzenbau* **5(Suppl.)**:118–124.

Harington, C.R., ed. 1992. The Year without a Summer: World Climate in 1816. Ottawa: Canadian Museum of Nature.

Holzhauser, H. 2002. Dendrochronologische Auswertung fossiler Hölzer zur Rekonstruktion der nacheiszeitlichen Gletschergeschichte. *Schweiz. Z. Forstwesen* **153**: 17–27.

Jarraud, M. 2005. Foreword. Increasing climatic variability and change: Reducing the vulnerability of agriculture and forestry. *Clim. Change* (Spec. Iss.) **1–2**:5–7.

Kalberg, S. 1994. Max Weber's Comparative-Historical Sociology. Cambridge: Polity.

Kaplan, S.L. 1976. Bread, Politics and Political Economy in the Reign of Louis XV, vol. 1. The Hague: Nijhoff.

Kates, R.W. 1985. The interaction of climate and society. In: SCOPE 27: Climate Impact Assessment. Studies of the Interaction of Climate and Society, ed. R.W. Kates, J. Ausubel, and M. Berberian, pp. 3–36. Chichester: Wiley.

Landsteiner, E. 2005. Wenig Brot und saurer Wein.: Kontinuität und Wandel in der zentraleuropäischen Ernährungskultur im letzten Drittel des 16. Jahrhunderts. In: Cultural Consequences of the "Little Ice Age," ed. W. Behringer, H. Lehmann, and C. Pfister, pp. 87–147. Göttingen: Vandenhoeck.

Le Roy Ladurie, E. 1972. Times of Feast, Times of Famine: A History of Climate since the Year 1000, transl. B. Bray. London: Allen and Unwin.

Luterbacher, J. 2000. Die Kleine Eiszeit (Little Ice Age, A.D. 1300–1900). In: Klimawandel im Schweizer Alpenraum, ed. H. Wanner et al., pp. 101–102. Zurich: vdf Hochschulverlag.

Luterbacher, J., D. Dietrich, E. Xoplaki, M. Grosjean, and H. Wanner. 2004. European seasonal and annual temperature variability, trends and extremes since 1500. *Science* **303**:1499–1503.

Luterbacher, J., E. Xoplaki, D. Dietrich et al. 2002. Reconstruction of sea level pressure fields over the Eastern North Atlantic and Europe back to 1500. *Clim. Dyn.* **18**:545–561.

Münch, P. 1992. Lebensformen in der Frühen Neuzeit. Frankfurt: Ullstein.

Persson, K.G. 1999. Grain Markets in Europe, 1500–1990: Integration and Deregulation. Cambridge: Cambridge Univ. Press.

Pfister, C. 1984. Das Klima der Schweiz von 1525 bis 1860 und seine Bedeutung in der Geschichte von Bevölkerung und Landwirtschaft. 2 vols. Academica Helvetica 5. Bern: Haupt.

Pfister, C. 1995. Im Strom der Modernisierung: Bevölkerung, Wirtschaft und Umwelt im Kanton Bern (1700–1914). Bern: Historischer Verein des Kantons Bern.

Pfister, C. 2001. Klimawandel in der Geschichte Europas: Zur Entwicklung und zum Potenzial der historischen Klimatologie. *Öster. Z. Geschichtswiss.* **12/2**:7–43.

Pfister, C. 2005. Weeping in the snow: The second period of Little Ice Age-type crises, 1570 to 1630. In: Kulturelle Konsequenzen der Kleinen Eiszeit: Cultural Consequences of the Little Ice Age, ed. W. Behringer, H. Lehmann, and C. Pfister, pp. 31–85. Göttingen: Vandenhoeck und Ruprecht.

Pfister, C., and H.R. Egli, eds. 1998. Historisch-Statistischer Atlas des Kantons Bern 1750–1995. Umwelt–Bevölkerung–Wirtschaft–Politik. Bern: Historischer Verein des Kantons Bern.

Post, J.D. 1990. Nutritional status and mortality in eighteenth-century Europe. In: Hunger in History: Food Shortage, Poverty and Deprivation, ed. L.F. Newman, pp. 241–280. Oxford: Blackwell.

Roy, E. 1982. Environment, Subsistence and System: The Ecology of Small-scale Social Formations. Cambridge: Cambridge Univ. Press.

Wanner, H. 2000. Vom Ende der letzten Eiszeit zum mittelalterlichen Klimaoptimum. In: Klimawandel im Schweizer Alpenraum, ed. H. Wanner et al., pp. 12–37. Zurich: vdf Hochschulverlag.

Wigley, T.M.L., N.J. Huckstep, and A.E.J. Ogilvie. 1985. Historical climate impact assessments. In: Climate Impact Assessment: Studies of the Interaction of Climate and Society, ed. R.W. Kates, J. Ausubel, and M. Berberian, pp. 529–564. Chichester: Wiley.

Worster, D. 1993. What is environmental history? In: Major Problems in American Environmental History, ed. C. Merchant, pp. 4–15. Lexington, MA: Heath.

Xoplaki, E., P. Maheras, and J. Luterbacher. 2001. Variability of climate in meridional Balkans during the periods 1675–1715 and 1780–1830 and its impact on human life. *Clim. Change* **48**:581–615.

APPENDIX 12.1 Assessing biophysical climate impact factors and their significance for grain price fluctuations.

Basic Assumptions

A model of biophysical impacts on grain markets was empirically assessed comparing biophysical climate impact factors (BCIF) for the Swiss Plateau to rye prices in Bavaria measured in grams of silver. Positive BCIFs are thought to agree with high prices and vice versa. Whereas selection of terms was guided by agronomical knowledge contained both in historical sources and present-day textbooks, the weighing of the terms was empirically done by fitting the impact factors to the curve of grain prices.

Computing Biophysical Climate Impact Factors

The basic climate data are monthly temperature (T) and precipitation (P) indices on a 7° classification (Pfister 1998):

3 = extremely warm/wet,	–1 = cold/dry,
2 = very warm/wet,	–2 = very cold/dry,
1 = warm/wet,	–3 = extremely cold/dry.
0 = normal,	

BCIFs are known to be nonlinear. Thus, temperature (T) and precipitation (P) indices were recoded as follows:

5 = extremely warm/wet,	–1 = cold/dry,
3 = very warm/wet,	–3 = very cold/dry,
1 = warm/wet,	–5 = extremely cold/dry.
0 = normal,	

The annual BCIFs were then computed and can be summarized as follows:

$$\begin{aligned}
&\text{IF}\left(T_{Mar}+T_{Apr}\right)<\left(T_{Mar}(N+1)+T_{Apr}(N+1)\right);\\
&\text{THEN}\,\frac{2\times T_{Mar}+3\times T_{Apr}}{-2};\\
&\text{ELSE}\,\frac{T_{Mar}(N+1)+2\times T_{Apr}(N+1)}{-2}\times\\
&\qquad\text{IF}\left(P_{Jul}+P_{Aug}\right)<-6;\\
&\qquad\text{THEN}\,\frac{3\times P_{Jul}+P_{Aug}}{-4};\\
&\qquad\text{ELSE}\,\frac{3\times P_{Jul}+P_{Aug}}{3}\\
&\qquad\quad+\frac{2\times P_{Sep}+3\times P_{Oct}}{5}\\
&\qquad\quad+\frac{3\times T_{Sep}+3\times T_{Oct}}{-2}\\
&\qquad\quad+\frac{2\times P_{Mar}+2\times P_{Apr}+P_{May}}{5}\\
&\qquad\quad+\frac{T_{May}+T_{Jun}+T_{Jul}+T_{Aug}}{4}
\end{aligned}$$

The first two terms require the most explication. The first term reflects spring temperatures. For grain prices, both the spring temperatures within a current meteorological year and those in the subsequent year need to be considered. The higher value goes into the equation. As spring temperatures are known to be significant for grain production, March and April temperatures are weighed.

The second term concerns precipitation patterns during the harvest period, which was from early July to early August. Extremely rainy conditions during this midsummer period had the most devastating impact and they had to be accordingly weighed.

The first and the second terms are then multiplied to simulate the summation effect, which resulted from the coincidence of a very cold spring and a very rainy midsummer in the same or in the subsequent year.

Four terms represent factors of minor importance: precipitation and temperatures in September and October, precipitation in spring and temperatures from May to September.

Results

For the years 1566/1567, 1629/1630, 1817/1818, and 1844/1845, breaks in the series of BCIFs proved significant. To establish the significance of these breaks, we utilized a statistical method known as *intervention analysis*, which Box and Tiao (1975) recommended for environmental time series that often show strong autocorrelations. Intervention analysis is an extension of autoregressive integrated moving average (ARIMA) modeling (Box and Jenkins 1976), a technique of time-series analysis in modern econometrics. According to this implicit model, average exposure shifted, although we only applied an intervention test in a stepwise fashion to the time as well as grain price series (Table 12A.1).

This result challenges the assumption of most economic historians: climatic shock on agricultural production was random. We have to assume that multidecadal periods of low and high BCIFs need to be distinguished.

Table 12A.1 Level of significance for tests of step changes in the mean level of biophysical climate impact factors (BCIFs) and rye prices in Munich. P-values greater than 0.05 are labeled as being not significant (n.s.).

	1566/67	1629/30	1678/79	1720/21	1766/67	1817/18	1844/45
BCIF	0.01	0.05	n.s	n.s.	n.s.	0.01	0.05
Rye prices	0.05	n.s.	n.s.	n.s.	0.05	0.05	0.05

References

Box, G.E.P., and G.M. Jenkins. 1970. Time Series Analysis: Forecasting and Control. Cambridge: Cambridge Univ. Press.

Box, G.E.P., and G.C. Tiao. 1975. Intervention analysis with applications to economic and environmental problems. *J. Am. Statist. Assn.* **70**:70–79.

Pfister C. 1998. Raumzeitliche Rekonstruktion von Witterungsanomalien und Naturkatastrophen 1496–1995. Zurich: vdf Hochschulverlaag.

13

Information Processing and Its Role in the Rise of the European World System

Sander E. van der Leeuw

School of Human Evolution and Social Change, Arizona State University, Tempe, AZ 85287–2402, U.S.A.

ABSTRACT

This chapter offers a millennial perspective on human history, a very global perspective that focuses on the underlying socionatural dynamics rather than on historical detail. It does so in the reverse order of what a historian would do. Whereas the latter would attempt to construct the present from the past, I have tried to reconstruct the past from my understanding of present-day dynamics. The chapter thus presents a millennial perspective on a centennial scale, a perspective that describes the period concerned in terms of large-scale dynamic generalizations. There is a distinct European bias, as very little attention is paid to what happened outside the European cultural sphere, except insofar as those events are relevant to Europe between A.D. 1000 and 2000.[1] Although this does not limit insight into the underlying dynamics in question, it does mean that the timescales are not relevant to other parts of the world.

The chapter will arbitrarily follow a rather "classical" division of the last millennium into periods of about two hundred years; this seems to fit some of the phenomena highlighted just as well as any other division, at least in Europe. However, the reader is forewarned about the biases this entails. The dynamics underlying what we normally observe as history occur simultaneously at an infinite number of different timescales, some of which are observable, while others remain hidden. These timescales stretch from several millennia to momentary, roughly from 10^4 to 10^{-2} years. As it is very difficult to address simultaneously more than two or three such scales, any theory assembled on the period under consideration will be profoundly underdetermined by our observations. Moreover, any such theory will be biased by our initial subdivision of the period, just as our choice of spatial scale will bias our ideas toward observations that can be made at the scale concerned. Even worse, we will not know how such choices have biased our ideas, as we make our choices before we have studied the phenomena concerned.

1 It is my contention, and that of several colleagues at the Dahlem IHOPE Workshop, that this approach is also valid for states, empires, and other large-scale social organizations elsewhere. To demonstrate this demands more space than is available in this volume. A study comparing the trajectories of Egypt, Mesoamerica, and the Roman Empire is, however, in preparation.

ROBUSTNESS, RESILIENCE, VULNERABILITY

In the context of this volume, which aims at initiating a cogent reflection on the sustainability of our way of life as it relates to the natural environment, it seems to me that as a social scientist, the most important dynamic I can highlight is the relative importance of societal processes in the combined socioenvironmental dynamic. In other words, if we are to study sustainability, we must answer the following question: When is society "strong" vis-à-vis the environment, and in which periods, if any, do environmental dynamics dominate the scene?

Before we do so, however, we must refine our conception of "strong" by distinguishing three states of a socioenvironmental system, which differ in the relative strength of its societal and environmental components. *Robustness* is the most recent of these terms. Its intrinsic meanings are still under (sometimes heated) discussion (cf. http://www.santafe.edu/robustness). In the present context, it seems to refer most usefully to the structural and other properties of a system that allow it to handle disturbances without adapting (i.e., without in any way durably changing either its structure or its dynamics). As defined by Holling (1973, p. 3), *resilience* refers to "the capacity of a system to absorb and utilize or even benefit from perturbations and changes that attain it, and so to persist without a qualitative change in the system's structure." Such a system may, however, take new external conditions into account by absorbing them into its mode of functioning (Holling 1986). The difference between the two concepts thus seems to lie in the extent to which (nonstructural) changes in the dynamics may be introduced into a system under the impact of changes in external circumstances. *Vulnerability* refers to instances in which neither its robustness, nor its resilience enables a system to survive without structural changes. In such cases, either the system does adapt structurally, or it is driven to chaos. All three terms express a temporary condition of the interaction between a system and its context. (cf. van der Leeuw 2001).

LONG-TERM TRAJECTORIES OF SOCIOENVIRONMENTAL DYNAMICS

In the remainder of this chapter I will argue that the efficiency and innovativeness of a society's information-processing apparatus are the best available indicators of that society's "strength" vis-à-vis its natural environment. To understand why that is so, we need to have a closer look at the evolution of the dynamics of interaction between societies and their environments, whether social or natural.

A useful way to look at these dynamics is to view them as a "risk spiral" (Müller-Herold and Sieferle 1998), driven by human societies' never-ending attempts to cope with risks. Although such attempts often succeed in reducing specific, frequently encountered risks, the efforts involved also create qualitatively new risks, some of which involve a larger spatial scale and/or a longer time frame. Such risks are often designated by the euphemism "unexpected consequences." Over time, as frequent, known risks are reduced, the "risk spectrum" shifts toward longer-term, unknown risks. For example, the increasing use of fossil fuels during industrialization, in particular after the 1950s, solved critical problems of food scarcity through impressive increases in yields per unit area, and food output per agricultural worker. These achievements, however, resulted in a host of new, longer-term risks including dependence on exhaustible resources, environmental problems caused by industrial agriculture (e.g., nutrient leaching, pesticide use, reduction of biodiversity), and global enrichment of carbon in the atmosphere leading to climate change. The need to solve such risks pushes the system further away from its original state, at increasing speeds. Ultimately, this can cause a conjunction of long-term risks to occur that would merit the term "crisis."

The need to keep this process under control, and to avoid such crises, has increasingly entailed the involvement of most human societies in the management of the environment. The present period has been characterized as the "Anthropocene," that is, a period in which humanity has taken effective control over the environment (Crutzen and Stoermer 2001). Why is this?

Although parts of the living environment operate at a very wide range of temporal scales—from the very short time frames in which viruses and bacteria can mutate to much longer timescales at which plants and animals change, or at which tectonics or erosion transform a landscape—many of the natural dynamics in a landscape occur relatively slowly in comparison with anthropogenic dynamics. Human beings can learn how to learn and, overall, human dynamics can often change more quickly than nonhuman dynamics. As a result, human beings generally begin by adapting themselves to the dynamics of their environment, but over the long term they modify these dynamics to suit themselves. They appropriate the environment by reducing its complexity in exchange for increasing the complexity of their societies.

Once this has been accomplished, however, there is no return: the people involved cannot stop investing knowledge and effort into the system that they have modified, because any reduction in effort will allow natural dynamics to take over and transform the environment into one to which the society is no longer adapted. Once a garden has been created out of a wilderness, one is bound to keep gardening. In addition, the more one has transformed the original wilderness, the more gardening there is to be done!

Therefore, long-term survival of any society depends on its capacity to solve the internal and external problems it encounters in a timely manner (cf. Tainter 2000). This includes adaptation of existing solutions and invention of new ones, the timely introduction of these changes into the societal dynamics, and the maintenance of social cohesion in the process. Over the longer term, solutions become more complex and depend upon the maintenance of ever-expanding expertise in an increasingly complex society to process the burgeoning information needed to keep the entire socioenvironmental system functioning. Hence, a system that is best able to innovate without fundamentally changing itself will be better equipped to cope with disturbances.

A THEORY OF SOCIOENVIRONMENTAL DYNAMICS

Any attempt to understand the circumstances under which the combined socioenvironmental system may be sustainable requires a theory about the dynamics that underpin societal cohesion. My starting point in outlining such a theory is the fact that humans differ from most other species in that they can learn and learn how to learn;[2] they can categorize, make abstractions, and hierarchically organize them; and they communicate between themselves by various kinds of symbolic means. Moreover, human beings have the capacity to transform their natural and material environment in a multitude of ways at various spatial and temporal scales. Our relations with our environment are part of the uninterrupted process of human learning, which may be seen as a positive feedback loop (Figure 13.1) that creates order out of our experiences of the world beyond us by isolating patterns, defining them in terms of a limited number of dimensions, and storing the latter in the form of knowledge. To summarize the feedback look in general terms: the more cognitive dimensions exist, the more problems can be tackled, and the more quickly knowledge is accumulated (cf. van der Leeuw and McGlade 1993; van der Leeuw and Aschan-Leygonie 2005).The result of this feedback loop is the continued accumulation of knowledge, and thus of information-processing capacity, which enables a concomitant increase in matter, energy, and information flows through the society, which in turn enables the society to grow.

Information, Information Processing, and Communication

I am using the concept of information in a very broad sense that differs considerably from both the everyday meaning of the word as well as from the term as it is used in Shannonian or other kinds of "information theory." It requires therefore some explanation.

We have seen that human learning involves the identification of patterns of all kinds, whether temporal, spatial, semantic, syntactic, or yet other. By identifying such patterns, we "organize" the world around us, infuse it with structure and meaning, make it possible for us to "understand" it and/or "know" things about it, intervene in it, etc. Such pattern identification is based on an interaction

2 The human capacity to process information is genetically encoded, but the information humans process and the ways in which they do so are not. Information is socioculturally and self-referentially developed and maintained.

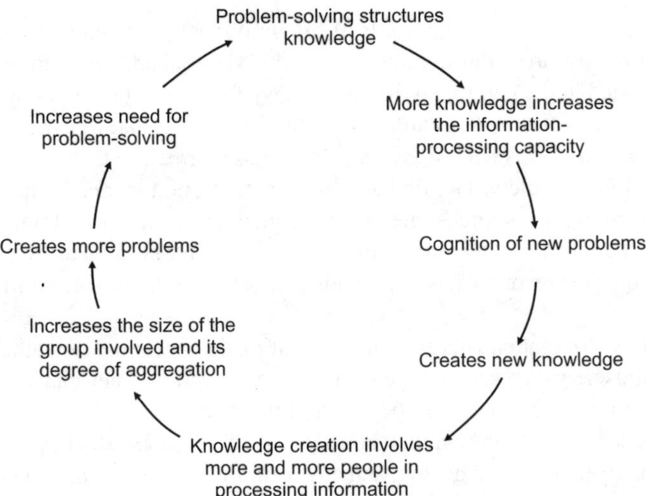

Figure 13.1 Depiction of human relations with their environment as a positive feedback loop.

between our minds and what we observe. In that interaction, more and more dimensions (e.g., color, shape, size, function) of the patterns are slowly but surely isolated by trial and error. Once they have been isolated and identified, perceiving them in another context (relative to other phenomena) provides information about these new phenomena, relating the new phenomena to the preexisting cognitive framework. *Information may thus be conceived as any signal that relates an observation to an existing set of meanings*, thus conveying significance.

In this context, what do we then mean by information-processing capacity? It includes of *all the available means an individual, group, or society has at its disposal to register and interpret information*, and thus to observe and understand what happens in its environment. It includes the sum total of the understanding, expertise, and skills of the people involved, including their technical and organizational means of solving problems, their means to maintain group cohesion, etc.

Such information-processing implies the use of the human capacities of learning and problem-solving, but also of communication. Such communication can be by any of a wide range of means. Most of these will spring readily to mind: example and imitation, gestures, spoken or written language, electronic transmission of speaking, images or other symbols, etc. However, it also includes less obvious means, such as the creation, transportation, and observation of artifacts (which may carry more than one message).

Human Societies as Flow Structures

Underlying this approach is the idea that human societies are flow structures; that is, their existence is dependent upon flows of matter, energy, and information which allow the needs of the individual participants to be met by distributing resources throughout the society. Material and energetic resources are identified in the natural environment, transformed by human knowledge in such a way that they are suitable for use in the society, and transformed again during use into forms with higher entropy. These forms can then be recycled or excreted by the society. The first type of transformation increases the information content of the resources; the second reduces information content. Thus, indirectly, the information content (or information value) is a measure of the extent to which the resource has been made compatible with the role it fulfils in the society.

The flows of matter, energy, and information do not follow the same pattern. Matter and energy are subject to the laws of conservation. They can be transmitted and stored, they may feed people and provide them with other necessary means of survival, but they cannot be shared or communicated. They can only be passed on from person to person, and any fortuitous constellation of people that only processes energy and matter will therefore immediately lose structure. In other words, flows of matter and energy alone could never have created durable human social institutions, let alone societies. Information, on the other hand, *can* be shared. Indeed, human societies are held together by expectations, by institutions, by worldviews, by ideas, by technical know-how, by a shared culture! Their various approaches to doing things distinguish them among other groups of people.

The sharing of information creates the channels through which information is processed. How do these all-important information-processing channels emerge? Through a process that could be called "alignment," that is, through the continued exchange of information that eventually allows different people to share perspectives, ways of doing things, beliefs, etc. That recursive communication process both facilitates shared understanding between individuals and draws more and more individuals into a network in which they can communicate more easily and with less effort and/or less risk of misunderstanding than they would experience with nonmembers of the network. There is thus a decided adaptive advantage to being part of such a network. Finally, when the recursive communication remains below a certain threshold, it keeps people out of a network because they cannot sufficiently maintain their alignment.

But that is not all. Shared information also creates the channels through which energy and matter flow, whether these channels are material (e.g., roads, cables, etc.) or remain virtual (e.g., exchange networks such as the *kula*). Sometimes the channels for information and matter and/or energy are the same, but this is not necessarily the case: electricity, petroleum, and coal are transported, processed, and delivered in different ways. The same is true of virtually all goods in everyday life that we do not collect or process ourselves. We conclude that the "fabric of society" consists of flows through multiple networks, held together by different kinds of (information) relations (e.g., kin, business, friendship, exchange, client–patron, power,), and transmitting different combinations of the three basic commodities.

SOCIETY–ENVIRONMENT DYNAMICS

Society uses information processing to ensure that the necessary matter and energy reach all of its members. Whereas matter and energy are found in the environment, information processing takes place in society. There is thus a feedback loop to consider, where matter and energy constitute the input into society and information is the output. Maintaining a society's growth requires a continued increase in the quantity of energy and matter flowing through society. Such growth is achieved through the identification, appropriation, and exploitation of an ever-increasing range of resources. At the most abstract level, therefore, the flow of information (structuration) into the environment enables the society to extract from that environment the matter and energy it needs to ensure the survival of its members. The dynamic is driven by the information-processing feedback loop which aligns an ever-increasing number of people effectively into a connected set of social networks, thus simultaneously increasing the degree of structuration of the society and the number of people involved.

For the whole to function correctly, the rates of information processing and those of processing energy and matter need to be commensurate. If too little information is processed, society loses coherence; people will act in their own immediate interests and the synergies inherent in collaboration will be lost. Thus, for a center to maintain its power, it must ensure that there is an information-processing gradient emanating outward from itself to the periphery of its territory. At every point in the trajectory between the center and the periphery, it must be more advantageous for the people involved to align themselves with what is happening closer to the center than with what is happening locally or further away.

Over time, such a gradient can only be driven by a continual stream of innovations emanating from the center toward the periphery. Such innovation is facilitated by the fact that the closer one is to the center, the higher the density of aligned individuals, and thus the more rapid is the information processing. One could state that the innovation density of such a system is thus always higher nearer the center. Innovations create value for those for whom they represent something desirable, but unattainable. The farther one is from the center of the system, the more unattainable the innovations are (because one is farther from the expertise that created the value). In general, therefore, the value gradient is inversely proportional to the information-processing gradient. Value, in turn, attracts raw materials and resources from the periphery. These raw materials and resources are transformed into (objects of) value wherever the (innovative) know-how to do so has spread, thus closing the loop between the two gradients. The objects are then exchanged with whomever considers them of value, that is, whoever cannot make them (as well or as efficiently).

An Example: The Expansion of the Roman Republic

To illustrate how this works in practice, let us look at the history of the Roman Empire (van der Leeuw and de Vries 2002). The expansion of the Roman Republic was enabled by the fact that, for centuries, Greco-Roman culture had spread from the Mediterranean to the north. It had, in effect, structured the societies in (modern) Italy, France, Spain, and elsewhere in a major way and led to inventions (such as money, the use of new crops, the plough), the building of infrastructure (towns, roads, aqueducts), the creation of administrative institutions, and the collection of wealth. Through an ingenious policy, the Romans aligned all these societies and made them subservient to their needs, that is, to the uninterrupted growth of the flows of matter (wealth, raw materials, foodstuff) and energy (slaves) throughout their territories. The Roman Republic and the Empire could expand as long as there were more pre-organized societies to be conquered. Once their armies reached beyond the (pre-structured) Mediterranean sphere of acculturation (i.e., when they came to the Rhine and the Danube), that was no longer the case and conquests stopped. Thereafter, a major investment in effort was needed in the territory thus conquered. This consisted of expanding the infrastructure (highways, *villae*, industries) and the trade sphere (Baltic, Scotland; but Roman trade goods have been found as far afield as India and Indonesia) in order to harness more resources. As large territories became "Romanized," they also became less dependent on Rome's innovations for their wealth, and thus their expectations from the Empire decreased. One might say that the "information gradient" between the center and the periphery leveled out, and this made it difficult to ensure that the necessary flows of matter and energy reached the core of the Empire. As the cost of maintaining these flows grew, in terms of maintaining a military and an administrative establishment (e.g., Tainter 2000; Crumley and Tainter, this volume), the coherence of the Empire decreased to such an extent that it ceased, for all intents and purposes, to exist. People began to focus on their own interest and local environment rather than on maintaining a central system. Other, smaller, structures emerged at the edges of the Empire and there the same process of extension from a core began anew, at a much smaller scale, and based on very different kinds of information. In other words, the alignment between different parts of the overall system broke down and new alignments emerged that were only relevant locally.

In such an overall flow-structure dynamic, cities play a major role. They function as demographic centers, administrative centers, foci of road systems, but above all they are the nodes in the system where the most information processing transpires. Put simply, they are the backbone of any large-scale social system. They operate in network-based "urban systems" which link all of them in a particular sphere of influence. Such systems have structural properties that derive from the relative position all the cities occupy on the information-processing gradient as well as in the communications and exchange networks that link them to each other (White 2006). Although the role of individual towns in such systems may change (relatively) rapidly (Guerin-Pace 1993), the overall dynamic structures are rather stable over long periods of time.

Because people congregate in cities, urban areas harness the densest and most diverse information-processing capacity. Not only does this relatively high information-processing capacity ensure that they are able to maintain control over the channels through which goods and people flow on a daily basis, but their cultural (and, thus, information-processing) diversity also makes them into preferred loci of invention and innovation.

This is best documented by referring to ongoing work in the EU-funded ISCOM project.[3] The ISCOM research team has convincingly demonstrated that innovation (as represented by the number of people involved in research, the number of research organizations, the number of patents submitted, etc.) has increased exponentially in recent years, in parallel with per capita income. It also scales superlinearly with the size of urban agglomerations (cf. Figures 13.2 and 13.3) (Bettencourt et al. 2004; Strumsky et al. 2005). The superlinear scaling of innovation with city size enables cities to ensure the long-term maintenance of the information gradient that structures the whole system. It is due to a positive feedback loop between two of any city's roles. The flow of goods and people transpires primarily through towns and cities. This means that urban people are confronted most intensely with information about what is happening elsewhere; this, in turn, enhances their potential for invention and innovation. However, these same connections enable them to export these innovations most effectively, thus exchanging a part of their information-processing superiority for material wealth.

The Evolution of the European World System

To make our theory—long-term sustainability of societies can be studied in terms of the evolution of innovativeness and information-processing efficiency—more convincing, let us now apply it to a case study using the last thousand years of western European history, as it presents a complete cycle from rural, very small-scale societies to a single, coherent global society. Such an application requires that we find ways to measure the evolution of these two parameters over time. As we cannot measure them directly for the sub-recent past, let alone for more distant periods, we will try to use a set of proxies. The proxies chosen have been based on two characteristics that seem to derive directly from innovativeness and information-processing efficiency: (a) societal cohesion

3 The "Information Society as a Complex System" (ISCOM) project is funded by DG Research of the European Union (contract no. IST-2001-35505) and coordinated by D. Lane (University of Reggio Emilia/Santa Fe Institute), D. Pumain (University of Paris 1/CNRS), S.E. van der Leeuw (Arizona State University/Santa Fe Institute), and G. West (Santa Fe Institute). The project focuses on the relationship between the dynamics of invention, innovation, and scaling in urban systems.

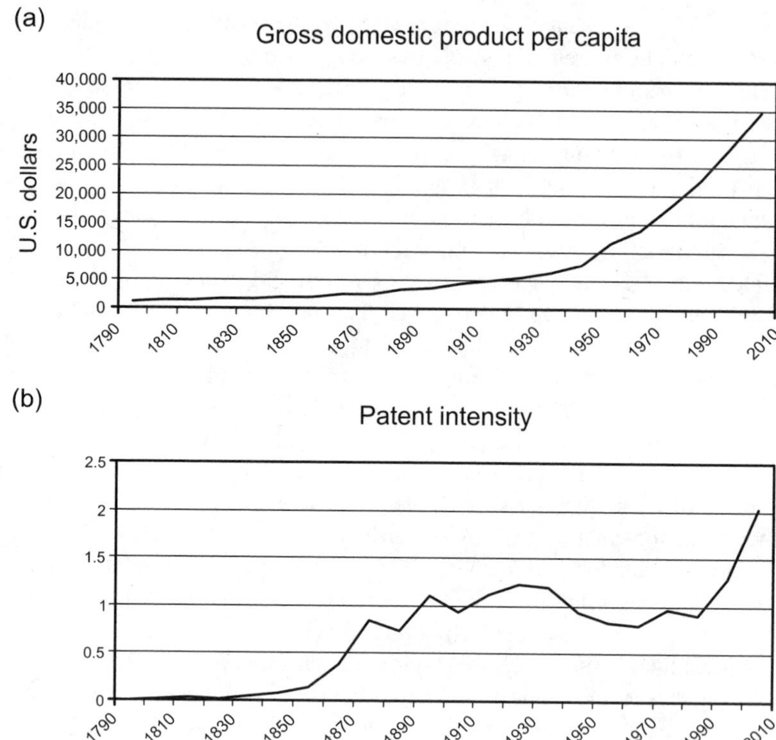

Figure 13.2 (a) Real per capita GDP and (b) number of patents per head of the population in the U.S. from 1800–2000. Thanks go to J. Lobo (Global Institute of Sustainability, Arizona State University) for preparing these graphs. The source of the data for patents counts is the U.S. Patent and Trademark Office (http://www.uspto.gov/); for population data, the source is the U.S. Census Bureau (http://www.census.gov/); for GDP per capita the source is the "Economic History Services" program jointly run by the University of Miami and Wake Forest University (http://www.eh.net/hmit/).

and (b) the flows of matter and energy that depend on the gradient from the center to the periphery of the system concerned:

- trade volumes and the spatial extent of trade flows are the most direct conceivable measure of energy and matter flows, and thus of the information-processing potential between the center and the periphery;
- the spatial extent of the territory pertaining to a system, relative to the territories of other such systems, as a measure of the area that the system can coherently organize;
- the degree of internal cohesion of the society as measured by the extent of monetization of its economy;
- the density and extent of road (and railroad) systems as a proxy for the density of information flows; for later periods this is to be replaced by the density and extent of telephone and other communication networks;

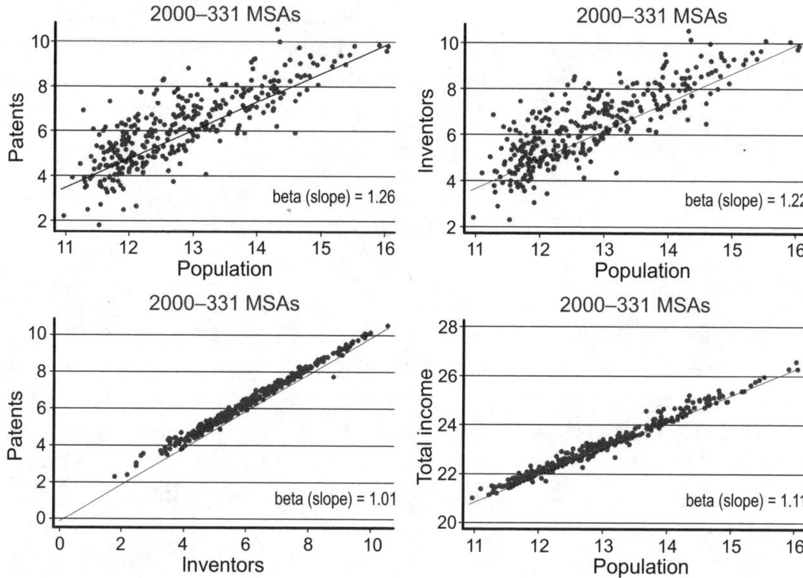

Figure 13.3 Scaling relationships (for the period 1980–2000) between various characteristics of 331 U.S. Metropolitan Statistical Areas (MSAs): (a) patents (as a proxy for inventive activity) versus population (superlinear); (b) number of inventors versus population (superlinear); (c) patents versus inventors (linear); and (d) total income versus population (slightly superlinear). Thanks go to J. Lobo (Global Institute of Sustainability, Arizona State University) for preparing these graphs. The patents and inventor data were assembled by Lee Fleming and Deborah Strumsky (Harvard Business School). Population, establishment, and income data were assembled by Deborah Strumsky from a variety of publicly available federal sources (Economic Censuses, County Business Patterns, and the Regional Economic Information System).

- the degree and gradient of wealth accumulation in the system, as measured for example, by the overall infrastructure level of the society, and the relative level of it in each town;
- the density and gradient of population aggregation across the system;
- the innovativeness of particular towns, regions, and periods.

Most of these cannot comparably be measured for each and every historical period and region, and are therefore of limited utility. Moreover, they constitute measures derived from different kinds of dynamics (although I believe that all fit within the overall dynamic outlined). They therefore operate at different temporal rates of change. As these proxies are all we have for the moment, I will try to do my best with them.

In view of their particular role in driving the overall dynamics, I pay particular attention to the state of the urban systems. During the whole of the millennium, we see a gradual strengthening of the urban (aggregated) mode of life, but this millennial tendency has its ups and downs, and manifests itself in different ways. A second long-term dynamic is that of European integration and expansion. All of these trends reflect different ways in which the European socioeconomic system strengthens itself vis-à-vis the external dynamics that it confronts, whether environmental or anthropogenic.

THE INDIVIDUAL PERIODS

The Dark Ages

In the last four centuries of the preceding millennium, after the end of the Roman Empire as such, we observe a weakening of society's structure and coherence when compared to the (admittedly high) level of Roman cohesion and structure. This period has been brilliantly described by Robert Lopez (1966). Essentially, between A.D. 600 and 1000, the whole of the fabric of society reached a higher level of entropy in western Europe, whereas in southeastern Europe the traditions of Greco-Roman urban culture were partially conserved. In the West, trade and long-distance contact diminished greatly, urban population dwindled (a famous example is that of Arles, which was for some time reduced to the perimeter of its Roman arena), and most villages were abandoned. All of society seems to fall back on immediate, local, survival strategies.[4] Much of Roman culture was lost (e.g., basic techniques as throwing pottery, building brick buildings, blowing glass). An exception is, however, formed by the Church, which through its estates and tithes was able to maintain some of the knowledge it inherited, especially in the arts of writing and bookkeeping. Because these were skills that the Church needed for its survival as an organization, it had the incentive to invest in their development. As most of the Roman provincial capitals became the See of a bishopric, this continued the concentration of such skills in an urban context.

Around the same time in northern Europe (from the shores of the Baltic and the North seas), one can observe a separate cultural sphere. It is essentially rural and military, and it enters the world system by raiding areas connected by the two seas involved, as well as through exchange in a few "entrepots" (e.g., Hedeby, Dorestad, and Quentovic; cf. Hodges 1982). Eventually these "Norsemen" settle in East Anglia, Normandy, and Sicily. Numismatic studies indicate that a considerable amount of their wealth was obtained from Russia, Byzantium, and (through these) from places further East. This wealth, however, did not in any way permeate society below its topmost tier.

4 We cannot definitively determine whether this period also saw any serious decrease in population, but irrespective of that, it seems that loss of skills is due to the (high) cost of maintaining those skills versus the (limited) local benefit obtained. Labor was needed for survival and locally one could not afford specialist artisans; the market (and thus the profit) to be obtained by artisans shrank with the withering of their lines of communication.

1000–1200: The First Stirrings

In his now famous thesis, Duby (1953) shows how, from about A.D. 1000, society in southern France began to rebuild itself "from the bottom up." Although the urban spatial backbone of the Roman Empire survived the darkest period, we see a completely new rural spatial structure emerge, even relatively close to the Mediterranean. In a couple of centuries of mostly local competition over access to resources, various small lords and princes strove to escalate the social ladder by conquering neighboring farms, villages, castles, and other positions of potential power. During that period, a new (feudal) social/hierarchical structure emerged. By means of access to resources, the most astute (who had the best information-processing and communication skills) were able to attract followers, notably by providing protection for peasants and thereby getting them to buy into the feudal system. That labor force in turn provided such lords with surplus that could be translated into the support of an army, etc.

From a systemic point of view, this was a period of oscillation between different entities, in which cohesion alternated with entropy even at the lowest levels. Ultimately the first larger spatial units are formed, bringing together the dispersed rural population under the "protection" of a hierarchy of feudal lords.[5] As this sociopolitical structure comes into being, more wealth accrues to the favored, and we begin to see the resurgence of an upper class, courtly, culture in the so-called Renaissance of the 12th century. This included tournaments, troubadours, and a wide range of other (mostly religious) artistic expressions.

Similar phenomena occurred in Lotharingia and the Rhineland, where a separate cultural sphere developed on both banks of the river, around commerce that linked Switzerland with the Low Countries. Further East, in Germany, the decay of whatever central authority the Holy Roman Empire exerted as well as the colonization of eastern Europe is observed. Concurrently, parts of Europe began to look outward: it is the time of the crusades against Islam.

We should, however, not overestimate the importance of long-distance commerce, towns, culture, and wealth. Southern Europe remained a mostly rural society of smallholders and serfs, whereas further north, there was hardly any sign of the 12th-century Renaissance.

In northern Europe, the trade connections forged in the period before A.D. 1000 led to the development of certain towns into commercial centers, later loosely federated into the Hanseatic League. These towns, however, remained essentially isolated islands in the rural countryside. During this period, North and South are still essentially separate cultural spheres.

---

5 The rate at which this happened varies for different regions and countries. France, Spain, and Britain moved more quickly, although in different ways, whereas Germany and Italy remained heavily fragmented for a considerably longer time than the first two. Eastern Europe was less structured from the outset and followed yet a third trajectory.

1200–1400: The Renaissance

Three major phenomena characterize this period: (a) the establishment of a durable link between the southern (Mediterranean) and the northern (Baltic and North seas) cultural and economic spheres, (b) the major demographic setback caused by the Black Death in the 14th century, and (c) the beginnings of the Italian Renaissance. The commercial and cultural links between South and North were initiated in the 11th and 12th centuries, essentially via the overland route from Italy and southern France via the Champagne to Bruges and Ghent in the Low Countries, where they connected with Britain and the Hanseatic trade system. In the 13th century, this corridor became essentially the mainstay of the continent-wide trading and wealth creation system. The wealth thus created enabled considerable population growth, both in the towns and in the countryside that fed them. Eventually, continued rural population growth drove the system in many areas to the limits of its carrying capacity, exploiting more distant and less fertile or less convenient areas.

It is difficult to assess how the three waves of bubonic plague epidemic, roughly one generation apart, affected the demographic structure in Europe. Opinions differ greatly, except on the fact that the impact was very uneven. In those areas where the impact was great, both the cities and the surrounding countryside were profoundly affected; people moved from the periphery into the traditionally more populous areas (where the plague had hit hardest), thus increasing both the level of aggregation of the population and its average per capita wealth. Other profound changes occurred in the cultural domain. The sudden major interruption led to a reevaluation of the role of religion, life and death, society and the individual, shaking society out of its traditional ideas and patterns of behavior.

These two major phenomena (the beginning integration of the two cultural spheres and the Black Death and its consequences) seem to have contributed to a localized "era of opportunity" in northern Italy and (a little later) in the Low Countries. In the highly competitive environment of the northern Italian city-states, cultural, institutional, technical, and economic inventions—whether homegrown or imported from the other shores of the Mediterranean (or farther afield)—were quickly adopted and led to a unique increase in information-processing gradient between these centers and the rest of the continent. Padgett (2001), for example, describes how financial and social innovations went hand in hand to transform the Florentine banking system, as it drew in increasing resources and invested them in an ever-widening range of commercial and industrial undertakings; this, in turn, transformed the practices in these domains. Many of the ideas developed were equally useful to the trading centers in the Low Counties, and these quickly followed Northern Italy in becoming very rich and powerful.

Semi-independently, we see the first stirrings of invention, innovation, long-distance trade, and urban development in the Iberian Peninsula. There, these two centuries were characterized by increasing cultural, military, and economic interaction between the Christian northern part and the Islamic southern section. In particular, Catalonia and Portugal developed long-distance links with very different cultural spheres.

The most striking phenomenon of these two centuries is the resurgence of cities. Power shifted from the castles and their lords to cities where trade was concentrated, wealth accumulated, and society reinvented. A bourgeoisie emerged, initially in the Italian and Northern European cities (including London), but a little later also in many French towns. Thus, the scene was set for most of western and southern Europe to cross another threshold on the way to becoming (for a while) the center of much of the world. This process reflects a systemic change. Whereas initially the (nonurbanized) geographic landscape was so homogeneous all across Europe that anyone could, in theory, grab local power and create a (small) principality or other unit, from this time onward, the opportunities of reaching the "top of the heap" were limited to certain geographic areas, where structural systemic changes (e.g., urbanization) had effectuated transformations enabling these areas to dispose of more and diverse resources as well as more effective information processing because they were linked into one or more system-level information flows.

1400–1600: The Birth of the Modern World System

This period was in many ways revolutionary, as it marked the final phase of the continent-wide transition from a rural barter economy, which was autarchic, to an urban, monetized economy where specialization and trade set the trend. Wallerstein (1972, 1980) describes it in the two volumes from which I have drawn the title of this section. In essence, the world of commerce and banking expanded across different political entities, different cultures, and different continents. Much of southern and western Europe, including Britain, Scandinavia, and the Baltic became an integral part of one and the same economic system.

Many rural areas saw their contacts with urban areas increase. These rural hinterlands had always provided the food and the raw materials (e.g., wool, wine, wood) for the industries and commercial activities of the towns. However, during this period, the flow in the opposite direction intensified. Rural areas began to use urban industrial products and trade goods themselves, and therefore became dependent on the cities. With the changing outlook on urban life, cities also began to look attractive to farmers in an overpopulated countryside that was continually disturbed and impoverished by marauding armies fighting out others' political conflicts. A major wave of rural emigration to these cities and towns followed, relieving the population pressure on the countryside and keeping the urban labor force cheap and thus enabling the urban bourgeoisie to increase their investments in, and profit from, industry. Consequently, a major industrial expansion followed the spread of commerce.

The driving force behind all of these changes was the European trade system, which was based in the cities. It put to work the enormous gains in information-processing capacity made during the innovative Renaissance. Through relatively unregulated commerce and industry, which profited from increased contrasts between the rich and the poor, Italian (e.g., Medici), German (e.g., Fugger), and later Dutch and British heads of commercial houses amassed enormous wealth, and used it to bankroll the interminable political conflicts and wars that disrupted the continent. In the process, they and their societies extended their control over much of the Western world. To expand and maintain that control, they created extensive, centralized networks for the gathering and communication of information, which included numerous spies in every important commercial, financial, and political center, as well as the first private courier services.[6] By investing in this information-processing infrastructure, these houses (and in their wake the political powers) explicitly acknowledged the fact that their power and influence was essentially based on their superior knowledge about what was going on in the world. Power is enhanced both by getting control over information flows and constructing new forms of information flow.

This is also the period of the first trade voyages to other continents, initially undertaken from Portugal and Holland and a little later from Britain and Spain. In the middle of the period concerned, the Americas were discovered, and Da Gama sailed to the Far East. By investing in further discoveries of these distant parts, the European trade system not only expanded its geographical scope, it also added a considerable new trajectory downstream from itself along the information-processing gradient. The huge, immediate profits made on the trade of spices and other extraneous goods more than compensated for the risks, even without the American gold and silver that quickly began to flow. More importantly, because of the physical distances involved, this long-distance trade opened the way for trading houses (and later the nation-states that took their place) to control for centuries an increasingly important and resource-rich part of the world.

This period can be seen as the heyday of the independent, city-based trading houses and of city power altogether. During these centuries, the likes of London, Amsterdam, Marseille, Lisbon, Seville, Genoa, and Venice had no competition from political overlords. On the contrary, the cities controlled the purse strings. It is the period with the steepest information gradient from the center of the European world system to its periphery, although on the European continent itself, that gradient began to level off a little bit as the cities in the hinterland, and eventually the territorial overlords, began to play the same game.

1600–1800: The Territorial States and the Trading Empires

Many attempted to get in this game—the Holy Roman Emperor, the kings of Spain, Portugal, England, Sweden, and France, not to mention innumerable

---

6 The first and foremost among them was run by the Vatican.

smaller princes, such as the Duke of Burgundy. However, although they possessed inherited legitimacy—or something approaching it—that stemmed ultimately from the Roman Empire, this did not pay the bills. On the contrary, the need to keep up a certain status was in itself a financial handicap until they could leverage their legitimacy against financial support from the rich middle classes, through the exchange of loans for taxes as their principal source of income. This, in turn, required them to guarantee peace and security (and optimal trading circumstances) within their territories.

Although the individual trajectories by which this was achieved in the major European countries differed, I think we can safely say that, at least in the case of Scandinavia, Holland, England, France, and the Iberian Peninsula, a degree of territorial integration and unity was achieved by A.D. 1600.[7] In view of what follows, it should be emphasized that in all of these cases, this unity (still) derived from the recognition that in these territories there was a single overlord, irregardless of whether he or she was called Emperor, King or *Stadholder* (the Dutch name for the office of the Princes of Orange, literally "the holder of the towns," since he was paid by the cities to guarantee peace in the Seven United Provinces).

It is not accidental that the regions in which such territorial integration was first achieved (Holland, England, and Spain) were also the regions that had the most extensive trade contacts outside of Europe. Long-distance trade guaranteed the steady stream of income that was necessary to pay for the standing armies and the infrastructure that eventually established the authority of the Prince over all others in the realm. Although the King of France was to a larger extent dependent on the riches of his own, predominantly rural, territory,[8] the internal pacification of the European kingdoms corresponded to the expansion of European trade overseas.

During this period, the foundations were laid for the large colonial Empires. Except in the case of Spain (which immediately proceeded to conquer and occupy the hinterlands of its coastal trading posts), this process initially involved the transformation of trading entrepôts (e.g., Batavia, Cape Town, Goa, Hong Kong, and others) into military strongholds, outposts of the European power involved. It went hand in hand with the intensification of trade, the introduction of more voluminous items (coffee, tea, cotton) with somewhat lower added value, the local production of necessities for the colony, the local transformation of some of the raw products, and a degree of out-migration from Europe to these colonies, in particular to the American colonies of Spain and Britain, as well as to British East India. As a result, both the European country and the colonies involved became economically dependent upon one another. Of course,

---

7 In Germany, Russia, and Italy, the process took much longer and did not come to completion by the end of the period we are discussing here.

8 Supplemented by income from Mediterranean, Rhenish, Burgundian, and Aquitaine commerce.

inevitably, the information gradient leveled out as the colonies were inhabited by Europeans who assimilated indigenous knowledge and (eventually) shared some of their knowledge with the local populations. In most instances, there were always more territories to be discovered, new products to be brought to Europe (or older products in greater quantities), and the mechanisms of trade and transport to be made more efficient: larger, faster ships to be constructed, natural (storms) and political (piracy) risks to be reduced, etc. In the case of the U.S., for example, toward the end of the period, the leveling of the information gradient led to the war of secession and the independence from Britain.

This same leveling off of the information gradient occurred in Europe itself, as more people learned about, and began to share in, the production of wealth and its benefits. The profits from the long-distance trade, as well as from the financial investments that it enabled, led to an important increase of the industrial base of the main European countries. This increase was for the moment achieved with essentially the same technological means as in earlier centuries: by involving more and more people in production and transformation of goods. At the same time, the tentacles of commerce and industry spread increasingly further into the rural hinterlands and all over the territory, aided by the improvement of road systems. That improvement, in turn, was triggered by a Prince's need to be able to communicate rapidly over long distances, as well as to move troops. This is particularly so in France, the largest of the territorial states, and the one with the smallest colonial Empire at the time. Ultimately, the leveling off of the internal European information gradient led to a major upset in the form of the French Revolution—a movement, in my opinion, that was born out of despair with the oppressive situation of the time as well as the perceived lack of any chance that things could get better in a country that had not exploited the opportunities offered by the establishment of a colonial Empire.

In summary, this period saw the transformation of a city-based economic system to one involving the entire territory of the state to a much larger extent. At the same time, many new partners made their entry into the system responsible for generating wealth. Among them were the political powers, which found ways to skim off some of the wealth in exchange for internal security. Still a larger, and growing, proportion of the population got involved in small-scale regional industries and commerce. The steep information gradient that was created during the last period thus leveled off in Europe, while the effect of the concomitant changes in the role and structure of the trade empires does the same at the periphery. Europe has thus become vulnerable!

1800–2000: The Industrial Revolution

The essence of the changes in this period is that at a time at which the overall structure of the European system had reached the point at which it was beginning to fray at the edges, the final phase in the isolation and control of energy as a resource and the concomitant mechanization of production and transport offered new opportunities for expansion. It did so by enabling the invention and implementation of a whole range of new technologies which reestablished the information gradient across Europe, as well as between it and its colonies.

The resultant shift was profound. Europe moved from being a zone in which internal consumption of high-value goods produced elsewhere generated most of the wealth, to one that mass-produced a wide range of goods for export and marketing throughout the rest of the world. To maintain this new system, Europe needed nevertheless to create wealth at the periphery of its sphere of influence so that local populations could acquire European goods. It did so primarily by extracting the raw materials from its colonies (India, the Dutch East Indies, Africa, South and Central America), which were then transformed in Europe itself into products to be re-exported to the colonies and sold. Thus, the status of the colonies themselves also changed: from areas that had produced goods having relatively little local value but very high value in Europe, to areas that mass-produced low value goods for export to Europe which in turn would either be sold in Europe or returned to the colonies for sale. In both cases, the added value to the colonies was minimal. This meant that a much larger investment needed to be made to control the colonies politically. This form of colonization resulted in an influx of large numbers of the colonial population into the European sphere as low-paid labor.

In Europe itself, this caused yet another fundamental shift in power. Until the Industrial Revolution, the large majority of the (rural) population had been so far removed from the process of wealth generation that it was relatively easy for the dominant classes to control the whole of the structure that generated their wealth. Now, for the first time, very large numbers of people became involved in the new wealth production system. Industrialization tied a very large working class to the (mechanized) production industry (e.g., coal mines, steel mills, textile factories) through low-paid, often dangerous, mass production jobs that gave little personal satisfaction and created much resentment. Concurrently, it provided major opportunities to a professional class that, through education, had mastered one of the many newly emerging technologies. Thus, the overall system involved ever more people, while exacerbating the social distance between them. Slowly but surely education came to be seen as the only way out of misery for large groups of the population, and some of the educated engaged the political battle to achieve this.

In summary, thanks to the invention of completely new technologies—both in the core and at the periphery—this particular form of the European economic system created much wealth. Ultimately, however, it created large groups of disenfranchised people, which undermined its very structure. Social movements were quick to emerge in the core (from about 1848) and persisted throughout this period up to the World War II. At the same time, those areas of Europe that had not been part of this system from the start aspired to create similar production and trade dynamics. This was initiated by the French in the 19 th century, occupying major areas of Africa and Southeast Asia that were of sufficient economic importance to provide the home country with considerable added wealth. In the latter part of the 19th century, Italy and Germany united into territorial states and then attempted to create colonial empires. Despite these attempts, however, the Germans and Italians were essentially too late and had to content themselves with the leftovers of the colonial banquet table. This fact contributed importantly to the causes of World Wars I and II that followed, as both countries sought expansion in Europe because it was denied them elsewhere.

Finally, it should be mentioned that during this period, the basis for the hegemony over large parts of the world, which Europe had thus far enjoyed, spread initially to include North America (in particular the U.S.) as well as Australia, Japan, South Africa, Central and South America, and, more recently Southeast Asia, China, and India. Europe or the U.S. are thus no longer in sole control of the information gradient responsible for the continued wealth creation, innovation, and aggregation of the world system, but must instead compete with these other regions.

THE SYSTEM'S IMPACT ON THE NATURAL ENVIRONMENT

Thus far, I have focused on the societal dynamics driving the socioenvironmental system. I have caricatured the environment as a potentially infinite source of material and energetic resources to be harnessed by society. In this section, I would like to correct that to some extent by discussing some of the overall long-term consequences of the above societal dynamics for its natural environment.

First, we must consider the nature of resources. In the most general sense, I would argue that everything in our natural environment is a potential resource. Such potential resources, however, do not play a significant role in the dynamics studied until they have been identified as a resource, and the cognitive and substantive infrastructure has been put into place to exploit it. This includes (but is not limited to) identifying for what purpose the resource can be used, how it is harnessed, how it is transformed into a form that makes it useful, how it can be applied to the purpose identified, and how the resulting product can be given the value that drives its distribution. Underlying the process is the need to align sufficient members of the society toward the purpose of using the resource, by giving them a role in the exploitation process, by providing them with the putative benefits of the resource, or in still other ways. With this in mind, I am not going to try and identify the history of individual resources and the impact of their exploitation on the natural environment. Rather, I am going to confine myself to a discussion of the main processes involved in the harnessing of an exponentially increasing number of energetic and/or material resources.

Such a discussion necessarily begins with a consideration of the evolution of the energetic resources at the disposal of European society. From this perspective, we can divide the period under consideration into two major parts. During the first of these, all energy used is due to current or recently stored sources: wind, human, or animal energy for transport, traction, or manufacturing; recently stored solar energy in the form of foods or fuel (wood, plants, or dung); or waterpower to drive mills or pumps. The second period sees the rapid increase of the use of fossil fuels (coal, oil, nuclear fuels) for all of these and many other purposes. The transition between these periods is, of course, of major importance, since it drove the Industrial Revolution. It is unique in the history of humankind, as it was driven—and enabled—by the unprecedented worldwide expansion of the European world system. From the perspective used here, it signals the victory of humankind over time in the domain of energy: it was the first occasion that societies became dependent on energy from resources millions of years old.

That event, however, is part of a trend that affected all material resources, which began thousands of years earlier when humankind first used stone, clay, and later metals to manufacture artifacts. The harnessing of each of these resources signified a major step in the development of humans' information-processing capability and entailed the invention and introduction of new concepts and innovative ways of doing things (cf. van der Leeuw 2000) as well as the reorganization of some of society's institutions. In some instances, it resulted in the reorganization of society itself. An example can be seen in the introduction of iron and steel-making technology during the middle of the first millennium B.C.; this led to the "democratization" of the capability to make high-quality weapons.

Technologies constitute the interface between society and its natural environment, as well as between the flows of information and those of energy and matter. The introduction of technologies plays a crucial role as a driver of social processes, but we must not forget that they also drive the expansion of interaction between humans and their environments by enabling the exploitation of ever more natural resources as well as their transformation into an increasing number (and kinds of) artifacts. Tim Ingold (1986) calls this process the *appropriation of nature*. Every time that a society durably intervenes in its natural environment (e.g., by converting a plant, animal, mineral, or other substance into a resource, or by building a dam, creating and maintaining a field, digging a mine, a canal or a pond, paving a road) it reinforces the co-dependency of the society and its environment by modifying and linking the natural and the societal dynamics. As a result, every such action makes the environment more dependent on disturbances of the natural order by human beings to maintain it in its current state, and the society more dependent on the natural resource to maintain itself in its current state.

If permitted to go on for a long time, this growing co-dependency leads to a reduction of the flexibility in the socioenvironmental system, because in the process, the dynamics become closely tied together. The system thus becomes hypercoherent (Rappaport 1976). For example, whereas in southern France about 10,000 years ago it took a period of *serious* climate deterioration *combined with* a *major* increase in the population's impact on the environment (due to the introduction of agriculture) to cause serious soil degradation, over the last few centuries, a *minor* climatic oscillation *or* a *minor* increase in demographic pressure have caused the same serious effect on the soil (van der Leeuw 2005). To put it in Holling's terms, such a hypercoherent system becomes "an accident waiting to happen," essentially because the different processes have become so overly interconnected that with time all flexibility disappears from the system (van der Leeuw 2000).

A reduction in flexibility, however, is not the only cumulative effect of such long-term socioenvironmental interactions. The other main effect is the emergence of unintended consequences, which lead in many cases to the occurrence of crises. This happens because the relationship between societies and their environments is asymmetric.[9] Not only do the two coupled systems have fundamentally different dynamics, but the society's perception of the environment and its impact on it is governed by the filter of human perception—and not the reverse. Human perception and subsequent interpretation of the environment is always simplified and incomplete. Moreover, it is often underdetermined by observations and overdetermined by a cognitive structure that was developed in the past. However, human interventions in the environment also enrich and modify the environment's dynamics in qualitative and quantitative ways that cannot immediately be observed. This asymmetry has important consequences for the way humans deal with the environment over the long term.

Take the following example. Cronon (2003) recounts how European colonists who arrived in the Americas began their adventure with a perception of what constitutes the natural environment that was determined by the area they left in Europe, rather than by the place where they landed in the Americas. Thus, they cut down certain species of trees that they perceived useful, without realizing that these were pivotal to the well-being of the New England forests. Through this process they transformed the natural forest dynamics, which eventually led to the forests' destruction. Unwittingly, they had shifted the "risk spectrum" of their actions with the local environment, and introduced new risks at unknown temporal scales.

In general, the effect of a series of human interventions in the natural environment is a reduction in the number of (known) minor disturbances, which gives the impression of increasing stability or control over the environment, as well as an increase in the risk of occurrence of less frequent, unexpected disturbances of unknown nature and scope (so-called "unintended consequences"). Over the long term, this may lead to a buildup of major unknown risks, or "time bombs."

---

9 This is expressed in the two words for environment in French: *milieu* and *environnement*. The asymmetry implies that what is degradation of the *environnement*, from an environmental perspective, can be a socialization of the *milieu*.

Once the density of such time bombs is sufficiently high, major crises are likely to occur.

Such crises have the impact of forcing the society to heavily restructure the socioenvironmental system under pressure or abandon its investment. In general, such reorganization entails increasing the diversity of resource use, production techniques, and products. This reduces the natural risks but enhances the complexity of societal management, and thus the risks due to the social dynamics.

Once a crisis looms, the time span available to find a solution is limited, and the outcome is dependent both on the nature of the solution found and the efficiency with which it is introduced. Too little is as useless as too late. Very often, whether a system survives or fails depends on minute differences in external conditions or in the nature of the dynamics. In this respect, differences in perception or social organization may be just as important as slight differences in precipitation, soil fertility, management efficiency, or product price.

Looking back over the past thousand years, this process is visible all around us. It is most evident, perhaps, in the unending quest to incorporate progressively more of the Earth's surface into the Western economic sphere—a quest that ultimately led to the colonial era of military occupation and, in the postcolonial period, to the widespread economic exploitation of the "developing" world. It is, however, also observable in the exponential increase in the number of different kinds of resources exploited, the investment in their exploitation, the new technologies developed to ensure that exploitation, and the massive intensification of such exploitation. In addition, it is behind the spread of infrastructure investment all over the surface of the Earth, whether for transportation of matter, energy, or (most recently) information.

Concomitantly, of course, our societies dispose of waste—solid, liquid and, more recently, gaseous and thermal—in such quantities that this threatens the Earth's natural systems, even in those areas where the wastes have not directly been "appropriated" by society (cf. Hibbard et al., this volume). Although we worry about this aspect of the socioenvironmental dynamic that drives our society, we tend to disassociate it from the input in raw materials of which it is the inevitabe result. More importantly, we tend to forget that the degree of structuration and complexity of our society is directly related to both. It would be difficult to maintain that degree of complexity without maintaining, in one form or other, the degree of material, energetic, and information throughput that it requires. The nature and form of the resources may change, but the flow has to be maintained!

THE SECOND-ORDER DYNAMICS

What is most striking, however, is that (notwithstanding its massive impact on the environment) the rise of modern Western society has not (yet) slowed down, either from a lack of resources or from insufficient or inadequate information processing. Each time a crisis has loomed, either a solution was found in time to continue expansion, or the slowdown was of relatively short duration (compared, e.g., to the slowdown at the end of the Roman Empire), or it affected only certain sectors of society. Once it was over, the lost terrain was regained very rapidly. Not only was this fortunate for our society, but it also hints at a very important aspect of the dynamics concerned.

To understand the significance of this observation, we have to look at how the dynamics of invention and innovation have themselves changed. In other words, we need to look at the dynamics of second-order change. When we do, we observe a series of inventive "explosions," or phases in which a considerable number of inventions and innovations quickly spread throughout large parts of society. Figure 13.4 illustrates a few of these, but there are many others in the same period (e.g., the introduction of bookkeeping, printing). In each case, such an explosion is triggered by a single invention (as in the case of the telephone, illustrated in Figure 13.5) which, in a relatively short time, is exploited in many different ways, reflecting the introduction of that invention as a "solution" to different, as yet unformulated (latent), "problems."

Two aspects of Figure 13.4 are of importance here. The first is the clear acceleration of the second-order process. The innovations succeed each other with increasing rapidity. Why? I would venture the following hypothesis: Most inventions are the result of "solving" a problem. They emerge when a new cognitive dimension is identified that "fits" an *almost* completely defined problem (i.e., a set of identified cognitive dimensions that is not yet experienced as "coherent").

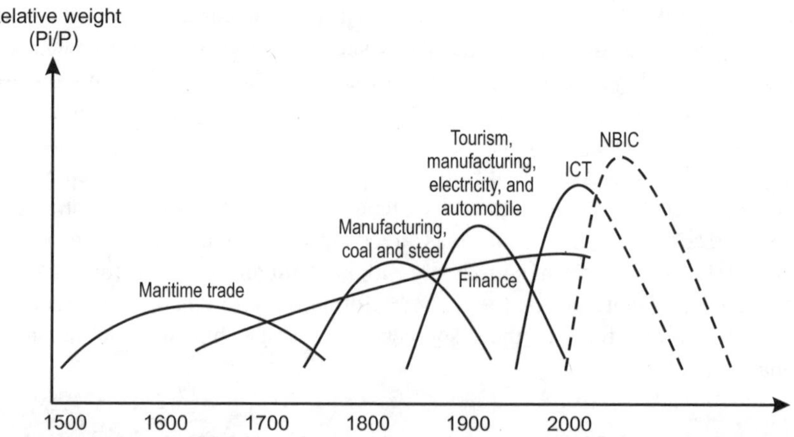

Figure 13.4 Innovation cascades in western Europe. The graph shows approximate dates and impact for several of the major innovation waves that have transformed Europe during the period A.D. 1000–2000. Used with kind permission of Denise Pumain (CNRS/Paris I, Paris, France).

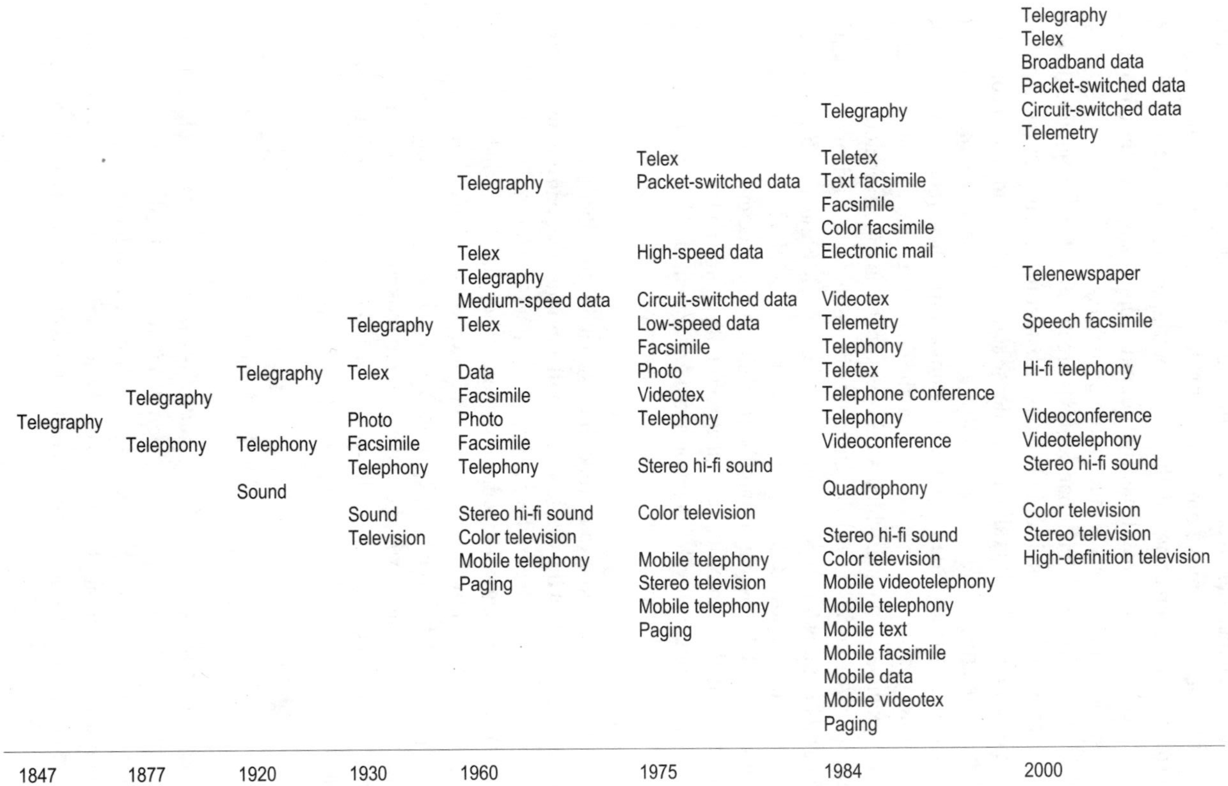

1847	1877	1920	1930	1960	1975	1984	2000
Telegraphy	Telegraphy Telephony	Telegraphy Telephony Sound	Telegraphy Telex Photo Facsimile Telephony Sound Television	Telegraphy Telex Telegraphy Medium-speed data Telex Data Facsimile Photo Facsimile Telephony Stereo hi-fi sound Color television Mobile telephony Paging	Telex Packet-switched data High-speed data Circuit-switched data Low-speed data Facsimile Photo Videotex Telephony Stereo hi-fi sound Color television Mobile telephony Stereo television Mobile telephony Paging	Telegraphy Teletex Text facsimile Facsimile Color facsimile Electronic mail Videotex Telemetry Telephony Teletex Telephone conference Telephony Videoconference Quadrophony Stereo hi-fi sound Color television Mobile videotelephony Mobile telephony Mobile text Mobile facsimile Mobile data Mobile videotex Paging	Telegraphy Telex Broadband data Packet-switched data Circuit-switched data Telemetry Telenewspaper Speech facsimile Hi-fi telephony Videoconference Videotelephony Stereo hi-fi sound Color television Stereo television High-definition television

Figure 13.5 Chronology of the innovation cascade triggered by the invention of the telephone.

Every once in a while, such a solution will theoretically fit a much wider class of latent problems. However, whether it will actually be seen to do so depends on the context in which the invention is either done or subsequently studied. If that context has a low cognitive dimensionality, then it most likely will not be applied to other situations, and will therefore not trigger a cascade of inventions. If the context, on the other hand, has high cognitive dimensionality, the emergence of a cascade of inventions is much more probable.

Earlier we discussed that urban contexts are much richer in cognitive dimensions than rural contexts, if only because urban contexts contain more people and a greater range of ideas all within a dense setting. This leads me to conclude that the more urbanized a society is, the higher are its chances to innovate (and the greater the society's dependency on innovation). There thus would seem to be an indirect causal relationship between the degree of urbanization of a society and its capacity to innovate, which would explain the observed acceleration in the rate at which innovation cascades emerge as being due to the urban explosion.

The other interesting aspect of Figure 13.4 is the role of finance. From our earlier arguments it follows that "value" plays a fundamental role in maintaining the various feedback loops that sustain the flow structure of society. Value is created in the realm of ideas (information) but attracts both energy and matter. Finance, of course, is the domain of value creation and management, and its rise was, until recently at least,[10] dependent on the added value generated by the nonfinancial innovation cascades. From that perspective, it is no surprise that each new innovation cascade creates more value (and more structure) than the last.

Does this mean that the whole system will continue if we can continue to find new sectors to add value to the economy in a timely manner? That would be a very positive conclusion, as we have argued earlier that the probability of new innovations, new innovation cascades, and new added value creation increases with population density and aggregation. Yet even if we do not want to conclude anything such as that, it would be interesting to look at the relation between innovation and value creation as the driver for today's demographic and urban explosion.

THE FUTURE

Although it is not, strictly speaking, part of my brief, I do want to close with a few remarks about the future. The period we have entered is in many ways unique in human history. For the first time, we are developing ways to isolate and control information. It took humanity tens of thousands of years to master matter, thousands of years to isolate and control energy, and now we are at the

10 Although in recent years the finance industry has itself become a generator of added value, I would argue that the link with value added in other domains is still essential to the system as a whole.

start of a period in which the third of the fundamental commodities of which our world consists, information, can be manipulated. The fact that this follows so closely on the heels of the Industrial Revolution is not accidental. It shows the strength of the invention and innovation dynamic that was set in motion a few centuries ago. It also shows how the concentration of information-processing capacity in urban areas has caused an exponential increase in innovative capacity.

A whole series of technologies are being developed (information technology, biotechnology, nanotechnology, and communication technology) and we have not yet reached our limits. Moreover, for the first time in human history, these technologies allow human beings to realize their capacity of reflexive intervention. It seems, therefore, that we may well be in a position to overcome a large number of the difficulties that we are facing today. If we are to do so, however, we need urgently to investigate and understand the dynamic of innovation itself, as it plays out in human societies. Without this understanding, we will never be able to control the direction "developments" will take, we will forever be running after the facts, and we will face and increasing number of unforeseen consequences of our actions.

One of the advantages of the archaeologist's perspective is that it makes us aware of the many times humanity has faced calamitous situations as well as huge risks to the survival of our environment and our species. Yes, in the nick of time, we have always been able to survive, thanks to our capacity to learn how to learn, and thereby accelerate our responses to dire circumstances. However, the archaeological record also documents collapse. I am therefore guardedly optimistic about the future. Future technologies will present us, more than ever before, with questions about what it means to be human. If the problems we face only require technological innovation for their solution, then there is reason for optimism. However, I question whether that is really the case.

ACKNOWLEDGMENT

All figures were prepared for this publication as part of the ISCOM research project.

REFERENCES

Bettencourt, L.M.A., J. Lobo and D. Strumsky. 2004. Invention in the city: Increasing returns to scale in metropolitan patenting. Working Paper 04-12-038. Santa Fe, NM: Santa Fe Institute.

Cronon, W. 2003. Changes in the Land: Indians, Colonists, and the Ecology of New England. New York: Hill and Wang.

Crutzen, P. and E. Stoermer. 2001. The "Anthropocene." *Glob. Change Newsl.* **41**: 12–13.

Duby, G. 1953. La société au XIe et XIIe siècles dans la region mâconnaise. Reprinted: Paris: Éditions de l'EHESS.

Guerin-Pace, F. 1993. Deux siècles de croissance urbaine. Paris: Anthropos.

Hodges, R. 1982. Dark Age Economics: The Origins of Towns and Trade A.D. 600–1000. London: Duckworth.

Holling, C.S. 1973. Resilience and stability of ecological systems. *Ann. Rev. Ecol. Syst.* **4**:1–23.

Holling, C.S. 1986. The resilience of terrestrial ecosystems: Local surprise and global change. In: Sustainable Development of the Biosphere, ed. W.C. Clark and R.E. Munn, pp. 292–317. Cambridge: Cambridge Univ. Press.

Ingold, T. 1986. The Appropriation of Nature: Essays on Human Ecology and Social Relations. Manchester: Manchester Univ. Press.

Lopez, R.S. 1966. The Birth of Europe [English Translation of: La Naissance de l'Europe, Paris, Arman Colin, 1962]. London: Evans-Lippincott.

Müller-Herold, U., and R.P. Sieferle. 1998. Surplus and survival: Risk, ruin and luxury in the evolution of early forms of subsistence. *Adv. Hum. Ecol.* **6**:201–220.

Padgett, J. 2001. Organizational genesis, identity and control: The transformation of banking Renaissance Florence. In: Networks and Markets, ed. J.E. Rauch and A. Casella, pp. 211–257. New York: Russell Sage.

Rappaport, R. 1976. Adaptation and maladaptation in Social Systems. In: The Ethical Basis of Economic Freedom, ed. I. Hill, pp. 39–79. Chapel Hill, NC: American Viewpoint, Inc.

Strumsky, D., J. Lobo, and L. Fleming. 2005. Metropolitan patenting, inventor agglomeration and social networks: A tale of two effects. Working Paper 05-02-004. Santa Fe, NM: Santa Fe Institute.

Tainter, J.A. 2000. Problem solving: Complexity, history, sustainability *Pop. Env.* **22(1)**: 3–41.

van der Leeuw, S.E. 2000. Making tools from stone and clay. In: Australian Archaeologist: Collected Papers in honour of J. Allen, ed. T. Murray and A. Anderson, pp. 69–88. Canberra: ANU Press.

van der Leeuw, S.E. 2001. "Vulnerability" and the integrated study of socio-natural phenomena. *IHDP Update* April:6–7.

van der Leeuw, S.E. 2005. Climate, hydrology, land use, and environmental degradation in the lower Rhone Valley during the Roman Period. *C.R. Geosci.* **337(1–2)**: 9–27.

van der Leeuw, S.E., and C. Aschan-Leygonie. 2005. A long-term perspective on resilience in socio-natural systems. In: Micro–Meso–Macro: Addressing Complex Systems Couplings, ed. H. Liljenström and U. Svedin, pp. 227–264. London: World Scientific Publ.

van der Leeuw, S.E., and B.J.M. de Vries. 2002. Empire: The Romans in the Mediterranean. In: Mappae Mundi: Humans and their Habitats in a Long-Term Socio-Ecological Perspective, ed. B.J.M. De Vries and J. Goudsblom, pp. 209–256, Amsterdam: Amsterdam Univ. Press.

van der Leeuw, S.E., and J. McGlade. 1993. Information, cohérence et dynamiques urbaines. In: Temporalités Urbaines, ed. B. Lepetit and D. Pumain, pp. 195–245. Paris: Anthropos/Economica.

Wallerstein. I. 1972. The Modern World System I: Capitalist Agriculture and the Origins of the European World-Economy in the Sixteenth Century. New York: Academic.

Wallerstein, I. 1980. The Modern World System II: Mercantilism and the Consolidation of the European World-Economy, 1600–1750. New York: Academic.

White, D.R. 2006. A qualitative network of the changes in Late Medieval European trade: A representational prelude to modeling macrosocial systems in the long 13th century. In: New Perspective on Innovation and Social Change, ed. D. Lane, D. Pumain, S.E. van der Leeuw, and G. West. Berlin: Springer, in press.

Left to right: Frank Hole, Richard Grove, Christian Pfister, Arnulf Grübler, Lisa Graumlich, Sander van der Leeuw, and John Dearing (not shown: Helmut Haberl)

14

Group Report: Integrating Socioenvironmental Interactions over Centennial Timescales

Needs and Issues

John A. Dearing, Rapporteur
Lisa J. Graumlich, Richard H. Grove, Arnulf Grübler, Helmut Haberl, Frank Hole, Christian Pfister, and Sander E. van der Leeuw

INTRODUCTION

This group report focuses on the means of improving our understanding of the dynamics of socioenvironmental change that occur over timescales of centuries, with specific focus on the past 1000 years. The term "socioenvironmental change" is used here to define the whole range of interactions that may connect together the climate, sociocultural, and ecological systems existing at any spatio-temporal scale. The overall importance in studying this timescale and period are as follows:

1. The past 1000 years were a period of substantial social and environmental change globally, exemplified by the growth of Islam, the European Renaissance, the 16th- to 17th-century scientific revolution, the rise of nation states, and the global exchange of European and Asian inventions and values. This period also encompasses the rise of colonialism, industrialization, global communications, accelerated global population growth, urbanization, significant modification of land use and land cover, and major climate change.
2. In this period, these changes fundamentally altered socioenvironmental dynamics that affect modern times, for example through the pervasive use of fossil fuels, the mass transport of people and goods, the development and spread of systems and technologies of information, and the

nature of globalization. Information, particularly, has been isolated as a commodity, increasingly transmitted and transported globally, and in increasingly shorter periods of time, in particular since the mid-19th century.

3. The seeds and drivers of both cumulative and systemic modifications of modern ecosystems and the Earth system are contained within this period. The past 1000 years include periods of crisis, vigorous growth, and dynamic stability that can place modern socioenvironmental interactions into a full and appropriate historical perspective. There is evidence that some socioenvironmental interactions are playing out on timescales of centuries and continue to the present day. We need to identify these contingent processes and physical legacies in order to understand fully current socioenvironmental changes.

Past, present, and future represent a continuum within which any idea of the future necessarily draws from experience. Historical narratives provide the only subject matter that allows us to reflect upon the future, particularly in terms of improving mitigation and adaptive strategies (Young, Leemans et al., this volume). For the past millennium, socioenvironmental interactions are normally described through a combination of data from documentary, instrumental, and natural archives. Compared with studies covering millennial timescales (Redman et al., this volume), there is generally less of an archaeological focus except in regions that lack written history. Continuous measurements are usually lacking before the 20th century, and have to be supplemented with the reconstruction of environmental change from analysis of historical documents as well as sedimentary, archaeological, and biological archives (e.g., Dearing, Battarbee et al. 2006a). The relatively imprecise or discontinuous nature of archival sources (proxy data) compared with modern data bases of census and instrument data and global observations means that the scope of studying centennial scales is constrained. Most data from before the mid-19th century reflect regional or local conditions; global or hemispheric time series extending back over 1000 years are currently only available for some atmospheric and oceanic properties (e.g., greenhouse gases). Therefore, scaling up local data to larger regions and the globe is a necessary, and major, challenge if centennial records are to be compared directly with 20th/21st-century records (Hibbard et al., this volume).

It should be noted that our research interest in studying phenomena of "global" change over centennial timescales within the last millennium comprises two distinct phenomena: (a) cumulative changes that are by definition localized (e.g., agricultural land clearing, or traffic congestion) but assume characteristics of global change because of their ubiquity across different ecological and social systems (cf. Turner et al. 1990), and (b) systemic changes operating at a planetary scale, like increasing trace gas composition of the atmosphere (cf. Steffen et al. 2004). It is only for systemic changes that construction of "global" indicators and data sets are really meaningful because the globalization of the human sphere has been a gradual process. However, while in principle most phenomena of interest within a millennial timescale perspective fall into the first category definition of cumulative global change, there are some important early precursors of increasing regional, and even global interdependence. The recent Syndrome Approach to global change (Schellnhuber et al. 2002) is an attempt to analyze regional and local settings that, though geographically different, are similar in terms of the interdependence between actors and the environment. Scaling down from global and hemispheric records in order to link them to local and regional socioenvironmental data sources represents further needs and challenges. This leads to an imbalance in how we actually understand past socioenvironmental interactions: the emphasis as we go back in time from the present focuses increasingly on local and regional case studies.

The focus in this report is often on Europe, which has not only provided the dominant socioenvironmental dynamic over the past 1000 years but probably provides the best combination of *currently available and accessible* archival data for the period. Although there is a voluminous literature dealing with climatic, societal, and ecological change over the past 1000 years, with notable exceptions (e.g., de Vries and Goudsblom 2002), too few attempts have been made to integrate systematically these three aspects of interconnected change. Therefore, the group aims were deliberately open-ended: to present perceptions, open questions, and controversies within the study of this timescale; to summarize some current trends in the field; and to suggest possible research projects. The focus was decidedly on critical and generic issues of methodology and understanding rather than historical review. Discussion was structured around a number of overarching topics and questions identified and introduced by members of the group. Thus the first section provides a possible schematic framework for studying socioenvironmental interactions and exemplifies through case studies some of the ways in which we derive insight from past records. This is followed by sections that deal with specific issues: understanding causality in complex systems; the role of technical innovation; and integrating frameworks based on information flow.

WHAT INSIGHTS CAN WE GAIN BY COMPARING HISTORICAL TRAJECTORIES OF SOCIOENVIRONMENTAL CHANGE?

Interactions and Scale

Socioenvironmental systems are complex entities that can be described by the nature of their intra- and inter-actions, and their temporal and spatial dimensions. Figure 14.1 provides a simple, but generic, schema to classify specific socioenvironmental systems according to the dominant intra- and inter-connections between natural forcings/climate, society, and ecosystems. The arrows

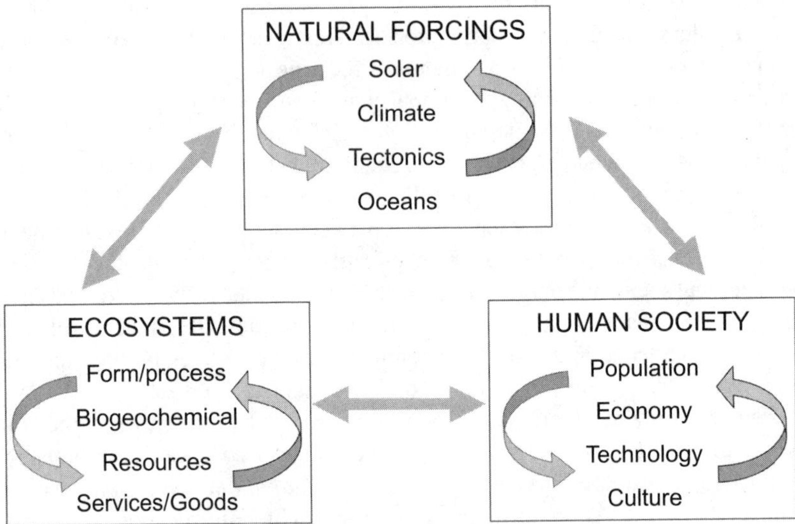

Figure 14.1 A schematic illustration of the potential interconnections between society, natural forcings, and ecosystems. Bi-directional arrows represent potential flows of energy, matter, and information between the three state systems that may define externally driven causality and feedback. Circular arrows within each box represent internal dynamical processes (Dearing 2006).

represent internal dynamics and sets of connections that may exist in both directions.

Using this schema it is possible to recognize numerous types of socioenvironmental conditions: impacts of climate on socioeconomic systems; indirect effects of climate on ecosystems that feed back on to social systems; direct impacts of human actions on ecosystem processes; and the potential feedback to climate through land cover change and atmospheric emissions. The value of the schema is that it recognizes a typology of interconnections in socioenvironmental systems that may vary through time at a specific location, and allows simple comparison between locations. The case studies summarized below illustrate the historical analysis for one set of conditions where evidence points to a strong climatic forcing of social behavior through the impact on ecosystem processes and agricultural productivity.

A full review of spatio-temporal variability is beyond the scope of this report, but Figure 14.2 exemplifies the range of detailed and high-resolution time series available for studying aspects of socioenvironmental systems during the last 1000 years at different spatial scales: demography, energy use, climate, crop harvests, and soil erosion. These particular time series illustrate the temporal quality and spatial scale of available data but are not necessarily causally linked. Indeed, Figure 14.2 serves to emphasize the importance of collecting parallel

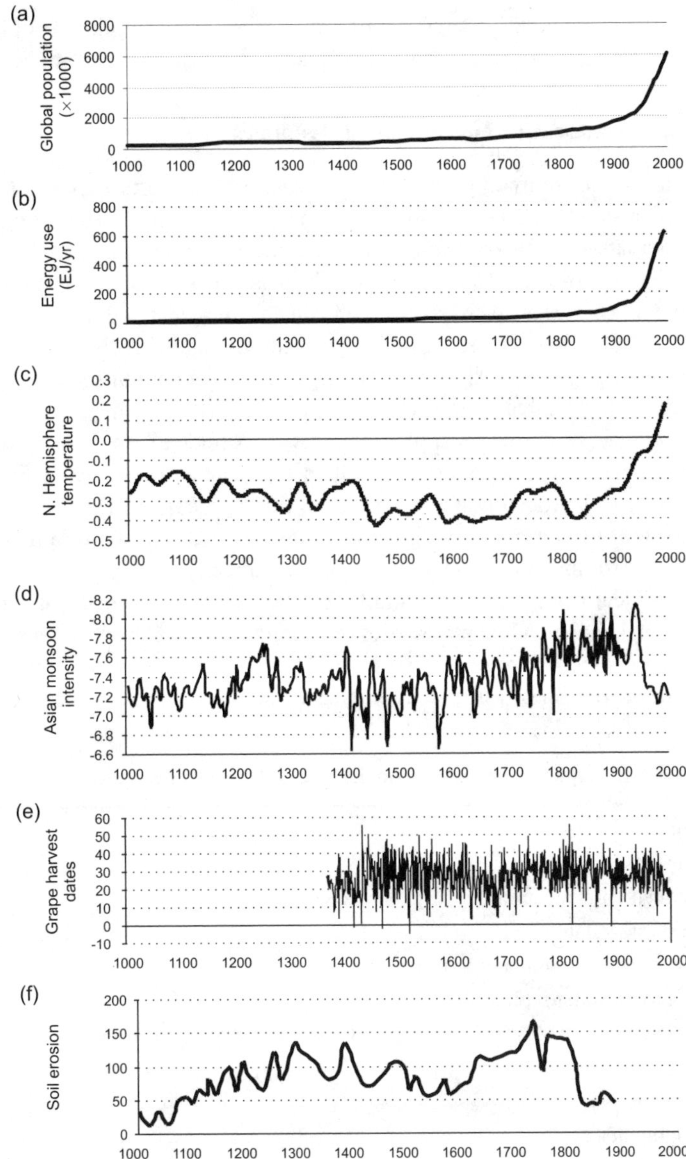

Figure 14.2 Selected centennial time series reflect the dynamic states of climate, ecosystems, and society, and the richness of data archives at this timescale. (a) global human population (Haberl 2006); (b) global human energy use (Haberl 2006); (c) Northern Hemisphere mean temperature as reconstructed from historical records, tree rings, and other proxy data (Jones and Mann 2004); (d) Asian monsoon intensity as indicated by oxygen isotope ratios in stalagmites in southern China (Wang et al. 2005); (e) yearly grape harvest dates from up to 18 villages in Burgundy; France (Chuine et al. 2004); (f) soil erosion rates in Germany as indicated by sediment cores (Zolitschka 1998).

records of independently dated data so that these can provide the optimum means for *testing* hypothesized causality, synchronicities, and teleconnections.

Centennial Timescales and Centennial Dynamics

To provide some common ground for discussing crises, perturbations, and trajectories, the group spent time considering the types of social and environmental processes that operate over centennial timescales, and the timing of important rates of change in the last 1000 years. Which social and environmental processes play out over centuries? And how have the time frames changed from the past to the present? This was seen as particularly useful for analyzing the origins of accelerated change in the 20th century. Categories of centennial-scale processes are divided into social and biophysical realms (Table 14.1), where the notion of "operating over centuries" is applied flexibly to exclude all processes from the upper list that clearly play out over less than 100 years or more than a thousand years. Major inflexions in rates of change of these processes (Table 14.2) were more difficult to compile, but agreement was reached that the timing should reflect a "saturating" effect and not the first instance.

What is clear from Tables 14.1 and 14.2 is the apparent shift in the balance toward societal change when major rates of change are considered. At least seven sets of major societal processes accelerated before the 20th century, in contrast with three in the biophysical realm. This difference may reflect the current

Table 14.1 Examples of centennial biophysical and societal dynamics.

Biophysical:
- Sea surface temperatures
- Centennial climate variability (e.g., Little Ice Age)
- Natural reforestation
- Recovery of soil depth and fertility
- Some groundwater replenishment
- Fluvial adjustment

Societal:
- Hegemonic rise
- Demographics
- Economic cycles (e.g., Braudel's *longue durée*)
- Invention and innovation—driven by need
- Social response and recovery from pandemics
- Complex problems with tradeoffs
- Some aspects of infrastructure
- Migration and transmission (human and human-assisted forms) crops/fauna
- Ideologies, memory times, and adaptive capacities

Table 14.2 Timings of some major inflexions in rates of change in biophysical and societal processes (Table 14.1) during the last 1000 years.

Biophysical

- Climate change/CO_2 concentrations: ~1800 (Moberg et al. 2005)
- Global increase in domesticated land: ~1900 (Steffen et al. 2004)
- Global loss of tropical rainforest: ~1925 (Steffen et al. 2004)

Societal

- Global population curve: 1800
- Individual and social energy usage: 1600–1950
- Speeding up of hegemonic rise and decline: 1450–1500
- Shortened social response and recovery to pandemics: 19th- to 20th-century medical intervention
- Invention and innovation speeded up/driven by supply—fertilizers: 1880
- Demographic transition and health care: industrialized 1800, developing 1950
- Ideologies (nationalism and fundamentalism): 1750
- The nature of archiving and literacy: printing press 1450
- Quest for economic growth, military: ~1650
- Perception of poverty and loss of hope: post-1950

unavailability of systemic historical data for many environmental processes or the incompleteness of Tables 14.1 and 14.2, but in any case reinforces the view that not all modern socioenvironmental change is rooted in the 20th century—some have seeds stretching back several centuries.

Theoretical Dynamical Frameworks

There are numerous published examples of well-researched crises and perturbations. Therefore, the group reviewed a framework for studying crises as the conjunction of processes with different dynamical timescales. As background (provided by Carole Crumley), it was useful to learn about the approach developed in the *Annales d'histoire économique et sociale,* founded by Marc Bloch and Lucien Febvre in 1928. Its main innovation was to shift the focus from writing problem-orientated analytical history to looking at human activity comprehensively. One of their most enduring innovations is the threefold means of dividing up the past:

- *Événement* (event): any portion of history that can be seen to have a clear beginning and end, and can be very short (a comet) or rather long (World War I).
- *Conjoncture* (concatenation): simultaneous arrival in time and space of trends, individuals, objects, etc. that each have histories, rhythms, and characteristics, but which together produce a notable effect.

- *Longue Durée* (long-term change) refers to trends, slow accretions, and other phenomena that build over long periods of time.

One example of a concatenation occurred when agrarian society created substantive links with industry. In Europe these links occurred in 1880 and 1950, and represented key phases of buffering capacity against climatic shocks. Expansion of maritime trade helped to avoid crises, as seen by the greater vulnerability of land-locked European countries (e.g., Austria) to climate events compared to, for example, the U.K. and Denmark. In 1880, railways allowed the import of grain from Russia and elsewhere into western Europe, without which additional climate-induced crises would have occurred. The Green Revolution in the mid-20th century strengthened further the dependency of agricultural productivity on industrial fertilizers.

Narratives of Socioenvironmental Change

In practice, narratives based on case studies dominate the means by which historical data and their trajectories are analyzed and communicated. The following example illustrates the insights that can be obtained through careful and rigorous analysis of independent climate and social records. Similar exercises have been done for different places and times, or could be undertaken in new research programs. The following paragraphs document the "Climatic anomalies and clusters of crises in Europe and Asia 1300–1700." Medium-term subsistence crises (2–40 years) may be superimposed upon processes that operate or vary over decades and centuries, such as solar variations and sea surface temperatures. The main human indicators for such crises are surges in grain prices (wheat, rye, rice), famine-mortality, a drop in birthrate, and domestic stock mortality causing multidecadal regional economic decline. Initially we discuss the significance of a cluster of extreme events in 14th-century Asia and Europe. We then discuss the impact of a longer cluster of extreme events from 1570 to 1700, a period which is sometimes referred to as the "Seventeenth-Century Crisis" in both Europe and Asia. We explore the evidence for exogenous impacts on agrarian societies based upon biomass production and consider the social and State response to these extreme events, although the evidence for the latter is necessarily somewhat impressionistic.

Fourteenth Century

In Europe the cluster of extreme events begins with the crisis of 1315–1317 that is related to Little Ice Age-type impacts (Pfister 2005). These are characterized by a combination of rainy autumns, cold springs, and wet summers. Such years are usually a consequence of major volcanic eruptions. The impact of the anomalous climate of 1315 carried on until at least the end of 1317. These are also the years of the first European plague epidemic of the 14th century

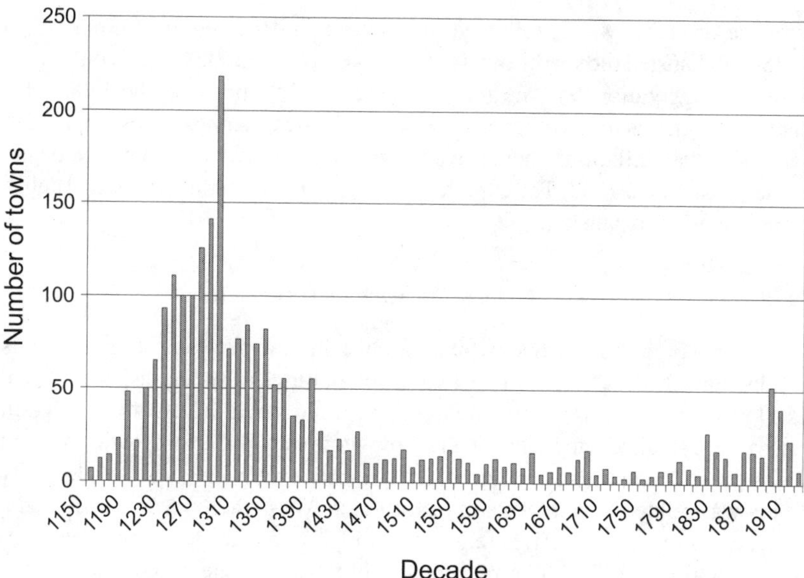

Figure 14.3 Foundation towns in Central Europe by decade (1150–1950). Values estimated from the graph in Stoob (1956).

(Gottfried 1983). The wider socioeconomic effects of this crisis were entirely unanticipated by society and outside the experience of two generations. It led directly to a breakdown in the previously vigorous phase of urban expansion (Figure 14.3). The period 1338 to 1350 includes the last huge locust invasion (1338), the "millennium flood" of 1342 (Germany, Switzerland, Netherlands), and a sequence of cold summers from 1345 to 1347, of which the summer of 1347 was the coldest of the millennium (Pfister 1988). From 1347 to 1350 there was the "Black Death," a pneumonic plague that was also associated with a "hemorrhagic disease" and high animal mortality (Scott and Duncan 2004). In 1348 a "millennium earthquake" in the Alpine area is known to have occurred. This unique cluster of extreme events undoubtedly undermined social confidence. One consequence was the seeking out of particular social groups as scapegoats for disaster, seen in terms of a major wave of anti-Semitic pogroms and systematic discrimination.

Turning to South Asia, the records of the Delhi Sultanate show prolonged, recurrent drought periods between 1296 and 1317. After 1300, there is evidence for prolonged drought and harvest failure in Upper Burma leading directly to the decline of the Pagan kingdom. After 1300 the lands of Pagana and Pakkoku became too arid for any kind of cultivation let alone that of rice. In the Delhi Sultanate, all attempts to maintain previously fixed levels of prices broke down completely after 1317, and the financial chaos that ensued rendered the state totally incapable of responding administratively to deficient harvests. A few years later the droughts which followed in uninterrupted succession from 1343 to 1345 so exhausted the sparks of vitality in his empire that Zia Barani could write without exaggeration that "the glory of the state and the power of the Sultan Muhammad from this time withered and decayed." These harvest failures helped to cause runaway inflation. There is some evidence that the drought period began earlier, possibly in 1333. The experience of recurrent droughts stimulated deliberate irrigation expansion.

1570 to 1720: The "Long Seventeenth-Century Crisis"

During this period in Europe two major clusters of anomalous climatic events are distinguished, which produce two different kinds of social response. From 1500 to 1570 crises were rare and not very pronounced. Fluctuations of grain prices were small and there was a phase of vigorous population growth. Around 1560, pre-14th-century population levels are regained. The crisis of the early 1570s was related to a succession of Little Ice Age-type impacts. It initiated a long period of frequent crises (e.g., 1585 to 1597, around 1610, and again between 1626 and 1629). The first cluster is related to the explosion of Billy Mitchell (Bougainville Island, Melanesia) in 1579/1580, Kelut (Java) in 1586, Raung (Java) in 1593, Ruiz (Colombia) in 1595, and Huaynaputina (Peru) in 1600 (Pfister 2005). The frequency and severity of "unnatural weather" and crises led to a surge of witch-hunts in weak states where no standing army was available to suppress the populist desire for scapegoats. This included Lorraine, the canton of Bern, and the southern German bishoprics (Behringer 1999). In general this period led to disorientation and pessimism. Many people thought that the end of the world was near.

The second period of crises culminated in the period 1688 to 1694 in western Europe (Lachiver 1991) and 1696/1697 in Northeastern Europe. Finland in particular experienced high mortality (Vesajoki and Tornberg 1994). The period of crises ends basically after the cold winter of 1708/1709. From 1650, States in western and central Europe increasingly responded to these crises by setting up regional grain stores. However, the effect of these measures is little understood. Distributing grain in times of crises beyond the capitals, within the entire territory, reflects a fundamental expansion in the notion of the role of the State itself toward a paternalistic regime.

By the 1570s, serious droughts had occurred again in South Asia, of a severity that had not occurred since 1410. From 1578 to 1579 a very severe El Niño initiated a pattern similar to that of the early 14th century, so that major famines were experienced at very frequent intervals from 1577 to 1710 throughout South and Southeast Asia. The most severe events were experienced simultaneously in India, Burma, and the East Indies, especially in 1614–1616, 1623/1624, 1629–1632, 1660–1662, 1665–1666, and 1685–1688. In the East Indies, the years 1624 to 1627 stand out as a period of severe drought. In Java, consecutive famines occurred between 1625 and 1627. In 1624, in Banten, stagnant water-courses stimulated malaria epidemics in which several thousands died. In the East Indies, El Niño droughts notably were associated with massive disease events, often leading to 50% mortality. During the El Niño droughts of 1629 to 1631 in Gujarat, up to three million people may have perished as a result. In addition, unprecedented out-migration led to permanent changes in the agrarian structure of western India. It is likely that this was the most destructive Indian famine of the early modern era. The events of 1629 to 1632 severely affected Burma, Bali, Siam, Java, and Central Asia. The climatic crises of the Long Seventeenth Century in Asia ended with the El Niño that caused the droughts of 1707 to 1709. They gave rise to massive social disruption, migration, and military conflict. They stimulated new kinds of property rights, taxation systems, and periods of runaway inflation. They also fundamentally increased the interdependence between regions of South and Southeast Asia, and directly caused an exponential growth in the trade of rice and other foods. Both in western Europe and India, the responses to extreme climate crises described above implied that the rising power of the State became more interventionist. This marks an important turning point in the evolution of the western state itself, setting the stage for later confrontations over the legitimate scope of states and governments.

Research Needs

Descriptions of such crises and perturbations within socioenvironmental systems abound in the literature, but what do we still need to know about the last 1000 years to improve our understanding of them? Table 14.3 describes some questions identified by the group as possible research priorities.

HOW DO WE MOVE BEYOND SIMPLISTIC NOTIONS OF CAUSALITY IN EXPLAINING PAST SOCIOENVIRONMENTAL SYSTEMS?

The background to this question lies with the growing awareness that the complexity of Earth and world systems, as represented by Figure 14.1, exists as a result of nonlinear interactions within and between systems. As a result it has been argued that adopting some form of "scientific realism" would be a suitable approach (Prigogine 1996). This position accepts the existence of nonobservable phenomena, structured and stratified systems with emergent properties, contingent relationships and prediction based largely on probabilities, and rejects a belief in direct and enduring relationships between cause and effect (Richards 1990). As Richards argues, the validity of assuming that falsification is possible in open systems is questionable and may be more appropriately supplanted by emphasis on the internal consistencies of theories and the explanations that

Table 14.3 Some data needs for studying past centennial crises and perturbations.

Biophysical

- Detailed and high-resolution climate histories for many parts of the world.
- Detailed and high-resolution reconstructions of environmental impacts on ecological, geomorphic, fluvial, limnic, and coastal systems at local or regional scales.
- Information about epizootics and the decline of domesticated animals, cattle, and draught power, a potentially multidecadal effect.
- Significance of environmental impact by all indigenous peoples.

Economic

- Role of capitalism on environmental change: where, when, who?
- Role of colonialism (and linked capitalist ideas) in changing environments (e.g., plantations, slaves).
- Role of credit systems in driving captialism and environmental degradation.
- Role of invention/innovation.
- Reasons for the acceleration and decelertion of economies.
- Effects of trade and specialization.
- Impacts of transport revolutions.

Social-cultural

- Drivers of agrarian–industrial conjunctions.
- Environmental history of the Russian Empire.
- Significance of religion for environmental impacts on governmental and individual attitudes.
- Knowledge transfer of successful environmental conservation and management scheme.
- What do we know about the changing levels of environmental knowledge?

derive from them. Thus, "the criteria for acceptance of a theory are not based on predictive success, but on explanatory power" (Richards 1990, p. 196).

Such an epistemological debate could be viewed as no more than a distraction derived from academic communities who rightly perceive the incompatibilities that exist between the *time domains* available for their theoretical, observational, monitoring, and modeling endeavors. This is particularly the case for studying environmental change over centennial timescales. Centennial perspectives are potentially important in providing descriptive explanations (cf. Deevey 1984), the means to observe the full pattern of environmental change that has preceded the present, and to aid the development and testing of predictive models and scenarios generation. But how do we maximize our opportunities to analyze centennial timescales in ways that generate an improved understanding of the dynamics of modern socioenvironmental systems? How do we tell the richer story of what has been a period of unprecedented growth, collapse and social demise?

Preindustrial Socioenvironmental Systems

One essential prerequisite to studying the complexity of human–environment interactions is not to assume at the outset that human activities are omissible. For early agrarian societies or pre-European agriculturalists, there may be a tendency or temptation to assume that human impacts on their immediate environment were insignificant. However, given the recent focus on the possible impact that early deforestation and paddy field farming may have had on global greenhouse gases (Ruddiman 2003), and the growing evidence for very early domestication of cereals stretching back to ~14,000 B.P. in western China (Yasuda 2002), and the wealth of evidence for industrialization and deforestation across Asia, Africa, and Europe during the Bronze and Iron Ages, it has to be argued that an a priori assumption that humans did not disturb their environment represents a flawed approach. A post-hoc hypothesis testing or modeling approach that seeks either to eliminate the effects of human actions or to quantify their impact is preferable. Thus we might query whether the *rate and nature* of the mid-Holocene desertification of the Sahara was entirely driven by the nonlinear switch in mid-Holocene precipitation caused by gradual changes in insolation (Claussen et al. 1999), or whether positive feedback on the vegetation cover through pastoralism, evidence for which is supported by the archaeological record (Hassan 2002), should also be considered. And if Holocene deforestation and wet, rice paddy cultivation cannot explain the whole pattern of atmospheric CO_2 and CH_4 concentrations, there still remains the unanswered question as to whether it was responsible for a measurable fraction of the increases.

Contingency and Timelines

A key and multifaceted concept in environmental change is contingency: the dependence of conditions on the operation of previous processes. This can take various forms, such as momentum/inertia, emergence and conditioning, and provides a powerful set of mechanisms for analyzing the links between the present and past (Dearing, this volume). Many European landscapes and ecosystems are clearly the products of previous agricultural regimes, but some have evolved along trajectories that may be essentially irreversible or have yet to reach a dynamic equilibrium. In a different context, house plans in the coastal areas of the Netherlands were developed in the 5th century B.C., but persisted well into the 20th century when the material constraints and world view, which had been responsible for their design, were no longer valid. Also, the legacy of the width of the U.K. and U.S. railway carriages, which in turn was a function of the width of coaches being driven in Europe over roads originally built in Roman times, is now seen in terms of the constraints on the size of NASA's rocket boosters!

One corollary is the idea that for any set of socioenvironmental conditions there may be a *critical length of time* before which events effectively become unconnected to the present, that is, contingency no longer operates. How do we ascertain the length of this timeline or determine whether, for practical purposes, it actually exists? From existing analyses we might argue that there are at least first- and second-order forcings of change. For example, to understand the rise and fall of a civilization like the Roman Empire it may be necessary to analyze the period well before the start: first-order forcings, in the shape of the economics and politics of the Roman Empire, are clear, but second-order forcings in the form of the agricultural revolution of the 8th to 5th centuries B.C. in the Northern Mediterranean, and the growth of population aggregation into towns outside the Roman area in the period after the 3rd century B.C., are necessary to gain a full understanding. A further example, the Anthropocene era (Crutzen and Stoermer 2001), also demonstrates the difficulty of defining the timing of the switch from nature-dominated to human-dominated global systems (Messerli et al. 2000). Was it the time when humans first had the technological capacity to modify the Earth system through cumulative impacts (e.g., the early to mid-Holocene: cf. Ruddiman 2003); the age of exploration and accelerated global connectivity (~1400); the beginning of the large-scale use of fossil fuels and steam technology (~1750–1850); the time when, according to paleoenvironmental analyses, global atmospheric pollution had risen above "natural" levels (~1800); or the unprecedented acceleration in resource depletion and ecosystem modification during the 20th century (~1950)? One conclusion is that to understand these phenomena we need to develop a multitemporal approach that is able to determine the appropriate temporal scales of analysis in interaction with the kinds of phenomena observed. When dealing with trajectories, it may be the case that the timescale of analysis has to be carefully chosen so that it unravels the apparent "moment" that can help reveal the underlying dynamics. Where there are clearly observable shifts in trajectory (acceleration or deceleration) future research needs to find ways of determining the causal chain. Were they responses to environmental crises, the result of a set of conjunctures, or a shift in values?

Dynamic Modeling of Socioenvironmental Phenomena

However detailed and penetrating, a full analysis of all available past records will not be able to generate alternative strategies for sustainable management *that can be tested* for their accuracy in informing the future over decadal–centennial timescales. The power of socioenvironmental reconstruction and narrative can only be utilized fully to inform alternative views of the future through simulation modeling. Case studies reinforce the nonlinear and contingent nature of the modern socioenvironmental system but also underline the challenge for simulation models whose goal is to extend socioenvironmental dynamics into the future: that is, to be functionally and dynamically realistic. At the heart of the challenge is the dichotomy in scientific methodologies: on one hand, they follow a reductionist path; on the other, they tackle systems holistically. Any method that promises to understand the holistic behavior of systems needs to include both fundamental rules and the means to synthesize these into emergent phenomena. Thus, we need simulation models that allow complex and macroscale emergent phenomena to arise from microscale interactions, with as few constraints as possible on spatial and temporal scales (Young, Leemans et al., this volume). The main requirements for socioenvironmental simulation models at local and regional scales are to capture the simultaneous growth of emergent phenomena over a variety of timescales and within spatially defined zones (Dearing 2006; Dearing, Battarbee et al. 2006b). In theory this can be achieved through models that simulate interactions between processes represented by fundamental rules.

Cellular automata (CA) models appear to satisfy many of these requirements. Cellular automata were originally created as toy models to simulate the complexity of hypothetical systems, but have now graduated to applications in the natural and social sciences. At their basis lies a spatially explicit landscape defined as a series of contiguous cells. Each cell has a number of rules that determine how neighboring cells will change. At each time step the state and conditions of each cell are updated to provide new states and conditions for the rules to operate on. Through continuous interaction, the rules generate emergent patterns and features, capturing along the way the feedbacks, time lags, and leads that prove so intractable to alternative methods. Complex and unpredictable behavior is typical of even simple toy models whose cells have rules for whether they should turn black or white according to the state of neighboring cells (Wolfram 2002).

Cellular automata models can be classified according to the level of functional rules used, the means by which and the timescales over which the model is validated, and the extent to which the activities of human agents and decision-making are made explicit. Tucker and Slingerland (1997) and Coulthard et al. (2002) have pioneered the use of mathematical biophysical cellular models in catchment hydrology with low-level rules, long timescales ranging from decades to millennia, but with limited inclusion of agents. Similar models have been developed for coastal zones. For example, Costanza and Ruth (1998) describe the use of the generic STELLA computing language to develop a simulation model of the Louisiana coastal wetlands. Set up with a spatial scale of 1 km^2, the model simulates the changing nature of the Louisiana coast over 50–100 year timescales as a function of management alternatives and climate variations. A similar approach has been adopted by Dearing, Richmond et al. (2006) in the CEMCOS model of estuarine sediment dynamics at a spatial scale of 2500 m^2. Wirtz and Lemmen (2003) use microbiological algorithms to simulate the centennial growth and distribution of human populations across the Neolithic transition, and find good correspondence with archaeological data. However, normally, the inclusion of human agents makes use of high-level rules and often a restricted history. For example, there are many models that simulate urban development. Wu and Martin (2002) model the potential growth of London as a function of land-use probability scores defined by proximity to, for example, transport networks, validating the model for 1991 and 1997.

There are ongoing developments that are likely to see improved CA-based modeling, through integration with GIS, macrolevel models and, in ecology, developing individual-based approaches (e.g., Gimblett 2002). With the increased availability of computer grid systems, processing power is unlikely to impose a major constraint. Perhaps most headway toward the development of integrated simulation models has been gained through the development of agent-based models (ABMs), particularly among the communities attempting to model interactions between economic agents in the development and adoption of new technologies, and changes in land cover and land use (e.g., IGBP–IHDP "Land Use and Cover Change" and "Global Land Project" programs). Validation has largely come through sequential maps of land cover derived from satellite imagery since the 1960s. ABMs combine a cellular model of a landscape with an agent-based model that introduces decision-making (Parker et al. 2002). Unfortunately, our current knowledge and theory do not allow us to define with any certainty the precise structure of the proposed socioenvironmental models. We probably have to recognize the importance of trial-and-error approaches, particularly in selecting the appropriate rules. Model development through experimentation at different process levels and spatio-temporal scales can be expected to advance theory significantly about decadal-centennial change in socio-environmental systems, but we are entering a phase of simulation model development where our highest level of certainty lies in the parallel histories that we can construct from historical case studies.

HOW DO TECHNICAL INNOVATIONS AND INVENTIONS MODIFY SOCIOENVIRONMENTAL INTERACTIONS?

From the earliest agrarian and urbanizing societies to the very present day, "technology" has been the main mediator between human societies and their respective environments. The impacts of technology on the environment are threefold: (a) negative impacts, which may be either direct (e.g., DDT buildup) or indirect (e.g., increased productivity leading to reduced consumer product costs and thus increased consumption); (b) positive impacts (e.g., environmental remediation, such as traditional "end-of-pipe" environmental effluent reduction technologies); and, finally, (c) technology as cognitive instrument to improve the understanding of socioenvironmental interactions (e.g., satellite observation systems, global climate models, etc.).

In its most fundamental definition, "technology" is a system of means to achieve particular ends that employs both technical artifacts (hardware, equipment) as well as (social) information. Information may be subdivided into "software," the know-how for the production and use of artifacts, and "orgware," the institutional settings governing rules and incentive structures for the development and employment of technologies (cf. Grübler 1998). These dual characteristics of technical artifacts and social information are important for a deeper understanding of how particular *systems of production and use*, which structure socioenvironmental interactions, emerge historically.

In addition, there is also a graduation in the evolutionary development sequence of technology. In the most fundamental (Schumpeterian) distinction there are four phases (Figure 14.4): invention (conception of a new idea), innovation (first market introduction of new idea, product, or process), niche market applications (that provide for essential experimentation and learning with new technologies), and (if successful in the earlier phases) pervasive diffusion. These are ultimately followed by decline and substitution by alternative technological solutions. This terminological distinction is of importance, as major impacts on socioeconomic–environmental interactions arise only when technologies are *widely* applied, that is, via technology diffusion, when individual technologies coalesce into entire dominant technology systems (Table 14.4).

Figure 14.4 also illustrates the two main aggregate hypotheses about the sources of technological change: "supply push" versus "demand pull." In fact, the literature has reached the conclusion that the suggested contrast between these seemingly opposed mechanisms of technological change is false (cf. Freeman 1994) because both factors are intrinsically intertwined, as are the seemingly sequential phases in a technology's life cycle. In other words there is broad consensus that the traditional linear model of innovation is dead. There is no automatic trickle-down effect across various phases of a technology life cycle. "Progress," therefore, requires continuous inputs via investments and appropriate incentive structures. Improved technology also requires gaining experience through actual market applications, as only these provide the essential learning

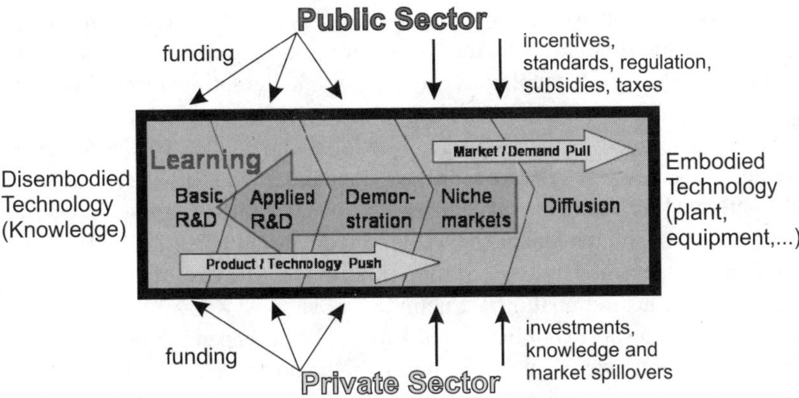

Figure 14.4 The black box of technology. A schematic illustration of main drivers and actors involved in technology development, deployment, and diffusion.

Table 14.4 Clusters of technologies and infrastructures characterizing particular periods of economic growth, which structure the interactions between socioeconomic systems and the environment. Dominant (top) and emerging (bottom) clusters that in turn become the dominant ones in the subsequent period. These technological/infrastructural clusters are embedded in concomitant social/institutional settings or "organizational styles" (Freeman and Perez 1988) and distinct core-periphery relationships in the development and use of technologies and resulting environmental impacts (not shown in the table; for a discussion see Grübler 1998).

1770–1830	1820–1890	1880–1945	1935–1995	1985–2050
Dominant				
Water power, ships, canals	Coal, railroads, steam power, mechanical equipment	Cars, trucks, trolleys, chemical industry, metallurgial processes	Electric power, oil, airplanes, radio, television, instruments and controls	Gas, nuclear, information, satellite and laser telecommunications
Emerging				
Mechanical equipment, coal, stationary steam power	Electricity, internal combustion, telegraphy, steam shipping	Electronics, jet engine, air transport	Nuclear, computers, gas, telecommunications	Biotechnology, artifical intelligence, space communication and transport

feedbacks (cf. Figure 14.4) to research and development. All these transitions between individual phases in a technology life cycle, augmented by feedbacks, explain why the process of technological development and diffusion is a long one. Development and diffusion times of many decades up to a century are characteristic for pervasive technology systems such as energy or transport infrastructures.

Equally important is to recognize that changes in technology are always intrinsically intertwined with economic and social processes. Social norms, incentives, and institutions structure the types of technologies that are developed and deployed. In turn the technologies in use enable changes in the form of social organization or the overcoming of resource and environmental constraints for societies. These dual feedbacks are an additional explanation why far-reaching changes in technology systems can take substantial periods of time and are often characterized by ruptures and "mismatches" between the various interacting components of an interdependent system (Freeman and Perez 1988). Historical research has identified five consecutive "clusters" or "waves" of interdependent technological/institutional/infrastructural innovations whose emergence and subsequent widespread application (diffusion) have shaped socioeconomic/environmental interaction since the onset of the Industrial Revolution (Table 14.4). Dominant technologies and infrastructures are paralleled by emerging systems that then become the dominant form in the "next wave"—a time-lag between core and peripheral technologies that is also mirrored in core–periphery relationships in the spatial extent of adoption of specific technology clusters. Dominant technology clusters may represent great barriers to change. Each cluster requires a specific organizational structure and this can often represent the largest barrier of all to change. Concepts of technological path dependency and lock-in are particularly relevant here (Arthur 1989).

There is wide agreement (e.g., Landes 1969; Mokyr 1990; Freeman and Louca 2001) that there have been profound changes in the social, economic, and institutional embedding of technology that have taken place over at least the past 300 years and that these have paved the way for the Great Acceleration of the 20th century (Hibbard et al., this volume). The major historical divide was the systematic development and application of new technologies at the origin of the Industrial Revolution, and which propelled Europe (and its offspring) from the position of a technological and economic laggard (compared to China or the Islamic world around 1000) to world leadership (Rosenberg and Birdzell 1986).

Historical Path Dependency and the Limits of Historical Contingencies

Energy and transport systems are particularly important *indicator technologies* for characterizing clusters and waves (Table 14.4). A sequence of transitions have characterized the evolution of fossil energy use since the onset of the Industrial Revolution, from systems dominated by renewable energy flows (wood) to systems dominated by coal, to their current dominance by oil and gas. Important consequences of these successive technological energy transitions were the possibility of a continued expansion of energy use in ever more diverse applications, rising levels of productivity and incomes, and evidently rising environmental impacts.

Figure 14.5 illustrates the historical evolution of rising energy use with reference to rising per capita incomes for the last 200 years for selected industrialized countries. It is interesting to note that while the overall relationship between growing energy use and economic development is pervasive, the development trajectories are diverse and punctuated by crises and turbulences. For example, the backward "snarl" for the U.S. during the 1930s is testimony to the Great Depression.

The differences in development trajectories span the extremes of "high intensity" (U.S.) to "high efficiency" (Japan). These originate in differences in initial conditions (e.g., resource endowments, relative prices) that lead to the adoption of particular technology and infrastructure development trajectories, which in turn influence, for example, settlement patterns and economic structures. Because of the cumulative nature of technological change, such development trajectories are persistent and maintain their momentum even when initial conditions (e.g., resource abundance) no longer prevail. This twin dependence on initial conditions and the development path followed has come to be known as "path dependency" and its resulting technological inertia as "lock-in." Path

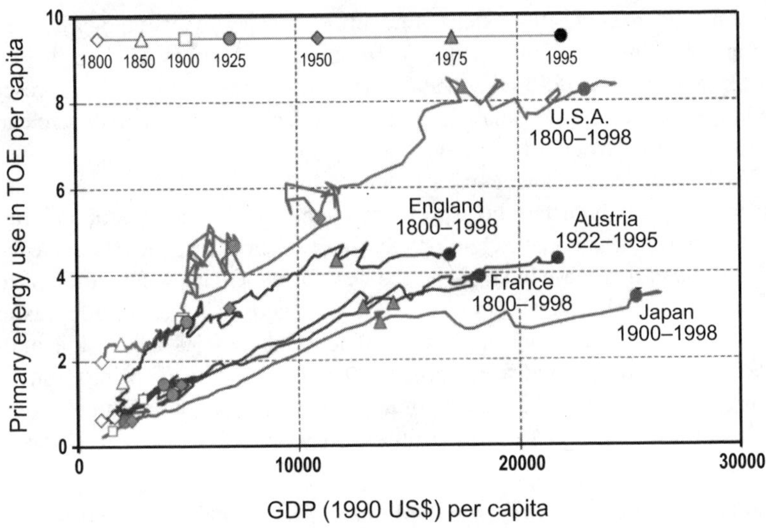

Figure 14.5 Per capita energy use versus income (GDP expressed at market exchange rates) for selected countries since 1800. TOE = tons oil equivalent. Adapted from Grübler (2004).

dependency has been identified in a variety of socioeconomic contexts and illustrates the value of having a deeper historical understanding of the pathways and conditions that have led to our current predicament.

However, Figure 14.5 also serves as a useful reminder that a historical perspective should not be misinterpreted as necessarily representing historical contingencies for the present and future, as even path-dependent phenomena can exhibit significantly different trends. For example, ever since the oil crises of the 1970s, growth in income has become progressively decoupled from growth in energy use. This break in trend was certainly not anticipated in the 1960s and 1970s, which from today's perspective were periods characterized by over-projections of future energy use and, as a result, ill-advised policies on perceived "silver bullets" of energy supply technology, such as the fast breeder reactor. Conversely, the true value of a historical perspective is not that it offers prescriptions for gazing into the crystal ball of an uncertain future, but instead that it informs us about the dynamics of change and the inevitable surprises and discontinuities of both social and technological nature that have shaped the interactions between socioeconomic, technological, and environmental systems.

Environmental Impacts beyond Air Emissions

The last few centuries have increasingly been characterized by clusters of technologies and infrastructures that enabled the large-scale utilization of new resources (Grübler 1998). The introduction of such new technologies has resulted in profound changes in socioeconomic structures, in the environment, and thus in socioenvironmental systems as a whole. Probably the most far-reaching change in socioenvironmental systems during the past 1000 years was triggered by the transition from the area-based energy system of agrarian societies to the fossil energy system of industrial society (Sieferle 2001; Smil 1992). Modern concerns about the impact of fossil fuels, which has followed the transition, focus upon atmospheric emissions of gases and particulates (e.g., IPCC 2001). But changes in energy use have had, and continue to have, far-reaching impacts beyond atmospheric pollution. This is by no means a historical, already completed process. Around the world the transition from a mainly biomass-based, agrarian pattern of energy use to fossil fuels is proceeding rapidly in most of the developing countries. This process is not only expected to have far-reaching impacts on global energy consumption and greenhouse gas emissions (Nakícenovíc and Swart 2001), it will also fundamentally transform socioenvironmental systems.

What can we learn from past experience and records of energy transitions about the likely impacts? Here we use the example of Austria 1830 to 2000 within its current boundaries (i.e., a reconstruction of Austria's current territory for the time before 1914) to underline major features of this transition. This period includes almost all of Austria's industrialization, as fossil fuel use was negligible at the beginning (<1%). Data show (Figure 14.6) that per-capita biomass use remained in a similar range throughout the period, while aggregate biomass use (and extraction from Austria's territory) increased considerably. Although some uses of biomass (e.g., fuel wood for industrial processes) were abandoned or at least greatly reduced in the process, per-capita biomass consumption was kept up by the transition to a diet more rich in animal food, which greatly increased the amount of biomass required in the food system. Figure 14.6 also shows that, until about 1950, industrialization was almost solely fueled by coal, while other energy carriers gained importance after that date.

The transition from the agrarian to the industrial energy system implies changes in resources used, in technology, in settlement patterns, in urban–rural relations, and in land use. During industrialization, fossil fuels are increasingly used and biomass loses its role as the main energy source of society. While the agrarian energy system relies almost exclusively on harnessing net primary production (NPP) for human energy needs, industrial metabolism relies on fossil fuels, the use of which is not constrained by land availability. The transition to an industrial energy system thus removes tight limits to growth imposed on agrarian societies by their dependence on the NPP of local ecosystems. The transition from biomass to fossil energy (and later nuclear energy, large-scale hydropower, and new renewable sources) was thus a precondition for the rapid economic (monetary and biophysical) growth in Austria, and much of Europe, from the 18th century onward (Sieferle 2001; Smil 1992). The role of land use in the socioeconomic metabolism thus changes qualitatively and quantitatively during industrialization. In an agrarian society, the agricultural sector as a whole must

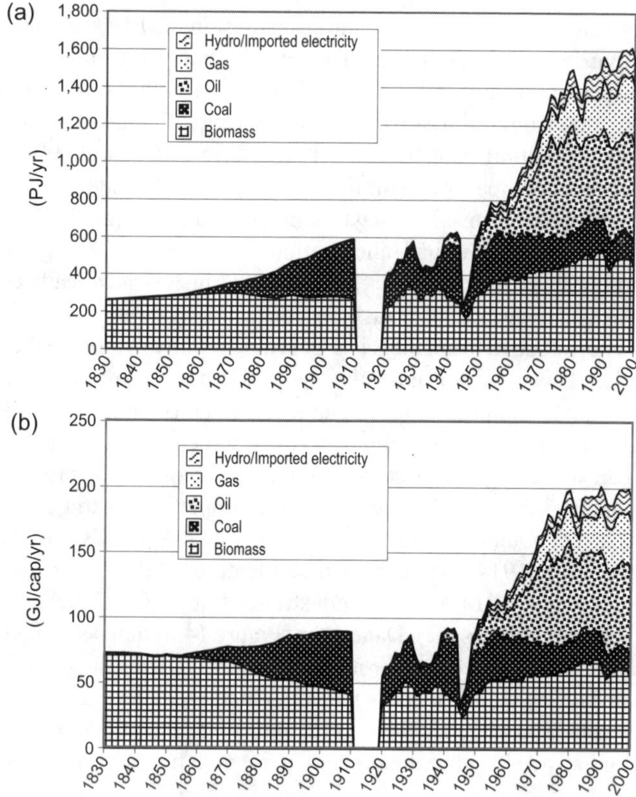

Figure 14.6 Domestic energy consumption of Austria 1830–2000, including biomass for human food, animal nutrition, etc. (a) Total energy consumption of Austria, (b) per-capita energy consumption (Krausmann and Haberl 2002; Krausmann, Schandl et al. 2003).

maintain a positive energy balance; that is, it must supply more energy to society than society invests into it. Under agrarian conditions, the energy return on investment (EROI) (Hall et al. 1986) must be between 5 and 10 (e.g., Giampietro and Pimentel 1991); otherwise an agrarian society cannot survive (e.g., Giampietro 1997). Industrial societies, by contrast, can afford to energetically "subsidize" their agricultural sector, and very often do so. For example, the aggregate EROI of Austria's agriculture was about 1 for the whole period from 1950 to 2000 (Krausmann, Haberl et al. 2003). Of course, many final products (e.g., meat, milk) have EROIs considerably below 1.

The removal of the tight limits to energy availability that were characteristic for agrarian societies was a prerequisite for the development of new agricultural technologies, such as artificial fertilizers, that allowed considerable increases in yields (Grübler 1998; Hall et al. 2000). Such increases in yields were a prerequisite for the growth in population and in per-capita consumption during industrialization. In the Austrian case this meant that the considerable increase in biomass use (Figure 14.6) could be harvested on ever-shrinking agricultural areas, allowing for a marked increase in forest area of about 22% during that time period, despite a rapid growth in land needed to accommodate infrastructure and urban areas (Krausmann and Haberl 2002). Fossil fuel use was also a prerequisite for a spectacular increase in agricultural labor efficiency, that is, the amount of food energy produced per agricultural worker, by more than two orders of magnitude. The growth of labor productivity was necessary to satisfy the growing demand of the manufacturing and services sectors for labor. The current pattern in industrial countries, where only 3–10% of the population works in the agricultural sector, while the industrial and service sectors have become dominant, would be impossible without this increase in agricultural labor efficiency.

Fossil and nuclear fuels as well as large-scale hydropower are highly concentrated and can be extracted with a high EROI (Hall et al. 1986). Fossil fuels can be conveniently stored, easily transported, and converted into useful energy with comparably high efficiency. This allowed the development of new transport systems with far-reaching consequences for the spatial organization of societies. Among others, it was a prerequisite for current patterns of urbanization, spatial organization of production and consumption, the spatial separation of different agricultural production processes, such as livestock rearing and cropland farming which resulted in a breaking up of formerly almost closed cycles of nitrogen, phosphorus, etc., and many other environmentally highly relevant phenomena (Krausmann, Haberl et al. 2003). Growth in transport volumes by far exceeds the growth in energy or materials throughput, and a considerable fraction of this growth is the inevitable consequence of urbanization (Fischer-Kowalski et al. 2004).

Among others, these changes also allowed an increase in local supply of, for example, food through imports from other regions or nation states. This means that at a local level, pollution, the use of ecosystems for production purposes, and even resource depletion, may be alleviated at the cost to distant locales. One consequence is that sustainability of local socioenvironmental systems cannot any longer be judged without looking at their "ecological footprint" in other parts of the world (Fischer-Kowalski and Erb 2003).

TOWARD AN INTEGRATIVE FRAMEWORK FOR STUDYING SOCIOENVIRONMENTAL SYSTEM RESILIENCE BASED ON INFORMATION FLOWS?

The Appropriation of Nature

Over the long term, landscapes with which humans interact become dependent on continuous human use; that is, the maintenance of the current state depends on their interactions with human beings. In the early Holocene, the human impact on the landscape was relatively light while social systems were heavily affected by natural dynamics. As people began to settle permanently and exploit their surroundings continuously, particularly through agriculture and stock-raising, landscapes became more "disturbance-dependent" and, as a result, less resilient to climatic events. When the sociocultural systems involved were unable to compensate for changes in natural systems, crises and even collapse of societies occurred. In the Southern Rhone Valley, for example, detailed research on the three interfaces between climate, society, and the environment has shown that in certain periods, heavy rainfall often created the necessary conditions for the erosion of soils. But whether erosion took place was contingent on the history of human exploitation, including the extent of soil degradation. As time proceeded, the level or extent of soil degradation increased to a point where progressively smaller oscillations in either climate or anthropogenic pressure created dramatic environmental crises in terms of erosion events. In such cases, the system has become hyper-coherent, similar to "an accident waiting to happen" (Holling et al. 2002).

What is the process underlying this familiar sequence of increasing loss of resilience and increasing vulnerability? One view, termed the "risk spiral" (Müller-Herold and Sieferle 1998), argues that the process is driven by human societies' never-ending attempts to cope with risks or exploit opportunity—a transformation of environmental complexity into social complexity. The key point is that while human actions often succeed in reducing specific risks, these efforts also create qualitatively new risks at a larger spatial scale and/or a longer time frame. For example, the increasing use of fossil fuels during industrialization has solved critical problems of food scarcity but these achievements have resulted in a host of new risks including, among others, the dependence on exhaustible resources, the environmental problems of industrial agriculture, and global increase of carbon in the atmosphere. Attempts to solve such problems run the risk of pushing systems further away from their original states. The process involves more and more management of the environment by human society and, importantly, although different parts of the living environment operate at a wide range of temporal scales, most of the natural dynamics in a *landscape* occur relatively slowly by comparison with human dynamics (cf. Tables 14.1 and 14.2). As a result, human beings initially adapt *themselves* to the dynamics of their environment, but over the long term societies' needs are best served by modifications to the environmental dynamics. Human societies thus become dependent on transformed, "colonized" systems, which require certain social institutions, especially those involved in the organized production, storage, and communication of knowledge (Fischer-Kowalski and Haberl 1998). However, like agro-ecosystems, domesticated, or even genetically engineered organisms, colonized systems may evolve rapidly in unexpected ways. If societies depend vitally on their proper functioning, to avoid being trapped in a risk spiral they face the challenge of keeping pace with knowledge production and transmission, and of adapting their social organization to the changes set in motion (Müller-Herold and Sieferle 1998).

At another level, our relations with our environment are part of the uninterrupted process of human learning, which may be seen as a positive feedback loop enabled by human information processing. This feedback loop creates order out of our experiences of the world beyond us by isolating patterns, defining them in terms of a limited number of dimensions, and storing the latter in the form of knowledge. The more cognitive dimensions exist, the more problems can be tackled and the more quickly knowledge is accumulated. The result of this process is the continued accumulation of knowledge, and thus of information-processing capacity, which enables a concomitant increase in matter, energy, and information flows through the society. In this context, information-processing capacity includes the total of the available means an individual, group, or society has at its disposal to register what happens in its environment, and to devise and implement responses to these observations.

Societies Depend on Processing Flows

One can state that societies depend on flows of matter, energy, and information that allow the needs of the individual participants to be met. But these commodities do not behave in the same way. Matter and energy are subject to the laws of conservation and entropy. They may feed people and provide them with other necessary means of survival, but they *cannot simultaneously be transmitted and kept*. They can only be passed on from person to person, and any group that only processes energy and matter will immediately lose structure. In other words, flows of matter and energy alone could never have created durable social institutions, let alone societies. Information, on the other hand, *can be shared*. Indeed, societies are held together by expectations, by institutions, worldviews, by ideas, by technical know-how—by a shared culture: shared patterns of information processing that can, when necessary, mobilize many individuals in a coherent manner!

These all-important information-processing channels emerge through a process that could be called "alignment," that is, through the continued exchange of information that eventually allows different people to share perspectives, ways of doing things, and beliefs. This recursive communication process both facilitates shared understanding between individuals, and draws more and more individuals into a network in which they can communicate more easily and with less effort and/or less risk of misunderstanding than they would experience with nonmembers of the network. There is thus a decided adaptive advantage to being part of such a network. Finally, when the recursive communication remains below a certain threshold, it keeps people out of a network because they cannot sufficiently maintain their alignment. In that case, societies' (different) ways of doing things maintain differences between different groups. The shared information also creates the channels through which energy and matter flow, whether these channels are material (e.g., roads, cables) or remain virtual (e.g., exchange networks such as the *kula* in the South Pacific). We conclude that the "fabric of society" consists of flows through multiple networks, held together by different kinds of (information) relations (kin, business, friendship, exchange, client–patron, power, etc.), and transmitting different combinations of the three basic commodities.

In order for the whole to function correctly, the rates of information processing and those of processing energy and matter need to be commensurate. If not enough information is processed the society loses coherence; people will act in their own immediate interests and the synergies inherent in collaboration will be lost. Thus, for an elite to maintain its power, for the society to remain a coherent entity, the society must ensure that there is an information-processing gradient outward from its center to the periphery of its territory, as in the case of the Roman Empire discussed by van der Leeuw (this volume; van der Leeuw and de Vries 2002). This does not only require a continuous stream of innovations, but also an increasingly extensive and efficient communication network. Throughout history, maintaining that gradient through continued innovation (increase in information-processing capacity) has been essential to the survival of every successful group of humans, whether a tribe, state, or empire; a pressure group such as a trade union; or a group of friends. The cumulative effects of the process can be seen both in the society itself and in its relations with the world around it. Internally, one observes an increasing formalization of social behavior, an increasing size and scope of social institutions, an increasing complexity of organizations, and an increasing range of artifacts and technologies. Externally, there is the increasing control over the known aspects of the environment, an ever wider range of resources used, social adaptation to more extreme natural circumstances, and the spatial homogenization of the environment. As space "shrinks," concomitant with the increase in the speed of communication, there is ultimately the inversion of the scalar structure of society, from a world of villages to a global village.

These lines of reasoning are supported by two empirical studies describing the evolution of society over the past 1000 years. The first is provided by Padgett (2001) for the 14th-century Florentine banking system. He tracked the changing patterns of personal and business relationships among 50,000 inhabitants. The relationships changed in nature over the time monitored, involving an ever wider group of people as the scope of Florentine trade expanded, so that less and less business was done "on a handshake." One sees how a single kin-based business network separates into different networks of kin and business relationships (though, of course, some people play a role in both). This led to increasing formalization of the relationships concerned, and eventually even to a series of financial inventions (among which was "double entry bookkeeping") to formalize the increasingly complex business transactions involved. These changes in turn triggered others extending more widely and transforming society as a whole, creating a new power structure, new customs, new ideas, and a trading empire.

The second comes from the modern world. The number of patents per head of the U.S. population increases nonlinearly with the expansion of the U.S. economic and political power, but it scales superlinearly with city size (the bigger the city, the higher the number of patents per head of the population) during the period of U.S. commercial domination (see Figure 13.2 in van der Leeuw, this volume). Of the many other phenomena tried, only proxies for innovation (patents, productivity, numbers of people devoted to research and innovation) scale superlinearly (Bettencourt et al. 2006). Thus, the larger the city, the larger the proportion of innovation-related people and structures. This seems to point to innovation as the driver for urban growth and the structuring of urban systems.

Crises

Crises reflect temporary incapacities of the information processing system to deal with the contemporary dynamics. They can be due to a sudden acceleration or deceleration of an internal or external part of the dynamics, causing a cascade of changes. There are times when societies are more vulnerable than at others, usually occurring after a period of perceived stability or plenty when the system pushes itself to the limits of adaptation. Agricultural communities that have expectations of continuously high yields may come too close to the carrying capacity to avoid the effects of extreme events, like drought. Likewise, the demise of empires has been linked to an inability to maintain material and administrative infrastructures (Tainter and Crumley, this volume). Internal societal processes seem in each case to push the society to the limits of its capacity to ensure the provision of sufficient energy and matter to all its members. This dynamic is closely related to the risk shift discussed above. Thus, successful agriculturalists leading a marginal existence, for example in arid lands, are familiar with oscillations in food supply and maintain their population well below the theoretical carrying capacity.

Therefore, one theoretical explanation of crises is that as the society solves the more frequent and better-known problems, its risk spectrum shifts toward unknown longer-term risks (problems) that are more likely to occur over centennial timescales. When the society has to confront its (unknown) accumulation of longer-term risks, it does so ultimately at a time when the system's flexibility is reduced because it has for some time adapted itself to particular circumstances, and when its overheads have increased due to the increasing coherence of the system mentioned earlier. The society will experience this as a crisis, a conjunction of phenomena to which it cannot adequately respond. Interestingly, this may shift the emphasis of concern whereby the "causes" of the incidence and impact of perturbations, which have occurred (and will occur) at all times, become secondary to determining the changes in the capacity required to respond to such perturbations.

Control over Information

The cognitive feedback loop that drives a social system creates an ever-expanding "cognitive space," but that space also packs the available cognitive dimensions ever more densely. This phenomenon seems to be responsible for innovative explosions that constitute the industrial and silicon revolutions and many others in the past. The nanotechnology–biotechnology–information–communications (NBIC) revolution that seems to define our current horizon is another case in point, with a particular twist. We have now isolated information, and we can potentially change ourselves. Theoretically, there will be opportunities to get out of the feedback loops that are currently destroying the environment. But we do not yet have enough insight into the dynamics of invention and innovation to control the emergence of the innovative cascade that is due. What will "being human" mean 50 years from now? The need to understand where (in conceptual space), or how, invention and innovation occur must be a first priority if we are to rise to the challenge. This does not imply that problems, such as the political process and its drivers, are ignored but it shifts the perspective to the need to encourage solutions leading to sustainability rather than encouraging a random pattern of inventions.

One way to approach this is to prioritize modeled simulations, scenario building, and testing against past records. This approach encompasses our current levels of understanding, uncertainty, and ignorance. It seems to be the only way we can, thus far, attempt to project into the future things we have learned from the past, if only in order to identify what we do not yet know. Identifying the range of problems that are likely to occur over various timescales, in local zones to the globe, would be a useful first step.

FINAL REMARKS

Socioenvironmental Systems

The group's main two messages are couched in terms of complex systems. First, today's socioenvironmental systems are complex and heavily contingent on the past. For some processes, the relevant timeline is relatively short but it is also clear that for others the trajectory extends back into multidecadal and centennial timescales, certainly before the 20th century. We cannot explain fully the present condition without recourse to the past, and it follows that society needs to assume that its actions today are reverberating, in climate terms but also in numerous other ways, into the centuries ahead. Second, descriptions of past socioenvironmental systems are an immensely rich resource that is currently under-exploited and under-researched. They provide analogs for the modern condition, baselines to provide perspective and targets, insight into system behavior and scaling, information on timescales involved in processes of change and transformation, and data for the development and testing of dynamic simulation models.

Interdisciplinary Integration

The members of our multidisciplinary group were able to experience at first hand the difficulties and rewards of interdisciplinary discourse. However, the experience served to reinforce awareness of the lack of integration that often exists between the research of the disciplines represented by the group, which included archaeology, anthropology, energy systems analysis, environmental history, historical ecology, and paleoecology. National and international agencies should continue to recognize the need for interdisciplinary research to be encouraged, developed, conducted, and funded within appropriately constructed integrative frameworks.

ACKNOWLEDGMENTS

We especially wish to thank Carole Crumley and Bert de Vries for their contributions to this report.

REFERENCES

Arthur, W.B. 1989. Competing technologies, increasing returns, and lock-in by historical events. *Econ. J.* **99**:116–131.

Behringer, W. 1999. Climatic change and witch-hunting: The impact of the Little Ice Age on mentalities. *Clim. Change* **1–2**:335–351.

Bettencourt, L.M.A., J. Lobo, and D. Strumsky. 2006. Invention in the city: Increasing returns to scale in metropolitan patenting. *Res. Pol.*, in press.

Chuine, I., P. Yiou, N. Viovy et al. 2004. Grape ripening as a past climate indicator. *Nature* **432**:289–290.

Claussen, M., C. Kubatzki, V. Brovkin et al. 1999. Simulation of an abrupt change in Saharan vegetation at the end of the mid-Holocene. *Geophys. Res. Lett.* **24**:2037–2040.

Costanza, R., and M. Ruth. 1998. Using dynamic modelling to scope environmental problems and build consensus. *Env. Manag.* **22**:183–195.

Coulthard, T.J., M.G. Macklin, and M.J. Kirby. 2002. Simulating upland river catchment and alluvial fan evolution. *Earth Surf. Proc. Landf.* **27**:269–288.

Crutzen, P.J., and E. Stoermer. 2001. The "Anthropocene." *Glob. Change Newsl.* **41**:12–13.

Dearing, J.A. 2006. Integration of world and earth systems: Heritage and foresight. In: The World System and the Earth System, ed. A. Hornborg and C.L. Crumley. Santa Barbara, CA: Left Coast Books, in press.

Dearing, J.A., R.W. Battarbee, R. Dikau, I. Larocque, and F. Oldfield. 2006a. Human–environment interactions: Learning from the past. *Reg. Env. Change* **6**:1–16.

Dearing, J.A., R.W. Battarbee, R. Dikau, I. Larocque, and F. Oldfield. 2006b. Human–environment interactions: Towards synthesis and simulation. *Reg. Env. Change* **6**:115–123.

Dearing, J.A., N. Richmond, A.J. Plater et al. 2006. Models for coastal simulation based on ceullar automata: The need and potential. *Phil. Trans. Roy. Soc. Lond.* A **364**: 1051–1071.

Deevey, E.S. 1984. Stress, strain and stability of lacustrine ecosystems. In: Lake Sediments and Environmental History, ed. E.Y. Haworth and J.W.G. Lund, pp. 203–229. Leicester: Leicester Univ. Press.

de Vries, B.J.M., and J. Goudsblom, eds. 2002. Mappae Mundi: Humans and their Habitats in a Long-Term Socio-Ecological Perspective. Myths, Maps and Models. Amsterdam: Amsterdam Univ. Press.

Fischer-Kowalski, M., and K.-H. Erb. 2003. Gesellschaftlicher Stoffwechsel im Raum: Auf der Suche nach einem sozialwissenschaftlichen Zugang zur biophysischen Realität. In: Humanökologie: Ansätze zur Überwindung der Natur-Kultur-Dichotomie, ed. P. Meusburger and T. Schwan, pp. 257–285. Stuttgart: Steiner.

Fischer-Kowalski, M., and H. Haberl. 1998. Sustainable development, long-term changes in socioeconomic metabolism, and colonization of nature. *Intl. Soc. Sci. J.* **158**:573–587.

Fischer-Kowalski, M., F. Krausmann, and B. Smetschka. 2004. Modelling scenarios of transport across history from a sociometabolic perspective. *Rev. Fernand Braudel Cent.* **27**:307–342.

Freeman, C. 1994. The economics of technical change. *Cambridge J. Econ.* **18**:463–514.

Freeman, C., and F. Louca. 2001. As Time Goes By: From the Industrial Revolutions to the Information Revolution. Oxford: Oxford Univ. Press.

Freeman, C., and C. Perez. 1988. Structural crises of adjustment, business cycles and investment behaviour. In: Technical Change and Economic Theory, ed. G. Dosi, C. Freeman, R.R. Nelson, G. Silverberg, and L. Soete, pp. 38–66. London: Pinter.

Giampietro, M. 1997. Linking technology, natural resources, and the socioeconomic structure of human society: A theoretical model. *Adv. Hum. Ecol.* **6**:75–130.

Giampietro, M., and D. Pimentel. 1991. Energy efficiency: Assessing the interaction between humans and their environment. *Ecol. Econ.* **4**:117–144.

Gimblett, H.R., ed. 2002. Integrating Geographic Information Systems and Agent-based Modelling Techniques for Simulating Social and Ecological Processes. Santa Fe Institute Studies in the Sciences of Complexity. Oxford: Oxford Univ. Press.

Gottfried, R.S. 1983. The Black Death: Natural and Human Disasters in Medieval Europe. New York: Free Press.

Grübler, A. 1998. Technology and Global Change. Cambridge: Cambridge Univ. Press.

Grübler, A. 2004. Transitions in energy use. In: Encyclopedia of Energy, vol. 6, pp. 163–177. Oxford: Elsevier.

Haberl, H. 2006. The global socioeconomic energetic metabolism as a sustainability problem. *Energy* **31**:87–99.

Hall, C.A.S., C.J. Cleveland, and R.K. Kaufmann, eds. 1986. Energy and Resource Quality: The Ecology of the Economic Process. New York: Wiley.

Hall, C.A.S., C.L. Perez, and G. Leclerc, eds. 2000. Quantifying Sustainable Development: The Future of Tropical Economies. San Diego: Academic.

Hassan, F.A. 2002. Holocene environmental change and the transition to agriculture in South-west Asia and North-east Africa. In: The Origins of Pottery and Agriculture, ed. Y. Yasuda, pp. 55–68. New Delhi: Roli.

Holling, C.S., L.G. Gunderson, and G.D. Peterson. 2002. Sustainability and panarchies. In: Panarchy: Understanding Transformations in Human and Natural Systems, ed. L.H Gunderson and C.S. Holling, pp. 63–102. Washington: Island.

IPCC (Intergovernmental Panel on Climate Change). 2001. Working Group I: Climate Change 2001: The Scientific Basis. Cambridge: Cambridge Univ. Press.

Jones, P.D., and M.E. Mann. 2004. Climate over past millennia. *Rev. Geophys*. **42(2)**: RG202 .

Krausmann, F., and H. Haberl. 2002. The process of industrialization from the perspective of energetic metabolism: Socioeconomic energy flows in Austria 1830–1995. *Ecol. Econ*. **41**:177–201.

Krausmann, F., H. Haberl, N.B. Schulz et al. 2003. Land-use change and socioeconomic metabolism in Austria. I. Driving forces of land-use change: 1950–1995. *Land Use Policy* **20**:1–20.

Krausmann, F., H. Schandl, and N.B. Schulz. 2003. Vergleichende Untersuchung zur langfristigen Entwicklung von gesellschaftlichem Stoffwechsel und Landnutzung in Österreich und dem Vereinigten Königreich. Stuttgart: Breuninger Stiftung.

Lachiver, M. 1991. Les années de misère: La famine au temps du Grand Roi 1680–1720. Paris: Fayard.

Landes, D. 1969. The Unbound Prometheus: Technological Change and Industrial Development in Western Europe from 1750 to the Present. Cambridge: Cambridge Univ. Press.

Messerli, B., M. Grosjean, T. Hofer, L. Nunez, and C. Pfister. 2000. From nature-dominated to human-dominated environmental changes. *Quat. Sci. Rev*. **19**:459–479.

Moberg, A., D.M. Sonechkin, K. Holmgren, N.M. Datsenko, and W. Karlén. 2005. Highly variable Northern Hemisphere temperatures reconstructed from low- and high-resolution proxy data. *Nature* **433**:613–617.

Mokyr, J. 1990. The Lever of Riches: Technological Creativity and Economic Progress. Oxford: Oxford Univ. Press.

Müller-Herold, U., and R.P. Sieferle. 1998. Surplus and survival: Risk, ruin and luxury in the evolution of early forms of subsistence. *Adv. Hum. Ecol*. **6**:201–220.

Nakícenovíc , N., and R. Swart, eds. 2001. Climate Change 2001: IPCC Special Report on Emissions Scenarios. Cambridge: Cambridge Univ. Press.

Padgett, J. 2001. Organizational genesis, identity and control: The transformation of banking in Renaissance Florence. In: Markets and Networks, ed. J. Rauch and A. Cassella, pp. 211–257. New York: Russell Sage.

Parker, D.C., T. Berger, and S.M. Manson. 2002. Agent-based Models of Land-use and Land-cover Change. Report and Review of an Intl. Workshop, Oct. 4–7, 2001. LUCC Report Series 6. Bloomington, IN: LUCC Focus 1 Office, Indiana Univ.

Pfister, C. 1988. Variations in the spring-summer climate of Central Europe from the High Middle Ages to 1850. In: Long and Short Term Variability of Climate, ed. H. Wanner and U Siegenthaler, pp. 57–82. Lecture Notes in Earth Sciences 16. Berlin: Springer.

Pfister, C. 2005. Weeping in the snow: The second period of Little Ice Age-type Crises, 1570 to 1630. In: Kulturelle Konsequenzen der Kleinen Eiszeit [Cultural Consequences of the Little Ice Age], ed. W. Behringer, H. Lehmann, and C. Pfister, pp. 31–85. Göttingen: Vandenhoeck.

Prigogine, I. 1996. The End of Certainty. New York: Free Press.

Richards, K.S. 1990. "Real" geomorphology. *Earth Surf. Proc. Landf.* **15**:195–197.

Rosenberg, N., and L.E. Birdzell. 1986. How the West Grew Rich: The Economic Transformation of the Industrial World. New York: Basic.

Ruddiman, W.F. 2003. The anthropogenic greenhouse era began thousands of years ago. *Clim. Change* **61**:261–293.

Schellnhuber, H.-J., M.K.B. Lüdeke, and G. Petschel-Held. 2002. The Syndromes Approach to scaling: Describing global change on an intermediate functional scale. *Integrated Assess.* **2–3**:201–219.

Scott, S., and C.J. Duncan. 2004. Return of the Black Death. Chichester: Wiley.

Sieferle, R.P. 2001. The Subterranean Forest: Energy Systems and the Industrial Revolution. Cambridge: White Horse.

Smil, V. 1992. Agricultural energy costs: National analysis. In: Energy in Farm Production, ed. R.C. Fluck, pp. 85–100. Amsterdam: Elsevier.

Steffen, W., A. Sanderson, P.D. Tyson et al. 2004. Global Change and the Earth System: A Planet under Pressure. Berlin: Springer.

Stoob, H. 1956. Kartographische Möglichkeiten zur Darstellung der Stadtentstehung in Mitteleuropa, besonders zwischen 1450 und 1800. In: Historische Raumforschung 1, vol. 6, pp. 21–76. Bremen: Akademie für Raumforschung und Landesplanung.

Tucker, G.E., and R. Slingerland. 1997. Drainage basin responses to climate change. *Water Resources Res.* **33**:2031–2047.

Turner, B.L., W.C. Clark, R.W. Kates et al. eds. 1990. The Earth as Transformed by Human Action: Global and Regional Changes in the Biosphere over the Past 300 Years. Cambridge: Cambridge Univ. Press.

van der Leeuw, S.E., and B.J.M. de Vries. 2002. Empire: The Romans in the Mediterranean. In: Mappae Mundi: Humans and their Habitats in a Long-Term Socio-Ecological Perspective. Myths, Maps and Models, ed. B.J.M. de Vries and J. Goudsblom, pp. 209–256. Amsterdam: Amsterdam Univ. Press.

Vesajoki, H., and M. Tornberg. 1994. Outlining the climate in Finland during the pre-instrumental period on the basis of documentary sources. In: Climatic Trends and Anomalies in Europe 1675–1715, ed. B. Frenzel, C. Pfister, and B. Glaeser, pp. 61–72. Stuttgart: Fischer.

Wang, Y., H. Cheng, R.L. Edwards et al. 2005. The Holocene Asian monsoon: Links to solar changes and North Atlantic climate. *Science* **308**:854–857.

Wirtz, K.W, and C. Lemmen. 2003. A global dyanamic model for the Neolithic transition. *Clim. Change* **59**:333–367.

Wolfram, S. 2002. A New Kind of Science. Champaign, IL: Wolfram Media.

Wu, F., and D. Martin. 2002. Urban expansion simulation of Southeast England using population surface modelling and cellular automata. *Env. Planning* **A34**:1855–1876.

Yasuda, Y., ed. 2002. The Origins of Pottery and Agriculture. New Delhi: Roli.

Zolitschka, B. 1998. A 14000 year sediment yield record from western Germany based on annually laminated sediments. *Geomorphology* **22**:1–17.

The Decadal Timescale: Up to 100 Years Ago

15

A Decadal Chronology of 20th-Century Changes in Earth's Natural Systems

Nathan J. Mantua

Climate Impacts Group, University of Washington, Seattle, WA 98195–4235, U.S.A.

ABSTRACT

Human activities were a major driver for secular changes in Earth's natural systems in the 20th century. Human-caused changes in the atmosphere included rapid increases in concentrations of greenhouse gases, aerosols (dust, sulfates, smoke, and soot), and low-level ozone, as well as late 20th-century decreases in stratospheric ozone. A rapid rise in global average temperatures has been attributed, in part, to the increased concentrations of greenhouse gases. As part of the climate warming, polar ice sheets and most glaciers shrank, and snow cover declined. The second half of the 20th century was especially notable for rapid ecosystem changes, including increases in the amount of freshwater impounded behind dams, increases in nitrogen and phosphorous used in fertilizers, conversion of forests to croplands, losses and degradation of mangroves and coral reefs, depletion in large marine predatory fish, and declines in global biodiversity.

Causes for decade- to century-scale variability in Earth systems can be traced to a variety of factors, many directly related to human activities, but also others arising from nonhuman aspects of the Earth system like natural climate variations. Decade- to century-scale variability in atmospheric circulation patterns contributed to prominent variability in regional climate and ecosystems. Decadal changes between pluvials (wet conditions) and drought were especially large in the Sahel and the continental United States. Prolonged regional drought was part of complex suites of changes in Earth systems that included vegetation change, increased dust loading in the atmosphere, and strong interactions between the land surface and atmosphere at regional to global spatial scales. Climate variations also contributed to variations in major marine ecosystems, especially in the North Pacific Ocean. In contrast, multidecade declines in Atlantic cod caused by overfishing were punctuated with a collapse in formerly productive cod fisheries in the early 1990s; however, these declines also coincided with major increases in the abundance of and fisheries for commercially valuable crab, shrimp, and lobster in many parts of the North Atlantic.

Climate modeling studies attribute decade- to century-scale variability in the 20th century to a combination of subtle changes in the intensity of the sun, human-caused increased concentrations of greenhouse gases and sulfate aerosols, human-caused stratospheric ozone depletion, and subtle changes in ocean surface temperatures. In some situations, feedbacks between the atmosphere and the Earth's surface (ocean temperatures and land surface characteristics that include snow, ice, and vegetation) appear to have amplified large-scale circulation-caused precipitation changes.

INTRODUCTION

In many respects, decade- to century-scale variability is variability at the human scale. A human generation, major infrastructure planning and investments, the length of a person's career and an individual's lifetime: each typically plays out over timescales of decades. Earth system changes over decadal timescales can therefore resonate with human systems in ways that lead to especially strong societal impacts. This chapter aims to provide an overview of key 20th-century decade- to century-scale changes in the Earth's natural systems due to both human and natural causes. Many components of the Earth system experienced dramatic changes in the 20th century. The recently published Millennium Ecosystem Assessment Synthesis Report concludes that over the past fifty years humans have changed Earth's natural systems more rapidly and extensively than in any comparable time in human history (MEA 2005). Decade- to century-scale variability in atmospheric circulation patterns also contributed to prominent variability in regional climate and ecosystems. Specific topic areas covered in this review are perspectives on global and regional climate, the hydrologic cycle, biogeochemical cycles, and ecosystems. Summary timelines are offered at the end to provide a means for integrating some of most prominent aspects of the 20th century's decade- to century-scale change in Earth system components.

INDICATORS OF GLOBAL CHANGE

Climate

Climate is a key part of natural systems, and fluctuations in climate influence the delivery of goods and services provided by natural systems. Surface water supplies, the biogeography of species, growing seasons, fish and animal populations, and consequently human welfare can all be significantly influenced by fluctuations in climate. Climate is simply defined as the statistics of weather. Among the most important aspects of climate is the fact that climate statistics are nonstationary: this means that the climate of a particular place or region depends on the period of data collection. Climate varies across a broad continuum of timescales in response to a wide variety of influences. For instance, subtle changes in the geometry of the Earth's orbit around the sun have been identified as a pacemaker for the glacial–interglacial cycles that have characterized Earth's climate over the past few million years. Century-to-century changes in the intensity of the sun have been linked with the multicentury Medieval Warm Period and subsequent Little Ice Age. More familiar, but equally dramatic, year-to-year shifts in global weather patterns are caused by El Niño and the Southern Oscillation, which are the results of naturally occurring interactions between the tropical Pacific Ocean and atmosphere.

Components of the Earth system that influence climate at decadal and longer timescales are the atmospheric composition, energy from the sun, atmosphere and ocean circulations, the hydrologic cycle, land cover/vegetation, and the cryosphere. I begin this section with a review of the prominent global indicators of decadal variations in Earth's climate and follow with a brief review of recurring regional patterns of decade- to century-scale climate variability. The National Research Council's *Decade-to-Century-Scale Climate Variability and Change: A Science Strategy* provides a comprehensive and widely accessible review of this subject matter (NRC 1998).

Surface Air Temperatures

Globally averaged surface air temperature records indicate two periods of strong global-scale warming in the past century: the first takes place from 1910–1945; the second begins in 1976 and continues to present (Figure 15.1). Global average surface temperatures were relatively stable from the late 1800s through 1910, and cooled slightly from 1945 to the mid-1970s. Over the 20th century, as a whole, global temperatures warmed by ~0.6°C ± 0.2°C (IPCC 2001). Analysis of various climate proxies for the Northern Hemisphere suggest that the rate of warming observed in the 20th century, as well as the magnitude of global temperatures from 1990 to 2004, is unprecedented in the past 2000 years (Moberg et al. 2005).

The spatial pattern of 20th-century warming was not uniform across the globe. The largest warming trends were in Alaska, western Canada, and Russia, with smaller warming trends over the global oceans and tropical land areas. Twentieth-century surface temperatures cooled slightly in parts of the Antarctic, parts of the Southern Ocean, and the North Atlantic Ocean just south and east of Greenland (IPCC 2001). The warming trends were also greatest at nighttime and in the wintertime.

Cryosphere

The cryosphere, the part of the Earth's surface that is perennially frozen or below the freezing point, is a crucial part of the Earth system. The most important influences on the cryosphere come from its interactions with the atmosphere, ocean, and land systems. Feedbacks between the cryosphere and other components of the climate system can yield especially high regional, and even global, climate sensitivity. For example, the *ice albedo* (surface reflectivity) feedback in

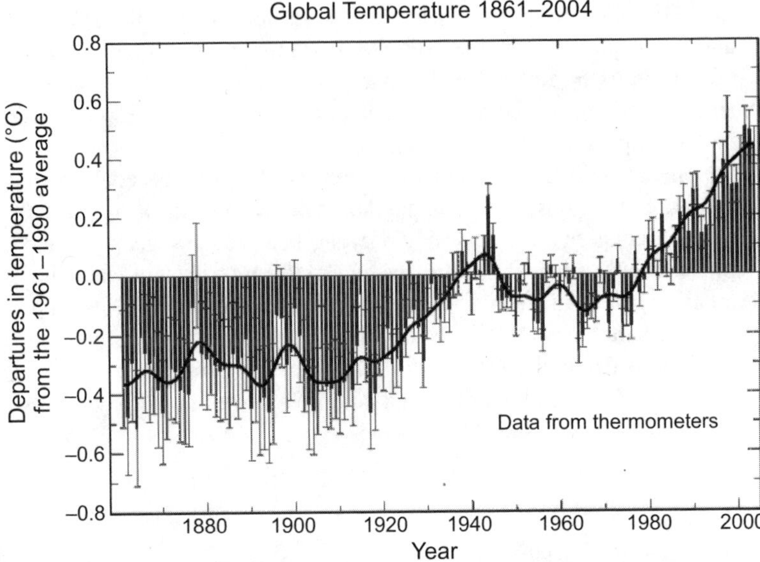

Figure 15.1 Globally averaged annual mean surface temperatures expressed as deviations from the 1961–1990 average. Annual averages are shown with the thin lines, a smooth version is shown by the heavy black line. Figure obtained from the Hadley Center's Climate Research Unit web-page on April 14, 2005 (http://www.metoffice.com/research/hadleycentre/CR_data/Annual/HadCRUG.gif). Reprinted with permission from the Met Office, © Crown 2005.

polar regions can lead to a high amplification of a small initial warming. Ice is highly reflective; thus melting ice increases the amount of sunlight absorbed, which raises surface temperatures, melts more ice, and leads to a positive feedback on the initial warming. Similar sets of feedbacks exists with snow cover and sea ice, such that an initially small amount of melt can start a rapid process of additional melting and surface warming (NRC 1998). Changes in permafrost, snow and ice cover are also ecologically important. Snow and sea ice insulate the surfaces below them from what can be extremely cold atmospheric temperatures. Melting permafrost can alter moisture infiltration rates in ways that lead to large changes in soil moisture and vegetation cover. In summary, the cryosphere directly influences climate through its role in the surface energy balance, and indirectly through its ecological role in biogeochemical cycles by influencing photosynthesis in terrestrial and marine plants and the decay of organic matter.

IPCC (2001) highlights the following cryosphere indicators of global climate change:

- Satellite data show that there have very likely been decreases of about 10% in the extent of snow cover since the late 1960s, and ground-based observations show that there has very likely been a reduction of about two weeks in the annual duration of lake and river ice cover in the mid- and high latitudes of the Northern Hemisphere over the 20th century.

- There has been a widespread retreat of mountain glaciers in nonpolar regions during the 20th century (only a small minority have been growing).
- Northern Hemisphere spring and summer sea-ice extent has decreased by about 10–15% since the 1950s. It is likely that there has been about a 40% decline in Arctic sea-ice thickness during late summer to early autumn in recent decades and a considerably slower decline in winter sea-ice thickness.

There were pronounced regional cryosphere changes in the latter half of the 20th century. Along the Antarctic Peninsula, the Wordie Ice Shelf and the Larsen Ice Shelf disintegrated in recent decades, dramatic indicators of a complex response to very strong regional warming trends of over 2.5°C since the 1950s. The Ward Hunt Ice Shelf, the Arctic's largest ice shelf located on the north coast of Ellesmere Island in Canada's Nunavut territory, fractured in September 2003 and released nearly all the freshwater from what was the Northern Hemisphere's largest ice-shelf lake. Snowpack changes were also prominent in midlatitude mountain areas, with trends to a widespread reduction in springtime snow pack and an advance in spring and summer snowmelt runoff timing documented for much of western North America for the last half-century (Mote et al. 2005; Stewart et al. 2005).

Variations in Key Components of Earth's Energy Balance

The Earth's energy balance boils down to two basic parts: the incoming radiant energy from the sun and the outgoing infrared radiation from the Earth's surface and atmosphere. The Earth's atmosphere, where clouds are absent, is nearly transparent to the visible radiation (light), which carries the bulk of the sun's energy. Clouds and aerosols (small particles) in the atmosphere and bright surfaces like snow cover reflect about 30% of the total incoming energy from the sun back to space, while the rest is absorbed by the Earth's surface. The absorbed solar radiation warms the Earth's surface, and because the Earth's surface is much cooler than the sun's, its heat is radiated upward to the atmosphere at much longer *infrared* wavelengths. Trace gases in the atmosphere, like water vapor (H_2O), carbon dioxide (CO_2), methane (CH_4), nitrous oxide (N_2O), and ozone (O_3), absorb and reemit some of the infrared radiation from the Earth's surface; half goes upward into space and the other half goes downward back to the Earth's surface, where it is again absorbed and reemitted back to the atmosphere. This natural greenhouse effect of trapped infrared energy warms both the atmosphere and the Earth's surface substantially; surface temperatures would be ~33°C cooler than they are today without the natural greenhouse effect. Changes in either the incoming solar energy absorbed at the surface or the atmospheric greenhouse effect directly influence the Earth's energy budget and alter Earth's climate.

Solar Radiation (Energy from the Sun)

The energy from the sun that reaches Earth is not constant. Solar influences on climate are related to several different factors that are known to vary over annual to decadal and longer timescales. There is a pronounced near 11-year cycle in sunspot numbers, with peaks corresponding to periods of peak solar intensity. Direct measurements find that 20th-century decadal-scale changes in solar output amount to less than 1% of the long-term average (Lean et al. 1995). Yet despite the subtle strength of these changes, decadal variations in 20th-century solar activity and aspects of Northern Hemisphere climate tracked each other quite closely (Labitzke and van Loon 1993).

Regional variations in cloud cover also affect the amount of solar radiation that is reflected back to space and absorbed at the Earth's surface. Decadal-scale variations with century-long trends toward increased cloud cover have been documented for Australia, North America, India, and Europe during the 20th century (McGuffie and Henderson-Sellers 1988).

Atmospheric Composition: Greenhouse Gases and Aerosols

During the 20th century, human activities rapidly altered the composition of the atmosphere to influence the Earth's energy balance. Atmospheric concentrations of human-made greenhouse gases—chlorofluorocarbons (CFCs), hydrofluorocarbons (HFCs), and perfluorocarbons (PFCs), as well as sulfur hexafluoride (SF_6)—are all the result of industrial development. Several of the important naturally occurring greenhouse gases (CO_2, CH_4, and N_2O) increased dramatically in the 20th century, especially since ~1950 (Figure 15.2; IPCC 2001). Carbon dioxide emissions from burning fossil fuels and converting forests into agricultural lands account for the observed rise in atmospheric CO_2 concentrations. Increased industrial activity and the use of nitrogen fertilizer are key factors contributing to the observed rises in N_2O. In contrast, causes for changing CH_4 concentrations are not well understood. The growth in atmospheric CH_4 concentrations began slowing in the 1980s, and CH_4 concentrations were basically stable from 1999–2002 (Dlugokenky et al. 2003). Based on measurements obtained from deep ice cores from Antarctica, the current CO_2 and CH_4 concentrations are higher now than they were at any time in at least the past 420,000 years (IPCC 2001).

Small airborne particles that influence the Earth's energy balance (i.e., aerosols) changed dramatically in the 20th century. Natural sources for aerosols include sulfur dioxide from volcanic eruptions, dust from desert areas, smoke from biomass burning, and salt from sea-spray. In the 20th century a growing

Indicators of the human influence on the atmosphere during the Industrial Era

(a) Global atmospheric concentration of three well-mixed greenhouse gases

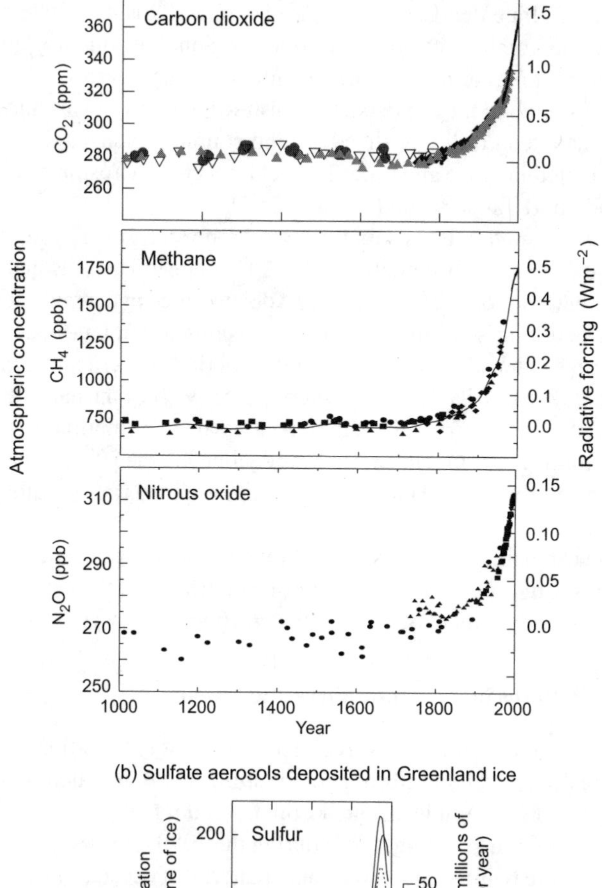

(b) Sulfate aerosols deposited in Greenland ice

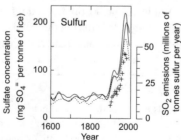

Sulfur

Sulfate concentration (mg $SO_4^=$ per tonne of ice)

200

100

0

50

25

0

SO_2 emissions (millions of tonnes sulfur per year)

1600 1800 2000

Year

Figure 15.2 (a) Indicators of human influence on the atmosphere over the past millennium. Figure obtained from the IPCC's (2001) Summary for Policy Makers. In panel (b) the +'s indicate estimates of relevant regional annual sulfate emissions contributing to sulfate deposited in Greenland ice (solid and dashed lines).

portion of atmospheric aerosols came from anthropogenic sources: (a) burning fossil fuels which added smoke, soot, and sulfuric acid to the atmosphere and (b) intensive farming practices that increased dustiness. For example, sulfuric acid deposition on Greenland ice rose dramatically with increased emissions from industrial activities in eastern North America from the 1950s until the 1990s (Figure 15.2). Sulfate deposition decreased in the 1990s as sulfate emissions were reduced to combat acid rain and other regional air quality problems in northeastern North America. Unlike greenhouse gases, aerosols have only a short residence time in the atmosphere. This means that aerosols and their impacts on climate tend to be localized around source regions and they are unevenly distributed in space and time. The net impacts of aerosols on climate are complex and poorly understood (see Box 15.1).

Ozone (O_3) is another important trace gas in the atmosphere whose concentrations have changed substantially in the 20th century. The seasonal development of an Antarctic "ozone hole" was first documented in the mid-1970s and is now understood to arise ultimately from emissions of human-made industrial gases known as chlorofluorocarbons. Stratospheric O_3 is crucial for life at the Earth's surface because it effectively filters out 95–99% of the harmful ultraviolet radiation that comes from the sun. Ozone is also an important greenhouse gas. While stratospheric O_3 concentrations declined in the 20th century, O_3 concentrations near ground level increased in smoggy areas as a consequence of hydrocarbon emissions from burning fossil fuels. In the formation of smog, this "bad" ozone can damage lung tissue and plants, and serves to increase locally the greenhouse effect in polluted areas. On a global average the net effect of total O_3 changes has been to increase slightly the greenhouse effect (see IPCC 2001).

Circulation Patterns in the Atmosphere and Ocean

At decade- to century timescales, regional temperature variations in the 20th century were not spatially uniform and not uniformly dominated by long-term warming trends. For example, the area from Labrador to the western Greenland Sea experienced strong cooling trends during the 1980s and 1990s when global temperatures were rising rapidly. Regional patterns of decadal temperature and precipitation variability are driven by decadal variations in regional

Box 15.1 Aerosol impacts on climate.

One direct effect of aerosols is to cool the local climate by reflecting incoming solar energy back to space. Increased aerosol concentrations also work to increase the number of water droplets in clouds, leading to smaller cloud droplets that reduce the rate at which water droplets fall out of clouds. A shift to smaller cloud droplets prolongs the lifetime of clouds so that they reflect more sunlight, thereby indirectly cooling local climate, and decreases the likelihood that droplets are large enough to fall to the ground before evaporating thereby leading to decreased precipitation.

atmospheric circulation patterns. The most prominent decadal climate patterns in the 20th century are described in more detail below. Some circulation patterns arise from atmospheric processes alone, but decade to century timescale variability is caused by factors external to the Earth's climate system (e.g., decade to century variations in the intensity of the sun or volcanic eruptions which put large amounts of aerosols into the atmosphere) or interactions between the atmosphere, the ocean, vegetation, and/or the cryosphere. Characteristics of recurring climate patterns are influenced by factors such as topography, the distribution of continents, snow cover and sea ice, land–sea contrasts, the size of ocean basins, and seasonally varying horizontal temperature gradients across distances of thousands of kilometers.

Decadal Variability of El Niño-Southern Oscillation

Probably the most widely recognized coupled ocean-atmosphere climate pattern is the El Niño-Southern Oscillation (ENSO). ENSO variability in the 20th century was most energetic at periods of two to seven years. Decadal variability of ENSO is manifest as eras of relatively high versus low subdecadal activity. The periods from 1900–1920 and the early 1960s to present have been notable for their high ENSO activity, whereas the period from the early 1920s to 1950s was notable for its relatively low ENSO activity. Warm episodes of ENSO (often called El Niño events) from the mid-1970s through the 1990s were more frequent, more intense, and more persistent, compared to the previous 100 years (IPCC 2001). Because ENSO has a global influence on patterns of storminess, rainfall, surface temperatures, as well as ocean conditions in much of the Pacific Basin, the decadal variability of ENSO can be detected in climate-sensitive aspects of Earth systems in many parts of the world.

The North Atlantic Oscillation

Each winter, over the North Atlantic, the Icelandic Low develops as a region of consistently low sea-level pressure and active storminess, while to the south the Azores High develops as a region of persistently high pressure and generally fair weather. In periods when the Icelandic Low is unusually deep, the Azores High tends to be unusually strong, and vice versa. This synchronous behavior of the Icelandic Low and Azores High pressure systems is referred to as the North Atlantic Oscillation (NAO).

A simple method for tracking the state of the NAO is provided by an index of atmospheric sea-level pressure differences between the Azores and Iceland. When the NAO index is positive, the region between the deep Icelandic Low and intense Azores High pressure cells is marked by strong westerly winds and frequent storms. As the core of the westerly winds nears the continent, it turns poleward bringing abundant precipitation and mild temperatures to northwestern Europe. The same circulation brings relatively dry conditions to southern Europe and the coastal lands bordering the Mediterranean Sea, thereby exacerbating drought conditions in what are typically already dry climates. Swings in the NAO also impact eastern North America and Greenland climate. While northwestern Europe faces milder and wetter winters during positive NAO periods, western Greenland and Labrador get abnormally cold and dry weather. Ice formation increases in the Labrador Sea, as does the transport of sea ice into the northeast Atlantic Ocean through the Fram Strait. Along the Atlantic coast of the United States, precipitation tends to increase while the frequency of cold air outbreaks and extremely cold temperatures is reduced. The opposite conditions hold with negative NAO periods.

Annual values of the wintertime NAO index from 1900–2003 are shown in Figure 15.3. There was a strong tendency for positive NAO winters from 1900–1915 and 1920–1930, negative NAO winters in the period from the early 1950s to the early 1970s, and an especially strong bias for positive NAO winters from the early 1970s through the late 1990s. The impacts of the recent period of strongly positive NAO winters have been pronounced: while glaciers around the world have been in a general century-long retreat, many glaciers in Scandanavia actually grew in recent decades due to the prevalence of positive NAO winters, which brought the region an abundance of precipitation; as previously noted, NAO-driven cooling trends over the North Atlantic Ocean were strong during the most recent period of strong global warming.

The Annular Modes, or Artic and Antarctic Oscillations

Over the past few years, numerous studies have highlighted the importance of a hemispheric pattern in the atmosphere that is centered over the Arctic, which includes the NAO as its regional expression over the North Atlantic. This pattern has been associated with an Antarctic twin, and both are now discussed under the label of *annular modes*. The Northern Annular Mode is called the Arctic Oscillation (AO), whereas the Southern Annular Mode is called the Antarctic Oscillation (AAO). The annular modes are the dominant patterns of atmospheric variability in their respective hemispheres, both patterns capturing the tendency for the polar vortex to intensify or weaken with an expanding or contracting ring of westerly winds over the polar latitudes. Positive phases of both the AO and AAO correspond to periods with an intensified polar vortex (over the Arctic and Antarctic, respectively).

An index for the AO is highly correlated with the NAO index, and decadal variations in the AO index are very similar to those in the NAO index. An interesting feature of the Southern Hemisphere's AAO index is the fact that it also shows a strong trend toward persistently positive values from the 1970s through the mid-1990s. Thompson and Solomon (2002) suggest that the trends in the AAO circulation pattern are linked to the seasonal development of the Antarctic

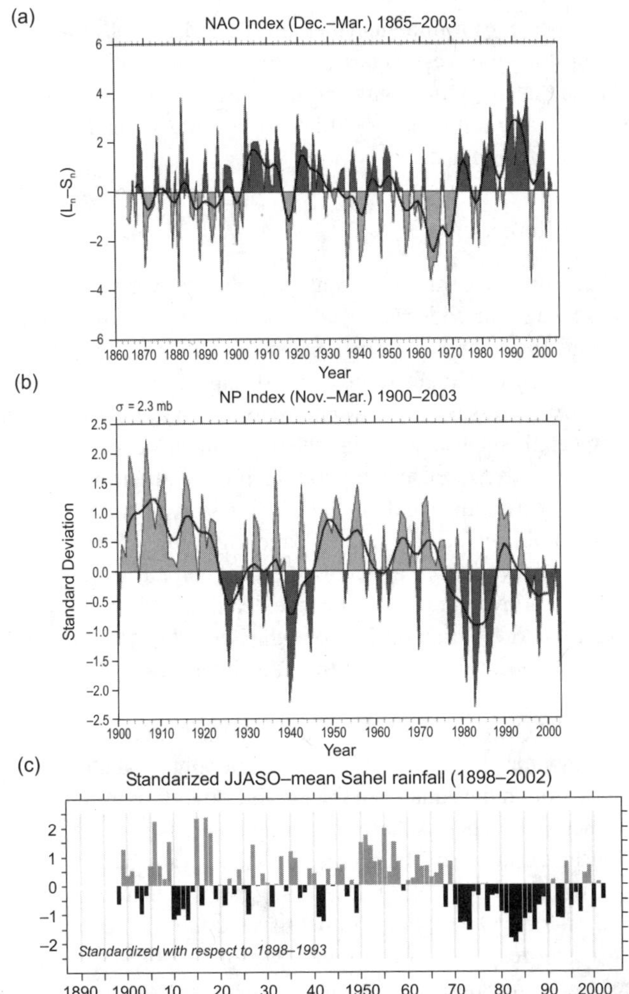

Figure 15.3 (a) Annual values for Hurrell's wintertime North Atlantic Oscillation (NAO) index. Dark gray shading indicates periods with a positive index; light gray depicts periods with a negative NAO index. The smooth black curve depicts 10-yr running average values. This figure was obtained from http://www.cgd.ucar.edu/cas/jhurrell/indices.html. (b) Annual values for Trenberth and Hurrell's wintertime North Pacific (NP) index, which tracks the intensity of the Aleutian Low (AL) atmospheric pressure cell. Dark gray shading indicates periods with a negative index (when the AL is especially deep), light gray for periods with a negative AL index. The smooth black curve depicts 10-yr running average values, which captures the multidecadal variability associated with the Pacific (inter)Decadal Oscillation, or PDO. This figure was obtained from http://www.cgd.ucar.edu/cas/jhurrell/indices.html. (c) Standardized annual wet-season (June–October) Sahel rainfall, 1898–2002. Image obtained from the University of Washington's Joint Institute for the Study of the Atmosphere and Oceans (http://www.jisao.washington.edu/data_sets). Figures reprinted with permission.

ozone hole that became prominent in the same period, as well as the observed intense warming along the West Antarctic Peninsula and cooling over the interior of the Antarctic Plateau. Others have argued that warming tropical oceans may have also contributed to the observed late 20th-century trends in both the AO (and NAO) and AAO indices.

Pacific Decadal Variability

The dominant pattern of surface climate variability over the North Pacific is characterized by a basin-wide rise or fall in sea-level pressures, with the center of this fluctuating pressure pattern located over the Aleutian Islands (NRC 1998). These pressure changes represent a strengthening or weakening of the Aleutian Low (AL) pressure cell. In some years and in some decades the wintertime AL is especially intense, while in others it is much weaker. Large-scale patterns of western North American temperature, precipitation, and river runoff are linked with AL variability, and in the 20th century the AL had prominent multidecadal variability (see the middle panel of Figure 15.3).

The multidecadal variations in many aspects of climate in the Pacific have received intense scrutiny over the past decade, leading to the notion first of a Pacific (inter)decadal oscillation (or PDO) in the North Pacific (Mantua and Hare 2002), and more recently to an even broader interdecadal Pacific oscillation (IPO) in both hemispheres of the Pacific (Power et al. 1999a, b). The climate variations of the IPO share many of the same spatial features as those of ENSO, showing near-symmetric patterns of ocean temperature and atmospheric circulation changes around the equator in the Pacific extending from the tropics into the midlatitudes of both hemispheres. Although there is no widely accepted explanation for the existence of the IPO, some theories suggest that it is simply a slowly evolving relative of ENSO. Others suggest that there is no preferred period of variation for IPO variations, rather that they simply reflect an integrated and filtered ocean response to a combination of ENSO influences and other higher-frequency atmospheric variations.

Whatever the cause for IPO variations may be, evidence for the impacts of multidecadal swings in Pacific climate and surrounding land areas are ubiquitous. Coastal ocean and land temperatures all along the Pacific coast of the Americas and in the tropical Pacific were mostly above average in the years from 1925–1946 and from 1977–1998, but mostly below average from 1900–1924, 1947–1976, and from 1999–2002. Moving from north to south along the Pacific coast, the warm periods also saw relatively wet conditions in coastal Gulf of Alaska, dry conditions in the interior of Alaska and for the Pacific northwest, wet in the extreme southern U.S., the Caribbean, and Central America, dry in northern South America, and wet conditions in much of Argentina and Chile. On the western side of the Pacific Basin, warm periods in the eastern Pacific Ocean are correlated with warm-dry conditions in eastern

Australia, and dry conditions in Korea, Japan, and the Russian Far East (Power et al. 1999b; Mantua and Hare 2002). Historical evidence also suggests that the multidecadal phase of the IPO influences the shorter timescale ENSO impacts on climate around the Pacific (Power et al. 1999a; Gershunov and Barnett 1998).

HYDROLOGIC CYCLE

Human Impacts

Globally, humans now use slightly more than 10% of the available renewable freshwater supply through household, agricultural, and industrial activities, although in some regions such as the Middle East and North Africa, humans use 120% of the renewable supplies with the excess obtained by pumping groundwater at rates greater than the rate of recharge. By the end of the 20th century, the construction of dams and other structures along rivers moderately or strongly affected flows in 60% of the large rivers systems in the world. Water removal for human uses has reduced the flow of several major rivers, including the Nile, Yellow, and Colorado rivers, to the extent that they do not always flow to the sea. As water flows have declined, so have sediment flows, which are the source of nutrients important for the maintenance of estuaries. Worldwide, sediment delivery to estuaries has declined by roughly 30% (MEA 2005).

Human impacts on hydrologic systems were especially profound in the latter half of the 20th century. Between 1960 and 2000, reservoir storage capacity quadrupled, and water withdrawals from rivers and lakes for irrigation or for urban or industrial use doubled. Large reservoir construction has doubled or tripled the average time that a drop of water takes to reach the sea (MEA 2005).

Impacts of Natural and Human-caused Climate Change

Climatically driven decadal-scale hydrologic variability is most evident in western Africa's Sahel and the continental U.S. (NRC 1998). Multiyear to multidecadal periods of relatively wet and dry conditions are evident not only in precipitation records, but also in stream flow and stored water in large lakes. Dramatic rises in the Great Salt Lake in 1983–1987 and the difference between the stream flows of the Colorado River during the first two decades of the 20th century (when they were exceptionally high) versus the period from 1999–2004 (when they were exceptionally low) demonstrate the ways in which hydrology integrates decadal climate variability. Extreme changes in vegetation and land-surface characteristics that accompanied the Dust Bowl in the 1930s and the Sahelian drought in the 1970s to 1990s highlight the potentially profound synergistic effects that decadal drought, vegetation responses, and human responses through land-use practices can exert on regional ecosystems and societal well-being.

Drought in the Sahel and Tropical Ocean Temperatures

A shift between persistent wet and drought conditions in the Sahel region of western Africa is among the most striking multidecadal regional climate shifts documented in the 20th century. From 1898 to 1949, rainfall in the Sahel exhibited variability at mostly subdecadal timescales. Sahel rainfall was consistently above average in the 1950s and 1960s, but the period from the early 1970s through 2002 brought a nearly unbroken string of low precipitation and severe drought (bottom panel of Figure 15.3). Climate modeling studies by Rotstayn and Lohmann (2002) and Giannini et al. (2003) support the hypothesis that global warming offset by multidecadal cooling trends in the Northern Hemisphere—the latter due to anthropogenic sulfate aerosol emissions in Europe and North America—were a proximate cause for the multidecadal Sahel drought, and that land-use and vegetation feedbacks amplified the incipient drought conditions. The basic chain of events proposed here starts with an equatorward shift in the seasonal belt of monsoon rains caused by a warming of the tropical oceans surrounding Africa. Subsequent drying of the Sahel reduced vegetation. Reduced vegetation led to a brighter land surface that reflected more sunlight and cooled the lower atmosphere. The dryer land cover also dried the lower atmosphere. The combined effects of cooling and drying the lower atmosphere further reduced regional precipitation. These positive feedbacks between land-surface changes and climate are thought to be important amplifying processes in many monsoon regions that help to sustain drought or wet periods.

Drought in the Sahel has been linked to broader-scale climate and ecological impacts over the Atlantic and Caribbean. Landsea and Gray (1992) link summertime west African rainfall with the frequency of intense hurricanes over the tropical Atlantic and the frequency of intense land-falling hurricanes on the eastern coast of the U.S. The Sahel drought has also contributed to a decadal increase in dust transported from Africa across the Atlantic to the Caribbean island of Barbados. Dust from the Sahel provides a radiatively significant aerosol loading to the atmosphere, locally reflecting sunlight and cooling surface temperatures. Minerals in the airborne dust from Africa add important nutrients for marine phytoplankton in iron-limited parts of the Atlantic Ocean, providing a link between drought, dustiness, and the global carbon cycle. Because Africa contributes about half of the global dust total today, these drought-related increases in dust provide a climate feedback that is much larger than the area of the drought region.

The 1930s Dust Bowl and Ocean Temperatures

Climate modeling studies implicate decadal fluctuations in tropical ocean temperatures as a proximate cause for the 1930s Dust Bowl era in North America (Schubert et al. 2004). In this case, a period with slightly cooler than average ocean temperatures in the tropical Pacific coincided with a period of slightly warmer than average ocean temperatures in the tropical Atlantic north of the equator. Once the ocean temperature-induced precipitation deficits were underway, land-surface feedbacks to the atmosphere amplified the drought conditions in the climate model simulations. In a related climate modeling study, Hoerling and Kumar (2003) identify the 1998–2002 coincidence of anomalously warm temperatures in the Indian Ocean and western tropical Pacific Ocean with cool temperatures in the eastern tropical Pacific as "the perfect ocean for drought" over much of the United States, southern Europe, and southwest Asia.

The Atlantic Multidecadal Oscillation and Drought in the U.S.

The Atlantic Multidecadal Oscillation (AMO) is the label given to observed changes in the average surface temperatures for the North Atlantic Ocean (Enfield et al. 2001). The AMO index is based on annual average ocean surface temperatures for the Atlantic Ocean from the equator to 70°N latitude. The AMO index was negative in all but one year from 1902–1925, positive in most years from 1926–1964, negative in most years from 1965–1994, and mostly positive from 1995–2004. Variability in the AMO has been linked with multidecadal changes in precipitation and stream flow in the continental U.S. (Enfield et al. 2001). McCabe et al. (2004) have shown that the two leading patterns of 20th-century decadal drought over the U.S. closely track the decadal changes in the AMO and PDO indices. Prolonged dry spells in the 1930s, the 1950s–1960s, and from 1996–2004 coincided with positive phases of the AMO. During these dry spells, changes in the phase of the IPO/PDO pattern corresponded with north–south shifts in drought areas such that drought concentrated in the northern U.S. for positive IPO/PDO periods but shifted to the southwestern U.S. during negative IPO/PDO periods.

Prolonged wet spells over much of the continental U.S. from 1905–1930, the 1940s, and from 1976–1995 coincided mostly with periods of cool North Atlantic Ocean temperatures (a negative AMO index).

ECOSYSTEMS

Human Impacts

Human-caused secular changes in ecosystems during the 20th century were profound. The recently released Millennium Ecosystem Assessment concludes that over the past fifty years humans have changed ecosystems more rapidly and extensively than during any comparable period of time in human history (MEA 2005; see Box 15.2). Even more alarming is the fact that many of the changes documented in key ecosystem metrics show rapid, nonlinear increases in rates of change that typically began around the mid-20th century. An accelerated rate of change in the last half-century was a characteristic of land converted to croplands, the loss or degradation of mangroves and coral reefs, the amount of

Box 15.2 Excerpts from the Millennium Ecosystem Assessment (MEA 2005, pp. 1–3).

Main Finding #1: Over the past fifty years, humans have changed ecosystems more rapidly and extensively than in any comparable period of time in human history, primarily to meet rapidly growing demands for food, freshwater, timber, fiber, and fuel. This has resulted in a substantial and largely irreversible loss in the diversity of life on Earth.

- More land was converted to cropland since 1945 than in the 18th and 19th centuries combined. Cultivated systems (areas where at least 30% of the landscape is in croplands, shifting cultivation, confined livestock production, or freshwater aquaculture) now cover one-quarter of Earth's terrestrial surface.
- Approximately 20% of the world's coral reefs were lost and an additional 20% degraded in the last several decades of the 20th century, and approximately 35% of mangrove area was lost during this time (in countries for which sufficient data exist, which encompass about half of the area of mangroves).
- The amount of water impounded behind dams quadrupled since 1960, and 3–6 times as much water is held in reservoirs as in natural rivers. Water withdrawals from rivers and lakes doubled since 1960; most water use (70% worldwide) is for agriculture.
- Since 1960, flows of reactive (biologically available) nitrogen in terrestrial ecosystems have doubled, and flows of phosphorus have tripled. More than half of all the synthetic nitrogen fertilizer, which was first manufactured in 1913, ever used on the planet has been used since 1985.
- Since 1750, the atmospheric concentration of CO_2 has increased by about 34% (from about 280 to 376 ppm in 2003), primarily due to the combustion of fossil fuels and land-use changes. Approximately 60% of that increase (60 ppm) has taken place since 1959.
- Humans are fundamentally, and to a significant extent irreversibly, changing the diversity of life on Earth, and most of these changes represent a loss of biodiversity.
- More than two-thirds of the area of two of the world's 14 major terrestrial biomes and more than half of the area of four other biomes have been converted by 1990, primarily to agriculture.
- Across a range of taxonomic groups, either the population size or range or both of the majority of species is currently declining.
- The distribution of species on Earth is becoming more homogenous; in other words, the set of species in any one region of the world is becoming more similar to the set in other regions primarily as a result of introductions of species, both intentionally and inadvertently in association with increased travel and shipping.
- The number of species on the planet is declining. Over the past few hundred years, humans have increased the species extinction rate by as much as 1,000 times over background rates typical over the planet's history (medium certainty). Some 10–30% of mammal, bird, and amphibian species are currently threatened with extinction (medium to high certainty). Freshwater ecosystems tend to have the highest proportion of species threatened with extinction.
- Genetic diversity has declined globally, particularly among cultivated species.

water impounded behind dams, water withdrawn from rivers and lakes (primarily for agriculture), increased flows of biologically available nitrogen and phosphorous in terrestrial ecosystems, and increased atmospheric concentrations of CO_2 and CH_4. The 20th century also brought major changes to biodiversity: in addition, either the population size or range or both of the majority of species is currently declining (MEA 2005).

Rapid declines in large predators have been documented for the world's oceans. At the global scale, fisheries data indicate a long-term downward trend in the trophic level (the level of an organism in the food chain) of marine fish catches over the last half-century. A spectacular collapse of Atlantic cod stocks in the late 20th century was a contributing factor to those declines; however, severe declines in other large predators (e.g., tuna, billfish, sharks, skates, and rays) have also been indicated by commercial fishing records (MEA 2005; Myers and Worm 2003).

At a regional scale, marine ecosystems in the North Atlantic experienced major human-caused changes in the 20th century. Atlantic cod populations off Newfoundland collapsed in the early 1990s after centuries of supporting commercial fisheries. More gradual declines in cod stocks in other parts of the Atlantic during the last few decades of the 20th century were mirrored by dramatic increases in many co-located crustacean populations. Worm and Myers (2003) examined Atlantic cod and northern shrimp catch and biomass estimates from nine different regions in the North Atlantic over the period 1960–2000 and found a strong negative correlation between the cod and shrimp series in eight regions, with the one exception being in the southernmost region in their study. The cod biomass series are positively correlated with water temperature, but the shrimp series are not correlated with water temperature. They argue that the overfishing of Atlantic cod led to both the widespread collapse of cod in the 1980s and the well-documented rapid increases in northern shrimp and many crab and lobster populations as the benthic crustaceans experienced significant declines in cod predation. A decade of sharp fishing restrictions and outright closures for cod has, thus far, failed to spur widespread signs of an Atlantic cod recovery.

Climate-related Decadal Variability in Marine Ecosystems

In addition to these profound secular changes in 20th-century ecosystems, there are striking examples of variability in ecosystems at decadal timescales. Kawasaki (1991) documents a remarkable coherence between 40–60 yr period fluctuations in sardine populations (and landings) off of Japan, California, Chile, and Peru. Equally remarkable in the 20th century was the fact that the sizes of two of the other largest coastal pelagic fish populations of the world—the Peruvian anchoveta and the South African sardine—varied directly out of phase with Pacific sardines. Mechanisms for this global synchrony in these widely separated fish populations are not understood, but most hypotheses include a role for global-scale climate forcing as the coordinating agent of change.

Many marine species in the subarctic North Pacific and Bering Sea have undergone strong decadal changes in the 20th century. From the 1950s through 1970s, crab and shrimp fisheries in the Gulf of Alaska and Bering Sea were highly productive; however, beginning in the 1980s those fisheries suffered steep declines which generally persisted through the 1990s. While Alaska's shrimp and crab fisheries collapsed in the 1980s and 1990s, the abundance and commercial landings of large predatory fish (particularly pollock, Pacific cod, halibut, and other flatfish) increased dramatically. Decadal changes in North Pacific phytoplankton, zooplankton, and Pacific salmon productivity have also been associated with decadal climate changes (for a more thorough review and list of references, see Mantua and Hare 2002).

Causes for these climate-induced changes in marine ecosystems are not well understood. Climate-induced changes in both marine food webs and stream conditions are important factors in the observed swings in Pacific salmon productivity (these changes are, of course, in addition to major changes wrought by humans via fishing, hatchery propagation, land- and water-use practices, and dam building in western North America). For the upwelling ecosystems of the eastern Pacific Ocean, climate-driven changes in the temperature, stratification, and depth of the seasonal mixed-layer has been linked with the availability of nutrients in near-surface waters where light is available for photosynthesis, the species composition of zooplankton, biogeographic shifts in pelagic fish and sea birds, changes in forage fish production, and changes in survival rates for various species of salmon and rockfish (Peterson and Schwing 2003; Chavez et al. 2003).

BIOGEOCHEMICAL CYCLES

Carbon Cycle

As previously noted, the atmospheric concentration of CO_2 has increased ~34% (from about 280 ppm in 1750 to 376 ppm in 2004) since preindustrial times, with ~60% of that increase taking place since 1959. The effect of changes in terrestrial ecosystems on the carbon cycle reversed during the last fifty years. Terrestrial ecosystems were on average a net source of CO_2 during the 19th and early 20th centuries (primarily due to deforestation and other land-use changes) and became a net sink sometime around the middle of the last century. Carbon losses due to land-use changes continue at high levels. Factors contributing to the growth of the role of ecosystems in carbon sequestration include afforestation, reforestation, and forest management in North America, Europe, China, and other regions; changing agriculture practices; and the fertilizing effects of nitrogen deposition and increasing atmospheric CO_2 (MEA 2005).

Feedbacks between terrestrial ecosystems, climate, and the global carbon cycle are likely to be important factors that will determine the strength of the terrestrial carbon sink in the future. For example, increasing heterotrophic (non-plant) respiration with increasing temperature as well as increasing CO_2 emissions due to increased fire frequency and insect attacks in forests driven by a warming and drying climate are likely to reduce the terrestrial ecosystems ability to store atmospheric CO_2 in a globally warmer climate.

Nitrogen and Phosphorus Cycles

The total amount of reactive (or biologically available) nitrogen created by human activities increased ninefold between 1890 and 1990, with most of that increase taking place over the past fifty years in association with increased use of fertilizers. More than half of all the synthetic nitrogen fertilizer (first produced in 1913) ever used on Earth has been applied since 1985. Human activities have now roughly doubled the rate of creation of reactive nitrogen on the land surfaces of Earth, and the flux of reactive nitrogen to the oceans increased by ~80% from 1860–1990. There are distinct regional patterns in human-caused changes in nitrogen flows, with fluxes from highly developed regions (including the northeastern U.S., the watersheds of the North Sea in Europe, and the Yellow River basin in China) having increased ten- to fifteenfold, while regions such as the Labrador and Hudson Bays in Canada have seen little change in nitrogen fluxes over that same time period (MEA 2005).

The use of phosphorus fertilizers and the rate of phosphorous accumulation in agricultural soils increased nearly threefold between 1960 and 1990, although the rate has declined somewhat since that time. The current flux of phosphorus to the oceans is now triple that of the background rates (MEA 2005).

SUMMARY

Twentieth-century decade- to century-scale variations in parts of the Earth system have been remarkable. As reported in the Millennium Ecosystem Assessment's Synthesis Report (MEA 2005), humans have caused extensive and rapid changes in a number of Earth system components. At the same time, decadal-scale changes have in some cases varied in a more quasi-periodic manner, demonstrating some indications for reversible change in parts of the Earth system. Recurring patterns of regional climate variability, generally in response to decadal variations in the atmosphere and/or ocean circulation patterns, have shown evidence for reversible change. However, the causes for decadal climate variability remain poorly understood.

Efforts aimed at synthesizing the causes for decade- to century-scale changes in Earth systems highlight the importance of feedbacks between Earth system components. For instance, the possible links between the industrial production of chlorofluorocarbons, stratospheric ozone depletion, increased greenhouse gases, warming tropical oceans, and trends in prominent high-latitude wind and weather patterns (associated with the AO and AAO circulation patterns) are high-priority topics in Earth systems research. Likewise, increased greenhouse gas concentrations, warming trends in tropical ocean temperatures, and seemingly subtle shifts in tropical rainfall patterns appear to be "teleconnected" (remotely connected) to the distribution of sustained drought over many parts of the world, including the Sahel, the continental U.S., Europe, southwest Asia, Australia, and the midlatitudes of South America and Africa. Drought itself is part of a complex suite of interactions involving feedbacks between the atmosphere, vegetation, and land-surface characteristics important for regional hydrology and ecosystems. On even longer timescales, feedbacks between ecosystems and the global carbon cycle have the potential to further alter climate by influencing the concentration of atmospheric greenhouse gases and the distribution of dust-borne aerosols. Increasing heterotrophic respiration as the result of higher temperatures and increasing CO_2 emissions, due to increased fire frequency and insect attacks in forests driven by a warming and drying climate, are likely to reduce the CO_2 storage capacity for terrestrial ecosystems.

A pair of summary time lines for decade- to century-scale changes in Earth systems are provided in Figures 15.4 and 15.5, respectively. Human activities caused substantial, rapid changes in key components of the Earth system in the last half-century (Figure 15.4). In many cases, rates of change accelerated from

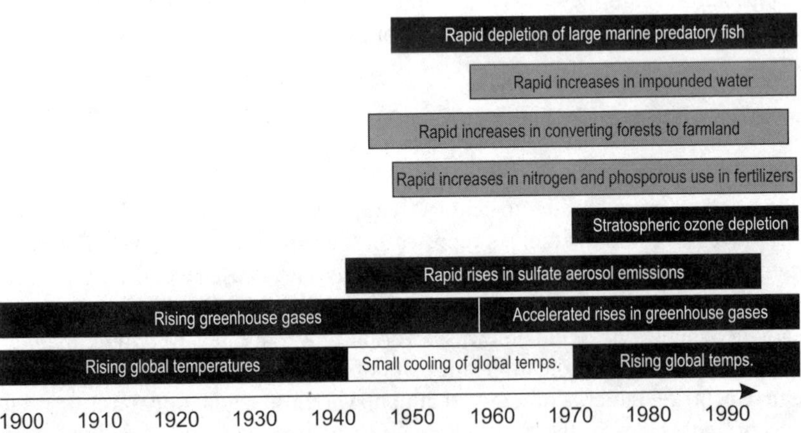

Figure 15.4 A summary time line for aspects of 20th-century secular trends in Earth's natural systems. Black boxes indicate rising global temperatures, rising concentrations of greenhouse gases and sulfate aerosol emissions, the depletion of stratospheric ozone, and the rapid depletion of large marine predatory fish. The white box indicates the midcentury period of cooling global temperatures. Gray shading highlights rapid increases in the amount of water impounded behind dams, the conversion of forests to farmland, and the increases in nitrogen and phosphorous use in fertilizers. Dates are meant to be approximate.

the beginning to the end of the 20th century. Decadal variations in the aspects of climate reported in this chapter are summarized in Figure 15.5. An important aspect of global temperature change was the phasing of decadal circulation patterns in the Pacific and Atlantic sectors. The prevalence of the negative phases of the AO/NAO and IPO/PDO from the mid-1950s to early 1970s contributed to the observed global cooling during that era, and the prevalence of the positive phases of these climate patterns in the late 1970s to mid-1990s contributed to the observed global warming in that era. Yet over the course of the entire 20th century, the indices used to track decadal climate patterns in the Atlantic and Pacific (AO/NAO and PDO/IPO indices, respectively) are uncorrelated. Likewise, the correspondence between the decadal variability in ENSO variance and the IPO/PDO are not correlated for the 20th century. In contrast, sustained drought in the Sahel and over the U.S. showed some correspondence with slowly varying ocean temperature patterns like that for the North Atlantic captured by the AMO index. Mechanisms linking the various decade- to century-scale patterns in Earth's climate are poorly understood, yet they are clearly of extreme importance for making projections of future regional climate changes. Presently, the

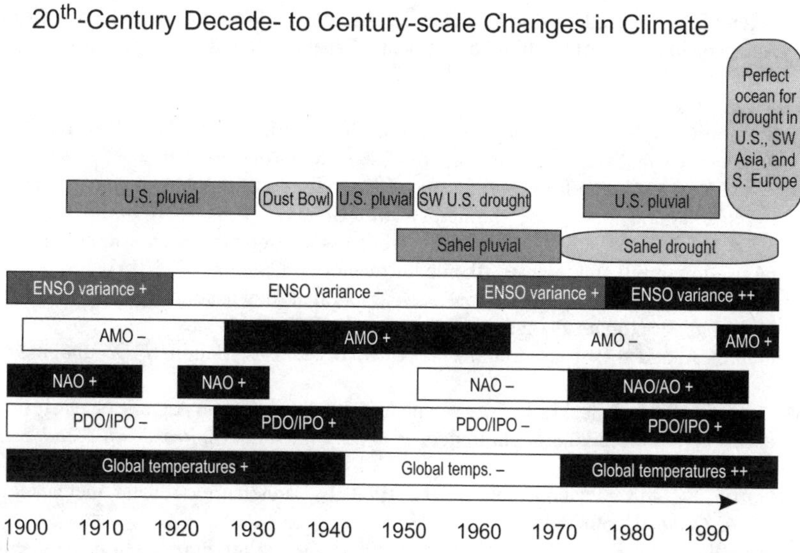

Figure 15.5 A summary time line for aspects of 20th-century decade- to century-scale changes in Earth's climate. Black boxes indicate rising global temperatures, periods with high ENSO activity, and positive (warm) phases of natural climate patterns (NAO/AO, PDO/IPO, and AMO). Here the NAO is treated as a regional expression of the AO, and the PDO is treated as the regional expression of the IPO. White boxes indicate the midcentury period of cooling global temperatures, the period of reduced ENSO activity, and negative (cool) phases of the climate patterns. Gray boxes highlight pluvials, whereas gray ellipses indicate droughts. Dates are meant to be approximate.

climate system models used to assess the impacts of changing concentrations of greenhouse gases and aerosols on climate generally do not provide a clear indication for how natural decadal climate patterns will behave in the future (IPCC 2001).

REFERENCES

Chavez, F.P., J. Ryan, S.E. Lluch-Cota, and C. Miguel Ñiquen. 2003. From anchovies to sardines and back: Multidecadal change in the Pacific Ocean. *Science* **299**:217–221.

Dlugokencky, E.J., S. Houweling, L. Bruhwiler et al. 2003. Atmospheric methane levels off: Temporary pause or a new steady-state? *Geophys. Res. Lett.* **30(19)**:1992.

Enfield, D.B., A.M. Mestas-Nunez, and P.J. Trimble. 2001. The Atlantic multidecadal oscillation and its relation to rainfall and river flows in the continental U.S. *Geophys. Res. Lett.* **28**:2077–2080.

Gershunov, A., and T.P. Barnett. 1998. Interdecadal modulation of ENSO teleconnections. *Bull. Am. Meteorol. Soc.* **79**:2715–2726.

Giannini, A., R. Saravanan, and P. Chang. 2003. Oceanic forcing of Sahel rainfall on interannual to interdecadal timescales. *Science* **302**:1027–1030.

Hoerling, M., and A. Kumar. 2003. The perfect ocean for drought. *Science* **299**:691–694.

IPCC (Intergovernmental Panel on Climate Change). 2001. Climate Change 2001: The Scientific Basis, ed. J.T. Houghton et al. Contribution of Working Group I to the Third Assessment Report of the Intergovernmental Panel on Climate Change. Cambridge: Cambridge Univ. Press.

Kawasaki, T. 1991. Long-term variability in the pelagic fish populations. In: Long-term Variability of Pelagic Fish Populations and Their Environment, ed. T. Kawasaki, S. Tanaka, Y. Toba, and A. Taniguchi, pp. 47–60. New York: Pergamon.

Labitzke, K., and H. van Loon. 1993. Some recent studies of probable connections between solar and atmospheric variability. *Ann. Geophys.* **11**:1084–1094.

Landsea, C.L., and W.M. Gray. 1992. The strong association between western Sahelian monsoon rainfall and intense Atlantic hurricanes. *J. Climate* **5**:435–453.

Lean, J., J. Beer, and R. Bradley. 1995. Reconstruction of solar irradiance since 1610: Implications for climate change. *Geophys. Res. Lett.* **22**:3195–3198.

Mantua, N.J., and S.R. Hare. 2002. The Pacific decadal oscillation. *J. Oceanogr.* **58**: 35–44.

McCabe, G.J., M.A. Palecki, and J.L. Betancourt. 2004. Pacific and Atlantic Ocean influences on multidecadal drought frequence in the United States. *Proc. Natl. Acad. Sci.* **101**:4136–4141.

McGuffie, K., and A. Henderson-Sellers. 1988. Is Canadian cloudiness increasing? *Atmos.Ocean* **26**:608–633.

MEA (Millennium Ecosystem Assessment). 2005. Millennium Ecosystem Assessment Synthesis Report: A Report of the Millennium Ecosystem Assessment. Washington, D.C.: Island. http://www.millenniumassessment.org/

Moberg, A., D.M. Sonechkin, K. Holmgren, N.M. Datsenko, and W. Karl. 2005. Highly variable Northern Hemisphere temperatures reconstructed from low- and high-resolution proxy data. *Nature* **443**:613–617.

Mote, P.W., A.F. Hamlet, M.P. Clark, and D.P. Lettenmaier. 2005. Declining mountain snow pack in western North America. *Bull. Am. Meteorol. Soc.* **86**:39–49.

Myers, R.A., and B. Worm. 2003. Rapid worldwide depletion of predatory fish communities. *Nature* **423**:280–283.

NRC (National Research Council). 1998. Decade-to-century-scale Climate Variability and Change: A Science Strategy. Washington, D.C.: Natl. Acad. Press.

Peterson, W.T., and F.B. Schwing. 2003. A new climate regime in northeast Pacific ecosystems. *Geophys. Res. Letts*. **30(17)**:1896.

Power, S., F. Tseitkin, V. Mehta et al. 1999a. Decadal climate variability in Australia during the twentieth century. *Intl. J. Climatol*. **19**:169–184.

Power, S., F. Tseitkin, S. Torok et al. 1999b. Australian temperature, Australian rainfall and the Southern Oscillation, 1910–1992: Coherent variability and recent changes. *Austral. Meteorol. Mag*. **47**:85–101.

Rotstayn, L.D., and U. Lohmann. 2002. Tropical rainfall trends and the indirect aerosol effect. *J. Climate* **14**:2103–2116.

Schubert, S.D., M.J. Suarez, P.J. Pegion, R.D. Koster, and J.T. Bacmeister. 2004. On the cause of the 1930s Dust Bowl. *Science* **303**:1855–1859.

Stewart, I.T., D.R. Cayan, and M.D. Dettinger. 2005. Changes towards earlier streamflow timing across western North America. *J. Climate* **18**:1136–1155.

Thompson, D., and S. Solomon. 2002. Interpretation of recent Southern Hemisphere climate change. *Science* **296**:895–899.

Worm, B., and R.A. Myers. 2003. Meta-analysis of cod-shrimp interactions reveals top-down control in oceanic food-webs. *Ecology* **84**:162–173.

16

Social, Economic, and Political Forces in Environmental Change

Decadal Scale (1900 to 2000)

John R. McNeill

History Department, Georgetown University, Washington, D.C. 20057, U.S.A.

ABSTRACT

This chapter provides a brief sketch of the scale and scope of environmental changes around the world over the past century or so. It also explains at greater length the evolving contours of global political economy which helped to shape environmental change. Over the past century a grand, global struggle took place between efforts to create autarkic national or imperial economies, championed at various time by Germany, Japan, and the U.S.S.R., and rival efforts to build an integrated, international economy, championed by Great Britain before 1914 and by the U.S.A. after 1945. While concentrating on this dimension of political economy, the chapter also treats modern trends in science and technology, in population and urbanization, and in energy use, all considered insofar as they related to environmental change.

INTRODUCTION

In this chapter, I wish to provide a sense of the social, political, and economic background to the environmental turbulence of the last hundred or so years by following two strategies simultaneously: (a) a narrative of the evolution of international political economy since the 1890s, and (b) summary treatments of driving forces behind environmental change operating on the decadal timescale. The first of these (section on GLOBAL POLITICAL ECONOMY SINCE 1890) may be too familiar to some readers to justify the time required to read it, but it is offered for those who are not already well acquainted with the subject. In it they may find historical developments that help explain components of the human–environment relationship. Thereafter, I present my thoughts on what is most important within recent human history for explaining the scale, scope, and variety of environmental change.

THE SCALE AND SCOPE OF ENVIRONMENTAL CHANGE (IN BRIEF)

Humankind has affected the biosphere in important ways since the harnessing of fire. In modern times, however, our power to alter nature grew exponentially with the onset of rapid population growth since the 18th century (and very rapid growth since 1945), the deployment of fossil fuel energy (from the 16th century, but very extensively since 1945), and the emergence of science and technology capable of decisive interventions (mainly since 1850). These developments were unsettling intellectually, socially, and environmentally, creating instabilities that fed upon one another. Durable and stable (not necessarily peaceful or pleasant) social and ecological orders cracked and a regime of perpetual disturbance ensued. Whether one dates this from the 18th century or, as I do here, from the 1890s, is a matter of perspective and emphasis.

Under these circumstances of tumult and flux, both cultural and biological evolution increasingly favored the adaptable as opposed to the well adapted. All this proved wrenching socially, cultural, and politically, as fortunes were reshuffled suddenly, not merely among individuals and families, but among states, societies, and cultures. States and societies survived if compatible with existing and emerging features of the global economy, such as growing integration and technological change, and with the shifting winds of international politics. Those states and societies not easily compatible with the various winds of change—the Ottoman Empire, the Qing Empire in China, the Russian Empire—were destroyed and either shattered forever (Ottoman Empire) or rebuilt in radically new forms (China and Russia).

States and societies entered an era in which economic and political competition was especially acute, and restraint in the interest of stability or the integrity of the biosphere seemed imprudent to almost all those in positions of power. Advocacy of anything short of maximum economic growth came to seem a form of lunacy or treason in the 20th century. And military preparedness, always a high priority of most states, commanded unquestioned support in most large countries, as in an age of total war the potential costs of unpreparedness appeared greater than ever.

Indeed, restraint in the interest of the biosphere was rarely given any consideration at all until the emergence of a more or less worldwide environmental movement in the 1960s. Although it had its precedents in many times and places, this was something novel, more general, more popular, and more sustained than prior conservation or preservation movements. It was provoked primarily by unbridled pollution, a Hegelian antithesis to industrialism and fossil fuels. However, it remained a weak antithesis, unable to command the political support necessary to challenge the regime of perpetual disturbance.

Thus it was that the global economy grew 14-fold in the 20th century, human population 4-fold, industrial production 40-fold, and energy use about 13-fold.

Nothing like this had ever happened before in human history. The mere fact of such growth, and its unevenness among societies, made for profound disruptions in both environment and society. Some of the ecological consequences of these trends are considered by Mantua (this volume) as well as in McNeill (2000) and Steffen et al. (2004).

(A MORE OR LESS CONVENTIONAL NARRATIVE OF) GLOBAL POLITICAL ECONOMY SINCE 1890

Readers familiar with the broad lines of world history in the 20th century may wish to skip over this lengthy section, although it provides an important context for the environmental changes of modern times. The emphasis here is on the evolution and growth of a unitary, interactive, global economy, which was not a foreordained matter but rather required the political defeat of rival visions and programs of how best to organize humankind and its relations with nature. Had the struggles within the political economy domain turned out differently, then the history of the biosphere would surely have been quite different from what it was.

War and Depression, 1890–1941: Visions of Autarky

In fits and starts, the various societies of the world had been forging economic, political, ecological, and other links among one another for millennia. The process gathered momentum with the oceanic navigations of the 16th–18th centuries and accelerated once more after about 1870 with technological changes in transport and communications such as the telegraph, railroad, and steamship (Osterhammel and Petersson 2005; McNeill and McNeill 2003).

The ongoing integration of the world's societies and economies after 1870 led many observers to suppose that war had become obsolete. In 1914 the outbreak of World War I shattered such fond hopes. The underlying reason for the war was the emergence of Germany as a great power in Europe, and the inability of the international system to accommodate an ascendant Germany. Prior to 1860 Germany existed only as a geographical expression; in a series of brief wars, however, it became a unified state by 1871. Its population (1871–1914) grew far faster than France's or Britain's. German urbanization and industrialization rates outstripped its neighbors' as well. This brought Britain and France, historic rivals, together after about 1900, and rival alliance systems soon formed involving all the major powers of Europe. The vigor of nationalism in European societies made war easier to accept, as did general ignorance of what war might entail.

In four years of unprecedented slaughter (1914–1918), an alliance between France, Britain, Russia, Italy, and eventually the U.S. narrowly prevailed over Germany, Austria–Hungary, and the Ottoman Empire. Military medicine had progressed to the point where doctors could keep gigantic armies free of epidemics long enough so that they could engage in the prolonged butchery of trench warfare. World War I killed about 9–10 million combatants in all. Millions more civilians died, from hunger, disease, or violence. European imperial interests ensured that the war spread to Africa, Asia, and the Pacific. The war became one of attrition, of both men and morale. All armies except the German underwent large-scale mutinies within 35 months after the onset of heavy casualties (Keegan 1976).

As an increasing number of men were fed into this meat grinder, governments concluded they had to mobilize every aspect of their societies for the war effort, a practice now known as "total war." War production became too serious a business to be left to the market, so official planning boards took over most decision making in industry, transport, and agriculture. Factories recruited millions of women to make weapons, munitions, and uniforms. Despite frustrating bottlenecks, it turned out that war economies, at least in some countries, could quickly ratchet up levels of production. This record of success offered an example of government management of national economies that proved appealing afterwards, in both war and peace.

The war shook the political and economic foundations of Europe and reverberated throughout the world. The peace treaties signed beginning in 1919 created new countries out of Austria–Hungary's multinational empire, shrank Germany slightly, and imposed a large indemnity upon Germany. It awarded some Ottoman territories to the victors through the mechanism of the League of Nations, an international body created in 1919–1920 to manage future conflicts and prevent war. What remained of the Ottoman Empire collapsed in revolution (1919–1924), when a secular, nationalist Turkey was born (see below). The Turks promptly abolished the caliphate of Islam, leaving the Muslim world without a religious head or political center for the first time in 1,300 years.

In some cases, the world of the victors came apart as well. By 1917 the strains of industrial warfare proved too much for Russia. Discontent fused into a revolution in the spring of 1917. A new, provisional government fatally chose to continue the war effort and, in the fall of 1917, fell in a *coup d'état* known as the Bolshevik Revolution, organized by Vladimir Lenin (1870–1924). Lenin made peace with Germany (March 1918) and won a Civil War (1918–1921), by using command economy methods he had admired in the German war effort. The country he created, the Union of Soviet Socialist Republics (U.S.S.R.), remained ever after indebted to methods of economic management pioneered by Germany during World War I.

Among the victorious Allies, others felt the strains of war too. Italy's government collapsed (1919–1922) giving rise to the fascist dictatorship of Benito Mussolini (1883–1945). Fascism repudiated democratic politics, vehemently rejected socialism, sought to mobilize the masses through a single political party, exalted the nation, and extolled martial virtues, and the purifying effects of war. Fascist movements and parties appealed especially to men who missed the emotional solidarity and camaraderie of military service.

To millions in Europe and around the world, Russia and Italy seemed to offer practical solutions to the troubles of the postwar age. The booms and busts of the market, which increasingly rippled throughout the world beyond anyone's control, meant constant risk of unemployment or, for peasants, of ruinously low prices for the food they sold. The globalizing economy, under reconstruction in the early 1920s, meant that people might lose their jobs because of a financial panic on the other side of the world, and peasants might go broke because of a bumper crop on another continent. This situation, while not entirely new, surpassed most people's understanding and accentuated grievances. With the public appetite for new economic and political arrangements whetted, communism and fascism found many supporters in Europe, and a few in China, colonial India, the U.S., and elsewhere. Part of the appeal of the Soviet and Italian models lay in the buffer they promised to provide against the buffeting of international markets. Both made economic self-sufficiency a high priority (although Italy never came close to achieving it).

During World War I, international trade fell off sharply. The combatants had ceased trading with one another and had methodically set about sinking one another's merchant ships, which submarines, a new weapon, made much easier. Thus shortages affected all participants, especially those most in need of imported food. Devising substitutes for lost imports became a crucial part of the war economies, but no one found a substitute for food. The U.S.S.R. and Italy, determined not to suffer such shortages again, made autarky—economic self-sufficiency—a peacetime policy. Although capital and trade flows recovered substantially in the 1920s, ambitions for autarky prevented the full resumption of the globalizing economy of the pre-1914 era.

The appeal and logic of autarky seemed to improve in 1929 when the reconstructed international economy collapsed. After 1924 American loans allowed Germany to pay some of its war reparations to Britain and France, and soon Germans, British, and French resumed their imports from around the world: the international economy flickered back to life. However, American loans dried up in 1928 when a stock market boom absorbed all available capital. Financing for exports from Latin America, Australia, and other agricultural regions evaporated as well, bringing on agricultural depression. Then the American stock market crashed in 1929, loans were called in, and banks and firms suddenly went bankrupt. The fact that banks were linked in worldwide networks of loans ensured that the New York crash and the subsequent failure of 40% of American banks ignited a chain reaction around the world. Soon millions of people lost their jobs in the industrial countries, and farmers around the world could not sell their crops. Given the memory of successful economic management during WWI, governments had to do something. They did.

They put up tariffs, quotas, and other obstacles to trade. They devalued their currencies, in the hopes of boosting exports and easing unemployment. They raised taxes and lowered public spending to balance budgets. They abandoned the gold standard, making all international transactions more difficult. In short, by seeking national escapes from the depression, they deepened it internationally. By 1932 the world economy had shrunk by a fifth, world trade by a quarter. Unemployment levels in hard-hit countries like the U.S., Canada, and Germany reached 20–35%. Food and raw material prices fell by half. Farmers and peasants around the world lost much of their incomes and, if in debt, often lost their lands. This was the Great Depression, which made millions hungry, insecure, and more inclined than ever before to contemplate novel social orders.

The most autarkic economies seemed to weather the storm best. Italy suffered less than other countries, and the U.S.S.R. also appeared to, although in reality it was undergoing an acute economic crisis of its own. Lenin's successor, Joseph Stalin (1879–1953), sought to extend the revolution to the countryside by replacing family farms with collective and state farms. This aroused bitter opposition, which Stalin answered with desperate brutality. About 10 million peasants were killed or starved to death in 1932–1933, and the Soviet economy contracted by perhaps a fifth. Stalin managed to keep all of this from the outside world, however, while trumpeting simultaneous—and very real—Soviet achievements in industrialization. Stalin aimed to end Soviet dependence on imported industrial and military goods. A planned national economy, which the U.S.S.R. became after 1928, could not easily mesh with the unpredictable swings in supply and demand characteristic of international markets. Autarky was the logical answer. State-sponsored industrialization also became national policy in Mexico, Argentina, and Brazil, without Stalinist terror. Japan's military government renewed the emphasis on heavy industry and agriculture, hoping to make the country self-sufficient in weaponry and food. Almost everywhere governments aspired to build more autarkic national economies.

This proved fateful. The most important consequence of the turn to autarky in the 1930s was that it made states expansionist, particularly those without overseas empires. To be free from imports that seemed ruinously expensive in peacetime and would (the experience of WWI suggested) be unreliable in war, it made sense to acquire more territory and resources.

In many countries, economic security seemed to require geographic expansion. Italy, Germany, and Japan all lacked crucial resources necessary to an autarkic economy, especially oil. Italy sought an empire in the Mediterranean and in the Horn of Africa, a quest that led to the conquest of Ethiopia in 1935–1936. Germany built an economic empire in southeastern Europe, luring smaller countries into exclusive trade agreements, and in 1938–1939 annexed Austria and Czechoslovakia. From 1931 Japan encroached on Chinese territory, especially in Manchuria, a region rich in coal and iron ore. It also tried to harness its colonies of Korea and Taiwan more tightly and, on a smaller scale, to tap the agricultural potential of the Micronesian islands it had acquired from Germany through the League of Nations after World War I. These quests for empire to make autarky more feasible brought on World War II.

On balance, the period from 1914 to 1940 saw a great disintegration of the world economy. As noted above, trade plummeted. Capital flows, never fully recovered from WWI, dwindled after 1929 and did not reach 1913 levels again until after WWII. The great era of migration (perhaps 100 million people took part in intercontinental migrations 1840–1920) also stopped, partly because some countries forbade emigration and partly because the largest recipient of immigrants, the U.S., adopted tight quotas in 1924. These were deep changes, and very disruptive to prosperity based upon free movement of goods, capital, and people. They undermined the politics of liberalism and shifted support to the idea that governments should direct economic life. In a nutshell, globalization after 1870 and war after 1914 generated resentments and fears which fed nationalism and the quest for autarky, thus undermining the cooperation and restraint necessary for a global economy, and for peace.

Resurgent Globalization: War and the Long Boom since 1941

The League of Nations proved unequal to the tasks it faced. The U.S.A., which had done the most to create it, refused to join, as did the U.S.S.R. until 1934. The League lacked the power to check the expansionist ambitions of Italy, Japan, and Germany, who formed a loose alliance in 1936, known as the Axis. The combination of their belligerence and the League's impotence obliged the U.S.S.R., France, Britain, and eventually the U.S. to prepare for war.

World at War, 1937–1945

World War II was, by a sizeable margin, the largest, most destructive struggle in history. It began with the Japanese assault on China proper in 1937. Combat lasted until 1945. The Japanese and the Germans had to prevail quickly if they were to win at all, because the Allies' economies could handily outproduce the Axis. The U.S.S.R. was preadapted to a war economy. It had used command economy methods from 1918 and economic planning since 1928. It specialized in hasty improvisation; its culture extolled frantic mobilization of the sort that built factories from scratch in a week, and encouraged (some might say obliged) women to work in field and factory. Its peasants and proletarians developed a high tolerance for Spartan conditions. The U.S. was also preadapted to a war economy by its "genius for mass production,"[1] its huge production runs, its tradition of entrepreneurial innovation and ruthless quests for efficiency. The U.S. created a huge military–industrial complex in one year, 1942, during which it alone outproduced the Axis in quantities of war materiel by about two to one. Despite much lower skill levels in its labor force and great disruptions due to lost territory, the U.S.S.R. outproduced Germany in every year of the war. The Allies had more raw materials, almost all of the world's oil, more men, and they eagerly recruited women for war work, even for combat in the Soviet case. In addition, thanks to clever decipherment, the Allies could read their enemies' secret codes.

1 Franklin Roosevelt's words. Hermann Goering derided American industry as capable only of producing razor blades.

Finally, the Axis never achieved anything like the degree of economic and strategic cooperation that the Allies, especially Britain and the U.S., managed. Thus once the Allies withstood the initial assaults of 1940–1941, their chance of winning improved daily. They won when the Soviet Army smashed its way into Berlin in May of 1945, and when American airmen dropped atomic bombs on Hiroshima and Nagasaki three months later (Overy 1995; Weinberg 1994).

World War II killed about 3% of the world's 1940 population, some 60 million people, including large numbers of civilians. The Germans and their accomplices murdered about 6 million Jews. Most of the fighting, and especially the dying, took place in eastern China and the western U.S.S.R. The U.S.S.R. lost perhaps 25 million to the war, two-thirds of them civilians, Poland lost 6 million (half of whom were Jews), Germany 4.5 million, Japan 2.4 million, Yugoslavia 1.6 million, and China maybe 15 million, although that figure is only a guess. The Americans, whose factories did so much to win the war, lost 290,000 in battle and 400,000 in all, about the same as Britain and Greece, rather less than France (Weinberg 1994).

The Cold War World

The U.S. emerged from WWII as the first global superpower. Imperial Rome and China had held unrivaled sway regionally, but never before had any state exercised such global reach. The U.S. had the world's biggest navy, it had (until 1949) a monopoly on nuclear weapons, it had (briefly) about half the world's industrial capacity, and nearly everyone owed its bankers money. It also enjoyed great prestige. Some of these advantages were simply the result of the prostration of the other major powers: the U.S. had not had its cities flattened, its industries blown up, its population decimated. But others derived from longer-term processes dating back to the 1800s. The U.S. had become the world's largest industrial power around 1890 and for the next 40 years was the "tiger" of the world economy, growing at 5–7% per year.

The key to American power, amply demonstrated in WWII, was the efficiency of its heavy industry, symbolized by the moving assembly line, which made production faster and cheaper. The symbiosis of mass production methods and a comparatively well-paid labor force amounted to a social contract (sometimes called Fordism) which undergirded American democracy after WWI. As millions of workers became consumers, businesses enjoyed larger and larger production runs, lowering their per-unit costs further, encouraging further consumption. Cheap energy and mass production made U.S. industrial firms the most competitive in the world. This allowed them to export successfully and to build subsidiary plants using American-style techniques overseas, which they did energetically in the 1920s. The Depression and WWII interrupted the emergence of American business on the world scene, but after 1945 American business enjoyed a double windfall. Industry in Europe and East Asia lay in ruins and the American government was prepared, as never before, to undertake management of the world economy.

Remembering the fiascos of peacemaking in 1919, the Allies in WWII insisted upon unconditional surrender so they could dictate terms to Germany and Japan. The Allies quickly quarreled, but nonetheless managed to reconstruct a stable order very quickly. The Americans did most of the managing. They sponsored and bankrolled a new slate of international institutions intended to make sure that the disasters of Depression and world war did not recur. The United Nations was intended to look after issues of world politics. Agreements and institutions were hatched to attend to the world's financial system (the Bretton Woods Agreement giving rise to the International Monetary Fund or IMF, and the World Bank), international trade (General Agreement on Trade and Tariffs, forerunner of the World Trade Organization), health (World Health Organization), and much more. The point was to succeed where the League of Nations had failed: to check nationalism, autarky, militarism, and other evils by enrolling countries in a set of clubs with rules. This new regime was intended to re-globalize the world but in a less anarchic way than had happened between 1870 and 1914, so that the resentments that experience had generated were not repeated (Ikenberry 2001).

Domestic policy in the U.S. and most of its allies harmonized with this general goal: it was designed to soften the blows inherent in participating in the international economy. This took various forms but usually involved commitments to full employment and/or unemployment insurance, pensions, and subsidies for farmers and some other primary producers (often the coal or oil industries). It obliged the state to regulate social and economic life, raise taxes to levels formerly reached only in wartime, and cultivate the arts of economic management.

This postwar arrangement nudged the world toward re-globalization. Information, money, goods, and technology all began to move around much faster than before 1945. Thanks to its power and prestige, the U.S. initially achieved a high degree of political and military cooperation among its friends. U.S. allies found the rewards of taking part in this arrangement usually outweighed the irksomeness of American arrogance. Thanks to the new rules, the efficiency of its industry, and some occasional help from the government, U.S. businesses after 1945 spearheaded a rapid—but partial—economic re-globalization.

Stalin mistrusted the whole enterprise. He remembered that the U.S., Britain, and Japan had tried to quash the Bolshevik Revolution by intervening in the Russian Civil War (1918–1921). He suspected that in WWII the Americans and British had delayed opening a second front (originally promised in 1942 but only delivered in June 1944 with the D-Day landings in France) so as to let the U.S.S.R. bear the brunt of fighting the Germans. His ideological training told him that capitalism was communism's mortal foe, and he was by nature mistrustful to the point of paranoia. A lifetime spent in the snake pits of revolutionary and Soviet politics had compounded his habitual mistrust.

Stalin, sensibly, wanted a postwar order that ensured that the U.S.S.R. would never relive the traumas of 1941–1945. In this respect, his goals were compatible with those of his erstwhile allies. However, his preferred means to this end were not at all compatible. His highest priorities were a weakened Germany and the creation of pliant buffer states between the U.S.S.R. and Germany. He did not care about safeguarding the world economy from depression, since he could not expose his planned economy to its chaotic swings anyway. When the Americans showed an inclination to rebuild Germany and Japan in their quest for a revitalized world economy, this looked to Stalin like conspiracy against the U.S.S.R. Stalin had the largest army in the world by 1945, and because much of it was in Eastern Europe, he was able to get the pliant buffer states he wanted, including the eastern third of Germany. He joined the United Nations in 1945 but refused to take part in the institutions erected to manage the world economy, and declined to accept any share of the money that the U.S. offered to Europe for reconstruction through the Marshall Plan, begun in 1947. Stalin used the tool he had—the Red Army—to consolidate control over eastern Europe, which included removing much of its industry to the U.S.S.R. The Americans, meanwhile, used one of the tools they had—money—to rebuild western Europe's industry and thereby create bonds that assured voluntarily pliant states, especially Britain and West Germany. They soon did the same in Japan, binding it economically and politically to the U.S. The U.S.–Soviet rivalry became the Cold War, in full career by 1948.

Soviet physicists and Chinese peasants improved Stalin's position in the Cold War in 1949. First, his physicists, with help from his spies, provided him with an atomic bomb, ending the American monopoly—a long stride toward military parity. Then, in October, a peasant army directed by the Chinese Communist Party, itself led by Mao Zedong (1893–1976), won a civil war and reunified China. The communists offered a radical program: peasants should dispossess landlords, women should be free from male oppression, Confucian hierarchy should be replaced by social equality, discipline and morality should prevail—and foreigners should go home. It featured an ideology that accommodated Marxist principles, as refracted through Lenin's writings and Soviet experience, to Chinese realities. Whereas Marx had complained of the "idiocy of rural life," and the Soviets had ruthlessly exploited the U.S.S.R.'s peasants, Mao proclaimed that peasants (of which China had many, including Mao's parents), and not merely proletarians (of which China had few), could be a revolutionary class. With this program, good military organization, excellent lieutenants, a little help from Stalin and a lot from the ineptitude of his enemies, Mao won control over mainland China (not Taiwan) by October 1949 (Mitter 2004).

The Chinese Revolution briefly created a communist bloc of a size to rival the American-led one. However, the Chinese and Soviets never achieved smooth cooperation or economic integration. They had doctrinal differences about Marxism. They had a history of intermittent Russo–Chinese friction dating back to the 1680s. They both had strong nationalist leanings and favored economic autarky. They had conflicting ideas about how to confront the capitalist enemy: Stalin and, after his death in 1953, his successors took a cautious approach, whereas Mao thought there were enough Chinese that he could afford to risk nuclear attack by the Americans. This cavalier position understandably scared the wits out of the Soviets.[2]

So, less than ten years after they had become allies in socialist solidarity, China and the U.S.S.R. had a falling out: this was the beginning of the end for the U.S.S.R. In 1968–1969 their armies clashed along their long frontier. Henceforth the Soviets felt they had to station a good chunk of their military forces along the Chinese border. Then, in the early 1970s China played the American card, initiating diplomacy with the U.S. which left the Soviets feeling encircled. All this added to military costs that the Soviet economy could ill afford to bear.

By about 1970 the Soviets had approached the limits of the possible within their command economy. They had shifted most of the peasantry into urban employment while mechanizing most agriculture. They had put women to work outside the domestic sphere on a large scale (obliging Soviet women to shoulder the double burdens of wage work plus domestic work, a powerful disincentive to childbearing). Both of these shifts raised productivity significantly, but these were cards that could only be played once. The Soviet system discouraged technical and organizational innovation because there was no competitive market for goods, no rewards for innovators. So the Soviets extended an industrial system built on coal and steel, the cutting edge of the 1870s, instead of constantly reinventing it. This made it technically and financially difficult to maintain a formidable military–industrial complex. Soviet agriculture missed out on crop-breeding developments that elsewhere ratcheted up yields, and by the late 1970s could not feed the Soviet population.

In the 1960s and 1970s Soviet rulers responded to these weaknesses by becoming a major exporter of Siberian oil and gas to western Europe. Moreover, cheap oil helped bond the eastern European satellites, who could not afford world energy prices, to the Soviet empire. When world oil prices tripled in 1973, and again in 1979, the U.S.S.R. briefly became a giant oil sheikdom, able to import consumer goods, new technology, and, gallingly, American grain. The Soviet economy became less autarkic and found itself once again vulnerable to the whiplash of international price fluctuations, its buoyancy dependent on oil prices beyond its control. The bubble burst when oil prices collapsed in 1984–1986, undermining Soviet finances.

Soviet political legitimacy suffered as well. New contacts and communications undermined the unofficial Soviet social contract. The U.S.S.R. had always

---

2 In 1957 Mao told Premier Nikita Khrushchev: "We shouldn't be afraid of atomic missiles. No matter what kind of war breaks out—conventional or thermonuclear—we'll win. As for China, if the imperialists unleash war on us, we may lose more than 300 million people. So what? War is war. The years will pass and we'll get to work producing more babies than ever before" (Khrushchev 1974, p. 255). Or so Khrushchev claimed: he had reasons to disparage Mao.

justified daily hardships with extravagant promises of a rosy future. This seemed plausible for a while, especially when the economy grew briskly. By the 1950s ordinary citizens knew they were materially better off than their parents and grandparents, and if they had to accept tight political control, that seemed acceptable to many. After the 1970s, however, it became increasingly obvious that the U.S.S.R.'s economy could not continue to deliver the goods. When the tight grip on information was relaxed, some people discovered via travel, movies, or West German television (which was accessible in East Germany) the degree to which they were doing without. New information changed their frame of reference: It no longer seemed so relevant that they were better off than their grandparents when they were visibly worse off than West Germans, Japanese, or Italians—people defeated in what Soviet citizens called "The Great Patriotic War." To make matters worse, the Soviets invaded Afghanistan in 1979 to install a client dictator, but ended up fighting a long, vicious, and unpopular war. By the mid-1980s almost no one believed in the Soviet dream anymore.

With state finance and legitimacy both crumbling, the leader who took over in 1985, Mikhail Gorbachev (b. 1931), desperately allowed more freedom of information and expression, hoping to revive the economy by exposing Soviet society to new ideas and technologies. A fire in a nuclear reactor at Chernobyl in 1986, initially covered up by Soviet officials, became an embarrassing international disaster, sprinkling radiation throughout the Northern Hemisphere—and underscoring the necessity of a more open society. However, openness let the genie of minority—and Russian—nationalisms out of the bottle and Gorbachev was not prepared to force it back in. His views narrowly prevailed over hardliners. In 1989 Gorbachev refused to intervene to prevent East Germans from fleeing to West Germany, and soon the entire set of buffer states had slipped from the Soviet orbit. The Soviets lost the Cold War for the same reason the Axis lost WWII: They could not create an interactive, cooperative, innovative, international economy to match the American-led one. They remained too wedded to the autarkic economy that Stalin had built. The reason they lost the Cold War *peacefully* was that Gorbachev prevailed over his elders and rivals (Kotkin 2002).

Decolonization

While the great powers engaged in hot and cold wars, they lost their empires. As of 1914 the industrial powers had acquired dominion over most of the globe, and in the 1930s Japan and Italy extended their empires into China and Ethiopia, respectively. Yet as early as 1918, the world's empires began to unravel under the strains of total war, the impact of insurgent nationalisms, the diffusion of information in general, and the arts of political mobilization in particular. By 1960 the technological and military gaps separating the weakest from the strongest were greater than in 1914, but even the weakest were becoming well enough organized to convince their colonial masters that imperialism did not pay.

WWI had broken up the Russian, Ottoman, and Austro–Hungarian empires. The Bolsheviks quickly reconstituted the Russian one, equipping it with new ideology and accepting different borders, but restoring the format of centralized bureaucratic empire. The Ottoman Empire vanished forever. The Austro-Hungarian Empire dissolved into four separate European states in 1919, each with its own restless ethnic minorities. Thus the initial burst of decolonization created only a few new states before it ended, and left plenty of problems for the future.

Ireland, or at least most of it, also won independence from Britain after World War I. Irish agitation for freedom dated back centuries, but during WWI Irish nationalists took advantage of Britain's heavy commitment in France and organized a rebellion in 1916. British troops suppressed it, but after the war the British government decided that maintaining a grip on Ireland would be too costly. So after much negotiation, the Republic of Ireland was born in 1922, leaving only the small territory of Northern Ireland, in British hands. This arrangement too left many problems for the future.

A second and larger burst of decolonization came between 1943 and 1975, liberating most of the world's colonies. The key to this process was the political restructuring of colonial societies. Original efforts at resistance to imperialism had been violent and often suicidal. The struggle required new weapons. Africans, Indians, and others took advantage of the opportunities made available by new communications and transport technologies. Some studied in Europe or the U.S. and learned about political struggles elsewhere. They formed political pressure groups, the first of note being the Congress (now the Congress Party), founded in India as early as 1885. These groups tried to harness the bonding power of nationalism, which worked well in ethnically homogenous colonies such as Vietnam or Korea. Elsewhere nationalists forged trans-ethnic political alliances, trying to create a nation where none had existed before.

The Great Depression of the 1930s made life harder in most colonies, raising rebellious sentiments. Colonial economies generally had a vulnerability that well-organized anticolonial nationalists could exploit. State revenues depended on exports of crops or minerals, which passed through the bottlenecks of railroad lines and ports. So when railwaymen or dockers went on strike, the colonial state risked insolvency, a fact eagerly exploited by insightful anticolonial nationalists as of the late1930s.

In many ways, WWII set the stage for the final drama. Millions of men from India, Indochina, Africa, and elsewhere had their horizons broadened by overseas military service. They heard American and British propaganda explain that the war was fought for freedom. They also learned modern military skills. Moreover, WWII completely destroyed the empires of Italy and Japan, freeing Ethiopians, Libyans, and Koreans from foreign control. It also weakened the finances and resolve of France, The Netherlands, and Britain. In the course of the war Syria and Lebanon became fully independent from France. In addition, after decades of boycotts, strikes, and marches against British rule, in WWII Indians made considerable blood sacrifice on behalf of the British, making it possible for them to negotiate their independence in 1947.

Elsewhere after the war, the French, Dutch, and British tried to reassert control, notably in Southeast Asia where the Japanese army had driven them out in 1941–1942. But rising nationalisms in Southeast Asia now required a stronger commitment than the colonial powers could afford. The Dutch gave up in Indonesia in 1949. Vietnamese forces humiliated the French in 1954, but because there the anticolonial nationalists were also communists, the U.S. gradually made France's war its own. This led to a long and bloody struggle before the Americans gave up in 1975.

In Africa, the war mattered as well, even if less so than in Asia. Again, the underlying issue was the development of political organization, skill, and commitment on the part of anticolonial nationalists. They raised the costs of continued imperialism beyond the level that France and Britain could stomach. At first, as in Southeast Asia, the colonial powers tried to strengthen their control, and invested in new infrastructure and economic development schemes. In this they sought cooperation from among Africa's educated elite, and also the blessing of the U.S., which had its doubts about continued European colonialism, at least where communists were not involved.

But events overtook these plans. A nasty and increasingly unpopular war in Algeria (1954–1962) finally convinced France to give up its empire in Africa. Algerian and Moroccan manpower had become essential to the French colonial forces due to the scarcity of young Frenchmen, so with the loss of its North African colonies France could not resist mounting pressures elsewhere for independence. The Suez crisis in 1956 exposed further weaknesses of France—and of Britain. In response to quarrels over the financing of the Aswan Dam project, Egypt seized the Suez Canal, hitherto managed by the British. Britain, France, and Israel mounted a surprise attack on Egypt, but the U.S. threatened to cut off the oil and dollars that kept Britain and France afloat, forcing a humiliating withdrawal. France and Britain needed full American support to withstand the challenges posed by anticolonial nationalism, and they clearly did not have it. Partisans of African independence took note and took heart, redoubling their efforts. By 1963 almost all of Africa had acquired its freedom, though in Mozambique and Angola nationalists had to fight on until 1975 against Portugal.

The next stage of decolonization came with the collapse of the U.S.S.R., noted above. Several small islands and enclaves remain colonies, dependencies, or UN trust territories today, although their combined population is tiny. China remains an empire of sorts, controlling Muslim and Tibetan populations in the western provinces that, if free to choose, would certainly go their own ways. So does Indonesia, an archipelago in which several populations would probably prefer to be free from Javanese control. However, unmistakably, the age of empire created by the great inequalities forged in the 18th and 19th century is now past. Indeed a political format that had been standard since the Bronze Age suddenly became (for the moment) an endangered species. The subordination of the weak to the strong now takes other forms.

The dismantling of empires since 1918 created well over 100 new countries. Most of them remained weak. The political skills that hastened decolonization did not easily translate into good government. Trans-ethnic unity proved fleeting. Successful industrialization required investment, skills, and markets that often scarcely existed. Rapid population growth made it hard to maintain living standards, especially in Africa. The most economically successful ex-colonies were both among the most brutally ruled (South Korea) and most gently (Cyprus, Hong Kong). They owed their good fortune to their comparatively well-educated populations, to ethnic solidarity that made nationalism a force for unity (with the exception of Cyprus), and in some cases to their strategic location that inclined the U.S. to favor them with both military protection and economic investment. A prosperous and stable South Korea and Taiwan, for example, helped in the struggle against communism in East Asia, so financial assistance and access to American markets suited U.S. interests there.

These new states faced tremendous pressures to industrialize and generate economic growth. Most of them subscribed to the general idea that the government should foment economic development through infrastructure projects, bringing on a golden age of dam, road, and port construction (ca. 1950–1990). In this quest they were ably assisted by aid agencies and development banks. Their economies generally did grow, but normally by little more than the rate of population increase. Postcolonial states were born in moments of optimism and high expectations of better days to come, which eventually made poor or even modest records of achievement seem unsatisfactory to the newly independent citizenries (Abernethy 2000).

The Long Boom and Re-globalization

Soon after World War II ended, the global economy entered its most remarkable era, growing 6-fold between 1950 and 1998. Indeed in the quarter century before 1973, the world's economy grew at nearly 5% per year, and 3% per year per capita. Yet even when economic growth slowed after 1973, it galloped faster than at any time before 1950 (Maddison 2003). Taken as a whole, this era is the most unusual in the history of economic growth, although many people, having experienced nothing else, now imagine it is normal.

It happened because of oil and energy, medicine and population growth, science and technology. It also happened because married women had fewer children, entered (and stayed) in the paid labor force in unprecedented numbers, and because farmers left the land at record pace for urban jobs. Each of these one-off social transformations added greatly to GNP numbers. More specifically, the long economic boom also happened because modern industrial techniques, such as electrified factories, spread to some very populous lands. The two great examples of this were Japan, where economic growth between 1950 and 1973 averaged nearly 10% annually, and China, where from 1978 to 1998 it averaged nearly 8% (Maddison 2003).

China's economic policies after the 1949 revolution included truly disastrous components that Mao's successors sought to reverse. Among Mao's notable economic initiatives were collectivization of agriculture and an industrialization scheme (the "Great Leap Forward"), which was intended to make China more autarkic and to allow China to surpass Britain in steel production—a target of symbolic importance to Mao. The Great Leap Forward generated vast tonnage of very bad steel and a famine that killed 20–40 million Chinese, as well as acute environmental damage (Shapiro 2001). After Mao's death in 1976 China reoriented itself, ended the collective farms, and liberalized rules in industry and trade. Shifting hundreds of millions of people into more productive work yielded spectacular results. Equally impressive growth took place among some far smaller populations of East Asia after 1980 (e.g., South Korea, Taiwan, Hong Kong, and Singapore). All together, this amounted to a real great leap forward in East Asia, in economic terms at least. Politically it raised the awkward issue of how the international system might accommodate a resurgent China.

The long boom also happened because of the re-globalization of the world economy, which yielded the usual returns from specialization and exchange. The international institutions created in the 1940s helped in this regard, as, perhaps more fundamentally, did the American commitment to open U.S. markets to the exports of Europe and East Asia. That commitment wavered, and in any case never extended to some sectors (e.g., agriculture), but it was a crucial political ingredient in the recipe of post-1950 world economic growth. Re-globalization intensified competition among firms, and the fortunate ones responded with innovations that dramatically raised productivity so that, for example, steelmakers increasingly could wring more steel out of the same amounts of iron ore, coal, and labor.

Trade played a big part in the long economic boom. In 1950 the share of goods produced for export was about the same as in 1870, and much lower than in 1913. But that share doubled between 1950 and 1973, and tripled by 1995 (Maddison 2003). The biggest increases took place in trans-Pacific trade, much assisted by the new technology of container shipping, invented in the 1950s but routine from the 1970s. Containers were big boxes of standard size that could travel by truck, train, or ship, and thus required far less handling, packing and unpacking, than had earlier methods. This reduced the time it took goods to get from Hong Kong to New York by two-thirds, cut labor costs sharply by putting legions of longshoremen out of work, and made the thorough participation of East Asia in the international trading economy much more practical.

The resurgence of capital flows eventually became a still bigger part of the long boom. The Marshall Plan was but a start. In the 1950s and 1960s, capital moved faster and in greater absolute volumes than ever before (although as a proportion of total investment, at lower levels than in 1870–1913). Capital flows accelerated when a great "financialization" of the world economy began in the late 1970s. The difficulties of the 1970s (slower growth, higher unemployment and inflation) were due primarily to higher energy costs associated with the tripling of oil prices in 1973 and again in 1979, itself mainly a political matter. But among the responses was an intellectual and policy shift toward liberalization of the rules governing capital flows, beginning in the U.S., the U.K., Chile, and a few other countries. This newly relaxed regulatory environment combined with new technologies to make it far easier to make money in finance than in production or trade. Telecommunications and networked computers drastically changed the ways of doing business, allowing instantaneous transactions all over the world at negligible costs. A tiny rise in an interest rate could cause the overnight migration of huge sums.

Whereas in 1870–1914 the great majority of international capital flows took the form of long-term investment in bonds, railroads, or factories, after 1980 it increasingly was short-term, restless capital looking for the opportunity of the moment—a currency or stock whose value was likely to rise. The volume of such capital flows increased exponentially, dwarfing the value of world trade by the late 1990s; every week foreign exchange markets handled business equivalent to the annual GNP of the U.S. In small and medium-sized countries, say Chile or Thailand, the ability of capital to flee instantly sharply restricted the range of policies open to them: if they departed from the preferences of investors they courted swift retribution in the form of declines in their currency's value and the price of their bonds. So they tried hard to conform to the wishes of mobile capital, balancing budgets and restricting the money supply. This was a perilous course, because it often meant firing government workers, trimming social programs, and other painful measures sure to arouse popular complaint. The choice between the Scylla of capital flight and Charybdis of popular unrest was eased, but not resolved, by the willingness of the IMF to lend money to governments that followed the new orthodoxy.

Although small in comparison to the growth in trade flows, and tiny compared to the oceans of capital sloshing around the world, the years after 1965 also saw a resurgence of international migration. Fast-growing economies with slow-growing populations imported labor. Poor regions with fast-growing populations needed to export it. When Europe's postwar recovery gathered speed after 1955, its industrial regions began to attract millions of laborers from southern Europe, Turkey, and North Africa. By the 1970s, even southern European countries became importers of people, mainly from North Africa, but also from South America and elsewhere. Europe, for 400 years an exporter of people, now attracted immigrants from far and wide. So did the newly rich oil states around the Persian Gulf, which imported laborers from Palestine, Pakistan, Korea, and elsewhere in Asia. By 1990 in Kuwait or Saudi Arabia immigrants did almost all the manual labor. Nigeria's oil boom (ca. 1975–1983) helped attract 2–3 million other West Africans. The U.S., Canada, and Australia—traditional recipients of immigrants—liberalized their racist policies in the 1960s, accepting skilled or rich people from almost anywhere, which in practice meant largely from East and South Asia. The U.S. quintupled its intake of legal migrants between 1965 and 1995, and at the same time received millions of clandestine immigrants, chiefly from Mexico and Central America. Canada doubled its immigration rate in the same years. In absolute numbers, the U.S. and Canada had more immigrants than at any time in their history by 2000, although in 1913 the proportion of immigrants to native-born was much higher (Hoerder 2002).

In general, cheap transport, cheap information about conditions elsewhere in the world, and relaxed quotas on migration encouraged scores of millions to uproot and try their luck elsewhere. By 2000 some 125 million people lived as immigrants, and the annual flow of legal migrants totaled about 2 million. Most, as in the past, were poor and unskilled, but a large minority had strong educations and marketable skills.

This accelerating swirl of migration helped ease the stresses of rapid population growth in places such as Algeria or El Salvador. It provided willing laborers in France or the U.S or Saudi Arabia, often in jobs that few native-born citizens would take. It took people from places where their labor would produce little to places where it produced more. In economic terms, it proved helpful to all but the laboring classes in recipient countries, whose wages were held down by the competition of immigrants. In cultural and political terms the great swirl brought new tensions. Most Britons did not welcome large numbers of Pakistanis and Jamaicans in their midst, and Algerians in France, Turks in Germany, and Filipinos in Kuwait also met cold receptions (Hoerder 2003).

Electrification of Information

The most important reasons behind the long economic boom were cheap energy and population growth. But new technology helped too. Transport technology mattered much less (despite containerization) than it had in the growth spurt of 1870–1913, while communications and information technology mattered more. Underlying this of course was cheap energy permitting the continuing electrification of the globe. Electrification and the development of new telecommunications and computer technologies reduced information costs, sometimes to the vanishing point. In 1930 a three-minute telephone call between London and New York cost $300, and in 1970 about $30. Then commercial satellites, optical fibers, computer microprocessors, and deregulation of telecommunication markets in the 1980s quickly cut the costs of such communication much further. By 2001 a transatlantic phone call could cost as little as 30 cents, and an e-mail exchange was just about free.

Computers first became useful during WWII for codebreaking and the first Internet exchanges took place within the U.S. military. Civilian Internet traffic began in the 1960s, but developed fast only after 1990 with the emergence of networked personal computers. The U.S. had 1 million personal computers in 1980, 10 million in 1983, and 44 million in 1989. By the early 1990s these were increasingly linked. This trend caught on faster than most earlier networked technologies, and by the 1990s the global electronic village was under construction. By 2000 the world had over a billion telephones (up from half a billion in 1980), all connected, and several hundred million computers with Internet access, and 1.6 billion web pages to choose from. Every minute, 10 million e-mail messages landed in the world's in-boxes.

The consequences of the electrification of information are hard to assess because the process is still in train. It has clearly played a large role in the "financialization" of the world economy. It has enriched the information-intensive service sector more than manufacturing and agriculture. It has strengthened the premium on education in the modern world, increasing the rewards for those who acquire schooling and shrinking the rewards for those who can contribute only a strong back or a nimble pair of hands. It has, so far, enhanced the status of English worldwide, because (as of 1999) 78% of Internet home pages were in English (Japanese came second with 2.5%). It changed the conduct of warfare for those who could afford it (mainly the U.S.), because satellites linked to computers allowed a level of precision with long-range weaponry previously impossible. But it created new vulnerabilities as well as new capacities. A malicious, skilled, and lucky hacker could raise havoc with air traffic control, city water systems, banks, and everything else that is computerized.

When inequality within and among societies grew sharply during the Industrial Revolution the effects had proved socially explosive. In the late 20th century such widening gaps occurred in an electronically unified world, one in which information circulated freely. So the world's poor are far more aware of their position than were their predecessors 150 years ago. They are unlikely to accept that position meekly, as the messages, spread so relentlessly via the electronic media, promote acquisition and consumption as the road to fulfillment. Conceivably, constant displays and occasional use of overwhelming force might suffice to protect the world's rich. So might updated versions of bread and circuses: the entertainment industry excels at supplying music, sports, and sex. It may be, of course, that information and communications technology will soon help narrow inequalities by making education cheaper and more accessible worldwide. But that, should it happen, will be a slow process during which we all face the dangers of glaring economic inequalities in a world of full disclosure.

In short, the surge of globalization between 1870 and 1914 generated inequalities and resentments that made nationalisms and war more likely. The war of 1914–1918 discredited nationalism (and war) for some, but deepened its appeal for others. The political and economic pressures of the postwar world, and especially the Depression, made nationalist autarkies seem sensible policy, leading to WWII. The new regime put in place after 1945 once again encouraged integration and globalization, helping generate unprecedented economic growth, and in ways that for a while reduced social inequality. After 1980 technology and policy combined to produce far faster globalization, this time combined with rapidly growing inequalities. This made it harder to conceal the ethnic frictions and economic cleavages that beset the world, and perhaps harder to control them.

TECHNOLOGY AND SCIENCE

Information Infrastructures

The great technological shifts in communications and transport in the 1800s (steamships, railroads, telegraphs) helped knit the world together more tightly, but some aspects of life remained untouched. The great distinction of the communications and transport technologies that shaped the 20th century (telephone, radio, television, movies, automobile, airplane, Internet) was that they altered the everyday lives of billions of people, enlarging their range of experience and their access to information. These new technologies democratized the transmission of information in prosperous countries, and until 1950 helped narrow gaps between rich and poor within them. But at the same time, because they conferred wealth and power upon those who used them, they helped widen the gap between the prosperous and the poor around the world. In 2000, about 60% of the U.S. population regularly used the Internet and about 35% in South Korea. But in Brazil only 6% did and in Nigeria less than 0.1%.

Roughly speaking, the technological transformations came in three waves. Although the telephone was invented in the 1870s, the automobile in the 1890s, and radio around 1900, they became widespread only in the 1920s, and even then mainly in the U.S. Their spread around the world came in fits and starts thereafter. The second wave came in the 1940s and 1950s, when television (invented in the 1930s) and commercial aviation became routine in the U.S. and soon in much of the rest of the world. Like telephones and radio, the airplane had been important in war well before it became commercially viable. The third wave, networked computers, originated in the 1960s but took off in the early 1990s. All of these new technologies worked as networks, which meant they were slow to catch on but once over a threshold they spread very rapidly. It made little sense to have a telephone before lots of other people had them, but once they did it made scant sense to be without one. Similarly, it made no sense to have a car before there were gas stations and passable roads, but once there were, having one proved irresistible to almost all who could afford one. So these technologies, once established, transformed daily life very quickly.

The cumulative effect of all these changes (and some others such as mass circulation newspapers) was to bombard people with new information, impressions, and ideas, and to allow more of them to travel further, faster, and more frequently than ever before. This proved disconcerting and disorienting, as well as seductive. It invited people to suppose that their circumstances need not be as they were, but could be improved—through emigration, revolution, education, hard work, crime, or some other initiative. With radio, movies, and television in particular, hungry illiterates could catch a glimpse (accurate or not) of how more fortunate people lived. This information, combined with massive urbanization (see below) inspired both ambition and resentment, providing potential recruits for an array of political movements.

Science

The new information infrastructures intensified competition in the marketplace of ideas. In some respects science underwent a shakeout. Fewer and fewer distinct ideas about, say, the origins of the Universe, the contents of the deep oceans, or the causes of disease, survived the 20th-century's rapid exchange of information and ideas. More and more people accepted a narrower range of ideas, and educated scientists came to accept a fairly consistent set of truths, whether they were based in Boston, Berlin, Bangui, or Beijing. The chief reason for this was the increasing global exchange of scientific data and ideas. International congresses and journals proliferated from the 1880s on. Young scientists from around the world went to study in Europe (Russia included) and North America. The power and prestige of Europe and the U.S. (and eventually Japan) were such that scientific ideas generated there swept the world.

The evolving scientific worldview had certain core features in common. First, the approaches to science that emerged in 17th-century Europe, emphasizing observation and experimentation acquired nearly unimpeachable authority, even in lands such as China with their own long and distinguished scientific traditions. Second, science increasingly abandoned the principle of timelessness and adopted evolutionary models of the natural world. Third, good science did not come cheap. Let's consider the last of these points.

In Darwin's day, science remained an occupation for gentlemen of leisure (or rarely gentlewomen) or academics. They typically worked alone, although thanks to post offices and scientific societies they communicated frequently with their fellow researchers. By the 1940s, the complexity and expense of modern science required systematic collaboration. The American effort to build an atomic bomb during WWII employed over 40,000 people (Rhodes 1986). In 1940 U.S. businesses employed 70,000 scientists in their research and development departments. In 1900, the two most scientifically advanced nations, Germany and Britain, had about 8,000 working scientists, but by 1980, the U.S. alone boasted over a million, and western Europe employed still more. After WWII showed what enormous funding and scientific manpower could do, governments and businesses increasingly bankrolled scientific research. While they were prepared to pay for a modest amount of pure science—disinterested inquiry about, say, the origins of the Universe—what they most wanted was applied science that would help build a better mousetrap—or, after the rise of biotechnology in the 1980s, a better mouse. Science could not do without state funding and, increasingly, states could not do without the technological fruits of science (Misa 2004).

The Marriage of Science and Technology

"The greatest invention of the 19th century was the invention of the method of invention." So wrote the English philosopher Alfred North Whitehead (1925, p. 98). What he observed bore far greater fruit in the 20th century than in the 19th. Throughout most of human history, science and technology had little to do with each other. Scientific changes altered ideas but not practices. Technological change came from tinkerers, people with little to no scientific education but with plenty of hands-on experience. In the course of the 19th century, science and technology grew closer together, a trend that sharply accelerated in the 20th century.

The wedding of science and technology took place in the late 19th century. With the imperial powers, chiefly Germany and Britain, competing to develop superior navies, governments began to organize scientists and engineers into teams directed to generate useful military technology. Scientific expertise gradually became a crucial component of military security. In the 1870s, industrial firms in Germany and the U.S. created their own research laboratories and maintained flocks of scientists assigned to solve particular problems. Chemical firms in particular developed ties with universities, financing research and assuring a stream of skilled graduates. As governments and firms became increasingly involved in funding science, the thrust of inquiry shifted toward applied science that could help win wars, improve health, and expand wealth. Almost all science, pure or applied, required expensive technologies and educational systems, which helped keep the international landscape of science dominated by a few countries, notably Germany until the 1930s and the U.S. thereafter.

After World War II the Americans built a sprawling scientific–industrial–military complex involving coordinated research and development by firms, universities, and the Pentagon. Britain had a much smaller one, dating back to the late 19th century. The Soviet Union embarked on a similar venture, with the additional feature of specially designated, often secret, science cities, where small armies of physicists and engineers lived and worked in near isolation from the broader society. The Cold War was the driving force behind these developments. But the implications of these investments in science and technology touched every aspect of life. Industrial chemists developed plastics, now ubiquitous. Solid-state physics research yielded the transistor in the 1950s, making radios cheap and portable. Plant geneticists created new strains of wheat, rice, and maize that, in favorable conditions, doubled or quadrupled crop yields (ca. 1960–1980), dramatically raising the world's food supply. The pace of invention and innovation accelerated as governments, universities, and businesses cooperated in supporting hordes of engineers and scientists.

The very success of 20th-century scientific and technological advance eventually brought public unease and suspicion, especially in those societies most transformed by it. Whereas at the beginning of the century the great Ernest Rutherford maintained that all good physics could be explained to a barmaid, a hundred years later no scientist supposed anything of the sort. Physics and genetics especially were well beyond lay understanding. Even the irreligious found some aspects of nuclear physics or genetic manipulation morally or politically repugnant, an attitude sometimes abetted by some scientists' apparent disinterest in the social consequences of their work. Even the suspicious were entirely dependent on the smooth functioning of immensely complex technological systems and continuing work of an elite cadre of highly skilled people (Misa 2004). And everywhere the marriage of science and technology contributed to the dynamism and instability of the social order and ecological regime.

POPULATION AND URBANIZATION

One of the most distinctive features of the 20th century was its population history, a tale of boom and partial bust, in which both science and social change played major roles.

Urbanization and population growth stands as the cardinal social change of the last century. For 5,000 years or more the typical human experience was village life, and human ideologies, institutions, and customs all evolved primarily in that setting. Now the majority human experience is that of city life with its anonymity and impersonal character. Past eras of urbanization, all slow and circumscribed compared to the modern one, put great pressure on reigning religions, ideologies, and worldviews as well as on standing political structures. Among the acute challenges of our time, it seems sure, is the process of social, political, psychological, moral, and ecological adjustment to life in the big city.

Population Growth

In 1900 the Earth supported some 1.6 billion people, about a fifth of them in China. By 2000, the total had quadrupled to 6.0 billion (China still accounted for a fifth, and India a sixth). Nothing like this had ever happened before, nor will it ever again. Most of this surge took place after 1960. The world's population growth rate peaked around 1970, at around 2% per annum. It has fallen, irregularly, ever since, and demographers expect it will slow to zero by 2050 or 2070 (Cohen 1995). What might happen after that is anyone's guess.

The main reason for this extraordinary burst was the transfer of successful death control measures to most populations on Earth. Before 1914 effective public health systems existed in only a few regions. But after 1950 vaccinations, antibiotics, and sanitation measures, the fruits of decades of scientific research, cut death rates everywhere. Life expectancy at birth, which globally had been less than 30 years in 1800 and about 35 years in 1900, reached 45 in 1950 and 67 in 2000. This constituted a radical change in the human condition. Japanese, currently the longest-lived people on earth, can expect to live twice as long as their great-grandparents' generation. Even the least long-lived, currently Botswanans, enjoy perhaps 20 extra years compared to their forebears in 1900. Worldwide, most of the progress in death control took place between 1945 and 1965. Until birth rates fell too (which has not happened everywhere), populations grew by as much as 4% per year in some African and Central American countries, fast enough to double in 16 years. In some countries, the transition from high birth and death rates to low ones took as little as 20 years, with a correspondingly modest leap in total population. South Korea, Taiwan, and Thailand all achieved this after 1960, and not coincidentally got comparatively rich in record time (Cohen 1995; Livi-Bacci 2001).

The tremendous human slaughter of the 20th century had little effect on population trends. If one adds up all the premature deaths from wars, genocides, state terror campaigns, and human-caused famines, the total comes to perhaps 180–190 million. This accounts for about 4% of the total deaths in the 20th century.[3] The acceleration of death from political causes did not nearly match the deceleration of death from public health measures and improved nutrition.

The inequality in population growth rates, so important in the 18th and 19th centuries, remained disruptive after 1890. Fertility stayed high in eastern Europe until 1914, which assured heightened tensions within the Russian and Austrian empires, despite high rates of emigration. But this was the twilight of rapid population growth in Europe. By 1920 almost every corner of Europe had reduced its fertility sharply, and the fastest growth now occurred in India, Latin America, and, after 1930 or so, in Africa. Whereas in Europe in 1900 emigration took away about one-third of the natural increase, in India, China, Latin America, and Africa emigration did not notably reduce pressures. Instead, population growth promoted political unrest, urbanization, and desperate state efforts to industrialize overnight. Africa's population history was especially dramatic—a 6- or 7-fold increase to roughly 750 million in the course of one century (Cohen 1995; Livi-Bacci 2001).

Urbanization

Exuberant growth of cities was another defining characteristic of the 20th century. In 1900 about 12–15% of humankind lived in cities, by 1950 some 30% did, but by 2001 more than half did. This also represents a great turning point in the human condition. Prior to 1880 cities everywhere were demographic black holes, mainly because of endemic childhood diseases but also because of recurrent epidemics. A big city, such as London in 1750, killed children and newcomers so quickly that it canceled half of the natural increase of all of England. But

3 See the arithmetic of Matthew White at http://users.erols.com/mwhite28/20centry.htm.

after the 1880s public health measures made city life safer, first through the provision of clean water. By the 1920s Chinese city-dwellers, for example, outlived their country cousins. Henceforth cities grew and grew, from natural increase as well as from in-migration.

In 1900 about 225 million people lived in cities. But by 2001 about 3 billion did, a 13-fold increase. Broadly speaking this surge to the cities came first in Europe, eastern North America, and Japan during their eras of industrialization (ca. 1850–1930). Britain was the first large country in which more than half the population lived in cities (by 1850), a proportion reached in Germany in 1890, in the U.S. by 1920, and in Japan by 1935. Next came the U.S.S.R. and much of Latin America, where state-sponsored industrialization helped propel a rush to the cities between 1930 and 1970. By the early 1960s the 50% urban threshold had been reached in both the U.S.S.R. and Latin America (taken as a whole). Chinese policy kept people in the villages until 1980, after which peasants stampeded to the cities in the largest, fastest urbanization the world has ever seen; China will probably surpass the 50% mark before 2010.

Cities became both more numerous and larger. Villages became cities almost overnight in regions where iron ore and coal could be found together, as in Germany's Ruhr region. Frontier settlement also brought forth great cities, such as Buenos Aires, Melbourne, or Chicago, each of which counted about 100,000 citizens in 1858, and more than half a million by 1900. (Chicago was then the fifth largest city in the world with 1.7 million.) Where industrialization occurred in a capital city, the growth was often greater still: Mexico City had a third of a million people in 1900, some 5 million by 1960, and 20–25 million by 2000. Indeed by 2000 sprawling megacities such as São Paulo, Shanghai, Cairo, and Delhi, each contained more people than the entire world did when agriculture was first invented, and about as many as Great Britain counted at the time of the Industrial Revolution.

This basic change in the human condition affected almost everything: morals, family, fertility, religion, identity, politics, ambitions, education, health, recreation, ecology, and much more, in ways that are as yet far from clear. In most places people had gradually worked out ways of living together in village settings, where every transaction took place in the context of personal encounters, where everyone's reputation was known to all, where customs had evolved to constrain and channel conflict. In urban settings such customs and constraints melted away, leaving only law, police, and moral education to discourage predatory behavior. A smorgasbord of social organizations sprang up to meet the needs of urban life, from street gangs to cults to neighborhood associations. But, as yet, no one had found a satisfactory moral code or a way to ensure smooth or stable social relations in urban settings.

Despite all its distresses, urban life offered great advantages. Opportunity for upward social mobility and more excitement for the young had always tempted migrants. Increasingly after 1890 cities also provided better access to education, health care, clean water, and electricity. After 1950 the easily acquired skill of driving a motor car could ensure an income without backbreaking labor. In many countries governments feared urban insurrections and consequently used their power to assure cheap food for the cities, sometimes by fixing prices, sometimes simply by confiscating it from the peasantry. All this helps explain the continuing magnetism of the cities.

Urbanization was also the main reason that world population growth began to tail off after 1970. In cities, at least where child labor is uncommon, children are costly to their parents for 15 or 20 years, whereas in rural settings, especially where there are goats or chickens to tend, children from the age of five or so are economically useful. In cities, girls are more likely to get formal education, and more educated women generally have fewer children. So wherever cities predominated, people within a generation or two abandoned the normal agrarian emphasis on fertility and had far fewer children. This became most apparent in Central Europe after World War I. Around 1930, Viennese reproduced so slowly that, without in-migration, the city's population would have declined by three-fourths within a generation. Berliners were almost as reluctant to bear children. By the 1970s urbanization (and city habits among the country folk) had spread so far that in Germany and Japan the national populations had sub-replacement fertility.[4] In Russia and Ukraine after 1980 birth rates fell, while death rates (especially for men) climbed, so that population decay set in very quickly.

This pattern, if it persists, means that cities are resuming their historic role as demographic black holes. Before 1880 they consumed population because their death rates were so high; after an interval of growth by natural increase, they began to consume population because their birth rates were so low. London today, as in 1750, would shrink without in-migration. (Today, indeed since the 1950s, London has drawn many migrants from the Caribbean and South Asia, not merely from the rest of Britain.) Will life in Lagos and Lima eventually make people as reluctant to bear children as has life in London? Will urban conditions always persist in discouraging people from parenthood? The answers are by no means clear. Unforeseeable developments in biotechnology, for example, could revise the entire procedures of reproduction and family life.

ENERGY HISTORY

One of the central reasons why human numbers could grow 4-fold in the 20th century, and why urban population could grow 13-fold, was the phenomenal success our species enjoyed in harnessing fossil fuels. Cheap and abundant

4 That is, less than about 2.1 children per woman over her lifetime. In such situations, population will eventually fall without immigration. It will not normally fall instantly, because usually there are enough people of childbearing age to keep the birth rate above the death rate for perhaps 10 or 20 years. The data on Vienna and Berlin are from Kirk (1946, p. 55).

energy, mainly in the form of fossil fuels, was the single most powerful reason why the 20th century was so environmentally turbulent.

The use of coal smashed old constraints on transport and industrial production. By about 1890, half the energy deployed around the world came from fossil fuels, mainly coal. In the 20th century the crucial development in energy history was the emergence of cheap oil.

The U.S. led the way in building its economy and society around oil. The first big gusher came in southeastern Texas in 1901, heralding a century of cheap energy. Big oil finds came later in Mexico, Venezuela, Indonesia, Siberia, and, by far the greatest of all, around the Persian Gulf, exploited heavily since the late 1940s. In global terms, the shift to oil came mainly between 1950 and 1973, during which time world oil production climbed from 10 million to 65 million barrels per day. These are, not coincidentally, the years in which world economic growth attained its crescendo. Oil production after 1973 rose more slowly, to 77 million barrels per day in 2003. Oil was useful not only for heating and electrical power generation (like coal). Together with internal combustion and jet engines it revolutionized transport. Oil transformed agriculture too, powering farm machinery and serving as the chemical feedstock for fertilizers, making it possible, for example, for 3% of Americans to feed all the rest. The rice, wheat, and potatoes that feed the world's population are made from oil as much as from soil, water, and photosynthesis.

With oil, and smaller contributions made by natural gas, hydroelectric power, and nuclear power, the world's energy harvest grew ever more bountiful after 1890. The average global citizen used 3–4 times as much energy in 2000 as in 1900. This average, of course, conceals great differences. Just as the industrial revolution widened inequalities of wealth and power around the world, so did the accelerating transition to high energy use. The ordinary Canadian or American, for example, by 1990 used 50–100 times as much energy as the average Bangladeshi. For those who enjoyed it, the transition to a high-energy society amounted to a great liberation from drudgery and muscular toil. It made people far more mobile, productive, and richer than their ancestors—or their contemporaries who did not share in the transition. In short, harnessing of fossil fuels, especially oil, simultaneously eased life for a large fraction of humanity and deepened the inequalities between the haves and have-nots (Yergin 1991; Smil 1994).

CONCLUSION

Our political institutions, which evolved over millennia to cope with other challenges, proved ill suited to large-scale but slow-moving environmental problems. The competitive international system impels states to maximize their wealth and power in the short run, assigning low priority to other concerns. Economic systems, whether capitalist or communist, encouraged similar attitudes and conduct. The arts of economic management, combined with technological change and social transformation, generated an enormous expansion of the world economy after 1890, despite the stagnation of 1914–1945. This put unprecedented strains upon the environment, both as a source of raw materials and as a sink for wastes.

The impetus for effective response to environmental ills came mainly from citizen agitation. That agitation typically focused on problems whose solution did not require any material sacrifice from the citizenry, nor much trust and cooperation across national boundaries. In rich countries in the 1980s, for example, it proved easy enough to reduce sulfur dioxide pollution from power plants or lead emissions from automobile exhausts by changing or altering fuels and engines. But few people desired the sacrifices that seemed necessary to check carbon dioxide emissions or fertilizer runoff. Environmental outcomes around the world reflected the preferences and compromises embedded in the prevailing political systems. The general approach in the 20th century was to make the most of resources, harness nature to the utmost, sacrifice ecological buffers and tomorrow's resilience if need be, and hope for the best, all while seeking to achieve the often elusive goals of economic prosperity and military security.

REFERENCES

Abernethy, D. 2000. Dynamics of Global Dominance. Palo Alto: Stanford Univ. Press.
Cohen, J. 1995. How Many People Can the Earth Support? New York: Norton.
Hoerder, D. 2002. Cultures in Contact: World Migration in the Second Millennium. Durham, NC: Duke Univ. Press.
Ikenberry, J. 2001. After Victory. Princeton, NJ: Princeton Univ. Press.
Keegan, J. 1976. The Face of Battle. New York: Viking.
Kirk, D. 1946. Europe's Population in the Interwar Years. Princeton, NJ: Princeton Univ. Press.
Kotkin, S. 2002. Armageddon Averted. New York: Oxford Univ. Press.
Krushchev, N. 1974. Krushchev Remembers: The Last Testament, transl. and ed. Strobe Talbot. Boston: Little Brown.
Livi-Bacci, M. 2001. A Concise History of World Population. Malden, MA: Blackwell.
Maddison, A. 2003. The World Economy: Historical Statistics. Paris: OECD.
McNeill, J.R. 2000. Something New under the Sun: An Environmental History of the Twentieth-century World. New York: Norton.
McNeill, J.R., and W.H. McNeill. 2003. The Human Web. New York: Norton.
Misa, T. 2004. Leonardo to the Internet: Technology and Culture from the Renaissance to the Present. Baltimore: Johns Hopkins Univ. Press.
Mitter, R. 2004. A Bitter Revolution. Oxford: Oxford Univ. Press.
Osterhammel, J., and N. Petersson. 2005. A Short History of Globalization. Princeton, NJ: Princeton Univ. Press.
Overy, R. 1995. Why the Allies Won. New York: Norton.
Rhodes, R. 1986. The Making of the Atomic Bomb. New York: Simon and Schuster.
Shapiro, J. 2001. Mao's War against Nature. New York: Cambridge Univ. Press.
Smil, V. 1994. Energy in World History. Boulder, CO: Westview.

Steffen, W., A. Sanderson, P.D. Tyson et al., eds. 2004. Global Change and the Earth System: A Planet under Pressure. Berlin and New York: Springer.
Weinberg, G. 1994. A World at Arms. New York: Cambridge Univ. Press.
Whitehead, A.N. 1925. Science and the Modern World. New York: Macmillan.
Yergin, D. 1991. The Prize: The Epic Quest for Oil, Money, and Power. New York: Simon and Schuster.

17

Integrated Human–Environment Approaches of Land Degradation in Drylands

Eric F. Lambin,[1] Helmut Geist,[1] James F. Reynolds,[2] and D. Mark Stafford Smith[3]

[1]Dept. of Geography, University of Louvain, 1348 Louvain-la-Neuve, Belgium
[2]Division of Environmental Science and Policy, Nicholas School of the Environment and Earth Science and Department of Biology, Phytotron Bldg., Duke University, Durham, NC 27708–0340, U.S.A.
[3]CSIRO Sustainable Ecosystems, Centre for Arid Zone Research, CSIRO, Alice Springs, Australia

ABSTRACT

Land degradation in drylands has wide-ranging impacts on human populations and environmental quality. Desertification is about biophysical and socioeconomic linkages and how they affect human welfare. It is driven by a limited suite of recurrent core variables, with strong causal factor synergies. Typical regional pathways of desertification are identified. Once certain thresholds are crossed, losses of ecosystem productivity cannot be compensated anymore by social resilience or government subsidies. A new paradigm on desertification that provides a framework to facilitate directed research effort is presented.

INTRODUCTION

Land degradation in drylands, which is referred to as desertification, impacts human populations (e.g., food security, health and well-being, sustainability) and environmental quality (e.g., dust storms, trace gas emissions to the atmosphere, soil erosion). Throughout human history, dryland degradation has been associated with the collapse or decline of a few ancient civilizations. Today, desertification is the subject of an international framework convention, the Convention to Combat Desertification (CCD).

Land degradation in drylands is still poorly documented, and its causes are hardly understood. Proponents of single-factor causation suggest various primary causes, such as irrational or unwise land mismanagement by nomadic pastoralists and growing populations in fragile semiarid ecosystems. Central to this understanding is the notion of "man-made deserts," that is, the human-driven, irreversible extension of desert landforms (Le Houérou 2002). Desertification has also been attributed to multiple causative factors that are specific to each locality, revealing no distinct pattern (Dregne 2002; Warren 2002). There has been a great deal of debate on whether the causes of desertification lie in the socioeconomic or biophysical spheres (human-induced land degradation vs. climate-driven desiccation). Irrespective of the causes of dryland degradation, its main consequences for human welfare are a decline in the level of well-being and an increase in vulnerability.

In this chapter we provide a brief overview of the nature and causes of land degradation in global drylands, with an emphasis on the interaction between human and natural dimensions of the problem. We then introduce a new paradigm on desertification that provides a framework to facilitate directed research effort and progress on this important global environmental issue.

THE NATURE OF DESERTIFICATION

The definition of desertification used by the CCD makes it clear that although biophysical components of ecosystems and their properties are involved (e.g., soil erosion, loss of vegetation), the interpretation of change as "loss" is dependent upon the integration of these components within the context of the socioeconomic activities of human beings. The CCD states that land degradation is the reduction or loss of the biological and economic productivity and complexity of terrestrial ecosystems, including soils, vegetation, other biota, and the ecological, biogeochemical, and hydrological processes that operate therein. The CCD's definition of desertification explicitly focuses on the linkages between humans and their environments that affect human welfare in arid and semiarid regions: land degradation in arid, semiarid and dry subhumid areas resulting from various factors, including climatic variations and human activities.

Unfortunately, the CCD definition of desertification is not amenable to easy quantification, especially as a single number or synthetic index. Most estimates of desertification are derived solely from either biophysical factors (e.g., soil erosion, loss of plant cover, change in albedo) or socioeconomic factors (e.g., decreased production, economic loss, population movements), but rarely both types simultaneously. Much of the confusion surrounding rates of desertification and regions affected could be eliminated by focusing on a small number of critical variables that contribute to an understanding of the *cause*, rather than *effect*, of desertification (Stafford Smith and Reynolds 2002). Management strategies to avoid desertification can be developed only after causes are understood.

Causes of Desertification

Geist and Lambin (2004) carried out a worldwide review of the causes of desertification based on 132 carefully selected case studies. Their aim was to generate a general understanding of the proximate causes and underlying driving forces of desertification, including cross-scalar interactions of causes and feedbacks, while preserving the descriptive richness of these case studies. Results showed that desertification is driven by a limited suite of recurrent core variables, with identifiable regional patterns of causal factor synergies.

At the proximate level, desertification is best explained by the combination of multiple social and biophysical factors, rather than by single variables. Dominating the broad clusters of proximate factors is the combination of agricultural activities, increased aridity, extension of infrastructure, and wood extraction (or related extractional activities), with clear regional variations. In particular, agricultural activities and increased aridity are associated together.

Agricultural activities include extensive grazing, nomadic pastoralism, and annual cropping. Livestock production activities slightly outweigh crop production as a cause of desertification, but both activities remain intricately interlinked in most of the cases. In comparison with pastoralism, agricultural expansion into marginal rangeland areas during wet periods often leaves farmers more seriously exposed to hazard when drought returns (Glantz 1994). Agricultural expansion on areas previously used for pastoral activities can also lead to overstocking on the remaining reduced rangeland. In addition, it can trigger soil degradation at sites that are not suitable, in particular, for permanent agriculture. Increased aridity is a robust proximate cause of desertification, through greater rainfall variability and prolonged droughts.

The extension of infrastructure associated with desertification is frequent primarily in cases from Asia, Africa, and Australia. Desertification is mostly linked to the development of water-related infrastructure for cropland irrigation and pasture development (reservoirs, dams, canals, boreholes). This in turn leads to a decrease in livestock mobility (Niamir-Fuller 1999). Water droughts are replaced with feed droughts. In the Asian and African cases, the buildup of irrigation infrastructure is associated with expanding human settlements, following an increase in food production and food security.

At the underlying level, desertification is also best explained by regionally distinct combinations of multiple, coupled social and biophysical factors, and drivers acting synergistically rather than by single-factor causation. A recurrent and robust broad factor combination implies the interplay of climatic factors leading to reduced rainfall, agricultural growth policies, newly introduced land-use technologies, and malfunctional land tenure arrangements which are no longer suited to contemporary dryland ecosystem management.

Climatic factors, mainly associated with a decrease in rainfall, are prominent underlying driving forces of desertification (Puigdefábregas 1998). They operate either through the (indirect) impact of rainfall variability via changes in land use or by directly impacting upon land cover in the form of prolonged droughts. Technological innovations are associated with desertification as frequently as are deficiencies of technological applications. Innovations mainly comprise improvements in land and water management through motor pumps and boreholes, or through the construction of hydrotechnical installations such as dams, reservoirs, canals, collectors, and artificial drainage networks. When applied, these developments often cause high water losses due to poor maintenance of the infrastructure. The disaster of the Aral Sea is an extreme case of such a perturbation (Saiko and Zonn 2000).

Among institutional and policy factors underlying desertification, modern policies and institutions are equally involved as are traditional institutions. Growth-oriented agricultural policies, including measures such as land (re)distribution, agrarian reforms, modern sector development projects, diffusion of agricultural intensification methods, and market liberalization policies, are important underlying causes of desertification as are institutional aspects of traditional land tenure such as equal sharing of land and splintering of herds due to traditional succession law, thus reducing flexibility in management and increasing the pressure upon constant land units. The collapse of necessary conditions for effective maintenance of common property grazing regimes (e.g., reduced capacity to prevent encroachment by other land uses) is another cause. The introduction of new land tenure systems, either through private (individual) or state (collective) management, is another important factor associated with desertification. Uncertain land tenure may arise from the overlapping of conflicting property-rights regimes, often leading to violent conflicts about land and thus reducing the adaptive capacity of herding and farming populations.

Often, the growth or increased economic influence of urban population triggers out-migration of poor cultivators and/or herders from high potential, peri-urban zones onto marginal dryland sites. Consequently, the sometimes rapid increases in the size of local human populations are often linked to in-migration of cultivators onto rangelands or in regions with large-scale irrigation schemes, or of herders onto hitherto unused, marginal sites, with the consequence of rising population densities there.

It is mostly the interactions between multiple causal factors that lead to desertification. A frequent pattern of causal interactions stems from the necessity for water-related infrastructures that are associated with the expansion of irrigated croplands and pastures. Typically, newly introduced irrigation infrastructures induce accelerated in-migration of farm workers into drylands and often stir more commercial–industrial developments as well as the growth of human settlements and related service economies. Irrigation infrastructure is often nested in a system of larger infrastructure extension related to regional economic growth. Commonly, road extension paves the way for the subsequent extension of irrigation, and (semi)urban land uses. In the developing world, underlying these proximate factors are national policies aimed at consolidating territorial control over remote, marginal areas and attaining self-sufficiency in food and clothing, with rice and cotton being the key irrigated crops.

Regional Pathways of Desertification

Dominant causative factors and feedbacks combined with environmental and land-use histories allow typical regional pathways of desertification to be identified (Geist 2005; Geist and Lambin 2004). In Central Asia, notably northern China, widespread increase in desert-like sand cover is the most spectacular outcome of desertification. It is linked to the exceptionally strong impact of socioeconomic driving forces such as centrally planned frontier colonization and (sometimes forced) population movements (Sneath 1998), but also to predominantly sandy soils and loess formations as well as the geological and climatic predisposition for desert formation of vast basins and plateaus. In ancient times and under various dynasties, climatic variations and destructive land uses operated in causal synergy so that oscillating desert margins became today's desertified land with the highest rates of dryland degradation worldwide (Genxu and Guodong 1999). Two central pathways of partly irreversible desertification in Central Asia are (a) the invasion of grain farming onto steppe grazing land, which triggered soil degradation as well as overstocking, and (b) large-scale hydraulic cultures into desert ecosystems that historically supported only localized traditional oasis farming (Zhou et al. 2002; Kharin 2003).

In contrast, a typically African pathway of desertification relates to the spatial concentration of pastoralists (through sedentarization) and farming populations around infrastructure nuclei, with the consequence of overgrazing, extensive fuelwood collection, and high cropping intensities ending up in vegetation degradation and soil productivity decline during periods of drought (Dube and Pickup 2001). "Beefing up" the fragile dryland ecosystems, with low or nil involvement of cropping, characterizes frequently the desertification pathways of Australia, and North and Latin America. Historically, these rangeland zones shared commonalities such as the rapid introduction by European settlers of exotic livestock species and commercial pastoralism into ecosystems that previously were not exposed to these uses. In both areas, dryland cover change happened in an episodic manner linked to shifts in rainfall and land uses, and rates of desertification had reached historical peaks in the late 19th and early 20th centuries, but have largely subsided since then (Pickup 1998).

Another common trajectory of dryland change is found in the Mediterranean basin of southern Europe. A millennia-old tradition of agropastoral land uses has removed next to all forest cover but favored a highly resilient *phrygana* (shrub) vegetation. A still valuable agricultural base is at risk only through the mechanization of farming on skeletal soils, inducing further soil erosion, and when grazing of remote mountain ranges is followed by devastating fires (Kosmas et al. 2000; Margaris et al. 1996).

Consequences of Desertification

From the socioeconomic point of view, most consequences of desertification (especially in pastoral systems) are a direct result of the decline in "productivity" or the capacity of the land to support plant growth and animal production. During early stages of desertification such losses are compensated by the social resilience of the local human populations, especially in developing countries, or by economic inputs from government (Vogel and Smith 2002). However, when certain thresholds are crossed, social resilience or government subsidies may not be enough to compensate for the loss of productivity, and this fuels a battery of socioeconomic changes that range from modifications in trade promoted by lower agricultural production to large population migrations (Fernández et al. 2002).

DISCUSSION

The meta-analysis of desertification cases identified "multiplicity" as the most common theme, as reported in empirically supported narratives of the impact on drylands of actors and activities: multiple agents; multiple uses of land; multiple responses to social, climatic, and ecological changes; multiple spatial and temporal scales in the causes of and responses to desertification; multiple connections in social and geographical space; and multiple ties between people and land in drylands (Geist and Lambin 2004; Rindfuss et al. 2003). The theoretical framework that best accounts for this complexity is system dynamics (Lambin et al. 2003), with special emphasis on the history of the system (initial conditions), heterogeneity among actors, hierarchical levels of organization, nonlinear dynamics caused by feedback mechanisms, and system learning and adaptation (Newell et al. 2005). The most important observation is that this complexity is associated with a limited number of recurrent pathways of desertification, which makes the problem tractable (Reynolds and Stafford Smith 2002). When identifying these pathways, it is important to recognize situations where traditional users of drylands become increasingly involved in wider worlds of business and commerce. It used to be the case that most pastoralists had few nonpastoral investments or ties. Increasingly, they are beginning to hold and maintain significant interests in other areas.

FUTURE DIRECTIONS

The simultaneous assessment of biophysical and socioeconomic drivers (and consequences) of desertification has been recognized as one of the most challenging topics for further research. Stafford Smith and Reynolds (2002) proposed the Dahlem Desertification Paradigm (DDP), which is unique in two ways: (a) it attempts to capture the multitude of interrelationships within human–environment systems that cause desertification within a single, synthetic framework; and (b) it is testable, which ensures that it can be revised and improved. The DDP consists of nine assertions (Table 17.1), which embrace a

Table 17.1 Assertions from the Dahlem Desertification Paradigm and some of their implications. From Stafford Smith and Reynolds (2002).

Assertions	Implications
1. Desertification always involves human and environmental drivers.	Always expect to include both socioeconomic and biophysical variables in any monitoring or intervention scheme. Model these interactions explicitly.
2. "Slow" variables are critical determinants of system dynamics.	Identify and manage for the small set of "slow" variables that drive the "fast" ecological goods and services that matter at any given scale. Do the same for social variables, especially those that signal a change in the values and interests of local people.
3. Thresholds are crucial, and may change over time.	Identify thresholds in the change variables at which there are significant increases in the costs of recovery, and quantify these costs. Seek ways to manage the thresholds to increase resilience.
4. Costs of intervention rise nonlinearly with increasing degradation.	Intervene early where possible, and invest to reduce the transaction costs of increasing scales of intervention.
5. Desertification is a regionally emergent property of local degradation.	Take care to define precisely the spatial and temporal extent of processes resulting in any given measure of local degradation. Do not try to probe desertification beyond a measure of generalized impact at higher scales.
6. Coupled human–environment systems change over time.	Understand and manage the circumstances in which the human and environmental subsystems become "decoupled." Facilitate change within institutional limits and encourage institutions to evolve.
7. Development of appropriate local environmental knowledge (LEK) must be accelerated.	Create better partnerships between LEK development and conventional scientific research. Employ good experimental design, coupled with effective adaptive feedback and monitoring.
8. Systems are hierarchically nested (manage the hierarchy!)	Recognize and manage the fact that changes at one level affect others. Create flexible but linked institutions across the hierarchical levels, and ensure processes are managed through scale-matched institutions.
9. A limited suite of processes and variables at any scale makes the problem tractable.	Analyze the types of syndromes at different scales and seek the investment levers that will best control their effects. Try awareness and regulation where the drivers are natural. Try changed policy and institutions where the drivers are social.

hierarchical view of land degradation and highlight key linkages between socioeconomic and biophysical systems at different scales.

The monitoring of desertification is an increasingly important development in the management of dryland areas. The establishment of long-term and rigorous monitoring programs is an effective way to assess the status of natural resources and the evolution of desertification processes. Increased research is dedicated to the development of easily accessible monitoring methods based on simple soil and vegetation indicators, and combining ground-based methods with remote sensing data (e.g., Pyke et al. 2002). The challenge, however, is to monitor and model not just biophysical indicators but the coupled human–environment system as a whole. This requires methods to link biophysical data to socioeconomic data on land managers, to capture the varying capacity of local agents to cope and respond to a decline in land productivity.

ACKNOWLEDGMENT

The authors are grateful for the financial support from the Science Policy Office of Belgium. This chapter has greatly benefited from ideas developed within several workshops of the Land-Use and Cover Change (LUCC) and Global Change and Terrestrial Ecology (GCTE) projects.

REFERENCES

Dregne, H. 2002. Land degradation in the drylands. *Arid Land Res. Manag.* **16**:99–132.

Dube, O.P., and G. Pickup. 2001. Effects of rainfall variability and communal and semi-commercial grazing on land cover in southern African rangelands. *Clim. Res.* **17**:195–208.

Fernandez, R.J., E.R.M. Archer, A.J. Ash et al. 2002. Degradation and recovery in socio-ecological systems: A view from the household/farm level. In: Global Desertification: Do Humans Cause Deserts?, ed. J.F. Reynolds and D.M. Stafford Smith, pp. 297–323. Dahlem Workshop Report 88. Berlin: Dahlem Univ. Press.

Geist, H.J. 2005. The Causes and Progression of Desertification. Aldershot: Ashgate.

Geist, H.J., and E.F. Lambin. 2004. Dynamic causal patterns of desertification. *BioScience* **54**:817–829.

Genxu, W., and C. Guodong. 1999. Water resource development and its influence on the environment in arid areas of China: The case of the Hei River basin. *J. Arid Env.* **43**:121–131.

Glantz, M.H. 1994. Drought Follows the Plow: Cultivating Marginal Areas. Cambridge: Cambridge Univ. Press.

Kharin, N. 2003. Vegetation Degradation in Central Asia under the Impact of Human Activities. Dordrecht: Kluwer.

Kosmas, C., S. Gerontidis, and M. Marathianou. 2000. The effect of land use change on soils and vegetation cover over various lithological formations on Lesvos (Greece). *Catena* **40**:51–68.

Lambin, E.F., H.J. Geist, and E. Lepers. 2003. Dynamics of land-use and land-cover change in tropical regions. *Ann. Rev. Env. Resour.* **28**:205–241.

Le Houérou, H.N. 2002. Man-made deserts: Desertization processes and threats. *Arid Land Res. Manag.* **16**:1–36.

Margaris, N.S., E. Koutsidou, and C. Giourga. 1996. Changes in traditional Mediterranean land-use systems. In: Mediterranean Desertification and Land Use, ed. J.C. Brandt and J.B. Thornes, pp. 29–42. Chichester: Wiley.

Newell, B., C.L. Crumley, N. Hassan et al. 2005. A conceptual template for human-environment research. *Global Env. Change* **15**:299-307.

Niamir-Fuller, M. 1999. Managing Mobility in African Rangelands: The Legitimization of Transhumance. London: Intermediate Technology Publ.

Pickup, G. 1998. Desertification and climate change: The Australian perspective. *Clim. Res.* **11**:51–63.

Puigdefábregas, J. 1998. Ecological impacts of global change on drylands and their implications for desertification. *Land Degrad. Dev.* **9**:393–406.

Pyke, D.A., J.E. Herrick, P.L. Shaver and M. Pellant. 2002. Rangeland health attributes and indicators for qualitative assessment. *J. Range Manag.* **55**:584–597.

Reynolds, J.F., and D.M. Stafford Smith. 2002. Do humans cause deserts? In: Global Desertification: Do Humans Cause Deserts?, ed. J.F. Reynolds and D.M. Stafford Smith, pp. 1–21. Dahlem Workshop Report 88. Berlin: Dahlem Univ. Press.

Rindfuss, R.R., S.J. Walsh, V. Mishra, J. Fox, and G.P. Dolcemascolo. 2003. Linking household and remotely sensed data: Methodological and practical problems. In: People and the Environment: Approaches for Linking Household and Community Surveys to Remote Sensing and GIS, ed. J. Fox, R.R. Rindfuss, S.J. Walsh, and V. Mishra, pp. 1–29. Boston: Kluwer.

Saiko, T.A., and I.S. Zonn. 2000. Irrigation expansion and dynamics of desertification in the Circum-Aral region of Central Asia. *Appl. Geogr.* **20**:349–367.

Sneath, D. 1998. State policy and pasture degradation in Inner Asia. *Science* **281**:1147–1148.

Stafford Smith, D.M., and J.F. Reynolds, eds. 2002. The Dahlem Desertification Paradigm: A new approach to an old problem. In: Global Desertification: Do Humans Cause Deserts?, ed. J.F. Reynolds and D.M. Stafford Smith, pp. 403–424. Dahlem Workshop Report 88. Berlin: Dahlem Univ. Press.

Vogel, C.H., and J. Smith. 2002. Building social resilience in arid ecosystems. In: Global Desertification: Do Humans Cause Deserts?, ed. J.F. Reynolds and D.M. Stafford Smith, pp. 149–166. Dahlem Workshop Report 88. Berlin: Dahlem Univ. Press.

Warren, A. 2002. Land degradation is contextual. *Land Degrad. Dev.* **13**:449–459.

Zhou, W.J., J. Dodson, M.J. Head et al. 2002. Environmental variability within the Chinese desert-loess transition zone over the last 20000 years. *Holocene* **12**:107–112.

Left to right: Diana Liverman, John McNeill, Paul Crutzen, Bruno Messerli, Kathy Hibbard, Will Steffen, Nate Mantua, and Eric Lambin

18

Group Report: Decadal-scale Interactions of Humans and the Environment

Kathy A. Hibbard, Rapporteur
Paul J. Crutzen, Eric F. Lambin, Diana M. Liverman, Nathan J. Mantua, John R. McNeill, Bruno Messerli, and Will Steffen

ABSTRACT

The complex dynamics of the human–environment relationship—from the deep past through the present into the future—provide a unique perspective in Earth system analysis. This chapter focuses on the last century, examining discontinuities, nonlinearities, thresholds, feedbacks, and lag effects in the human–environment relationship. Environmental responses to human activities, which include changes in knowledge, science, and technology as well as feedbacks through population dynamics, energy, institutions, and political economies, are presented as a linked human–environment system. In this system it is clear, particularly since the 1950s, that a tension exists between the momentum of past changes and the potential for future variable rates of changes. The 20th century can be characterized by global change processes of a magnitude which never occurred in human history. The strength of the links and associated feedbacks between the human and environmental components (positive and negative) operate over variable time and space domains. Indicators of the significant decadal processes are presented to aid understanding of the rapid changes in the human–environment system since the 1950s. It is suggested that by accounting for the discontinuities, thresholds, and surprises inherent in a complex systems approach, insight into the mechanisms of the human–environment relationship can be gained to allow testable hypotheses for the future of the Earth system.

INTRODUCTION

Of the timescales that reach from the deep millennial past to the forecast of the future, the decadal scale provides the most detailed insights into human interactions within the Earth system. Human activities were major drivers for secular changes of the Earth's environment in the 20th century (McNeill 2000). Humanity is facing an increasing number of global-scale problems, and, as these problems become more evident, there are increasing calls for a change in how we manage our affairs and our relationship with the environment (Costanza et al. 2005). The primary finding reported by the Millennium Ecosystem Assessment (MEA 2005, p. 1) is that "over the past 50 years, humans have changed ecosystems more rapidly and extensively than in any comparable period of time in human history, largely to meet rapidly growing demands for food, fresh water, timber, fiber, and fuel. This has resulted in a substantial and largely irreversible loss in the diversity of life on Earth." The trends in carbon dioxide (CO_2) emissions and associated temperature changes also suggest a rapid acceleration of human impacts on the atmosphere over the last 50 years. These and many other changes demonstrate a distinct increase in the rates of change in many human–environment interactions as a result of amplified human impact on the environment after World War II—a period that we term the "Great Acceleration." Although many of the trends are associated with the degradation of the environment, we identify some important factors that have the potential to reduce human impact on the environment in the future (i.e., decelerating tendencies) as well as important nonlinearities, thresholds, and time lags that provide momentum for continued human impacts on the environment well into the 21st century.

Since 1950, dramatic changes and even switches in the rates and feedbacks of human enterprise and its associated environmental signatures have occurred. The primary human processes that have caused altered rates of change (acceleration or deceleration) can be broadly defined as changes in: (a) human knowledge, science, and technologies, (b) energy systems development, (c) human populations and their demography, (d) production and consumption, and (e) political and economic structures and institutions (i.e., political economy).

We frame our discussion of recent decades of the 20th century in the human–environment system around three main approaches:

1. the application of a systems approach to human as well as environmental components of the human–environment relationship, with an emphasis on feedbacks (see section on SYSTEMS APPROACHES TO DECADAL DYNAMICS);
2. the tension between those processes that decelerate or mitigate the human impact and those that carry momentum or inertia, such that changes will continue for decades or will be difficult to reverse;
3. the possibility of discontinuities and nonlinearities in the human–environment system, which can lead to rapid and surprising shifts in the state of the Earth system.

Human and Environmental Precursors to the Great Acceleration

Humans have altered local and regional ecosystems for centuries through the domestication of plants and animals and the manipulation of fire. Large-scale alterations have been evident since at least the 1500s, with the 20th century acting as a pivotal point in the relationship between humans and the environment

Table 18.1 Scale of environmental change from the 1890s to 1990s and some of the driving forces behind the change (after McNeill 2005).

Scale	Coefficient of Increase
Freshwater use	9-fold increase
Marine fish catch	35
Cropland	2
Irrigated area	5
Pasture area	1.8
Forest area	0 (i.e., 20% reduction)
CO_2 emissions	17
Lead emissions	8
Cattle population	4
Driving Forces	
Population	4-fold increase
Urban population	13
World economy	14
Industrial output	40
Energy use	13

(Table 18.1). One of the most vivid examples from the early 20th century was the Dust Bowl drought (1931–1939) in the Canadian prairies and midwestern United States, where a decade of low annual rainfall and poor farming practices resulted in crop failures, a massive loss of topsoil, continental-scale dust storms, social disruption, and large-scale emigration of farming families out of the affected areas. The negative impacts on the affected population were exacerbated by its coincidental occurrence with the much larger-scale Great Depression (late 1920s–1930s), as economic opportunities for dislocated people were extremely limited (McNeill 2000; Worster 1986). Unlike the human-induced accelerating degradation trends of the Earth system (discussed below), events like the Dust Bowl were episodic. Although topsoil loss was essentially irreversible, improved rainfall conditions and better farming practices allowed for some revitalization of the midwestern farming economy in subsequent years. In other cases of collapsing resource systems (e.g., the collapse of the California sardine fishery, discussed below), the resultant degradation of ecosystem services[1] prompted a relocation of people and infrastructure changes, and subsequent economic activities actually revitalized the economy of the region impacted by the loss.

Since the 1950s, there has been a Great Acceleration in the scope, scale, and intensity of mutual impacts on the human–environment system. The Great

[1] The goods (e.g., food, timber, fresh water) and services (e.g., nutrient cycling, climate regulation, recreation) that humans acquire directly or indirectly from ecosystems.

Acceleration occurred when it did partially because prior constraints on economic growth, population growth, and technological change were removed. In the three decades prior to 1945, the world economy struggled as a result of two world wars (and a host of smaller ones) and the Great Depression (1929–1939), which diverted or stymied investment in technological change. Economic growth from 1914 to 1945 proceeded more slowly than it had during 1874 to 1914. War increased the mortality rates in combatant countries and, although these effects were short-lived, both war and economic depression dampened fertility in many populations.

World War II was pivotal in many respects: armies of scientists were enlisted in closely directed projects, most notably the building of nuclear weapons. This practice continued after the war and was instrumental in coordinating the work of scientists and engineers; remarkable results were obtained, especially in the energy, chemical, and agricultural industries. During the Cold War (from ca. 1947–1991), competition between the U.S. and the U.S.S.R. led to advances in knowledge, science, and technology. The Russian space launch of Sputnik 1 in 1957 was most notable in this regard.

Economically, the Allied victory in 1945 precipitated the spread of a liberal market economy that linked North America, western Europe, and Japan more closely than ever before through trade, political alliance, and investment. The autarkic economies of the 1930s vanished, except in the Soviet Union and some developing countries. This restructuring of economies led to record growth in the postwar decades (1946–1973). It also accelerated the spread of new technologies, some of which had marked impacts on human–environment interactions (e.g., DDT which was used around the world in anti-malaria campaigns, the hybrid seeds of the Green Revolution, or the implementation of assembly-line technology in manufacturing).

The dominant characteristics of the world economy since 1945 have been growth and integration. In addition to the postwar economic recoveries of Europe and Japan, the world economy received an additional boost through the increasing participation of the densely populated regions of east and southeast Asia: Taiwan in the 1960s, Korea in the 1970s. Thereafter, during the 1980s, several southeast Asian countries and China developed export-oriented economies and recorded very high growth rates. Political decisions taken by the successors of Mao Zedong after 1976 were crucial in removing constraints on China's economy. The end of the Soviet system after 1991 had similar effects in some eastern European countries and Russia.

Although large areas, including the Soviet Union and China, were not immediately affected by, or were representative of the shifts in political economy of the 1950s, they are currently experiencing a phase of dramatic regime shifts. Similar changes are occurring in developing societies: Brazil and Argentina have initiated massive investments in science, which previously was inhibited by debt crisis.

Explanatory Frameworks

To understand the dynamics between humans and the environment, we can contrast and employ several explanatory frameworks. Social science has a range of often competing theories and explanations for the political, economic, and environmental history of the second half of the 20th century.[2] In some cases, especially as relates to economics, these theories have informed political ideologies or have been applied in policy.

For example, a fundamental divergence in economic theory after World War II saw John Maynard Keynes arguing that economic cycles and slumps were integral to economic growth and that state intervention was required to moderate the effects in the form of welfare programs and active monetary policy. This theory was widely adopted after 1945 through extensive state intervention in economies, including industrial and agricultural subsidies and government-owned enterprises, which in many cases drove growth and environmental degradation. However, Keynesian policies also provided a model for state investment, public ownership, and an interventionist command-and-control model that led to the growth of environmental management from about 1970.

Alternatives to the Keynesian model are the theories of classical and neoclassical economics associated with Adam Smith and Milton Friedman. These theories argue that the economy functions most efficiently in a free market with less government intervention and have underpinned neoliberal policies of reduced government intervention and free trade since about 1980.

A third set of theories and practices are associated with political economic explanations of the dynamics of world system—attributable to the work of Karl Marx and Emmanuel Wallerstein—in which capital's drive for profit exploits labor and nature unsustainably unless moderated by the state or social protest. One solution that allows continued accumulation of profits is global expansion (e.g., through colonialism or trade) and the subsequent creation of new markets and commodities (through establishing private property over the commons and natural systems as well as through advertising). According to this approach, the Great Acceleration would be associated with the unbridled expansion of capitalism and its exploitation of resources and people. One version of Marxism was translated into the Russian and Chinese communist systems, which promoted collective resource ownership under heavy state control.

Each of these theories is reflected in paradigms that seek to explain the social drivers and responses to human–environment interactions over the last 50 years. Thus there is a contrast between those who argue for a strong role of the state in environmental management, those who favor market mechanisms as the most effective environmental protection, and those who see no need for any controls on environmental impacts.

2 For a comprehensive review, the reader is referred to the *International Encyclopedia of the Social and Behavioral Sciences* (Smelser and Baltes 2001).

Social science seeks to theorize specific relationships, such as those between population and environment, technology and environment, and human attitudes and environment. In terms of population, there is a fundamental disagreement between Malthusians, who see population as a primary cause of environmental problems (Ehrlich 1968; Malthus 1803), and those who perceive population to be a resource that can be used to manage the environment more efficiently and sustainably (e.g., Boserup 1965). Technology has been viewed as the main cause of environmental degradation (Commoner 1971) or as the solution to environmental problems (Ausubel 2002).

These different approaches can confuse and complicate the explanation of the Great Acceleration and its environmental impacts unless care is taken to assess the empirical evidence for the human role in environmental change. We must consider the possibility that explanations may work in some places and times, but not in others.

From a decadal perspective, we synthesize below the significant change, evident globally and regionally, through a systems approach of nonlinear processes and associated lags and feedbacks in the human–environment relationship.

THE GREAT ACCELERATION

Indicators

The Great Acceleration can be characterized through a suite of global indicators that show dramatic increases over the last few decades (Figure 18.1). For example, human populations have increased 4-fold in the 20th century (McNeill 2005; Table 18.1) and, further, have increased from 1972 to 2000 from 3.85 to 6.1 billion (UNFPA 2001). Over the second half of the 20th century, human activities have expanded rapidly and have led, for example, to nonlinear increases in global atmospheric CO_2 concentrations (Figure 18.1b). Although some correlations can be found between population and CO_2 growth rates in the 20th century (Figure 18.1a, b), increased CO_2 is also associated with a rapid rise in per capita consumption by a relatively small fraction of the human population. The rapid accumulation rate of atmospheric CO_2 has placed the Earth system in a domain with unknown consequences and feedbacks that are unprecedented over at least the last 730,000 years (Petit et al. 1999; EPICA 2005). Coincident with increased atmospheric CO_2 concentrations, average global temperatures have risen as a result of long residence times of CO_2 and the "greenhouse effect" (Figure 18.1c) (IPCC 2001).

Nitrogen serves as a further indicator (Figure 18.1d). The use of nitrogenous fertilizers, made possible through the Haber–Bosch process (Box 18.1), has negatively impacted global terrestrial and freshwater ecosystems. The rate of resource extraction and harvest is a direct result of increased population and technology. In terms of marine fisheries (Figure 18.1e), the FAO (2000) reports that

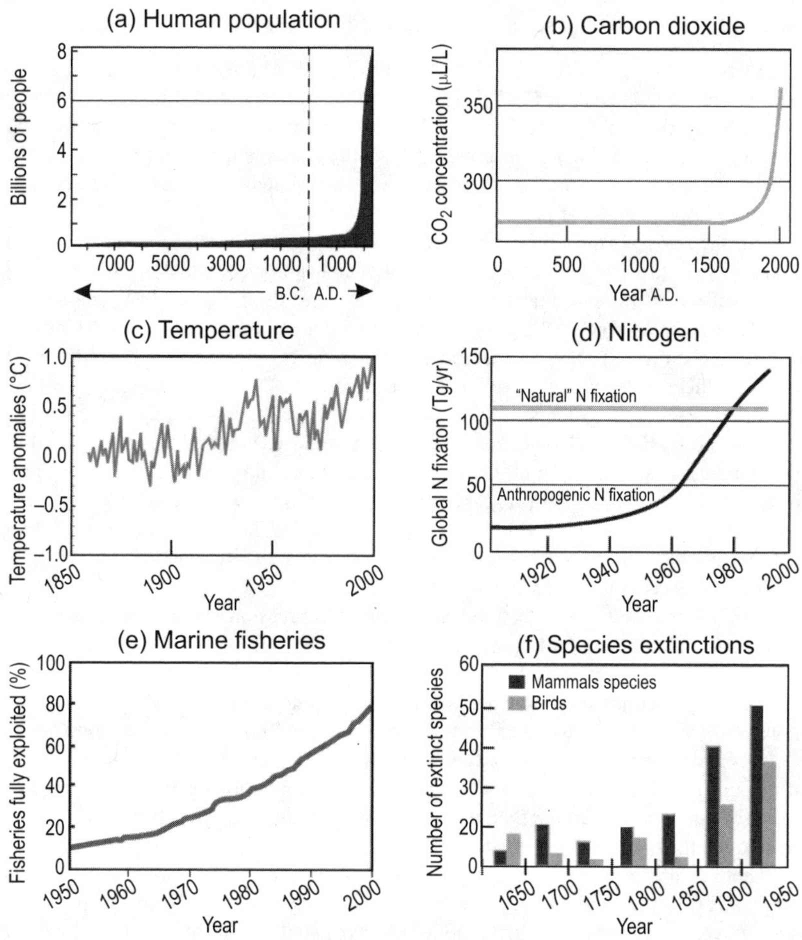

Figure 18.1 Global indicators of the Great Acceleration. (a) Human population (WRI 2003); (b) CO_2 (Globalview–CO_2, 2005); (c) temperature (Vitousek 1994); (d) nitrogen (Vitousek 1994); (e) the state of the world's fisheries and aquaculture (FAO 2000); (f) the scientific basis for conserving biodiversity (Reid and Miller 1989).

up to 80% of possible fish populations have been exploited. In addition, the exponential extinction rate (Figure 18.1f) of species highlights unforeseen and unpredictable consequences of human activities coupled with resource degradation.

Other indicators most certainly exist, which points to the necessity of understanding the causal relationships between human–environment interactions in an effort to provide a framework for hypothesis testing and projections of Earth system dynamics.

Box 18.1 The Haber–Bosch ammonia synthesis and the global environment.

The German chemist, Fritz Haber (Dec. 8, 1868–Jan. 29, 1934), began experimenting with ammonia synthesis in 1904. Like many others, he recognized that crop yields were chiefly limited by nitrogen availability. Although guano (a natural fertilizer high in nitrogen and phosphorus obtained from the excrement of seabirds) had been imported from South America, crop yield was insufficient, and need increased as Germany approached war pre-1914.

Nitrogen exists in the atmosphere in limitless supply, and thus the challenge was to find a way of extracting it. In 1908, Haber patented his process, which stipulated that under sufficient pressure and temperature, and using the right catalyst, ammonia could be synthesized ($N_2 + 3H_2 \rightarrow 2NH_3$). This process reprepresented a milestone in industrial chemistry, because it enabled nitrogen products (e.g., fertilizer, explosives) to be produced independently from natural sources.

In 1910, under the guidance of Carl Bosch, the Haber process was made commercially viable by the firm BASF. The process, however, was not used to produce fertilizer, but rather explosives for the German military. Haber contributed to the war effort by developing the use of chlorine gas as a military weapon; it was first used at Ypres in 1915, an event that is reputed to have led Haber's wife Clara, herself a chemist, to commit suicide at the dinner honoring him for his efforts. In 1918, Haber received the Nobel Prize in chemistry for ammonia synthesis.

World Wars I and II, and the intervening Great Depression, diverted the development and implementation of nitrogenous fertilizers. By 1950, however, ca. 5 million tons had been produced worldwide. In 2000, production stood at 80 million tons. Major increases in crop yields were possible due to the generous applications of nitrogenous fertilizer and other technical changes in agriculture. For some staple food crops (e.g., cereal grains), yields doubled and quadrupled (Mosier et al. 2004). Through the use of nitrogenous fertilizers, it is estimated that nearly 2 billion people more are able to be fed per annum. Without their use, 30–35% more cropland would need to be cultivated—a tall order in a crowded world.

There is, however, a downside to the use of nitrogenous fertilizers. Runoff from croplands flow into streams, lakes, and seas and has negatively impacted aquatic ecosystems (e.g., by nourishing algal blooms). Source: Smil (2001); http://en.wikipedia.org/wiki/Fritz_Haber.

Explanations

As our discussions reveal, ultimately a myriad of complex, intertwining socioeconomic trends underpin the Great Acceleration (see also McNeill, this volume, and McNeill 2000). Here we highlight a few of the most important socioeconomic factors as a precursor to a discussion of how the human–environment relationship changed during the Great Acceleration.

Population

Global populations began a long secular increase in the mid-1700s, with the maximum rate of increase occurring around 1970 with a current growth rate of about 77 million people per year. Since then, fertility rates have been generally declining; however, there is tremendous ethnic, social class, and regional variation, primarily in Asia and Africa. Contemporary rates of population growth are largely attributable to increased medical and human health knowledge, science and technology, as well as the reduction of the impact from epidemics and infant mortality on global population size. In contrast, historic epidemics and pandemics significantly limited population growth on a global scale.

As population increased, the ratio of energy cost per unit of production was substantially reduced. In other words, energy costs for manufacture became cheap relative to labor. For the U.S., this occurred in the early 1920s; for Japan, in the 1960s to 1970s; for Europe, in the late 1950s. The opening of the big "elephant" oil fields around the Persian Gulf in the late 1940s was one of the drivers behind decreased energy costs, together with the large hydro/nuclear projects between 1930 and 1980. In general, the greatest changes occurred between 1920 and the 1970s, with the peak observed in the 1950s (Pfister 1995).

Science and Technology

The breaks, regime changes, and/or points of inflection in accelerated knowledge, science, and technology are much less well defined. Gradual increases that set the stage can be found from the origins of the modern scientific method in the Renaissance period of the 16th or 17th century (e.g., Galileo). As the 19th century progressed, sponsored science (industrial R&D) developed in Europe and North America from 1850–1880. After World War II, "big science"[3] increased greatly. Between 1945 and 1960, science and technology was contained (with few exceptions) within national borders. In many countries, "Science Foundations" were first established in the 1950s or 1960s. As of 1960, the scientific community began to recognize the need for better international scientific cooperation and communication. This evolved rapidly thereafter (Box 18.2), and the internationalization of science and technology contributed strongly to the Great Acceleration.

Political Economy

Dramatic shifts in political and economic structures as well as in institutions have occurred since 1950. In terms of the influence this had on human–environmental impacts, some of the most significant examples include:

- The establishment of the Bretton Woods Institutions (World Bank, International Monetary Fund) in 1944 and the United Nations in 1945: These institutions played important roles in (a) programs of economic

3 "Big science" is a term used to describe a series of changes in science that occurred in industrial nations during and after World War II. It usually implies big budgets, big staffs, big machines (e.g., the cyclotron at the Lawrence Livermore Laboratory), and big laboratories.

Box 18.2 The internationalization of science.

As of 1960, the scientific community recognized the need for better international scientific cooperation. The International Biological Programme was founded as a result; it was subsequently replaced in 1971 by UNESCO's Man and Biosphere Programme (MAB), because the interactions between vegetation and biodiversity could not be understood without taking into account the impact of human activities. Within MAB, 14 interdisciplinary projects were established, covering all of the main global biomes and special topics (e.g., urban areas, costal zones, mountains, and small islands). Most fascinating was the focus on the interface between humans and the biosphere (i.e., the human–environment relationship). Initially, the challenge was to find a common language and an integrating methodology, for at that time interdisciplinary cooperation between natural and social scientists was quite unknown, or even unnecessary, for many disciplines. In 1972 the first international United Nations Conference on the Environment took place in Stockholm. This led not only to the foundation of the United Nations Environment Programme (UNEP), but also provided new stimuli for the scientific research community and supported better cooperation in the domain of human–environment interactions.

After 1980 the scientific community was forced to take on new responsibilities and create new structures and organizations that would follow the rapidly growing process of economic globalization. In 1986 the International Council for Scientific Unions (ICSU) established the International Geosphere–Biosphere Programme (IGBP), and some years later the International Social Science Council (ISSC) followed by founding the International Human Dimension on Global Environmental Change Programme (IHDP). The World Meteorological Organization (WMO), together with UN organizations and ICSU established the World Climate Research Programme (WCRP), and UNESCO and ICSU supported the foundation of DIVERSITAS, a program designed to study and monitor biological diversity. In addition to these four global research programs, ICSU and several UN organizations initiated the Global Observing System Programme for land, climate, and ocean with a focus on the changing environment. In a first joint action, ICSU and ISSC, together with the engineering sciences, prepared an Agenda of Science for Environment and Development into the 21st Century (UNCED), which was discussed in Vienna, November 1991, and presented at the conference in Rio de Janeiro, June 1992. Even if several statements were not included in Agenda 21 (UN 1993) for political reasons, it was an impressive manifestation by the global scientific community for the global Agenda 21. The Rio conference was an important event for the internationalization or globalization of science, and was reinforced by conventions on climate change, biodiversity, and combating desertification. As a result, the International Panel on Climate Change (IPCC) was established in 1988; its findings have stimulated scientific and political discussions ever since.

The interaction between human activities and natural environment became so important, that the UN Secretary-General called for a Millennium Ecosystem Assessment in his report to the UN General Assembly in 2000. Subsequently, 1360 experts from 95 countries worked on this worldwide evaluation of the human impact on the environment. All together, in only three decades, the human–environment relationship has become recognized as a fundamental problem for humanity, and the scientific community was able to create the necessary global programs to follow the rapidly growing processes of globalization into the 21st century.

development, which altered human–environment interactions through, e.g., the building of large dams, (b) programs of overall investment and debt, which increased industrialization or required the adjustment of economic and environmental policy, and (c) the creation of global environmental management organizations such as UNEP and regimes that produced conventions on issues such as climate and biodiversity (see section below on *Governance*).

- The Cold War and its conclusion in 1989: The Cold War indirectly impacted the environment in many ways, including the ramifications from nuclear weapons programs and the diversion of resources to conflict and defense. The collapse of the Soviet block in 1989–1991 produced a distinct reduction in environmental impacts as industrial activity dropped.
- A period of protectionism and heavy state intervention in economies followed by a period of (neo)liberalization, reduced government, and free trade (phased in different regions): The shift from protectionism and state involvement to free trade has been associated with a multitude of environmental impacts, including those associated with subsidized industry and export crops. Free trade is seen by some either to promote environmental protection as incomes increase or to degrade the environment as polluting firms seek looser environmental regulations (see section below on *Globalization*).
- Decolonization and the spread of democratic governments to many regions of the world have been associated with (a) the difficulties of establishing environmental governance in newly independent but economically dependent nations and (b) the growth of environmental movements under less repressive political regimes.
- Incorporation of new commodities and property rights into the economic system through privatization of un-owned, communally or state-owned resources, and the marketing of nature (e.g., biodiversity, genetic material): The environmental impacts of privatizing and pricing nature are a topic of debate. Some view the economy and private property as the most efficient way to protect the environment; others see loss of common property resources and nonmarket values as a threat to environmental sustainability (Liverman 2004).

How Has the Great Acceleration Changed the Human–Environment Relationship?

Rapid changes in population, migration, energy use, knowledge, science and technology, and political economies have all interacted at various rates and magnitudes over the past 50+ years. The nature of these interactions is described below using three examples of major changes: globalization, urbanization, and governance. Emergent properties of interactions from these examples can be characterized through the globalization of trade, consumption, and transportation, urbanization, and the decentralization of political economies.

Globalization

Globalization is a term used to describe a set of processes that has transformed the world over recent decades and has two faces. One consists of increasing poverty, hunger and food insecurity, and water scarcity–sanitation–health problems. The second face is based on the rapid increase in flows of goods, services, ideas, and people through the massive expansion of trade, the global reach of financial services and digital communication, the rapid diffusion of consumer cultures and scientific information, and the growth of tourism, migration, and transportation links between countries.

The impacts of these processes on human–environment interactions are extremely complex and somewhat controversial. In brief, the debate about trade and environment suggests that the general increase in flow of goods around the world has:

- increased pollution from transport;
- resulted in expanded production in some regions and declines in others with associated environmentally damaging impacts;
- produced better environmental practices in some regions with strong environmental management, increased wealth, or where corporations bring in high standards for reputation or efficiency reasons;
- produced environmental degradation where economic growth is not accompanied by environmental enforcement and where production seeks out lower environmental standards.

The globalization of financial services has produced environmentally damaging development projects and debts that may drive resource overuse, but it has also been associated with institutional innovations that reduce environmental impact, such as sustainable development policies of the World Bank or climate risk avoidance and lower carbon preferences by investors. Digital communication was initially expected to reduce travel and paper use and to facilitate the diffusion and linking of scientific ideas, environmental activism, and technologies; however, business travel and paper production have increased as never before. The unprecedented increase in communication capabilities served as a double-edged sword: In developed countries, and increasingly in developing nations, digital technologies and access to computers has reached ordinary households, providing unparalleled access to information around the globe. Similarly, an increasingly global media has been blamed for the spread of environmentally damaging consumption preferences and the loss of indigenous traditional knowledge of environmental stewardship. Finally, the movement of people around the globe—as migrants and tourists—has increased pollution from transportation. For example, human populations have increased 4-fold in the 20th century (McNeill 2005; Table 18.1) and, further, have increased from 1972 to 2000 from 3.85 to 6.1 billion (UNFPA 2001). More significantly, however, it has raised the risks of disease transmission. Increased rates of trade and transportation have interacted with the environment through ecological exchange and invasions (Box 18.3). There has been a "homogenization of the biosphere" (H.A. Mooney, pers. comm.) through monoculture and other human activities.

Cultural and social values have been altered through accelerated rates of consumption. In the U.S., Europe, and Japan, rapidly increasing consumption per capita started around the 1920s, but became globally significant after World War II. Several processes have affected humans and natural resources in the 20th century across various scales and sectors. Accelerated use of resources has led to divergent changes in the environment. In general, this "consumption of the new" has led to higher standards of production globally with more rapid flows of information (and misinformation). Globalization of consumption and finance has had direct impacts on the environment through debt crises. Countries that must repay loans are less likely to attend to problems of soil and land degradation, water quality, and air pollution issues.

The global human system crossed several thresholds in network connectivity over the last decades and, indeed, the last century. Two different timelines for transport and communication reflect technological changes (but not exclusively). The first major change in transportation was steam-driven technology, which occurred around 1840. Railroads and steamships reduced transport costs and had subsequent impacts for human migrations. The second was the

Box 18.3 Globalization impacts on the environment along the U.S.–Mexico border.

The interaction between processes of globalization and the human–environment relationship can be clearly seen on the Mexican side of the U.S.–Mexico border, where trade liberalization, migration, and resource privatization have had significant material impacts on the environment. The early (1963) establishment of urban free-trade zones for in-bound manufacturing using lower-cost Mexican labor (the Maquila Program) and the development of irrigation systems placed pressures on the capacity of air and water systems. These pressures generally accelerated after the signing of the North American Free Trade Agreement (NAFTA) in 1994, which increased the number of factories, the production of water-intensive export crops, and vehicle traffic in the border zone (Sanchez 2002). In addition, population and consumption increased as a result of migration to the border area, urbanization, improved housing, and consumption cultures; this raised per capita water use, household energy consumption, the use of automobiles, and waste generation. Some attempt was made to mitigate the environmental impacts of NAFTA through the establishment of new trilateral environmental monitoring (CEC) and bilateral infrastructure (BECC) institutions, as well as improved environmental law and management in Mexico. However, NAFTA institutions and Mexican environmental agencies have lacked funding and capacity, and have been overwhelmed by the political and economic imperatives of trade and development. In response to both export and domestic demands, the general expansion of resource development in Mexico over the last 50 years has produced a dramatic increase in CO_2 emissions and agricultural chemical use (Liverman et al. 1999).

development of road networks, trucking, and automobiles, which happened in the U.S. in the 1920s and was adopted thereafter in other parts of the world. The technology and energy costs associated with this were discussed earlier. Aviation (routine from the 1950s onward) represents the third stage of transportation change. After 1950, transport has been flooded with options. A timeline for changes in communication technology begins with the printing press (invented 1450s), leading up several centuries later to the telegraph (widespread by the 1850s) and telephone (1920s). After the 1980s, computing and the Internet have provided unprecedented global-scale information and communication capabilities.

Urbanization

Urbanization has dramatically alterred the relationship between human societies and the environment. Urban populations have a different perception of "nature" than rural residents. While rural (farming) populations have a more direct knowledge of nature, the contact of urban dwellers with nature is more indirect. However, urban populations generally have a broader scientific understanding of the functioning of ecosystems, due to greater access to education and information. Land use in urban settings is generally thought to be more efficient, given economies of scale. Urbanization has been associated with a decrease in fertility rate and, therefore, contributes to a slowdown of world population growth (see section on EVIDENCE OF DECELERATION). Urban centers are the home of "nodal" interactions that contribute to rapid developments in knowledge, science, and technology. However, urban lifestyles tend to raise consumption expectations with wealth and access as important contributing factors. Urbanization affects ecosystems through the transformation of urban–rural linkages (i.e., the urban "ecological footprint"). Large-scale urban agglomerations and extended peri-urban settlements fragment large landscapes and threaten various ecosystem processes through near-complete reliance on importing material goods and unsustainable resource use (e.g., drinking water).

Governance

The last 50 years has brought several major changes in governance that have accelerated or altered environmental change in important ways. Some of the most significant changes over the last 10–15 years are associated with the shift from the centrally planned economies of the Soviet block and China to more open markets and entrepreneurial strategies. Initially, the sudden fall of the Soviet block actually caused a decline in pollution, as inefficient state-run heavy industry was closed. This, however, has now been replaced by renewed economic growth, especially in the energy sector. Currently there is much greater market freedom and opportunity for economic growth in China and the former Soviet Union. The implications of the resultant economic expansion for the environment include an increase in greenhouse gas emissions as well as the contamination of freshwater bodies and estuaries. Significant air pollution and water pollution with associated health problems, which were serious under the old systems, have intensified in China and are reemerging in post-Soviet Russia. Under certain circumstances a dramatic reduction in the provision of ecosystems services may occur as increasingly more resources are appropriated for humankind. In short, the results of these political and institutional reforms will be even more globalization, urbanization, consumption, and trade than otherwise would have occurred, and will be accompanied by the associated impacts on the environment from local to global scales.

Another significant governance shift is the decentralization of environmental management to local government, nongovernmental organizations (NGOs), and the private sector (WRI 2003). This occurred in some countries without the parallel decentralization or transfer of funds and capacity to implement environmental protection. In others the involvement of local people and outsiders has led to more sensitive and legitimate management of the environment (discussed further below).

EVIDENCE OF DECELERATION

As stated, rates of change have markedly accelerated over the past 50 years. Nonetheless, emergent properties of the Great Acceleration have also contributed to decelerating trends in the human–environment relationship. Indicators of decelerating trends have been observed, but it is not clear whether these trends are sustainable within considerable momentum of the Great Acceleration itself.

Indicators

Some indicators of decelerating trends include declining fertility and birth rates, the development of environmental institutions and governance, global threats to human health, technologies that have increased energy efficiency, and value changes that have expanded the range of and incentives for consumption patterns.

Declining Fertility

During the 20th century, global population increased as never before, yet after 1970, the rate of growth slackened from a maximum of almost 2.0% per annum to 1.4% by the year 2000. This occurred despite post-1970 gains in life expectancy and continued reductions in mortality, and can be attributed to rapid declines in fertility: from an average of 5 children per woman in 1955 to 2.65 children per woman in 2005. This decline is significant, especially in east Asia. By

2005, in sub-Saharan Africa as well as in parts of south Asia, fertility decline progressed quite unevenly.

Fertility decline is a complicated issue and may have occurred in different contexts for a variety of reasons. One may, however, safely state that (a) urbanization played a major role almost everywhere; (b) government policy played a modest role in several countries and a strong one in China, where rules imposed in 1978 placed strong penalties on couples who had more than one child; and (c) in many societies and cultures, the emancipation of women, improved health services, and an increase of the economic standard of living played crucial roles. Demographers are confident that fertility decline will continue over the decades ahead, resulting in a world population of about 9 billion in 2050, up from 6.5 billion in 2005, but below earlier projections of more than 10 billion (Lutz 2006).

Improved Environmental Governance

The emergence of environmental governance and institutions has occurred at many levels. For most developed countries, the inflection point is associated with the occurrence of environmental disasters (e.g., mercury contamination at Minimata, Japan, DDT poisoning of birds, and oil spills at Santa Barbara, U.S.) and the emergence of environmental movements in the 1970s, which led to the establishment of environmental agencies and laws to protect water, air, ecosystems, and public health. The 1972 UN Conference on the Environment in Stockholm and the 1992 Earth Summit in Rio are frequently identified as key events that signaled the emergence of environmental policy and sustainable development on national and international agendas. Internationally there were specific attempts to halt environmental degradation through the signing of international treaties on issues such as acid rain, ozone (Montreal Protocol), climate change (UN Framework Convention on Climate Change), and biodiversity (Convention on Biological Diversity).

National governments have initiated environmental management strategies that include aggressive attempts to control air and water pollution. Since the 1970s there has been an evolution of NGOs and civil societies' interest in the environment to influence environmental management.

International environmental organizations and regimes have also played major roles including, but not limited to, UNEP, UNFCCC, as well as the Montreal and Kyoto protocols. The ozone hole, or the complete loss of ozone during late winter and early spring in the lower stratosphere over the Antarctic, came as a total surprise when it was first reported by scientists of the British Antarctic Survey in 1985 (see Box18.4).

Technologies

As with the Great Acceleration, decelerating trends can by driven by changes in technologies, such as improved water treatment, air pollution controls, and the

Box 18.4 The ozone hole.

First reported in 1985 by scientists of the British Antarctic Survey, the ozone hole (i.e., the complete loss of ozone in the lower stratosphere during late winter and early spring over the Antarctic) is due to a combination of natural and anthropogenic feedbacks. The appearance of sunlight after the long polar night combined with cold temperatures allow the formation of ice particles on whose surfaces highly reactive chlorine radicals (Cl and ClO) are formed which destroy ozone. The primary source of these radicals is photodissociation of the chlorofluorocarbons (CFCs). These conditions for rapid ozone destruction ($CFCl_3$ and CF_2Cl_2) are met over Antarctica—the furthest place away from the locations in the Northern Hemisphere where CFCs are actually released into the atmosphere. The destruction of ozone depends on the square of the chlorine content in the stratosphere, which explains the rapidly growing nonlinear destruction of ozone.

Since 1996, CFCs have no longer been produced in the developed world. Somewhat optimistic estimates suggest that the chlorine content of the stratosphere may first return to pre-ozone hole conditions by the middle of 21st century. However, these estimates do not account for the cooling of the stratosphere by growing concentrations of CO_2, which will delay the recovery of ozone. For example, the 2005 hole was larger and deeper than the holes that formed when the discovery was made. However, the situation would be much worse if the Montreal Protocol had not come into force.

de-materialization and de-carbonization of some economies. There are many examples of such technology-driven deceleration: Motor cars have become more fuel-efficient and the amount of emissions per kilometer driven has dropped significantly. Aircraft have also steadily become more fuel-efficient. Many household appliances have been made more energy-efficient. In terms of the production of energy itself, some forms of renewable energy, such as wind power, are now reliable producers of electricity and have been integrated into national power grids in many areas of the world. In some countries, the economy has continued to grow although the amount of physical material flowing through the economy has not; this has led to a partial decoupling of economic growth and environmental impacts.

Improved management techniques can also power the trend toward deceleration. For example, the agricultural industry is continually improving its efficiency. Much less fertilizer is now being used to grow crops, based on an improved understanding of the agro-ecological system. Less water is also being used to irrigate the same amount of cropland, which, when combined with improved fertilizer use, still has a high crop yield (often termed precision farming). Management of traffic in urban areas is also being streamlined by using network approaches to the flow of vehicles, leading to further gains in fuel efficiency.

Governments that have large debt crises also contribute to deceleration via reduced consumption. Their financial burden, coupled with corruption and weak governmental structure, inhibits technological advance; often investments are made in prestige projects without a sense for development problems and the need for science. The drivers behind these contributions to deceleration are negative in terms of socioeconomic development but are likely to be temporary.

Paradoxes within Larger Changes

The Great Acceleration presents various paradoxes, for example, in the ways in which aerosol production has balanced greenhouse gas warming. Increasing atmospheric greenhouse gas concentrations (CO_2, CH_4, N_2O, O_3) exert a radiative climate forcing of 2.7 ± 0.3 w m^{-2}. By the end of the 21st century, this could cause climate warming which, according to the IPCC (2001) under a future scenario, may be as high as 5.8°C on a global average. The observed global average temperature increase of 0.6°C is lower than might be expected for several reasons. First, considerable lag effects are part of the climate system due to various buffering factors. Thus, the climate system is not in equilibrium with additional temperature increases "in the pipeline." Second, aerosols (particulate matter) play a major role in dampening climate warming, as they cool the climate by partially reflecting solar radiation back into space and increase the Earth's albedo (brighter reflectance) by making clouds more reflective. Recent studies emphasizing the role of aerosols cannot, however, be accurately assessed. An analysis by Crutzen and Ramanathan (2003) suggests that about 50% of potential greenhouse gas warming is being masked by aerosol cooling; higher estimates of the aerosol cooling effect have been published (e.g., Anderson et al. 2003; Andreae et al. 2005).

The opposing roles of greenhouse gases and aerosols on the Earth's climate pose a dilemma for policy making. Cleaning up air pollution by reducing aerosol emissions to the atmosphere (which is desired) may actually aggravate global climate warming, if concurrent efforts are not undertaken to reduce CO_2 and other greenhouse gases at an accelerated rate.

An additional paradox can be found in full cost accounting and deceleration. A major part of the problem with environmental degradation lies in the fact that many of the costs associated with economic production are external to the market system. In terms of the environment, the market sends out the wrong signals, and producers and consumers engage in "over-consuming" environmental resources. One way to correct this problem and create the appropriate deceleration of environmental degradation is to "internalize" these external costs. This could be done by taxing environmental degradation (i.e., Pigovian taxes), establishing cap and trade systems, or estimating the true external costs of companies and publishing that information for use by investors (i.e., http://www.trucost.com). For example, organic food production involves much lower external costs than conventional agriculture. If the full external costs of conventional agriculture (water pollution from overuse of chemical fertilizers, pesticides, soil erosion, etc.) were internalized and incorporated into the price, there is no doubt that conventionally grown food would become much more expensive. This could precipiate a significant shift in consumption patterns, from conventional to organic food (Bernow et al. 1998), which would lower environmental externalities and decelerate the loss of ecosystem services from this source. Likewise, an economy-wide effort to internalize environmental externalities would lead to dramatic lessening of environmental impacts and enhance the chances for sustained deceleration.

Sustainability of Deceleration

Are the observable trends in deceleration sustainable? There is an inevitable tension between deceleration trends and the momentum of the Great Acceleration. Signs of global deceleration are indicated by political and social protection of lands and afforestation through the abandonment of unproductive croplands. Timber consumption in the U.S. has declined (for a description of the paradox in forestry management, see Box 18.5). Energy consumption has declined or stabilized in Japan and parts of Europe, but not in other parts of the globe. In addition, freshwater consumption has decreased in Australia and the U.S. As mentioned earlier, the Montreal Protocol for the protection of the stratospheric ozone layer is an example of an institutional response to environmental pressures. What drives these deceleration trends?

It is primarily through the interaction of social behavior with knowledge and the resultant feedbacks where we find the beginnings of sustainable deceleration. For example, access to information on and the ability to perceive and understand environmental consequences have improved in human populations. Establishment of environmental regulations and market-based incentives contribute to the motivation and perceptions of agents and social behavior. Environmental regulations and markets for ecosystem services are growing. There is also the capacity to implement change: new institutions and the ability to invest in clean technologies have increased, for instance, in production and consumption patterns and technological change. The issues, however, are much more deeply rooted. Environmental degradation has consequences for political stability and undermines the production and sustainability of ecosystem goods and services. A glimpse into the future might suggest that the globalization and internationalization of institutions through international policy will also function as future drivers (e.g., the World Trade Organization and technology transfer). In the near term, however, small steps of sustainable deceleration are underway.

SYSTEMS APPROACHES TO DECADAL DYNAMICS

The Great Acceleration is clearly the result of a number of interacting factors, and thus it is useful to identify chains or sequences of events that are part of the dynamics of the system (Figure 18.2). Here we briefly cast the factors discussed above into a systems dynamics framework and focus on the interactions among the factors.

Box 18.5 Forestry in the Pacific northwest.

In May 1991, Federal District Judge William Dwyer issued a landmark decision which declared that the U.S. Forest Service violated the National Forest Management Act by failing to implement an acceptable management plan for the northern spotted owl. His decision prohibited logging across the spotted owl region until the Forest Service implemented an acceptable plan. An injunction that blocked logging in the northern spotted owl habitat affected 17 national forests in Washington, Oregon, and Northern California. The short- and long-term consequences for the rural economy in many areas of the Pacific northwest were devastating. As many as 135 mills were closed, which resulted in an unemployment rate of up to 25% in some small communities. The mill closings affected cutters, loggers, and truck drivers as well as other businesses that provided services to them.

The decision did increase forest density, providing a temporary sustainable habitat for the spotted owl in old growth. However, as a result of over 100 years of fire suppression in this area, these forests were potential tinderboxes.

In 2002, catastrophic wildland fires resulted in the loss of hundreds of thousands of forested hectares in the state of Oregon alone. Although the effective federal block on logging may have salvaged some spotted owl habitat, enactment of a single prescription (e.g., the Bush administration's "Healthy Forest Act") does not and will not provide effective environmental husbandry and management. Conceptually, many different approaches to environmental or human interactions may stabilize the system; however, significant reduction or only one approach may produce fluctuations that ripple throughout human–environment systems, thereby increasing vulnerability and reducing resilience and sustainability.

Resource economists counter that the "collapse" of the small timber economy coincided with one of the biggest economic booms in the history of the northwest, which was almost entirely concentrated in urban/suburban areas as a result of a high-tech boom. High-tech workers immigrated into the region, creating an increase need for the construction and service industries. It is difficult to ascertain, however, whether this was the direct result of policy and human intervention on forestry practices or simply a fortuitous coincidence for the greater metropolitan areas. It is, however, highly likely that the remaining logging communities—which stretch from northern California to Washington and remain hamstrung with regard to their profession—do not share in the economists' perception.

We propose that a critical causal factor in the Great Acceleration was the change in knowledge systems and worldviews traceable back to the Enlightenment, an era typified by rational thinking and which provided the basis for the technological explosion of the 20th century. This change brought about improvements in hygiene and health care, which supported population growth, as well as new technologies for accessing energy sources, especially fossil fuels. Population growth and cheap, convenient energy led to the development of a powerful production and consumption system. These three factors—population, energy, and production/consumption systems—form a feedback loop, with

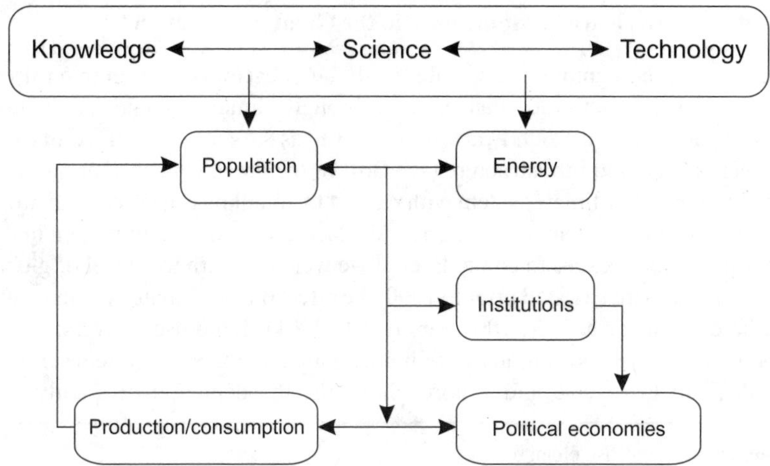

Figure 18.2 A systems approach to understanding the linkages, interactions, and feedbacks in human–environment relations. Note that there are no specific positive or negative feedback indicators between the representative states. There can exist, at any given time, a simultaneous increase or decrease in, e.g., production and consumption that can impact both population and political economies.

a constantly improving knowledge base supporting the acceleration of the feedback loop.

One additional feature is required to complete this first-order picture of the system driving the Great Acceleration: the change in political economy, with the liberalization of national and world markets, the integration of parts of the Third World into the global economy, and the abandonment of centralized economic systems in the former U.S.S.R. and China. These changes were associated with new institutions at the local, national, and global scales.

The powerful feedback loops embedded in Figure 18.2 have led to significant impacts on the natural environment (Boxes 18.4, 18.5). However, the system has also the potential to support deceleration, in terms of the importance of knowledge/science/technology to act as an accelerator or brake. For example, the perception that human activities can cause widespread and significant environmental changes has been around for a while, at least since the middle of the 19th century (e.g., Marsh 1874/1970). Currently, we are developing a new understanding of the place of humans in the Earth system, with enhanced scientific activities to understand global change and detailed assessment projects to communicate this new understanding to decision makers and the public. These trends in perception, coupled with the types of technological advances described above, may lead to a future, substantial deceleration of human impacts on the Earth system. The fundamental question regarding the system depicted in Figure 18.2 is whether there is a strong enough self-regulated feedback loop in the system to slow and eventually reverse the momentum of the Great Acceleration.

What Is the Evidence of Momentum in the Great Acceleration?

Tremendous momentum is associated with the Great Acceleration in terms of the impacts and activities that emerge through climate change, population growth, land use, and groundwater usage as well as the sunk-cost effects of capital, technology, and infrastructure (see Box 18.6). There is inertia, or momentum, built into the climate system with respect to greenhouse gas accumulations because greenhouse gases emitted over the last few decades will remain in the atmosphere for decades to come. Even if we were to return to a level of global CO_2 emissions that existed prior to 1990, the effect on the climate system would not be evident for at least 100 years (IPCC 2001). Land-use conversion has long-term legacies as well: losses of biodiversity and other ecosystem services are difficult to reverse and restore. Similarly, the depletion of quality and

Box 18.6 Land-use change.

As long as humans have controlled fire and domesticated plants and animals, they have cleared forests to extract a higher value from the land. About half of the ice-free land surface has been converted or substantially modified by human activities over the last 10,000 years. Undisturbed (or wilderness) areas represent 46% of the Earth's land surface. Although forests covered about 50% of the Earth's land area some 8000 years ago, they occupy only 30% today. Cropland has increased globally from an estimated 300–400 million ha in 1700 to 1500–1800 million ha in 1990. Finally, pasture lands for grazing have increased from around 500 million ha in 1700 to around 3100 million ha in 1990.

Concerns about transformations of the land surface emerged in the research agenda on global environmental change several decades ago, when it was realized that land-surface processes influence climate: Land-cover change modifies surface albedo and thus surface–atmosphere energy exchanges; terrestrial ecosystems are sources and sinks of carbon and thus affect the global climate via the carbon cycle; and land cover controls the contribution of local evapotranspiration to the water cycle.

A much broader range of impacts of land-use/cover change on ecosystem goods and services has now been identified. Of primary concern are impacts on biotic diversity worldwide, soil degradation, and the ability of biological systems to support human needs. Land-use/cover changes also determine, in part, the vulnerability of places and people to climatic, economic, or sociopolitical perturbations. When aggregated globally, land-use/cover changes significantly affect central aspects of Earth system functioning. All impacts are not negative though, as many forms of land-use/cover changes are associated with continuing increases in food and fiber production, in resource use efficiency, as well as in human wealth and well-being.

Historically, humans have increased agricultural output mainly by bringing more land into production. The period after 1960 witnessed a decoupling between food production increase and cropland expansion. The near doubling in world food production from 1961 to 1996 was associated with only a 10% increase of land under cultivation; however, large increases in the amount of irrigated cropland and in the global annual rate of nitrogen and phosphorus fertilization also resulted.

quantity of fossil groundwater contributes to the momentum of the Great Acceleration.

Population also affects momentum as the delayed impacts of changes in fertility and mortality take decades to be realized in terms of population growth rates. Similarly, the costs of investment in current infrastructure—large dams and energy plants, technological "lock-ins" such as transportation, production, and capital investments of polluting technologies—cannot easily be reversed. There is inertia with how accounts and investments are made because of loans and planning horizons, which often last 30 years or more. The paradigm of growth and the culture of consumption continue to be dominant and are difficult to change. Value systems emphasizing consumption as an indicator of well-being (e.g., GDP) are still working their way into the developing world through the globalization of communications. These and other indicators of the multiple, interacting factors that contribute to the continued momentum of the Great Acceleration help position us to ask whether the human–environment system can self-regulate at the global scale.

Observations of Self-regulation in the Human–Environment System

This section begins a transition from understanding the indicators and factors explaining the Great Acceleration and subsequent decelerating trends to the implementation of a systems approach to the human and environment relationship which might involve self-regulation. Are there examples of rapid changes in human values and activities that are directly attributable to environmental degradation, which show the capacity for intentional or accidental self-regulation? One example of a rapid feedback from energy, institutions, and political economies can be found in the Netherlands. In the 1960s, local air pollution from the burning of coal—the primary energy source in the Netherlands—posed a major health threat to the population. In the early 1970s, large indigenous reserves of natural gas were discovered, and within five years, all households and industry switched from coal to natural gas (see Figure 18.2).

There are theoretical arguments concerning self-regulation, such as the Gaia hypothesis (Lovelock 1979), the market-based arguments of neoclassical economics, the population collapse theories of the Malthusians, and even Marxist theories that imply controlled exploitation to ensure that capitalism does not destroy the resources needed to sustain it (O'Connor 1998). The Gaia hypothesis suggests that living organisms must regulate their planetary environment; otherwise the forces of physical and chemical evolution will render it uninhabitable. Arguments in favor of a self-regulating, human–environment system at the global scale are countered with visions of overshoot and collapse (e.g., Meadows et al. 1972) and with resource constraints, such as running out of fossil fuels. Neoclassical economics would suggest that once the costs of environmental degradation outweigh the benefits, alternatives that are more environmentally benign will emerge.

The primary difference between the positive and negative theoretical constructs is the effectiveness of the feedbacks in the system. For instance, if demographers are correct and global population stabilizes, and this is linked to improved environmental technologies in 50 years, can overshoot and collapse be offset? Is the small niche of decelerating trends powerful enough to expand and provide a counterbalance to the Great Acceleration?

There is scant evidence to support the notion that we have reached a point where existing decelerating trends suggest a path of self-regulation in global-scale human–environment interactions. The magnitude and number of accelerating trends in environmental degradation are simply too great. The human–environment relationship also encumbers large institutional processes with formidable inertia. The healing of the ozone layer, although slow, is an example of deceleration. It may be, however, that a systems collapse at a regional level (e.g., sub-Saharan Africa) may trigger an inoculum in the human–environment system and contribute a vital feedback to sustainability. Alternatively, early signs of deceleration of the human impact on the environment may be sustained in coming decades.

Self-regulation is assisted by several factors including environmental monitoring, increased awareness, and better institutions (Lambin 2005). First, the ability to gather, process, understand, and communicate information on the state of the environment and on alternative resource management practices is constantly improving, thanks to, for example, global change scientific programs, Earth observation satellites, progress in simulation modeling of complex systems, the increasing attention paid by media to the issue, and the penetration of sustainability concerns into the policy maker and corporate worlds. This enables society to make increasingly accurate diagnoses of the causes and impacts of, and solutions to, environmental change. Policy makers also become aware that environmental degradation undermines the ability to maintain high standards of living and has consequences for political stability. Second, the sources of behavior of agents—environmental attitudes, deeply held values, and knowledge systems but also institutions—are being gradually transformed to integrate sustainability concerns more effectively. This takes place via new environmental regulations, market-based instruments applied to ecosystem services, and various social movements. Third, the capacity to implement change in environmental management via the provision of appropriate physical, technical, and institutional infrastructure is also growing as a result of research and development efforts: new environmental institutions, clean technologies, and slow changes in consumption patterns.

Changes in environmental management to implement sustainable development policies may be hampered by conflicts of interest groups. In summary, we do have some evidence of decelerating trends. It is not yet clear, however, whether the decelerating trends currently observed in some developed countries will become global in scope or sustainable for a long period of time.

EARTH SYSTEM CONTEXT

Placing the human–environment relationship in an Earth system context requires an understanding of the feedbacks, nonlinearities, and vulnerabilities. How does this changing human–environment relationship affect the Earth system, and at which critical points? Representative examples of how changes in the human–environment relationship have either amplified or dampened Earth system processes provide some insight into these intersections.

Intersection of Environmental Variability/Change and Socioeconomic Change

Amplifying Effects

The socioeconomics of fisheries are highly dependent on climate. Therefore, they provide good regional examples of amplification of Earth system processes by human activities. McEvoy (1986) provides an insightful case study of the 1940s collapse of the California sardine fishery. In McEvoy's conception, fisheries consist of three strongly interacting elements: (a) law—a system of social control; (b) economy— production, price, and profit; and (c) nature—the ecosystem that produces the target fish stocks.

From this perspective, the health of a fishery is indicative of healthy interactions between each of the three component parts. For the California sardine fishery, a productive economy consisting of local fishers, large corporate processors, and a network of distribution markets (primarily for chicken feed) developed around the intense commercial harvest of highly abundant sardines from the California Current. With abundant and productive sardine populations in the late 19th and early 20th century came a rapid expansion of California's sardine economy. Harvest capacity soon rose to match the peak productivity of the sardine ecosystem. What was not well understood in the early 20th century was the sensitivity of sardine productivity to decadal variations in climate and the environmental conditions in the California Current system. In the 1940s, what had been favorable environmental conditions for high sardine productivity were replaced with unfavorable environmental conditions, and sardine productivity dropped significantly. When the regulatory agency (the California Department of Fish and Game) started to observe signs of reduced productivity in the sardine population, recommendations for reduced harvests were met with intense opposition from politically influential economic stakeholders in the sardine industry. The breakdown in the three-part fishery system was soon followed by intense overfishing, which resulted in a rapid depletion of the sardine population; this led, in turn, to an economic collapse of the formerly productive sardine industry. This regional-scale example gained global significance when the infrastructure (purse seiner fleet[4] and the onshore processing facilities) from California's suddenly defunct sardine fishery was largely relocated to Peru in the 1950s. By the 1960s, the Peruvian anchoveta fishery had become the world's largest commercial fishery. Here, too, the consequences of an unforeseen collision between the dynamic changes in the ecosystem and a slowly responding economy and regulatory system were observed, when the Peruvian anchoveta fishery collapsed in the early 1970s. Remnants from the collapse of the California sardine fishery are still evident in the California Current ecosystem today; the slow recovery in the abundance of California sardines has yet to reach productivity levels of the early 20th century.

From a terrestrial perspective, drought, drying, and their interactions with political economies have had significant impacts on Earth system dynamics, particularly in strongly seasonal arid and semiarid regions of the tropics and subtropics (e.g., the Sahel, Australia, and the southwestern U.S.; see Lambin et al., this volume). Experience in Australia and elsewhere suggests that land deregulation is greatest during and immediately after drought, particularly in areas that are heavily settled. More recently it has become apparent that people find it difficult—if not impossible—in the short term to distinguish between drought and drying. When drying rather than drought occurs, production systems have to be recalibrated, if they are not to collapse. Further terrestrial examples include tropical deforestation, leading to drying and further loss of forest and widespread biomass burning on peatlands, such as occurred in Indonesia during the 1997–1998 ENSO as well as in 2002 and 2004.

Urban examples include severe heatwaves, such as in France in 2003, following which a possible increase in the use of air conditioners in the future could lead to increased energy consumption and thus to increased emissions of CO_2, which in turn could contribute to a further temperature increase globally. Another example concerns the switch from coal to natural gas energy systems in London.

An example from the cryosphere suggests that the increased melting of glaciers due to higher temperatures alters runoff dynamics with consequences for water resources. As populations increase water use and freshwater resources become inadequate, desalinization efforts will have to increase. This will create higher demands for more energy to secure the water, and thus increase greenhouse gas emissions, leading to future warming (see Box 18.7).

Damping Effects

One class of damping effects is the disappearance of opportunity and space. For instance, building dams eventually slows down because all of the best sites have been taken. The transformation of prairie and grassland to cropland becomes

---

4 Throughout the last half century, the purse seiner has been the most commonly used net on West Coast fishing vessels. Purse seiners capture surface fish by encircling them with a long net and drawing or "pursing" the bottom of the net.

Box 18.7 The melting cryosphere.

Glaciers, snow, and permafrost are sensitive indicators of climate change because, in many places, they are close to the temperature at which they will melt. Durations and depths of snow are especially important because snow responds annually and seasonally to any climate oscillation and can be precisely identified and monitored by remote sensing methods.

The cryosphere exists in all climatic zones, from the polar and high-latitude regions with large ice- and snow-covered areas to lower latitudes with relatively small spots in mountain areas. The changing cryosphere can be measured and compared because similar methods and techniques are used all over the different climatic zones; this is not possible with areas that contain highly diverse vegetation patterns.

Regional examples of the melting cryosphere in the second half of the 20th century show the following characteristic features:

- Arctic region: rapid disappearance of sea ice.
- Western U.S.: 80% of the water used for agricultural, industrial, and domestic purposes originates from high-elevation winter–spring snowpacks.
- Andes: extreme retreat of glaciers in Bolivia is creating serious problems for irrigation in the Altiplano and water supply for La Paz.
- Himalayas: extreme retreat of glaciers led to the formation of ice and moraine dammed lakes; outbreaks of dammed lakes could become a disaster for the populated adjacent valleys.
- Alps: in the extremely warm summers of 1947 and 2003, only those rivers with a glacial regime had a high enough flow to support full hydroelectric power production; all other rivers experienced an extremely low flow, which negatively impacted not only agriculture and domestic use but also the groundwater tables.

self-limiting because most potentially arable lands are already under cultivation. When humans run out of one resource, the system can be either dampened or amplified through intensification of resource management. An example of such an amplifying effect is the combustion of coal if hydroelectric power becomes limiting or not sustainable.

The Kuznets curve, which gives a relationship between economic wealth and environmental damage, is an empirical representation of a form of damping in human–environment systems (Figure 18.3). This curve shows increasing environmental damage with economic wealth up to a point at which the curve turns over and, as economic wealth increases further, a society can then afford to invest in environmental protection, and environmental damage subsequently decreases. The Kuznets curve, however, is scale dependent and somewhat institution dependent and does not seem to work for all pollutants. The rapid accumulation of wealth without institutional regulations excludes environmental benefits (or benefits have not been observed). There is no consensus on whether this is effective at large scales, however, and in various socioeconomic contexts.

Competitiveness and the free market interacting with climate variability and enhanced degradation are a final example of damping effects. In the African

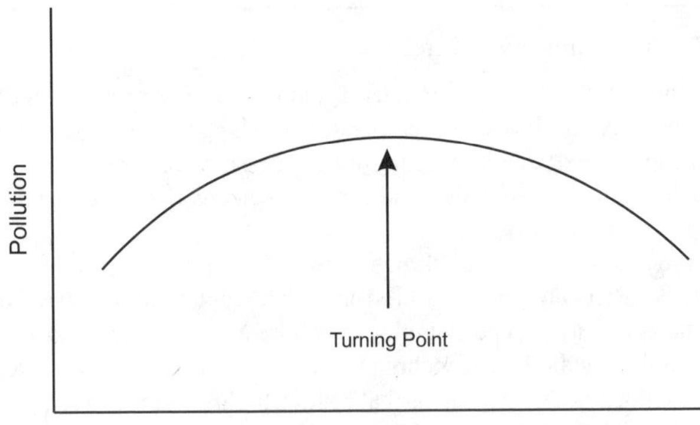

Figure 18.3 Example of an environmental Kuznets curve: an inverted U-shaped relationship between pollution and per capita income (Cole 1999).

Sahel, diversification through a mixture of livestock production and cropping provides increased flexibility in a market economy where price varies with climatic conditions (increased price of food crops and decreased price of livestock during drought years, and vice versa in wet years).

Nonlinearities

There are many examples of surprises and nonlinearities in the Earth system. Discovery of the ozone hole itself came as a complete surprise, because the ozone layer in the stratosphere was thought to be very stable. The chemical reactions that drive the formation of the ozone hole show strong nonlinearities, with some rates related to the fourth power of the concentrations of the reactants. This complex chemistry is also typified by teleconnections and lag effects, with long-term implications up to the end of this century, at least. In addition, a near disaster was averted, during the development phase of refrigerants and propellants (in the 1920s), when chlorine compounds were chosen instead of bromine compounds. Although bromine was more efficient than chlorine, it was also more expensive, and thus, an administrative decision was taken in favor of chlorine—a serendipitous decision as it turns out, for bromine compounds are also about 100 times more reactive in the atmosphere than chlorine ones. Had bromine been used, the results would have been much worse: there would have been a much deeper ozone hole over the entire planet during all seasons.

Human awareness was also nonlinear with respect to acid rain (detected in 1890). Acidification of lakes poses another example of nonlinearity, due to the buffering capacity. Basically, as the H_2SO_4 loading on a lake increases, the pH goes down but in a highly nonlinear way. Once the buffering capacity is exceeded, the drop in pH is rapid. Effects of acid can be highly differentiated with dying forests as secondary effect. Although remediation through reduced sulfur emissions appears to have ameliorated the problem, lingering problems exist with lakes that are still acidic. Lessons learned are dependent on a range of socioeconomic factors. China is rich in coal reserves that are high in sulfur content. There is the potential for China to use its coal for energy, but this will result in the same problem experienced by North America and Europe through lake acidification and forest mortality. The current institutional pressures for carbon sequestration, however, may offset the pressure for increased burning of coal. In addition, if China and India can obtain cleaner fuels (e.g., as oil and natural gas) at reasonable cost, it may be possible to minimize a potentially severe acid rain problem in Asia.

The nexus between rapid developments in science and technology coupled with fundamental cultural belief systems is arguably the greatest potential nonlinearity in the contemporary human–environment system. For example, genomics can have major implications for human–environment systems through rapidly developing knowledge coupled with feedbacks from fundamentalist elements of society. Genomics could have the potential for major gains in human health, but research could also be blocked by those in societies holding strong religious beliefs against such work. Also, endocrine disruptors could be leading to the observed decrease in human sperm counts, and thus partly to drops in fertility. This trend, however, may be counteracted by other developments in the biosciences. The Green Revolution offers another example, where scientific and technological developments led to rapid gains in food production but also to vastly increased consumption and pollution. Environmental groups have opposed much of this, arguing for alternative means of food production.

Signs of Dangerous Thresholds in the Human–Environment System

The relationship between humans and the environment may be approaching dangerous and irreversible thresholds. Some of the indicators include inequality of information, increased climate sensitivity, and degradation of ecosystem services including water resources and energy transitions (see Box 18.8).

1. *Inequality, wealth, and the information revolution*. There is a growing inequality between the rich and the poor coupled with the information revolution. The poor are now more aware that they are much poorer than the wealthy, and this has become a source of civil unrest and resentment as well as possible conflict.
2. *Increase in climate sensitivity*. It appears as though the climate system may be more sensitive to human forcing than earlier thought. One of the main factors is increasingly long temporal lags in impact assessment and political response, leading to the sense that we have already crossed over a threshold to irreversibility. It can be argued that the feedback to the atmosphere is too slow from political systems and institutions. In addition,

Box 18.8 Climate sensitivity and human activities.

Climate sensitivity is defined as the increase in global mean temperature that can be expected from a doubling of atmospheric CO_2 concentration above the preindustrial level. Climate sensitivity coupled with the assumed trajectory of greenhouse gas emissions through this century has largely determined the projected global mean temperature in 2100. The range of projected temperatures in 2100 has remained remarkably constant over the past decade, between about 1.4° and 5.8°C (IPCC 2001). However, there are a number of lines of recent evidence that challenge current thinking on climate sensitivity and suggest that climate may be more sensitive to human perturbation than earlier perceived (Schellnhuber et al. 2006):

- *Inertia in the climate system*: There is now good evidence (Barnett et al. 2005) that significant amounts of heat are being stored in the surface waters of the ocean. This implies that even if greenhouse gas emissions could be reduced sharply over a short period of time, the climate would continue to warm for decades into the future.
- *Model parameterization*: Many of the critical processes in General Circulation Models (GCMs) are characterized by parameters that are quantified with "best-guess" values based on the current level of process understanding. Sensitivity studies, in which several of these parameters are varied within plausible limits, give a broader range of sensitivities than previously acknowledged. The low end of the range (about 1.5°C) is about the same as before; there is a significant "tail" at the high end of the range with sensitivities much higher than ca. 6°C (Schellnhuber et al. 2006).
- *Interactive carbon cycle feedbacks*: Including an interactive carbon cycle, that is, carbon dynamics that respond to the changing climate, increases the amount of CO_2 in the atmosphere in 2100 over the scenario projection by 50 to 200 ppm (Friedlingstein et al. 2006).
- *Aerosols*: Aerosol particles in the atmosphere have complex radiative properties but in general act to cool the Earth's surface by scattering some of the incoming radiation. Quantitative estimates of this cooling effect suggest that aerosols have masked up to half of the warming that we should have experienced for the current level of greenhouse gases in the atmosphere. This means that as aerosol emissions are reduced over the coming decades, in response to the health impacts of local and regional air pollution, there will be an enhanced rate of increase of warming at the Earth's surface (Andreae et al. 2005).
- *Cryospheric change*: Melting sea ice and decrease of winter snow cover can accelerate warming by changing the albedo of the Earth's surface. A recent assessment of the impacts of climate change in the Arctic show that these processes are already under way and could accelerate in the second half of this century (ACIA 2004).

Taken together, these factors provide a strong argument that climate sensitivity is probably higher than we have assumed over the past decade. By contrast, it is difficult to find counterarguments or feedback processes which suggest that climate sensitivity is actually lower than we currently believe.

it is conceivable that the Greenland ice sheet is already "lost": this translates to about 6 or 7 meters of sea-level rise coupled with atmospheric feedbacks and increased freshwater inputs to the oceans (implying a potential trigger to flip the thermohaline circulation). This and other nonlinear thresholds are highlighted by Clark et al. (2004) (Figure 18.4).

3. *Degradation of ecosystem services*: The Millennium Ecosystem Assessment provides strong evidence that there has been significant loss of biodiversity, keystone species, and ecosystem services with unknown consequences. The MEA states (2005, p. 1) that "there is *established but incomplete* evidence that changes being made in ecosystems are increasing the likelihood of nonlinear changes in ecosystems (including accelerating, abrupt, and potentially irreversible changes) that have important consequences for human well-being. Examples of such changes include disease emergence, abrupt alterations in water quality, the creation of 'dead zones' in coastal waters, the collapse of fisheries, and shifts in regional climate." We may soon reach a threshold where ecosystem collapse is imminent at local and regional scales. It is unclear what (not if) the consequences will be for the continuity of the human–environment system into the future. A major concern is the dynamics of biogeography and future transition to the new ecosystems that cannot provide the services required by human societies.
4. *Water resources*: The number of people living under water stress is expected to rise over the next several decades, especially in arid and

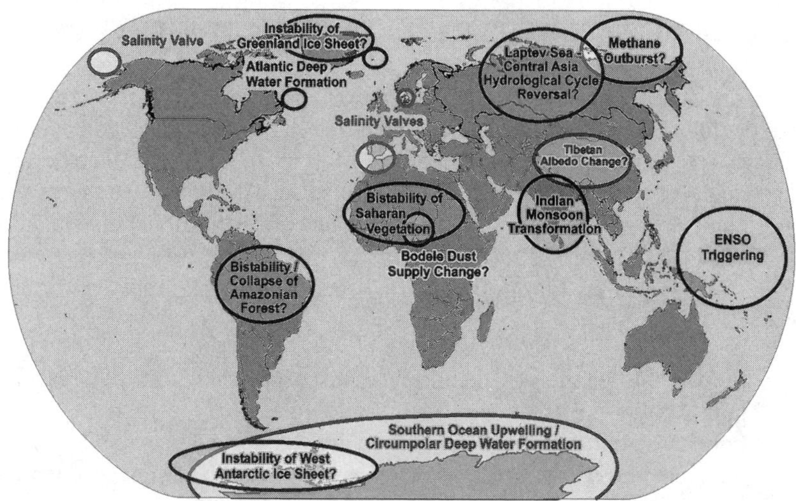

Figure 18.4 Critical region analysis of hot spots or switch and choke points; an early attempt at identifying parts of the Earth where changes at the regional scale can cause significant changes in the functioning of the Earth system as a whole (Clark et al. 2004).

semiarid systems (ca. 40% of land surface) (Vörosmarty et al. 2000). Changes in water supply can show threshold/abrupt change behavior, for example, when changes in rainfall are amplified by 2–3 times in terms of runoff. This happened in Perth, Australia, around 1970 when a 15% drop in rainfall resulted in a 50% reduction in the water flow into the city's reservoirs, necessitating construction of a desalination plant (Steffen et al. 2006).

5. *Energy Transitions*: Over the past 100 years, energy systems have undergone a transition from solid fuels such as coal to oil and gas, as well as from the distribution of high-quality processed fuels (e.g., liquids, gases and electricity) to dedicated energy infrastructure grids (Flavin and Lenssen 1994). However, the next 100 years or so are unlikely to unfold as a simple extrapolation of these past trends (Grübler 2004). Although there is considerable agreement that high-quality processed fuels (e.g., liquids, gases, electricity) supplied to the consumer via energy grids will continue to grow, there is much uncertainty surrounding energy projections to 2100. The link between income level and quantity and type of energy is changing, large differences in regional development pathways exist now and will likely increase, and transitions toward more efficient and environmentally friendly energy sources are far from complete. Technology and innovation will be key, but are highly uncertain, leading to the possibility of surprises and sharp shifts in the energy transition process.

A 20th-CENTURY PERSPECTIVE ON THE FUTURE

Since one goal of understanding human–environment relationships is to predict into the future, we test our understanding of the last few decades by imagining that we were based in 1905 and trying to forecast the human–environment relationship over the next century. One notable difference in 1905 is that there were large numbers of people who still had little contact with the global system. This level of isolation is rare in 2005.

In 1905, if you were European or North American, you might see the world as characterized by:

- Remarkable technological and economic progress with very few misgivings about progress.
- Political stability based on colonial administration and a sense that among the great powers, war had become obsolete because national economies had become too interconnected.
- Modest anxiety about urban pollution and landscape change in Europe, America, and Australia but very little conservation awareness.

- For the great majority of people in industrial countries—workers, bosses, professionals—pollution was seen as an inevitable consequence of what makes employment and a comfortable lifestyle.
- A power system operated by elites that did not include women, developing countries, or other marginal groups.

Technologies that are common in 2005 were almost inconceivable in 1905, including contemporary computing and electronics, biotechnology, and space flight. For many people, decolonization, universal suffrage, and mobility were unattainable and the devastations of World War I and II as well as the Cold War were unimaginable.

The message we learned by placing ourselves in 1905 is that the human part of the Earth system may be highly unpredictable. We doubt that a 1905 working group could have predicted the Great Acceleration, because in 1905, this would have been a science fiction scenario. For example, we now understand that in reality, there is an exponential decay of reliability of predictions with time. This is because there is an increasing probability of high impact/low probabilistic events with time. The least predictable process can feedback strongly in a nonlinear fashion to "more" predictable processes in an unpredictable manner. The suggestion for providing projections of the future is not to predict but rather to produce scenarios. Similarly, understanding recent decadal events through accelerating and decelerating interactions of the human–environment system provides some hindsight and insight for future projections and scenarios.

We propose a hierarchy of predictability. Biophysical processes (incoming solar radiation, ocean circulation, Arrhenius functions) are probably the easiest component of the human–environment system to predict, followed by population/demographics, technological and scientific change, and geopolitical change. Values and social movements are the least predictable variables in the human–environmental equation.

The remarkable technological advances of the Great Acceleration exacerbated the momentum and inertia in the human–environment system. Tension between deceleration and inertial or momentum processes make it difficult to offset the large machine of global change. At the same time, observations underlying nonlinearities, discontinuities, and thresholds in the biophysical world make system predictability almost impossible. We begin the 21st century in a very volatile situation.

REFERENCES

ACIA (Arctic Climate Impact Assessment Report). 2004. Arctic Council and International Arctic Science Committee. http://www.acia.uaf.edu/

Anderson, T.L., R.J. Charlson, S.E. Schwartz et al. 2003. Climate forcing by aerosols: A hazy picture. *Science* **300**:1103–1104.

Andreae, M.O., C.D. Jones, and P.M. Cox. 2005. Strong present-day aerosol cooling implies a hot future. *Nature* **435**:1187–1190.

Ausubel, J. 2002. Maglevs and the vision of St. Hubert. In: Challenges of a Changing Earth: Proc. of the Global Change Open Science Conf. Amsterdam, July 10–13, 2001, ed. W. Steffen, J. Jäger, D. Carson, and C. Bradshaw, pp. 175–182. Berlin: Springer.

Barnett, T.P., D.W. Pierce, K.M. AchutaRao et al. 2005. Penetration of human-induced warming into the world's oceans. *Science* **309**:284–287.

Bernow, S., R. Costanza, H. Daly, and R. DeGennaro. 1998. Ecological tax reform. *BioScience* **48**:193–196.

Boserup, E. 1965. The Conditions of Agricultural Growth. London: Allen and Unwin.

Clark, W.C., P.J. Crutzen, and H.J. Schellnhuber. 2004. Science for global sustainability: Toward a new paradigm. In: Earth System Analysis for Sustainability, ed. H.J. Schellnhuber, P.J. Crutzen, W.C. Clark, M. Claussen, and H. Held, pp. 1–28. Dahlem Workshop Report 91. Cambridge, MA: MIT Press.

Cole, A. 1999. Limits to growth, sustainable development, and environmental Kuznets curves: An examination of the environmental impact of economic development. *Sustain. Dev.* **7**:87–97.

Commoner, B. 1971. The Closing Circle: Nature, Man, and Technology. New York: Knopf.

Costanza, R., W. Steffen, L. Graumlich, K. Hibbard, and D. Schimel. 2005. *Glob. Change Newsl.* **64**:19–22.

Crutzen, P.J., and V. Ramanathan. 2004. Atmospheric chemistry and climate in the Anthropocene: Where are we heading? In: Earth System Analysis for Sustainability, ed. H.J. Schellnhuber, P.J. Crutzen, W.C. Clark, M. Claussen, and H. Held, pp. 265–292. Dahlem Workshop Report 91. Cambridge, MA: MIT Press.

Ehrlich, P.R. 1968. The Population Bomb. New York: Ballantine.

EPICA (European Project for Ice Coring in Antarctica). 2004. Eight glacial cycles from an Antarctic ice core. *Nature* **429**:623–628.

FAO (Food and Agricultural Organization of the United Nations). 2000. The State of the World's Fisheries and Aquaculture. Rome: UN.

Flavin, C., and N. Lenssen. 1994. Power Surge: Guide to the coming energy revolution. Washington, D.C.: Worldwatch Institute.

Friedlingstein, P., P. Cox, R. Betts et al. 2006. Climate–carbon cycle feedback analysis, results from the C^4MIP model intercomparison. *J. Climate*, in press.

Globalview–CO_2, 2005. Cooperative Atmospheric Data Integration Project. http://www.cmdl.noaa.gov/ccgg/globalview/co2/

Grübler, A. 2004. Transitions in energy use. In: Encyclopedia of Energy, ed. C.J. Cleveland, vol. 6, pp. 163–177. Amsterdam: Elsevier.

IPCC (Intergovernmental Panel on Climate Change). 2001. Climate Change 2001: The Scientific Basis. Cambridge: Cambridge Univ. Press.

Lambin, E.F. 2005. Conditions for sustainability of human–environment systems: Information, motivation, and capacity. *Glob. Env. Change* **15**:177–180.

Liverman, D.M. 2004. Who governs, at what scale, and at what price? Geography, environmental governance, and the commodification of nature. *Ann. Assn. Am. Geogr.* **94**:734–738.

Liverman, D.M., R.G. Varady, O. Chavez, and R. Sanchez. 1999. Environmental issues along the United States–Mexico Border: Drivers of change and responses to citizens and institutions. *Ann. Rev. Energy Env.* **24**:607–643.

Lovelock, J.E. 1979. Gaia: A New Look at Life on Earth. Oxford: Oxford Univ. Press.

Lutz, W. 2006. Global demographic trends, education and health. In: Global Environmental Change and Human Health, ed. M.O. Andreae, U. Confalonieri, and A.J. McMichael. Rome: Pontifical Acad. of Sciences.

Malthus, T.R. 1803/1992. Essay on the Principle of Population. 5th ed. Cambridge: Cambridge Univ. Press.

Marsh, G.P. 1874/1970. The Earth as Modified by Human Action: A New Edition of Man and Nature. St. Clair Shores, MI: Scholarly Press.

McEvoy, A.E. 1986. The Fisherman's Problem: Ecology and Law in the California Fisheries, 1850–1980. Cambridge: Cambridge Univ. Press.

McNeill, J.R. 2000. Something New Under the Sun: An Environmental History of the Twentieth-Century World. New York: Norton.

McNeill, J.R. 2005. Modern Global Environmental History: A Turbulent and Dramatic Scenario. IHDP – Update. *Newsl. IHDP Glob. Env. Change* **2**:1–3.

MEA (Millennium Ecosystem Assessment). 2005. Ecosystems and Human Well-being: Synthesis. Washington, D.C.: Island. http://www.millenniumassessment.org/

Meadows, D.H., D.L. Meadows, J. Randers, and W.W. Behrens III. 1972. The Limits to Growth: A Report for The Club of Rome's Project on the Predicament of Mankind. New York: Universe.

Mosier, A., J.K. Syers, and J.R. Freney, eds. 2004. Agriculture and the Nitrogen Cycle. Scope 65. Washington, D.C.: Island Press

O'Connor, J. 1998. Natural Causes: Essays in Ecological Marxism. New York: Guilford.

Petit, J.R., J. Jouzel, D. Raynaud et al. 1999. Climate and atmospheric history of the past 420,000 years from the Vostok ice core, Antarctica. *Nature* **399**:429–436.

Pfister, C. 1995. Das 1950er Syndrom. Bern: Haupt.

Reid, W.V., and K. Miller. 1989. Keeping Options Alive: The Scientific Basis for Conserving Biodiversity. Washington, D.C.: World Resources Institute.

Sanchez, R.A. 2002. Governance, trade, and the environment in the context of NAFTA. *Am. Behav. Sci.* **45**:1369–1393.

Schellnhuber, H.J., W. Cramer, N. Nakícenovíc, T. Wigley, and G. Yohe, eds. 2006. Avoiding Dangerous Climate Change. Cambridge: Cambridge Univ. Press.

Smelser, N.J., and P.B. Baltes, eds. 2001. International Encyclopedia of the Social and Behavioral Sciences. Oxford: Elsevier.

Smil, V. 2001. Enriching the Earth. Cambridge, MA: MIT Press.

Steffen, W., G. Love, and P. Whetton. 2006. Approaches to defining dangerous climate change: A southern hemisphere perspective. In: Avoiding Dangerous Climate Change, ed. H.J. Schellnhuber, W. Cramer, N. Nakícenovíc, T. Wigley, and G. Yohe. Cambridge: Cambridge Univ. Press.

UN (United Nations). 1993. Agenda 21: A Blueprint for Action for Global Sustainable Development into the 21st Century. United Nations Conf. on Environment and Development (UNCED), Rio de Janeiro, June 3–14, 1992. New York: UN.

UNFPA (United Nations Population Fund). 2001. The State of the World Population. Footprints and Milestones: Population and Environmental Change. UN: New York.

Vitousek, P.M. 1994. Beyond global warming: Ecology and global change. *Ecology* **75**:1861–1876.

Vörösmarty, C.J., P. Green, J. Salisbury, and R.B. Lammers. 2000. Global water resources: Vulnerability from climate change and population growth. *Science* **289**: 284–288.

Worster, D. 1986. Dust Bowl. New York: Oxford Univ. Press.

WRI (World Resources Institute). 2003. World Resources 2002–2004: Decisions for the Earth: Balance, Voice, and Power. Washington, D.C.: World Resources Institute.

The Future

19

Scenarios

Guidance for an Uncertain and Complex World?

Bert J. M. de Vries

Netherlands Environmental Assessment Agency (MNP), 3720 BA Bilthoven, The Netherlands and Copernicus Institute for Sustainable Development and Innovation, Utrecht University, 3508 TC Utrecht, The Netherlands

ABSTRACT

Scenarios are novel tools to explore an increasingly uncertain and complex world. Their increasing use reflects the limitations of the control paradigm as well as the rise of democratic pluralism in values and behavior. In conjunction with simulation models, the scenario approach has been applied in several large global change projects. To draw lessons for future scenario projects, this chapter reviews two projects: TARGETS and SRES. To improve the next round of global change scenarios, suggestions are made. First, we must acknowledge the need to deal adequately with scientific knowledge and the insights from a branch like "complexity science." Second, people's values must be taken explicitly into consideration. Simulation games and policy exercises are helpful in this respect.

INTRODUCTION

Having been asked to write about scenarios, I interpret this within the focal question for the 96th Dahlem Workshop: How can we best use human–environment systems history and models to generate plausible future scenarios that can integrate with various policy, decision making, and stakeholder communities? I will—and only can—do this within the context of my personal experience in global change modeling, in energy and greenhouse gas emission scenario construction, and in investigating past and present aspects of the search for sustainable development (Rotmans and de Vries 1997; Nakícenovíc et al. 2000; de Vries and Goudsblom 2002; MNP 2004).

In this contribution I will first briefly describe the scenario method and interpret it in the historical context. Next, I discuss some global change-related scenarios on long-term (up to 2100) global developments, with an emphasis on model-based scenario construction efforts and the perspective of sustainable development. In particular, I use the TARGETS model-based scenarios and the emission scenarios for the Intergovernmental Panel on Climate Change (IPCC). I then give some suggestions for improvement, focusing on uncertainty, complexity, values, and participation. I end with some statements for discussion. Throughout this chapter I have been parsimonious with references, on request.

THE SCENARIO METHOD

The scenario method has become rather widespread over the last few decades. The many practitioners have given a large array of definitions.[1] Common elements are:

- Scenarios are a tool for [better] [strategic] decision making.
- The scenario method emphasizes the construction of alternative futures in order to prepare for divergent plausible futures.
- To this purpose, existing mental models should be challenged, and qualitative ("storytelling," narrative) as well as quantitative ("modeling") approaches are to be used.
- It is important to know for whom scenarios are made and for which purpose. Credibility, legitimacy, and creativity are important aspects, then, of process and product.
- Scenario construction is a training in finding key trends, recognizing prevalent myths, and imagining attitudes of key players (see Box 19.1).

One should not make more than three or four scenarios because people cannot handle more due to cognitive limitations. The identification of the driving forces (i.e., what makes it going), of predetermined elements (in particular slow changing variables), and of critical uncertainties provide the structure or *logic* of a scenario (Schwartz 1991). It is claimed that the scenario method presents people with multiple perspectives on the world, which is an alternative or at least a complement to the conventional languages of business and science in addressing the often complex and ill-structured questions of today's world.

Of course, long before the advent of scenarios, governing elites were interested in anticipating future events and in strategy development. Influential advisers to the ruling elite have earned their place in history. Often, priests were at the forefront of such endeavors as a conduit to allow the gods to participate in judging, legitimizing, and rationalizing. Usually, the objective was sustaining and/or expanding power within a rather well-defined organizational context. There have always been those who challenged the rulers with opposing views of what the future could or should be—rebels, visionaries, prophets. Thomas More

1 See, e.g., van Notten et al. (2003) for a recent overview of the scenario literature and scenarios.

Box 19.1 Some related concepts.

Scenarios are, it seems, halfway myth and plan. A *myth* is "the way things are" as people in a particular society believe them to be (Schwartz 1991), often unconsciously. *Strategy* is the art of deliberately recognizing major trends, establishing one's own course of action, and translating this into practical *plans*. Part of scenario construction is the process of *visioning*. "Visioning means imagining, at first generally and then with increasing specificity, what you really want...not what you have learnt to be willing to settle for. Visioning means taking off all the constraints of assumed 'feasibility' " (Meadows et al. 1992). As such they express the ethos of their times (Heilbroner 1995).

offered his Utopia as a visionary critique, Karl Marx offered a rationale for the demise of capitalism, Jules Verne expanded technical possibilities far beyond the known options, and religious leaders have promised mixtures of catastrophe and salvation. Such views were rooted in deviant valuations and interpretations of the present, which often provoked violent oppression.[2] Nowadays, there are hundreds of individuals and organizations who offer their view of the future, which ranges from alerts and warnings to technological paradise and fundamentalist doom.[3]

How should we interpret the scenario method in historical perspective? Let me try a concise answer. With the emergence of science and technology as the "modern worldview," old centers of authority were increasingly challenged and new and powerful actors appeared on the scene: scientists, entrepreneurs, corporations, citizen groups (or today's nongovernmental organizations, NGOs). The legitimacy of decision making had to change. Science and technology began to offer tools for deeper and more rational forms of control and management: physical, as in factories and transport and nowadays ecosystems, as well as mental, as in offices and media.[4] Useful and influential as this control paradigm may have become, it has also met its limitations in the past decades—a development sometimes associated with postmodernism. With the spread of education, communication, and democratic forms of governance and with the ever more visible unintended—and for many undesirable—consequences of this "mechanization of the world," what used to be seen as unprecedented success has been increasingly challenged. Simultaneously, science has become aware of the limitations of its "Newtonian paradigm" and is offering novel perspectives

2 Castells (2000) discusses such forces at work in the information age. He distinguishes three sources of identity: legitimizing by the powers-that-be, resistance against it by those suppressed, and solidarity for a common project for societal reform. Examples of the last are ecological and feminist movements.

3 Modern alerts can often be identified from the titles: *Limits to Growth* (Meadows et al. 1972), *Social Limits to Growth* (Hirsch 1976), *Limits of Organization* (Arrow 1974), *Limits to Competition* (Group of Lisbon 1995), and *Limits to Certainty* (Giarini and Stahel 1993).

4 Illustrative examples are the use of optimal control techniques to analyze management options in large societal systems.

in the study of complex systems such as the climate, ecological, and human–environment systems.[5]

I argue that the scenario method, with its explicit consideration of uncertainties, multiple perspectives, and stakeholders, is only a logical next step in this development. Naturally, the modern-day equivalents of the centers of authority—the corporate and government planning institutions—are among the first to apply such novel methods. Resistance comes from (some) engineers and social scientists. The former do not subscribe to the nonscientific approach, preferring to adhere to the control paradigm instead; the latter may have built their careers on mimicking mechanical science methods or fear invasion of their domain by latter-day scientists. Politicians may also show signs of dislike, as they will always be tempted to prefer command and control over participation and pluralism. Let us look at this closer through some examples.

GLOBAL SCENARIOS

Global [Change] Modeling

One of the predecessors in the search for more sustainable forms of human activity is the book *The Limits to Growth* published in 1972 by Meadows et al. At the request of a group of industrialists—the Club of Rome—a number of scientists at MIT in Boston constructed a system dynamics model of the global system and showed that humanity would face a future full of catastrophes unless there were drastic changes in the mechanisms that cause exponential growth in population and economic production.[6] In subsequent years, the *Limits to Growth* analysis was severely criticized for various reasons: no regional dynamics, in particular the rich–poor gap; no or insufficient price response dynamics; too pessimistic resource base estimates; and no or insufficient technological progress. Several attempts were made in the 1970s and 1980s to eliminate these perceived shortcomings, leading to a variety of global change models. Integration became a buzzword and global change modeling became part of *integrated assessment modeling* (IAM). Variation in input assumptions became established practice, and the resulting outcome was no longer called "model run" but "scenario."

Simultaneously, the quest for sustainable development intensified with publications from the UN Environment Program (UNEP) and the World Bank's World Development Reports. Government and business circles became more deeply involved, partly in response to expanding NGO activities. The "good

5 Although the signs of this change are manifold, Funtowicz and Ravetz's (1990) introduction of the concept of postnormal science is widely recognized as a milestone.

6 In 1992 a new edition of the book appeared with the title *Beyond the Limits* (Meadows et al. 1992). The authors concluded that many trends in the period from 1970–1990 confirmed the claims made in *Limits to Growth*. In 2004 the book *Limits to Growth: The 30-year Update* was published (Meadows et al. 2004).

intentions" of all those concerned about environmental degradation and social and economic inequity were confronted with the involvement of practitioners of "the dismal science": economics. Macroeconomic modelers entered the scene and issues such as the cost-efficiency of policy instruments and the trade-offs between development and environmental and social objectives became prominent. For instance, Duchin and coworkers (1994) did an extensive analysis of the macroeconomic feasibility of the objectives suggested in the UN report, *Our Common Future*, of the UN World Commission on Environment and Development (WCED), also called the "Brundtland Report" (WCED 1986). It was concluded that the positive effects of technological adjustments such as recycling and energy efficiency are insufficient to realize the development aspirations sketched in the Brundtland Report on a sustainable basis.

In the early 1990s, the first attempts at scenario analysis became more widely publicized as a somewhat belated and modified continuation of the famous *The Year 2000: A Framework for Speculation on the Next Thirty-Three Years* by Kahn and Wiener (1967). Schwartz (1991) discussed the scenario approach, giving practical advice on how to do it and illustrating it with real-world examples in multinational companies. Hammond (1998) published a scenario study that reflected several years of discussion within the Global Scenario Group and explicitly considered regional diversity (http://www.gsg.org).[7]

I tend to have most affinity with and expectations about scenarios that combine storytelling and modeling.[8] The storytelling part consists of carefully constructed narratives, built around interpretations of past and current observations and trends. The modeling part implies an attempt to introduce consistency and sharpness by quantifying certain parts of the narrative on the basis of available (statistical) data and formalized (mathematical) relationships. I will briefly discuss two projects in which simulation models were used to construct storyline-based scenarios: the TARGETS project (Rotmans and de Vries 1997; de Vries 2001) and the Special Report on Emission Scenarios (SRES; Nakícenovíc et al. 2000; IPCC 2001) for the IPCC. I leave out other interesting scenario developments which have taken place in the last decade, such as:

- The Millennium Institute's Threshold 21 (T21) model, an offshoot of the Global 2000 Report to U.S. President Carter. This is a system dynamics one-country model which had the objective of providing a generic tool for the exploration of sustainable development strategies (see http://www.threshold21.com).
- The International Futures (IFs) model, a tool designed to help understand the state of the world and develop strategies for desirable futures. Rooted

7 This largely qualitative research was supported with the PoleStar model, a country-based accounting framework aiming to provide transparency and consistency.

8 Labeled "computer-aided storytelling," this has somewhat reluctantly been taken up in economic science, too.

in earlier global models (e.g., WIM and GLOBUS), it emphasizes the analysis of the potential for conflict between nations (see http://www.du.edu/~bhughes/ifs.html).

- The IMAGE model, developed since the early 1990s for climate change-related research and broader global change scenarios (Alcamo et al. 1998) (http://www.mnp.nl/image). It is a typical IAM of intermediate complexity and has been used extensively in the Global Environmental Outlook 3 in the construction of scenarios within the SRES framework (http://www.unep.org).
- Two other interesting, recent scenario studies are the VISIONS project (http://www.icis.unimaas.nl), which used novel software tools and participatory methods to develop three regional scenarios for Europe, and the Millennium Ecosystem Assessment (2005), which contains four scenarios to explore the causes and consequences of global ecosystem degradation.

The Targets Project

The TARGETS model is an IAM on global change at a high level of aggregation. It was the result of a 5-year project (Rotmans and de Vries 1997; de Vries 2001) on global change at the Dutch National Institute for Public Health and the Environment (RIVM). The project's main objective was to operationalize the concept of sustainable development, using a systems dynamics approach. Pressure–State–Impact–Response (PSIR) chains and transitions were used as guiding concepts.[9] The system was conceived of as a collection of well-defined subsystems interacting with each other—population and health, water, land and food, energy, and biochemical cycles. Subsystems were described on the basis of meta-models, that is, simplified "expert models" of the long-term dynamics, integrated horizontally (between subsystems) and vertically (modeling human behavior on top of the environmental dynamics "substrate").

The resulting world model, TARGETS1.0, was a meta-model, not an expert model, representing in a coherent and systematic way the various insights from scientific disciplines on the functioning of the Earth system in order to frame sustainability issues and provide a context for debate. A series of global scenarios for the 21st century have been constructed with the model, using the cultural theory of Thompson et al. (1990) to make coherent interpretations and assumptions of how the world fits together and how it should be run. Cultural theory combines insights from cultural anthropology and ecology in distinguishing cultural perspectives, based on the degree to which individuals behave and feel themselves part of a larger group of individuals with whom they share values and beliefs (the "group" axis) and the extent to which individuals are subjected

9 A more recent, similar integrated Earth system model is the Global Unified Metamodel of the BiOsphere (GUMBO), with a focus on the dynamics and values of ecosystem services (Boumans et al. 2002).

to role prescriptions within a larger structural entity (the "grid" axis). The resulting four perspectives are related to their position along these two axes: the *hierarchic* (high on both), the *individualist* (low on both), the *egalitarian* (high in "group," low in "grid"), and the *fatalist* (low in "group," high in "grid"). It was then assumed that utopian futures unfold if both the worldview ("how the world functions") and the management style ("how the world is managed") are in agreement.[10] If not, dystopian futures will unfold.

From a scenario point of view, the findings from the TARGETS model experiments can be summarized in these utopia–dystopia terms:

- Continuing large growth in population and material welfare will cause ever more pressure on the natural environment, but this can be managed for the next 100 years if new and powerful technologies—geared to more efficient resource use—are put into practice and if natural ecosystems—flora and fauna—are not [too] vulnerable. This would lead to an *individualist utopia*—a highly managed, high-tech world with a materialist outlook.
- If these conditions cannot be met, the future will develop in less attractive and dystopian ways. The burden of environmental degradation may be shifted onto the weakest members of society and an impoverished world of islands of extreme wealth in the middle of mass suffering and crime may evolve.
- It is also possible that the egalitarian ideal of frugality will gain ground so that economic growth will be less or less energy and material intensive. This could be triggered through ecological disasters or "grass-roots" environmental initiatives, violence resulting from large and visible income gaps, changes in consumer preferences and lifestyles, rebirth of spiritual/moral movements, or a combination of all these. The pressure on the environment in such an *egalitarian world* will then be much less, meaning that catastrophes are avoided or at least anticipated and hence better handled.
- It is very well possible that the future contains a mix of elements of these utopias and dystopias but that the mainstream unfolds along the lines of the hierarchic perspective. Many of the present trends will continue. If this happens on the basis of correct knowledge, a *hierarchic utopia* will unfold with incremental policy measures on the basis of institutional expertise, conventions, and control. If these conditions are not met, "overshoot and collapse" will occur in more-or-less serious forms because of waiting too long for more knowledge/expertise or because governments get bogged down with bureaucratic and regulating ineffectiveness and corruption.

---

10 See de Vries et al. (2002) for such an approach with respect to past human–environment interactions in Himalayan and Alpine villages.

> The only consolation in the face of such an oncoming disaster is perhaps that the emergency actions are well organized.

What did we learn from the TARGETS project? In retrospect, it can be summarized:

- A "one-world" IAM like TARGETS can be made to reproduce most of the available statistics on the system Earth since 1900. This is largely a form of calibration—validation in a more rigorous sense is difficult because many of the model variables are nonobservable aggregates and many of the presumed relationships are generalized forms of dynamic processes observed at the local/regional scale. Hence, such models are useful for framing issues about long-term population–resource–environment developments, which lends a more of synthesizing and conceptual rather than hypothesis-testing value. A next step has to be to disaggregate in representative regions and develop generic dynamic models on the basis of past records. Such an approach has, for instance, been attempted by Wirtz in his simulation of the Neolithic transition (de Vries et al. 2002).
- The TARGETS model experiments showed that the most widely published result of the 1972 *Limits to Growth* report—overshoot and collapse—is one of the many possible outcomes, namely the one in which egalitarian environmentalists are right about the finiteness and fragility of the natural system and the power is in the hands of those who act upon the opposite assumption.
- The project made clear that transcending disciplinary boundaries in knowledge is difficult. Scientists from different disciplines use different methods and concepts—and they are attached to them. There is a surprising lack of generalized "stylized facts" knowledge in the areas in-between the disciplines (e.g., about water–energy and food–health links). Integration remains a challenge but it would help if more research were done on issues in-between specialized disciplines and if novel methods, such as looking at the system in terms of networks, were applied.
- Constructing "perspective-based scenarios" invites people to take part in the discussion and makes people's values visible in the debates about where the world might and/or should be headed. Although much effort was put into communicating the insights with good visualization techniques and interactive use,[11] participatory use of the model was not very successful due to lack of a rigorous method and serious time and budget constraints. Moreover, there was a marketing problem (which has only grown since then). In communicating with the media, I got the impression that announcing doomsday for April 30, 2019 would have made a more

---

[11] For this purpose, a new simulation environment, M, for model development and interactive communication was developed at RIVM (http://www.m.rivm.no/info/ftp.htm).

lasting impression than the rather sophisticated risk-oriented results of the TARGETS approach (see also Meadows, this volume).

I will come back to possible solutions to these shortcomings.

The Special Report on Emission Scenarios (SRES)

The threat of human-induced climate change has rapidly become one of the most prominent environmental issues on the global agenda. In 1988, to assess the causes and consequences of human-induced climate change, the IPCC established three working groups. They published reports in 1990, 1996, and 2001 (http://www.ipcc.ch; see also Nakícenovíc et al. 2000; IPCC 2001 and Hosoda et al. 2000). The 1990 report contained one of the first emission trajectories published, the so-called IS92 scenarios. With almost no endogenous relationships and lack of qualitative, descriptive detail, one could hardly speak of scenarios. In 1997 the IPCC established a team to develop a series of standardized reference emissions scenarios (SRES). The objective of this team was to review existing emission scenarios and to revise the earlier IPCC IS92 emission scenarios. The new scenarios should project future greenhouse gas emissions from all sectors, without considering specific climate policies and their impact on emission reductions. They should provide a baseline or benchmark against which climate policy scenarios (mitigation, stabilization, adaptation) could be evaluated.

The team decided to use different *storylines* or narratives, each describing in qualitative terms how the future could evolve. These narratives were the basis for the input assumptions to be used by the six modeling groups involved. Each group (two from the U.S.A., two from Japan, and two from Europe) used its own energy model, which of course caused large problems of harmonization.[12] Most models were significantly better than the models used for the earlier emission scenarios: more regions/countries; inclusion of energy efficiency, renewable options, and fuel trade; learning-by-doing dynamics; linkage with land use, land cover, and material flow models to evaluate the role of carbon sequestration and traditional and modern biofuels; and linkage with macroeconomic models to analyze leakage and rebound effects. The time horizon was set at the year 2100, countries were clustered into four regions, and 13 emission sources were taken into account.

Early in the process it was decided to organize the storylines in four *scenario families* based on divergence along two axes: (a) whether the world will continue on the path of globalization or reverse in a more protectionist regionalization of economic, cultural, and political blocks, and (b) whether the prevailing attitude of people will be toward material welfare and high-tech

12 Fourteen modeling groups have interpreted some or most of the narratives into their models. This resulted in over 40 different emission paths but for communication purposes only one model was selected to represent one of the narratives. The SRES scenarios were used in the third assessment of IPCC (2001).

consumerism or tend toward social and environmental quality-of-life aspects. This led to the four scenarios which have become known as A1, B1, A2, and B2 after the SRES team had decided to choose neutral names. In daily practice, A stands for economic and B for environmental, 1 for global and 2 for regional. Evidently, the storylines leave ample room for divergent interpretation and emphasis even within a scenario family.[13] Figure 19.1 shows the two axes and the names of scenarios constructed by various individuals and organizations over the last decade as arranged by me.

The SRES project has accomplished a lot, thanks to the tremendous efforts of all participating groups and project leaders. It can be argued that SRES was trendsetting in its attempt to merge the quantitative and qualitative (story-scenario model), to use multiple logics and perspectives, and to simulate at regional/local scale. A major achievement has been that modelers from various backgrounds as well as policy analysts and decision makers have been and still are confronted with each other's expertise, ignorance, and values. The scenario approach has contributed significantly to this outcome. Another advantage of the narrative approach, widely used, is that scientists working on other aspects of the climate change issue, such as mitigation or adaptation potential, can begin to build on these storylines (e.g., Hosoda et al. 2000). For instance, in a B1 future, the world is quite able and willing to introduce effective and efficient emission reduction and help potential victims in adapting; the A2 future, however, would offer a very different prospect.

Obviously, such an effort will not—and cannot—be without omissions and errors. An entire volume can be devoted to a discussion about the merits and shortcomings of the SRES approach and the resulting scenarios. I will thus confine myself to a couple of critical comments.

1. The four proposed leading storylines—indicated as *marker scenarios* in the modeling context—give too much room for divergent interpretations. For instance, globalization (1 vs. 2) was (and is) not a well-defined process, measured by economists as an increase in the flows of goods and capital, discussed by demographers in terms of migration, described by Di Castri as *"une diversité globalement uniforme: plus de diversité locale...mais la meme partout"* (Theys 1998) but hardly considered at all from a cultural point of view.[14] Similarly, the distinction between economic versus environmental (A vs. B) is not at all clear. I have come

13 The storylines are described in the SRES report but they have grown significantly in quantity and diversity of interpretation (e.g., Nakícenovíc 2000). There are clear links with the previously discussed TARGETS scenarios: whereas the A1 world reflects the market- and high-tech-orientation of the individualist, the B1 and possibly even more the B2 worlds would be driven by egalitarian values.

14 In fact, the "cultural clash" aspect of a storyline, notably the A2 scenario, was taboo just like the prospects of terrorism or collapse due to famine, disease, and mismanagement. Such avoidance may be symptomatic for UN processes.

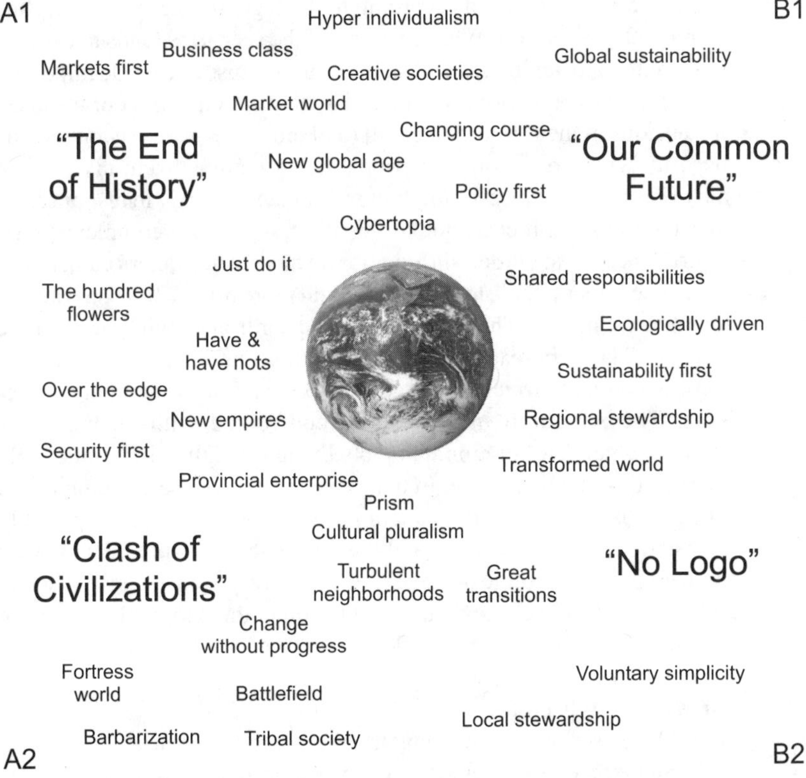

Figure 19.1 Depiction of two axes and the names of scenarios constructed by various individuals and organizations over the last decade. (Image from Visible Earth, NASA.)

across a variety of dichotomies: market versus government, deregulation versus overregulation, competition versus coordination, efficiency versus equity, consumer tech versus green tech. It will be helpful to bring in aspects of governance and technology more explicitly.

2. In my experience, confusion about how to interpret the storylines occurred because key assumptions from high-level aggregate empirical relationships ("stylized facts" or meta-models) were not well understood. To mention a few of the ones used: net population growth declines with income and, directly and indirectly, with globalization; economic growth and energy intensity are bell-shaped functions of income; and globalization in the form of less trade barriers—operationalized by lower transport costs—increases economic growth via higher rates of capital and technology transfer (de Vries et al. 2000). Many more such assumptions are hidden and/or implicit and lifted from country to region level; I will come back to this.

3. The world will not unfold according to the logic of one single storyline over a 100-year period. Whenever one of the scenarios tends to become dominant, opposing forces will start to compromise and erode important features of such a worldview and lead to new directions.[15] For instance, the tensions in the A1 narrative will (and did) show up as rising income disparity with large groups of people being marginalized; tax evasion and the associated "race-to-the-bottom" dynamic; partly in response, increasing transaction costs due to litigation, regulatory complexity, and security measures; burden shifting, for instance when deteriorating public service forces people to spend more and more private time on waiting, communicating, etc. The B1 future will be confronted with the negative aspects of large bureaucracies (e.g., inertia, inefficiency, and corruption), which may drive it into more market and/or protectionist directions. The A2 future may well be confronted with such serious environmental deterioration that globally and socially/environmentally oriented forces will emerge. In that sense, a serious shortcoming of the SRES scenarios is that (a) no attempt was made to include the possibly false underlying assumptions and the associated risks in terms of environmental and economic damage, and (b) the response in terms of social and political feedbacks toward one of the other three logics has not been considered. This, of course, reflects the absence of an explicit sociocultural dynamic over and above some rather simple economic and demographic "stylized facts."
4. A fourth and serious shortcoming in SRES was, in my view, the fact that the computer models became leading in what had been envisaged. Translating storylines into model assumptions is expectedly a procrustean process. All of the models used had (and still have) a bias in terms of process relationships and data reliability which may make them somewhat realistic to project the future economy–energy–emission path for the U.S. or Europe; this causes, however, serious distortions and pseudo-insights if applied, for instance, to 800 million Chinese or Indian farmers in 2030 in a setting of traditional culture and economic protectionism. This has been shown to be especially devastating for the B2 scenario: the models cannot realistically cope with, for instance, energy demand-price or renewable energy penetration rate dynamics in a diverse, equity- and environment-oriented world.[16] To some extent the models became a legitimization not for how world regions are but how they should be, thus

---

15 This has been illustrated with the TARGETS model in "the battle of perspectives" (Janssen and de Vries 1998).

16 This was one of the reasons that the B2 scenario quickly became a middle-of-the-road dynamics-as-usual scenario, without fulfilling its promise of offering think-space to alternative visions, values, and rules. In turn, the resulting middle-of-the-road emissions path led those who were not interested in the storyline to interpret it as "the" medium scenario.

representing (inadvertently?) the ideology of the Washington consensus (World Bank, IMF) that dominated the 1990s.

These comments hopefully indicate that it makes no sense to assign probabilities to the scenarios. SRES rightly states that *there is no single most likely, "central" or "best-guess" scenario*. In a way, all four scenarios are highly implausible. Once the system boundaries are enlarged to include the human system and not just industrial and power plants or pollutant flows, the only way forward is to refine and adjust continuously the storyline logic and improve the model structure, data, and applications accordingly.[17]

HOW CAN GLOBAL CHANGE SCENARIOS BE IMPROVED?

In essence, my critique and suggestions for improvement boil down to four keywords: *uncertainty*, *complexity*, *values*, and *participation*. We must address explicitly and scientifically uncertainty and complexity. A new epistemology is needed. We should incorporate people and their values in the process of scenario construction and use. Participatory methods, such as simulation games and policy exercises, are necessary complements.

Uncertainty and Complexity

Most of the scenario analyses discussed thus far are about (parts of) an extremely complex socionatural system: Earth. The merits and limitations of the classical science and engineering methods—in the vein of the control paradigm—have become ever more evident in the course of the 20th century. Still, many issues relevant for global change are clouded by uncertainties and controversies, giving rise to divergent or even antagonistic interpretations of what happened and expectations of what might and should happen.

Philosophers of science (epistemology) have proposed that uncertainty and ignorance be acknowledged and identified and have suggested ways to do so (Funtowicz and Ravetz 1990). Science itself, notably mathematics and systems theory, has contributed to new ways of addressing complex systems, introducing notions such as self-similarity and self-organization, chaos and catastrophe, bifurcations and resilience. Economic science is picking up these advances in branches such as behavioral economics and ecological and evolutionary economics, and is introducing ideas about transaction costs and technology transitions, for instance. As a result and in combination with the advent of strong computers and satellite data, our insights into the evolution of socionatural systems

---

[17] This, of course, in no way dismisses modelers of the task to do careful sensitivity analyses and evaluate the importance of uncertainty in assumptions for the objectives and policies at hand.

are expanding enormously.[18] How can these developments be incorporated in the generation of useful, plausible scenarios for humankind? Let me briefly mention four avenues.

The first concerns *uncertainty analysis*. The models used for (global) scenarios are usually not scientific in terms of practical validation (comparison of simulated results with observational data) and conceptual validation (scrutinizing the concepts and theoretical laws of the system under consideration).[19] Various plausible but sometimes contradictory explanations of phenomena can be constructed. In the discussion on climate change impacts and mitigation options, there are, for instance, important uncertainties regarding size and cost of energy resources and regarding temperature response to rising greenhouse gas concentrations. The quest is for a balanced use of sensitivity and uncertainty analytical tools, ranging from multiregression Monte Carlo techniques to Delphi expert panels (see htttp://www.nusap.org). The outcomes can—and should—guide the construction of scenarios.

The second is *model improvement*. The usual way to achieve this is to add detail (e.g., higher spatial resolution, more countries, etc.). However, it may be more important to improve the meta-models that link the processes in the various subsystems. To mention a few in a question form:

- *Population*: How do aging and income disparities interact with the prospects for economic growth (health costs, savings rate, proneness for radicalism and revolt, female employment)? Which value changes are possible in a world where more than half of human activity takes place in China and India?
- *Economic*: What is the link between physical and monetary fluxes and should *mer* (market exchange rates) or *ppp* (purchasing power parity) be used for intercountry comparison in a distant future? Are the transitions to a service economy and dematerialization universal phenomena? Which role is played by the shadow economy, including organized crime, and how is it related to trade and tax regimes? In which ways might climate change feed back on social and economic conditions?
- *Technology and resources*: Is there such a thing as a "long wave" in economics, with nano- and biotechnology and robotics spurring the next one? How effective are government R&D programs and what are the determinants of technology transfer? Which land-use/cover constraints are to be expected and by which mechanisms are land-change processes mediated? Are large oil deposits associated with dictatorship because both weapons

18 See, e.g., Steffen et al. (2004) for a recent overview of "Earth science" insights; http://www.pages.sunibe.ch for research on past human–environment interactions.

19 I have suggested the concept of "strong knowledge" to indicate the knowledge which is or can be gained from repeated experiments in controlled subsystems. Most knowledge about human systems is, in this sense, relatively weak. Of course, the law of large numbers can be applied to lead to rather certain probabilistic statements.

and legitimacy can be bought (the "resource curse")? A rigorous testing of social science hypotheses is welcome here, although it will never eliminate the uncertainties.[20]

The next involves *issue (re)framing*. One of the permanent challenges in global change modeling is to handle adequately different scales in a nested dynamical hierarchy. Ecological research has explored this issue deeply (see, e.g., Berkes et al. 2003). An interesting framework is the Syndrome Approach (Petschel-Held et al. 1999). *Syndromes* are archetypical nonsustainable patterns of civilization–nature interactions. It has been suggested to distinguish between utilization, development, and sink syndromes. The persistent structural properties of a region, such as its biogeography, determine the disposition for certain syndromes. With high disposition, exposition factors such as natural catastrophes or social revolt can activate a syndrome and cause a downward spiral of degradation collapse. Human history provides illustrative examples (de Vries and Goudsblom 2002). Constructing generic dynamic models for such well-defined syndrome areas and using novel techniques such as network theory and multi-agent simulation constitute, in my view, a crucial next step.

The fourth avenue is *logic structuring*. From an even broader perspective, one of the more difficult parts of scenario analysis is to create relevant logics for the complex parts of a system. A helpful framework is the cultural theory which introduces three "active" solidarities according to which people manage their social and natural environment (cf. TARGETS approach and VISIONS project). These three cultural biases, it is claimed, provide the requisite plurality if one is to incorporate items (e.g., fairness, risk, and innovation) into scenarios in a structured way. As Michael Thompson (pers. comm.) puts it: "*Uncertainty, far from being an unproductive desert that will only bloom when we have managed to irrigate it with knowledge, is a resource: something that is all the time being colonized and fought over by contradictory certitudes.... Visions of the future function at the individual and collective level as 'final' causes: attractors and repellers...that variously define in the here-and-now what is technically possible, socially desirable, and morally acceptable*." The dynamic interactions between adherents of different worldviews can be a way of addressing surprise and discontinuity because much "catastrophic change" may be the result of a rising discrepancy between what (a majority of) people believe and what they experience. After all, the unexpected is often the consequence of limited or false information and of restricted or perverted exchange of views and values—as, for instance, recent analyses of the Iraq war show.

Evidently, in each of these items, novel ways to model human–environment systems are to play an important role—without any further consideration I mention the importance of spatial dynamics with cellular automata and the complex

20 See Ross (2001) for an example of empirical testing of hypotheses on the role between oil richness and democracy.

adaptive systems and multi-agent simulation modeling approaches. Local and regional land-use/cover scenarios are the avant-garde in this endeavor.

Involving People: Values and Participation

In the recent Sustainability Outlook of the Dutch Netherlands Environmental Assessment Agency (MNP), the decision was made to operationalize sustainable development as the continuation and expansion of qualities of (human) life. Which qualities, for whom and how long? It is a continuous balancing act between the prevailing values of (a group of) people on the one hand and available knowledge and skills to satisfy these on the other. It plays itself out between the ends and means, each having its own kind of rationality and methods. To do this, we used the results of value surveys in the Netherlands, which indicated that people's value orientations can meaningfully be grouped in 8 clusters, and a framework for questioning people about their expectations and desires about the future—almost identical to the SRES framework. We found that people are concerned about large-scale environmental disruption, social inequity, and loss of social cohesion and that only a small minority perceives a globalizing, high-tech high-growth world (say, A1) as possible and desirable.[21] This suggests that the dominant view in (neoclassical) economics of humans as maximizers of discounted individual utility with perfect foresight represents only one form of rationality; it has to be complemented with others (cf. Jager et al. 2000).

The Sustainability Outlook process was initiated to involve people explicitly as stakeholders in the discussion on their local, regional, and global future. Telling a story and supporting it with the best available scientific expertise and models while acknowledging the uncertainties and controversies is, in my view, the way forward. In this way, people become involved with their values and interpretations, and policy for change can become legitimate. This is also part of the previously mentioned issue (re)framing and logic structuring. A framework proposed during the Dutch COOL process is shown in Figure 19.2, which shows how the degree of consensus about values and consensus in knowledge largely determine the problem context. A similar framework, but with different jargon (added in italics), is used in business schools, I discovered later. This brings me to the last point.

Another approach to address ill-structured problems is gaming simulation (Duke and Geurts 2004). Opening people to multiple perspectives is a lofty goal but may easily result in an exchange of views and opinions of bewildering complexity and unclear and quickly forgotten outcomes. One reason is, of course, time and money constraints; another is the lack of a structured setting. When a clear organizational setting is absent, as is often the case with global change

[21] An interesting finding in this context is that Europeans are on average positive about their quality of life but see it only vaguely related to economic performance and see no conflict between environment and competitiveness (Special Eurobarometer, February 2005 215/Wave 62.1).

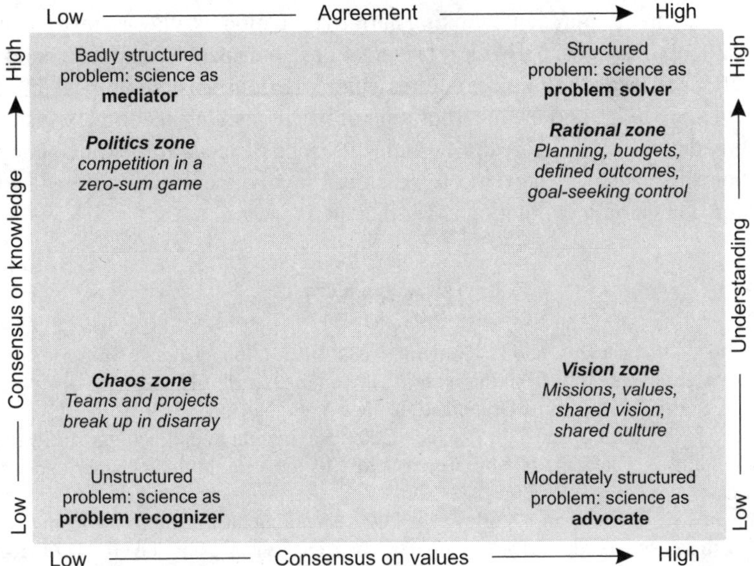

Figure 19.2 A framework proposed during the Dutch COOL process depicting the degree of consensus about values and consensus in knowledge that largely determines the problem context.

issues before the stage of concrete strategies and plans, there is an urgent need for an organizing framework to be able to structure, focus, and prioritize. Gaming simulation can be very helpful. Equally important is the use of simulation games and policy exercises in an educational setting, as this is an effective way to teach the art of scenario construction in a participatory learning mode.[22] Interactive scenario construction is one of the methods that may gain ground as the infrastructure (e.g., Internet, simulation tools) further improves.

CONCLUDING STATEMENTS

The art of scenario construction and use is spreading. This is good news because it is one of the more promising avenues to combine qualitative insights on human–environment interactions from the social sciences, in the form of narratives, anecdotes, and analogs, with the more formal and quantitative models of the natural sciences. However, experience with long-term global change scenario construction indicates that there are still significant barriers to overcome. Integration across disciplines in concepts, data, and methods as well as transparent and interactive communication of scientific insights are two important areas for improvement. To improve the effective construction and use of scenarios, a

[22] For example, the simulation games Stratagem and FishBanks Ltd. (http://www.sustainability institute.org/tools_resources/games.html).

more systematic approach in linking narratives to models, via system dynamics causal loop diagrams, network representations, and spatially explicit social interactions is needed. This also requires a more rigorous epistemological foundation of knowledge and the incorporation of recent insights in complex systems. To broaden the realism, creativity, and legitimacy of scenarios, explicit consideration of people's values and a structured involvement of people as stakeholders via gaming simulation is another step forward.

REFERENCES

Alcamo, J., R. Leemans, and E. Kreileman, eds. 1998. Global Change Scenarios of the 21st Century: Results from the Image 2.1 Model. Oxford: Elsevier.

Arrow, K.J. 1974. Limits of Organization. New York: Norton.

Berkes, F., J. Colding, and C. Folke, eds. 2003. Navigating Social-ecological Systems: Building Resilience for Complexity and Change. Cambridge: Cambridge Univ. Press.

Boumans, R., R. Costanza, J. Farley et al. 2002. Modeling the dynamics of the integrated Earth system and the value of global ecosystem services using the GUMBO model. *Ecol. Econ.* **41**:529–560.

Castells, M. 2000. The Information Age: Economy, Society, and Culture. Oxford: Blackwell.

de Vries, B.J.M. 2001. Perceptions and risks in the search for a sustainable world: A model-based approach. *Intl. J. Sust. Dev.* **4**:434–453.

de Vries, B.J.M., J. Bollen, L. Bouwman et al. 2000. Greenhouse-gas emissions in an equity-environment- and service-oriented world: An IMAGE-based scenario for the next century. *Tech. Forec. Soc. Change* **63**:2–3.

de Vries, B.J.M., and J. Goudsblom, eds. 2002. Mappae Mundi: Humans and their Habitats in a Long-Term Socio-Ecological Perspective. Myths, Maps and Models. Amsterdam: Amsterdam Univ. Press.

de Vries, B.J.M., M. Thompson, and K. Wirtz. 2002. Understanding: Fragments of a unifying perspective. In: Mappae Mundi: Humans and their Habitats in a Long-Term Socio-Ecological Perspective, ed. B.J.M. De Vries and J. Goudsblom, pp. 257–299. Amsterdam: Amsterdam Univ. Press.

Duchin, F., and G.-M. Lange. 1994. The Future of the Environment. Oxford: Oxford Univ. Press.

Duke, R., and J. Geurts. 2004. Policy Games for Strategic Management: Pathways into the Unknown. Amsterdam: Dutch Univ. Press.

Funtowicz, S., and J.R. Ravetz. 1990. Uncertainty and Quality in Science for Policy. Dordrecht: Kluwer.

Giarini, O., and W. Stahel. 1993. Limits to Certainty. Dordrecht: Kluwer.

Group of Lisbon. 1995. Limits to Competition. Cambridge, MA: MIT Press.

Hammond, A. 1998. Which World? Scenarios for the 21st Century: Global Destinies, Regional Choices. Washington, D.C.: Island.

Heilbroner, R. 1995. Visions of the Future: The Distant Past, Yesterday, Today, and Tomorrow. Oxford: Oxford Univ. Press.

Hirsch, F. 1976. Social Limits to Growth. Cambridge, MA: Harvard Univ. Press.

Hosoda, E., et al. 2000. Long-term scenarios and socioeconomic development and climate policies. *Env. Econ. Policy Stud.* **3**:65–290.

IPCC (Intergovernmental Panel on Climate Change). 2001. Climate Change 2001: Mitigation. Summary for Policy Makers and Technical Summary of the Working Group III Report. Cambridge: Cambridge Univ. Press.

Jager, W., M. Janssen, B. de Vries, J. de Greef, and C. Vlek. 2000. Behaviour in commons dilemmas: *Homo economnicus* and *homo psychologicus* in an ecological-economic model. *Ecol. Econ.* **35**:357–379.

Janssen, M., and B. de Vries. 1998. The battle of perspectives: A multi-agent model with adaptive responses to climate change. *Ecol. Econ.* **26**:43–65.

Kahn, H., and A.J. Wiener. 1967. The Year 2000: A Framework for Speculation on the Next Thirty-Three Years. New York: Macmillan.

Meadows, D.H., D.L. Meadows, J. Randers, and W.W. Behrens III. 1972. The Limits to Growth. New York: Universe.

Meadows, D.H., D.L. Meadows, and J. Randers. 1992. Beyond the Limits: Confronting Global Collapse, Envisioning a Sustainable Future. Post Mills, VT: Chelsea Green.

Meadows, D.H., J. Randers, and D.L. Meadows. 2004. Limits to Growth: The 30-year Update. London: Earthscan.

Millennium Ecosystem Assessment. 2005. Millennium Ecosystem Assessment Synthesis Report. Washington, D.C.: Island. http://www.millenniumassessment.org/

MNP (Netherlands Environmental Assessment Agency). 2004. Kwaliteit en toekomst: Een verkenning van duurzaamheid. Bilthoven: MNP/Natl. Institute of Public Health and the Environment (RIVM).

Nakícenovíc, N. 2000. Global greenhouse gas emissions scenarios: Five modeling approaches (Spec. iss.) *Tech. Forec. Soc. Change* **63(2&3)**.

Nakícenovíc , N., J. Alcamo, G. Davis et al. 2000. Special Report on Emissions Scenarios (SRES) for the Intergovernmental Panel on Climate Change (IPCC). Cambridge: Cambridge Univ. Press.

Petschel-Held, G., A. Block, M. Cassel-Gintz et al. 1999. Syndromes of global change: A qualitative modelling approach to assist global environmental management. *Env. Model. Assess.* **4**:295–314.

Ross, M. 2001. Does oil hinder democracy? *World Pol.* **53**:325–361.

Rotmans, J., and B. de Vries. 1997. Perspectives on Global Change: The TARGETS Approach. Cambridge: Cambridge Univ. Press.

Schwartz, P. 1991. The Art of the Long View: Planning for the Future in an Uncertain World. New York: Wiley.

Steffen, W., A. Sanderson, P. Tyson et al. 2004. Global Change and the Earth System: A Planet under Pressure. Berlin: Springer.

Theys, J., ed. 1998. L'environnement au XXIe siècle [The Environment in the 21st Century]. Paris: Germes.

Thompson, M., R. Ellis, and A. Wildawsky. 1990. Cultural Theory. Boulder, CO: Westview.

van Notten, P., J. Rotmans, M. van Asselt, and D. Rothman. 2003. An updated scenario typology. *Futures* **35**:423–443.

WCED (UN World Commission on Environment and Development). 1986. Our Common Future. The Brundtland Report. Geneva: WCED.

20

Evaluating Past Forecasts

Reflections on One Critique of *The Limits to Growth*

Dennis L. Meadows

Laboratory for Interactive Learning, Durham, NH 03824, U.S.A.

ABSTRACT

In early 2005 after completing four years of study, an international network of 1,500 specialists produced a massive study, The Millennium Ecosystem Assessment. The principal findings in their reports generally confirm the long-term forecasts made over three decades earlier in a 1972 book, *The Limits to Growth* by Donella Meadows and her colleagues.

The main features of the *Limits to Growth* projections have also been supported by other scientific studies during the past decade. They are also confirmed increasingly in the media by the widespread and growing concern about global problems related to oil depletion, water scarcity, agricultural soil deterioration, fisheries collapse, species loss, and climate change.

Despite this corroboration, the initial *Limits to Growth* report was met with a storm of criticism that has persisted among many decision makers down to the present. In addition, a general survey of policy makers, at least in the United States today, would probably show they are either totally uninformed about the substance of the *Limits to Growth* forecasts, or that they have a vague and general impression that the report was disproved long ago and is no longer worthy of serious consideration.

What explains this contrast? There is basic confirmation of the *Limits to Growth* scenarios within the fields of physical and biological science while there is general disdain and disregard within the fields of economics and policy. This is a question of central importance for the Dahlem Workshop, which hopes to produce a body of analysis that will have a different fate. The question can be usefully addressed by examining several of its corollaries.

What vocabulary is useful for describing forecasting efforts and their products? What is the role of scenarios like those prepared by the *Limits to Growth* team? Why are the *Limits to Growth* conclusions still debated and criticized, after they have been generally confirmed by physical and biological developments on the planet? What would be the criteria and the procedures relevant for someone who is honestly trying to evaluate forecasts like those presented in *Limits to Growth*? What motivations prompt those who criticize such work? What lessons can we learn that will raise the quality and the impact of any forecasts that emerge from this Dahlem Workshop process?

In this chapter I first present a brief glossary of terms useful in describing and evaluating efforts to foretell the future. Then I summarize the main features of the original *Limits to Growth* forecasts and compare those forecasts with scientists' current understanding of the global situation as summarized most recently in the Millennium Ecosystem Assessment. I examine one early and influential criticism of the *Limits to Growth* report—a major magazine article by Peter Passell and two colleagues, academic economists in the U.S. I list the various ways in which these three authors sought to dispute our analyses and discredit our team. I offer a conceptual perspective that is useful for understanding the role of forecasts in the evolution of social policy. Finally, I provide a list of seven concrete steps the IHOPE participants could take when releasing their report. The goal of all this is to provide the 96th Dahlem Workshop participants with information and perspectives that will help them increase the value and the impact of their work to develop better foundations for describing the integrated history and the future of people on Earth.

GLOSSARY OF TERMS FOR DISCUSSING EFFORTS TO FORETELL THE FUTURE

Numerous names for efforts to foretell the future are used commonly as synonyms. Among them are: conjecture, prediction, prophecy, scenario, prognostication, augury, divination, projection, prognosis, and forecast. There are, however, subtle differences among these terms. An influential futurist in the social sciences, Frenchman Bertrand de Jouvenel (1964, 1967), used the term *conjecture*. For this workshop I was asked to write a paper entitled, "Testing Past Predictions." But professionals in the field have come to prefer the word *forecast*. It is used both as a verb and a noun. It is the term I normally use in describing my own research, and it is the word I will use in this chapter.

Forecast has three connotations: uncertainty, assessment of the current situation, and preparation for action. The term, *prediction*, has a simpler meaning. It designates an effort to describe what will exist at some point or during some interval in the future. One predicts an eclipse of the moon, but one forecasts the weather. This use of the term *forecast* became widely used in response to the monumental report to OECD by Erich Jantsch (1967) after he conducted a worldwide study of formal methods used by industry and the military to anticipate the diverse futures that will emerge to affect their success.

There are different reasons for forecasting, and one must be explicit about the goals of a specific effort before making any evaluation of it.

> The first thing to recognize in judging the worth of predictions…is that predictions can serve quite different purposes (Iklé 1967, p. 733).

I find it useful to differentiate among three objectives. Some forecasts are *philosophical*. I could, for example, forecast how an epidemic of the Marburg virus in Africa would follow a different course, if there were no motorized transport in the region. Were I to make such a forecast, I would have no impression whatsoever that I could change any of the factors that are included in my analysis, nor could I imagine that the outcome, whatever it turns out to be, will have any personal implications for me. A wonderfully informative and insightful forecast of this sort, especially relevant to the IHOPE participants is available in Harrison Brown's insightful portrait of humanity's long-term interaction with technology and the natural environment (Brown 1954). Iklé called these forecasts, "Predictions for the purpose of entertainment."

Some forecasts are *defensive*. I could, for example, forecast that global oil prices will rise to over $100/barrel in the coming decade. In making this forecast, I have no impression that I can change the conditions that will affect oil price, but I certainly believe that the forecasts, if they are accurate, will permit me to take actions that will reduce the negative effects on me and my household of such high prices.

As a result of this defensive forecast, I might move to a new home in the city center to eliminate my reliance on an automobile, or I could decide to install solar collectors on my house to provide hot water at prices unaffected by the cost of petroleum, or I could install better thermal insulation in my house, or I could do other things to reduce the difficulties that will come with much higher oil prices. I might even decide to invest in the stocks of oil companies that have large petroleum reserves. Iklé called these forecasts, "predictions to overcome indecision."

Some forecasts are *proactive*. I suppose the Dahlem Workshop is designed to give us better foundations for this kind of statement about the future. I could, for example, forecast that intensive use of herbicides in the yard of the house where I live will eventually lead to chemical contamination of the drinking water I pump from an underground aquifer. In this instance, if I act early enough, I can change the physical conditions upon which I predicated my forecast and thereby avoid the result and the negative consequences I foresaw. I can avoid the use of chemicals on my grass, or I might move my well to a distant location. For this category Iklé coined the term, "guiding predictions." As he wrote, these are made to "describe the consequences of a course of action and of some of its alternatives so that we can shape the future more to our liking."

The consequences of an inaccurate forecast in each of the three situations are very different. The criteria for deciding if the forecast was useful are also very different, and each of the forecasts could possibly be quite useful, even if it turned out to provide an inaccurate picture of the future.

For example, I have made some philosophical forecasts that turned out to be useful, because they led me to learn about relevant issues, meet interesting people, or read through important literature. In some cases, where I worked together with a group, the forecasting process was useful, because it forced us to identify our differences and resolve them. Note that these criteria of utility are largely unrelated to the ultimate numerical accuracy of the forecasts we produced.

A defensive forecast could turn out to be useful, because it prompted actions that led to valuable results, even though the world moved down a different path than was anticipated. At my university in the fall of 1999 we made a defensive forecast that abbreviated date codes in many computer programs would disrupt essential research and administrative functions at the turn of the millennium. After the fact we must admit that the Year 2000 problem was mainly a bust. However, the forecast led my university colleagues to make many investments and changes in their computer systems that have been subsequently useful, even without any Year 2000 Armageddon.

It could be that the proactive forecast I made was extremely useful, even correct, although ultimately not a single molecule of herbicide ever entered my drinking water. Why? Because I was motivated by the forecast to take effective, preventive action. I changed the conditions upon which the negative forecast had been based.

> …it appears that a forecast must be correct to be of value. However, in the situation where the decision maker has partial control over the outcome, forecasting then becomes more complicated. If the decision maker is given a forecast of a particular event that he considers undesirable, he may use what control he has to forestall the event. To the extent that he is successful, he invalidates the forecast (Martino 1973, p. 28).

> It may be argued that the utility of a forecast is determined by how favorably it affects the decisions made by the forecast user. This approach makes it possible for a wholly inaccurate forecast to be a very good forecast…it may be suggested that the definition of forecast-system validity be changed to reflect the utility of the forecasts generated rather than their accuracy (Gerjuoy 1977, p. 37).

Another important phrase is "*time horizon*." This is synonymous with "time scale" as that expression is used in the IHOPE conference documents. Time horizon is the interval of time over which we are concerned about the character and evolution of a system. This workshop explicitly differentiates among millennial, centennial, and decadal intervals. This is useful, because the types of relationships involved in change over a decade are very different than those that cause change over periods of a hundred or many thousands of years.

It is important to differentiate between *static* and *dynamic* forecasts. The first address factors that do not change over the chosen time horizon. To be useful, static forecasts should provide quite precise numerical estimates of numerical values for the key elements. Dynamic forecasts address factors that do change their value significantly over the relevant time period. With dynamic forecasts the value more often resides in accurate portrayal of behavior patterns. In this meeting the issues of interest require dynamic forecasts; our concern is with alternative patterns of behavior not with the precise numerical value of some factor in the distant future.

Three more terms are useful as we identify our conceptual approach to the factors we consider in thinking about the future. There are *excluded* factors; they are totally omitted from the analysis. They simply do not appear in the discussion.

Exogenous factors are treated by us as if they are external to the system. We recognize that they may change their value over time, and that their change will affect the system of interest. But exogenous factors are assumed not to be influenced by the evolution of the system. For example, population growth is often an exogenous factor in forecasts of future climate. We all know that climate will have important impacts on both birth and death rates. But since we do not understand them well, we treat population as an exogenous factor. We make assumptions about how it will change in the future. We feed into our models the values we assumed. We represent the impact of population on important variables within our analysis. But we do not attend to the effects any variable in our model might have on the change in population.

There are also *endogenous* variables. These influence other parts of the studied system, and their own behaviors are influenced by some of those other parts.

Finally, it is important to categorize the method used in developing a particular forecast. There are many formal techniques for converting information about the present into images of the future. Erich Jantsch's book (1967) describes several: econometrics, morphological analysis, trend extrapolation, and scenario writing are just a few of the formal methods he describes. The precise methodological differences among these various procedures are not especially relevant here. But it is extremely important to understand the terms that Jantsch used to differentiate between two fundamentally different philosophies of forecasting: *normative* and *extrapolative*.

The first, normative, develops a desirable image of the future and then traces back, step by step, to the present the sequence of events that would need to occur for that image to be realized. If you think the goal is compelling enough to call itself into realization, then the step-by-step sequence becomes a forecast. The second, extrapolative, analyzes the numerical values and the momentum of current factors and extends them out to some point or interval in the future.

The World Order Models Project (WOMP) was an important normative forecasting effort. The project originated in the 1960s among North American academics. Guidance came from Saul Mendlowitz, a faculty member in law at Rutgers University. It dealt with the specifics of culture and history in different nations of the world. WOMP participants described preferred futures and worked to understand plausible and effective transition strategies. Many books and reports were produced within the WOMP. The first was by Richard Falk (1975).

The 1976 book by Amílcar Herrera and his colleagues is another excellent example of a normative global modeling project. In their work they developed an image of an attractive global society in the year 2060 and then created a model to determine what policies would be required to achieve it.

> The model presented here is quite explicitly normative. It is not an attempt to discover what will happen if present trends continue but tries to indicate a way of reaching a final goal, the goal of a world liberated from underdevelopment and misery. It does not pretend to be "objective" in the sense of being value-free as

> generally understood. It portrays a conception of the world shared by its authors and to which they are deeply committed (Herrera et al. 1976, p. 7).

Using the terms I have explained in this brief glossary, I hope you will now understand precisely my meaning when I say that the book, *Limits to Growth*, reported on the results of a proactive, extrapolative forecasting effort that examined the dynamics of population and industry for a 200-year time horizon, 1900 through 2100. It treated population, industrial capital, natural resources, pollution, and other important global variables as endogenous variables.

THE ORIGINAL FORECASTS OF THE *LIMITS TO GROWTH* TEAM

There were many forecasts during the second half of the last century that led to major impacts. For example, there were scenarios of potential disruption from the Year 2000 effect, extrapolations of ever-growing stratospheric ozone holes, weather forecasts, statements about the imminent peak in global oil production, and declarations that Saddam Hussein would use weapons of mass destruction. Among the influential forecasts were those made in our book, which sold millions of copies in about 30 languages.

The first chapter of *Limits to Growth*, immediately listed our three main conclusions (quoted below from Meadows et al. 1972, p. 24):

1. If the present growth trends in world population, industrialization, pollution, food production and resource depletion continue unchanged, the limits to growth on this planet will be reached sometime within the next one hundred years. The most probable result will be a rather sudden and uncontrollable decline in both population and industrial capacity.
2. It is possible to alter these trends and to establish a condition of ecological and economic stability that is sustainable far into the future.
3. The sooner the world's people decide to strive for this stability, the greater will be their chance of success.

Left unsaid in the third conclusion is the notion that waiting too long will make it impossible to avoid collapse.

Notice that the model used to derive the *Limits to Growth* forecasts, World3, did generate thousands of numerical values for hundreds of parameters for the period 1900 through 2100. However, we did not use World3 to make point predictions. We used it to determine the main behavioral tendencies of the global system. Although most of the media focus was on the precise numerical paths of variables in the model, the authors paid no particular attention to them.

> We have deliberately omitted the vertical scales and we have made the horizontal time scale somewhat vague because we want to emphasize the general behavior

modes of these computer outputs, not the numerical values, which are only approximately known (Meadows et al. 1972, pp. 123–124).

We maintained this orientation and revalidated the main conclusions in the two subsequent updates to the 1972 book (Meadows et al. 1992, 2004).

There was only one important change. In 1972 we concluded that global population and industrial activity were still below the levels that could be supported indefinitely on Earth. By 2004 it was clear to us that they had grown above sustainable levels. So in 1972 the main objective seemed to us to be finding ways of slowing down physical expansion on the planet. In 2004 the main objective had become getting physical flows that are propelled by population and industry back down below the carrying capacity of the planet.

The change in 2004 did not result from discovery of mistakes in our initial analysis; it reflected the population and industrial growth that had occurred during the three decades that elapsed between our first and our third editions.

Leave that change aside. Here I will focus on the 1972 analysis, for it is that book that affords us three decades of perspective. That perspective supports two conclusions. First, the initial analysis was basically correct. Second, the *Limits to Growth* did not finally have any discernible impact on the trends it found to be so threatening. I believe, looking back over the decade of the 1970s, that our forecast was extremely useful. It generated hundreds of studies, papers, conferences, and books on long-term global futures. Thousands of people today genuinely feel that reading our book profoundly changed their understanding of global trends; many of these people also feel that our forecasts caused them to change the field of their studies and their career focus. Nevertheless, as we make clear in our third edition, our initial study did not prompt any fundamental changes in the policies that govern growth in population and industrial activity and that are driving this planet to major ecological disruptions.

In the remainder of this chapter I briefly substantiate the first conclusion and then reflect on the reasons for the second. I conclude by reflecting on the past three decades and giving some guidelines that might help the Dahlem IHOPE network be less influenced by the criticisms that its work is certain to provoke.

RECENT CONFIRMATION FROM THE BIOLOGICAL AND PHYSICAL SCIENCES

In 1972 it had not yet occurred to the vast majority of humanity that there was any chance our species could expand its physical demands on the planet to levels that would cause serious, irreversible damage to key ecosystems. However, through the 1980s and 1990s this awareness began to dawn. Concern about global pesticide use and recognition of the stratospheric ozone hole were among the early harbingers, and they stimulated more systematic analyses. These studies generated an enormous new body of data, theory, and literature. Most of it focused on specific issues in particular locations and on policies of interest to a narrow group of decision makers. Thus broad views and general consensus were slow to emerge. There were, however, at least two major efforts to develop a holistic perspective based on biological and physical science that could gain the support of a large group of individuals from many nations and disciplines. Both of them provided confirmation of our initial forecasts.

In 1992 a group comprised of more than 1,600 scientists, including 102 Nobel laureates, from 70 countries, issued a manifesto, *World Scientists' Warning to Humanity.* It said:

> Human beings and the natural world are on a collision course. Human activities inflict harsh and often irreversible damage on the environment and on critical resources. If not checked, many of our current practices put at serious risk the future that we wish for human society and the plant and animal kingdoms, and may so alter the living world that it will be unable to sustain life in the manner that we know. Fundamental changes are urgent if we are to avoid the collision our present course will bring about (http://www.worldtrans.org/whole/warning.html).

This statement attracted substantial attention in Europe and Japan; it evoked little interest in the United States. The same may be said about a more recent effort.

Between 2001 and 2005 a group of approximately 1,360 experts from 95 countries worked in loose collaboration to assess the consequences of ecosystem change for human well-being and to establish the scientific basis for actions needed to enhance the conservation and sustainable use of ecosystems.

Their work, summarized in *The Millennium Ecosystem Assessment Synthesis Report* (MEA 2005) reported many conclusions. But, generally speaking, we can say that the report fully supports the main results of the *Limits to Growth* project.

In the MEA among the five drivers of change in ecosystems and their services, population change (which means growth) and change in economic activity were the first two to be mentioned. The MEA documented many instances in which these drivers have already taken consumption of ecosystem services above sustainable levels:

> Ecosystem services that have been degraded over the past 50 years include capture fisheries, water supply, waste treatment and detoxification, water purification, natural hazard protection, regulation of air quality, regulation of regional and local climate, regulation of erosion, spiritual fulfillment, and aesthetic enjoyment. The use of two ecosystem services—capture fisheries and fresh water—is now well beyond levels that can be sustained even at current demands much less future ones. At least one-quarter of important commercial fish stocks are over harvested. From 5% to possibly 25% of global freshwater use exceeds long-term, accessible supplies....Some 15–35% of irrigation withdrawals exceed supply rates and are therefore unsustainable (MEA 2005, p. 6).

Note that the MEA did not examine nonrenewable resources, such as petroleum, which are, by their nature, used unsustainably today.

The MEA mainly examined the past and present. It did, however, give some basis for anticipating the nature of global change over the next fifty years. In the process, in carefully couched language, it also substantiated the notion that collapse is possible, even likely.

> Approximately 60% (15 out of 24) of the ecosystem services examined during the Millennium Ecosystem Assessment are being degraded or used unsustainably, including fresh water, capture fisheries, air and water purification, and the regulation of regional and local climate, natural hazards, and pests.... The challenge of reversing the degradation of ecosystems while meeting increasing demands for their services can be partially met under some scenarios that the MA has considered but these involve significant changes in policies, institutions and practices, that are not currently under way (MEA 2005, p. 1).

Of course some details of the original *Limits to Growth* forecasts remain in doubt. But the global community of biological and physical scientists no longer disputes that physical expansion in population and industrial activities are causing irreversible damage to important earthly ecosystems.

So why can many skeptics still claim the opposite:

> Our species is better off in just about every measurable material way. And there is stronger reason than ever to believe that progressive trends will continue past the year 2000, past the year 2100, and indefinitely (Simon 1997, p.1).

> In 1972, the Club of Rome published *Limits to Growth* questioning the sustainability of economic and population growth....None of these developments has even begun to occur. So the Club of Rome was wrong (ExxonMobile 2002).

Thomas Kuhn (1962) long ago disabused us of the notion that paradigms, overarching views of the world, evolve in a steady and objective fashion through reasoned consideration of each new piece of information. It is quite the opposite! Those, scientists or policy makers, who have found one view of the world to be useful first ignore and then, when that becomes impossible, bitterly fight against information that might suggest a different view is more valid or useful. Think of the Catholic Church condemning Galileo to life imprisonment for his suggestion that the universe does not revolve around the Earth.

The clash between those subscribing to the established paradigm and those proposing an alternative may ensue for decades, even if the new paradigm is more valid or useful and even if the material stakes are relatively small. Or it may be brief. It depends on the rate at which disconfirming evidence can be marshaled and on the culture and ethics of the disputants related to evidence and debate.

When the paradigm involves systems that evolve slowly over decades; when there are enormous ego, financial, or political benefits at stake; when the establishment controls the channels of communication and sets the rules of the battle; the fight between proponents of an obsolete and a new paradigm can be long and rancorous. In the case of Galileo, it was only 350 years after his death, in 1992, when the Church finally admitted that errors had been made in his1633 trial.

The Church has no monopoly on obstinacy. The contemporary debates between those who believe or disbelieve that human activities are causing climate change, or between those on both sides of the question of when global oil production will peak, illustrate the heat generated by paradigm clash.

If even relatively objective, physical phenomena, like climate change and oil production, can stir hard feelings and debate, imagine how long it will take us to achieve consensus on the need for the physical dimensions of population and industry to cease their expansion and decline back down to sustainable levels.

I will illustrate the motivations and the strategies at work in a clash of this sort by describing an early and influential criticism of the *Limits to Growth* forecasts.

THE NEW YORK TIMES BOOK REVIEW OF *LIMITS TO GROWTH*

Shortly after our book appeared, *The New York Times Book Review* Sunday magazine published a general critique of our work and of two earlier books by Jay Forrester (Passell et al. 1972). That review was authored by three young academic economists (Peter Passell, Marc Roberts, and Leonard Ross), who claimed no shred of expertise in either the biological or the physical sciences. None of the three had ever managed an enterprise that developed new technologies. They had no reputation in forecasting, and, indeed, did not offer any of their own forecasts in their review except to predict that, "A virtually infinite source of energy, the controlled nuclear fusion of hydrogen, will probably be tapped within 50 years."

Two of them had a conflict of interest, since they were about to release a book, *The Retreat from Riches: Affluence and its Enemies*, which would lose its rationale and, presumably, a large share of its potential market, if the *Limits to Growth* thesis were accepted as valid.

In their review the three critics made numerous statements about our research that were demonstrably false, they ascribed goals and opinions to us without even once citing an excerpt from our text that would substantiate their claims, and they resorted frequently to *ad hominem* attacks that impugned the expertise, the integrity, and the motivations of the *Limits to Growth* team members, even though the review authors had never met any of the authors.

In short they produced a polemic in the guise of science, a piece of work so shoddy that today Dr. Passell, now the editor of a modest house publication for the Milken Institute, would certainly feel compelled to discard it in the nearest waste bin, if it were submitted for publication in his journal.

Yet, and this is the important part of the story, the piece was accepted and widely referenced, and many of its mistakes have been incorporated into the popular myths about *Limits to Growth*. How could that happen? The answer is simple: The review confirmed the views of the economic and political establishment.

The *New York Times* review was particularly vitriolic, but the goals, the strategies, and the mistakes of the three authors were representative of many hundreds of critiques written during the 1970s in response to our work. Thus it is useful to address briefly the work of Passell et al. (1972) before turning to the question of major interest here: What strategies could we employ in the IHOPE effort to achieve a better outcome when it eventually starts to make statements about the future on the basis of its work.

The *Times* review was written in 1972. However, it remains a useful illustration of the tactics and the mistakes that any IHOPE effort will eventually confront, if it challenges prevailing views. So it is useful to examine the article in some detail.

There are three important features of the Passell text:

1. The authors actually did not read the entire book. They saw reports of the work, decided that it was a challenge to their own reputations and paradigms, and set out to destroy its credibility in whatever way possible. As a result, they made many objectively false statements about the contents of the book. They may have done this through calculation, carelessness, or through ignorance; it makes no difference.
2. Peter Passell and his colleagues ignored our statement of goals and imputed to us the objectives they would have had, if they had done a global modeling project.
3. The authors of the review also endeavored to impugn the goals, the expertise, and the integrity of the *Limits to Growth* authors. They did this with no objective evidence through a process we might label, "proof through insinuation."

It is clear that Passell and his colleagues did not read the entire book. Whatever examination of the text actually took place was for the purpose of finding support for a set of preexisting, critical views. This sounds harsh, but I cannot otherwise explain their errors. The following is not an exhaustive list, but it does illustrate the range of the literal mistakes in their *New York Times* manuscript all subsequently carried over into their book, when they took the review and incorporated it verbatim into their text.

In the following I cite from the review and then quote relevant portions of the *Limits to Growth* text.

Passell et al: "Factors the researchers believe influence population and income are boiled down to a few dozen equations."

Limits to Growth: "The relationships discussed above comprise only three of the hundred or so causal links that make up the world model" (p. 121). Plus the precise diagram of the entire model is reproduced on pages 102 and 103; it shows over 150 different factors, each of which would have to be represented by one or more equations.

Passell et al.: "World reserves of vital materials (silver, tungsten, mercury, etc.) are exhausted within 40 years."

Limits to Growth: The flow diagram of the model, and our text, makes clear that World3 does not distinguish among nonrenewable resources. "We plot one generalized resource that represents the combined reserves of all nonrenewable resources" (p. 94).

The nonrenewable resources are all lumped together in one stock in our model, implicitly assuming enormous possibilities for substitution. And in every one of the 13 scenarios published in the book in the year 2010, at least 50%, and more often 80%, of the resources originally assumed to exist in the year 1990 are still available for use.

On this issue Passell and his colleagues mistakenly attributed to our model a set of statistics that we indicated clearly came from a U.S. Bureau of Mines' publication. The statistics had absolutely no relation to our model in any way.

Passell et al.: "It is no coincidence that all the simulations based on the Meadows world model invariable end in collapse."

Limits to Growth: The principal point of the book lies in the two scenarios shown on pages 165 and 168. They do not collapse. Indeed they are labeled Stabilized World Model I and II. I suspect Passell and his colleagues did not get that far in the book, because they had already written their review.

Passell et al.: "It is disconcerting to note that one earlier variant of the world model, which does manage to avoid collapse, is not even discussed in *Limits*.

Limits to Growth: See above; we present and discuss at length two noncatastrophic possibilities.

Passell et al.: "In the *Limits* version…crowding actually increases birth rates."

Limits to Growth: There is nowhere in the *Limits* text, nor in any of the book's six diagrams that portray the causes of birth rates, any suggestion that crowding directly influences birth rates. The precise flow diagram of the model directly contradicts that statement.

Passell et al.: "Its message is simple: Either civilization or growth must end, and soon."

Limits to Growth: There is no justification for this statement in the book. Indeed we speak at length about the options that would be available after population and industrial capital stocks had stopped expanding. "If society's time horizon is as long as 70 years, the permissible population and capital levels may not be too different from those existing today" (p. 173). "An equilibrium defined in this way does not mean stagnation…corporations could expand or fail, local populations could increase or decrease, income could become more or less evenly distributed. Technological advance would permit the services provided by a constant stock of capital to increase slowly." "No one can predict what sort of institutions mankind might develop under these new conditions" (p. 174).

"It seems possible, however, that a society released from struggling with the many problems caused by growth may have more energy and ingenuity available for solving other problems." To emphasize that point we quoted from another economist, John Stuart Mill, who wrote in 1857, "It is scarcely necessary to remark that a stationary condition of capital and population implies no stationary state of human improvement. There would be as much scope as ever for all kinds of mental culture, and moral and social progress; as much room for improving the Art of Living and much more likelihood of its being improved."

More important than the literal mistakes was the total misunderstanding of our goals and important results. Economists, most especially those in the U.S. in the late 20th century, work and think with essentially linear models. Their goal is precise prediction of numerical values in the near future. They have notorious difficulty with delays, nonlinearities, and unquantified variables. So variables for which there is not an extensive data series are generally omitted from their forecasts. That is one reason their forecasts have such poor accuracy when called upon to predict variables that lie far outside the period for which they have data.

Passell and his colleagues assumed, or at least implied, that we had the same goals and constraints, even though they acknowledged that we were using a different forecasting method. In fact we frequently acknowledged the impossibility of making point predictions, and we recognized the imperfect nature of our data.

THE PURPOSE OF THE WORLD MODEL

"In this first simple world model, we are interested only in the broad behavior modes of the population-capital system. By *behavior modes* we mean the tendencies of the variables in the system to change as time progresses. The output graphs reproduced later in this book show values for world population, capital, and other variables on a time scale that begins in the year 1900 and continues until 2100. These graphs are *not* exact predictions of the values of the variables at any particular year in the future. They are indications of the system's behavioral tendencies only" (Meadows et al. 1972, pp. 92–93).

Passell et al: "*Limits* pretends to a degree of certainty so exaggerated as to obscure the few modest (and unoriginal) insights that it genuinely contains."

Note that Passell and his coauthors do not quote any of our text to prove this assertion. In fact, they do not quote any of our text anywhere in their review. Were they to do so, it would call their assertions into question. Here is what we actually said:

> "Can anything be learned from such a highly aggregated model? Can its output be considered meaningful? In terms of exact predictions, the output is not meaningful.... The data we have to work with are certainly not sufficient for such forecasts, even if it were our purpose to make them" (Meadows et al. 1972, p. 94).
>
> "We hope that by posing each relationship as a hypothesis, and emphasizing its importance in the total world system, we may generate discussion and research that will eventually improve the data we have to work with" (Meadows et al. 1972, p. 121).
>
> "If decision-makers at any level had access to precise predictions and scientifically correct analyses of alternate policies, we would certainly not bother to construct or publish a simulation model based on partial knowledge (Meadows et al. 1972, p. 122).
>
> "We have deliberately omitted the vertical scales (on our computer plots) and we have made the horizontal time scale somewhat vague because we want to emphasize the general behavior modes of these computer outputs, not the numerical values, which are only approximately known" (Meadows et al. 1972, p. 123–124).

Because our text does not justify the review's many criticisms, finally the three authors resort to character assassination.

Passell et al.: "*The Limits to Growth* in our view, is an empty and misleading work. Its imposing apparatus of computer technology and systems jargon conceals a kind of intellectual Rube Goldberg device—one which takes arbitrary assumptions, shakes them up and comes out with arbitrary conclusions that have the ring of science."

> "Less than pseudoscience and little more than polemical fiction."
>
> "The *Limits* researchers jigger the assumptions just enough to eliminate this noncatastrophic possibility." [A note to readers with English as their second language: "jigger" in this context means "to manipulate or rearrange (figures, etc.) especially so as to mislead."]
>
> "Still, *The Limits to Growth* might be excused in spite of its lack of originality and scent of technical chicanery." [Note: "chicanery" in this context means "the use of trickery in debate or action; deception."]

I often disagree with assumptions and conclusions that I read in the papers of my colleagues. But when that happens, I start from the assumption that our varying perceptions arise from honest differences in paradigms, or discrepancies in our knowledge, or varying levels of expertise. It even occurs to me, when I see someone else with a different point of view, that I might be the one making a mistake. I wonder how Passell, Roberts, and Ross came to conclude that the authors of *Limits* were engaged in deliberately lying and misrepresentation of their research. Their review does not give any indication of the basis for this view.

Whatever the origins of their views, the attributes of their review—misrepresentation, misunderstanding, and malevolence are characteristic of those within the establishment who feel threatened by potential for a major paradigm shift. We can expect similar responses to any forecasts developed by the IHOPE project. So it is useful to reflect on the past reviews by Passell and others in a similar vein, and decide what could be done to reduce their frequency and their impact.

STRATEGIES FOR COUCHING A FORECAST

Let me immediately acknowledge that this is the weakest part of this chapter. I have not thought much about these issues, and I have certainly not been very effective in blunting criticism of my own work. Nevertheless I may have some insights that are useful:

1. Be aware! Do not imagine you are entering a scientific exercise in which the best theories and data will inevitably and quickly win. Scenarios that portray futures for the complex global system will challenge the perceived vested interests of important players. Ego gratification and self-importance, political influence and financial income are at stake. When people see threats to these, they do not respond as scientists eager to improve their knowledge; they fight back and attempt to destroy the credibility of the messenger.
2. Be explicit and repetitious about the limitations of your analysis and the focus of your work. Be careful to label your speculation as that and to advance your conclusions in ways that can be fully justified by your data and analysis.
3. As part of your analysis describe an explicit set of time-series data that could be monitored in the future to gauge whether or not the world is moving in the directions you forecast. State explicitly what you would expect to see happening over time, if you are right. Indicate explicitly what trends or events would cause you to decide that important parts of your forecast had been in error.
4. Wherever possible, muster a large and impressive group of people to endorse your findings and recommendations. This strategy was, for example, employed by Goldsmith (1972) in his *Blueprint for Survival* in 1972 and by the Union of Concerned Scientists (1992).

5. In any event, take pains to relate what you have done to earlier work that generally supports your views. Where possible include many citations to earlier work that supports your conclusions.
6. In your report acknowledge that some readers will disagree; spell out in explicit detail what they should present in place of your findings, if they wish to promote other views.
7. Respond immediately and in kind to those who criticize your work. If they have made legitimate and useful criticisms, acknowledge those and revise your analysis. If they have resorted to distortions and character assassination, point that out explicitly in publications that will come to the attention of their professional peers. Or, better yet, avoid publication delays and take your case to the web.

Will these steps mean that others evaluate your forecasts on their own scientific merits only? Of course not—not if you are working in areas where much is at stake. But it will perhaps gain you a longer period of measured consideration by those who genuinely wish to improve their understanding.

REFERENCES

Brown, H. 1954. The Challenge of Man's Future. New York: Viking.

de Jouvenel, B. 1964. L'art de la conjecture. Monaco: Editions du Rocher.

de Jouvenel, B. 1967. The Art of Conjecture. London: Weidenfeld and Nicolson and New York: Basic.

ExxonMobil. 2002. Limits to growth? *Wall Street J.* July 25.

Falk, R. 1975. A Study of Future Worlds: Preferred Worlds for the 1990s. Old Tappan, NJ: Free Press.

Gerjuoy, H. 1977. Validity of forecasting systems. In: The Study of the Future: An Agenda for Research, ed. W.I. Boucher, pp 33–37. WDC Stock No. 038–000–00327–5. Washington, D.C.: U.S. Govt. Printing Office.

Goldsmith, E.A., R. Allen et al. 1972. Blueprint for Survival. Boston: Houghton Mifflin.

Herrera, A.O., H. Scolnik, G. Chichilnisky et al. 1976. Castrophe or New Society? A Latin American World Model. Ottawa: Intl. Development Research Centre.

Iklé, F.C. 1967. Can social predictions be evaluated? *Daedalus* **96**:733–758.

Jantsch, E. 1967. Technological Forecasting in Perspective. Paris: OECD.

Kuhn, T. 1962. The Structure of Scientific Revolutions. Chicago: Univ. of Chicago Press.

Martino, J.P. 1973. Evaluating forecast validity. In: A Guide to Practical Technological Forecasting, ed. J.R. Bright and M.E.F. Schoeman, pp. 26–52. Englewood Cliffs, NJ: Prentice Hall.

Meadows, D.H., D.L. Meadows, J. Randers, and W.W. Behrens III. 1972. The Limits to Growth. New York: Universe.

Meadows, D.H., D.L. Meadows, and J. Randers. 1992. Beyond the Limits. White River Junction, VT: Chelsea Green.

Meadows, D.H., J. Randers, and D.L. Meadows. 2004. Limits to Growth: The 30-Year Update. White River Junction, VT: Chelsea Green.

Millennium Ecosystem Assessment. 2005. Millennium Ecosystem Assessment Synthesis Report. Washington, D.C.: Island. http://www.millenniumassessment.org/

Passell, P., M. Roberts, and L. Ross. 1972. Review of *The Limits to Growth. NY Times Book Rev.* April 2.

Simon, J. 1997. The State of Humanity. Malden, MA: Blackwell.

Union of Concerned Scientists. 1992. World Scientists' Warning to Humanity. Cambridge, MA. The text is available at: http://www.worldtrans.org/whole/warning.html

21

Integrated Global Models

Robert Costanza,[1] Rik Leemans,[2] Roelof M. J. Boumans,[1] and Erica Gaddis[1]

[1]Gund Institute for Ecological Economics, Rubenstein School of Environment and Natural Resources, The University of Vermont, Burlington, VT 05405–1708, U.S.A.
[2]Environmental Systems Analysis Group, Department of Environmental Sciences, Wageningen University, 6700 AA Wageningen, The Netherlands

ABSTRACT

Integrated global models (IGMs) attempt to build quantitative understanding of the complex, dynamic history and future of human–environment interactions at the global scale. There is now a 30-year history of this approach. Over this period, computer simulation modeling has become a well-accepted technique in scientific analysis, but truly integrated simulation models—those that address the dynamics of both the natural and human components of the system and their interactions—are still relatively rare, and those that do this at the global scale are even rarer. This chapter provides a survey of past experience with IGMs to serve as the basis for discussion about their role in the IHOPE project. We analyze seven IGMs in some detail, comparing and contrasting their characteristics, performance, and limitations. The integrated global data base that IHOPE will create can greatly spur the development, testing, and application of IGMs. At the same time, the development of IGMs can greatly facilitate thinking about what data needs to be collected. IGMs, therefore, will play a central role in the IHOPE project and deserve careful consideration.

INTRODUCTION

There is now a relatively long history of integrated global modeling using computer simulations, starting in the 1970s with the World2 (Forrester 1971) and World3 models (Meadows et al. 1972; Meadows and Meadows 1975). Since then the field has expanded, owing partly to the increasing availability and speed of computers and to the rapidly expanding global data base that has been created in response to increased interest in global climate change issues (Meadows 1985; Meadows et al. 1992; Nordhaus 1994; Alcamo et al. 1998; Rotmans and de Vries 1997; IPCC 1992, 1995, 2001; Boumans et al. 2002; Meadows et al. 2004). Collectively, the applications of integrated global models (IGMs) constitute a relatively well focused and coherent discussion about our collective future. As Meadows pointed out:

> Global models are not meant to predict, do not include every possible aspect of the world, and do not support either pure optimism or pure pessimism about the future. They represent mathematical assumptions about the interrelationships among global concerns such as population, industrial output, natural resources, and pollution. Global modelers investigate what *might* happen if policies continue along present lines, or if specific changes are instituted (Meadows 1985, p. 55; italics added).

This chapter evaluates IHOPE's needs for further developing IGMs by identifying pressing questions to be answered by IHOPE, and presenting and discussing already available IGMs. We will also identify some of the gaps and challenges. Needless to say, this analysis is based on our personal views and limited experiences. The out-of-IHOPE emerging innovative IGMs could well have capabilities to deliver insights far beyond our current imagination.

THE NEED FOR IGMs

In today's world, human impacts on ecological life-support systems are increasingly complex and far-reaching (Gallagher and Carpenter 1997). In this world, the emphasis needs to shift from addressing problems in isolation to studying whole, complex, interconnected systems and the dynamic interactions between the parts. Complexity implies that the whole is significantly different from the simple sum of the parts and that scaling (the transfer of understanding across spatial, temporal, and complexity scales) is a core problem. Incorporating both biophysical and social dynamics makes these problems impossible to address from within the confines of any single discipline.

To address these problems, we must supplement our rapidly growing body of data and analysis about the natural world and humanity's role in it with a significantly expanded emphasis on synthesis of this information (Pfirman and the AC-ERE 2003) so that it can be better communicated and used by both the scientific community and society. This synthesis can take many forms, but they all include some form of modeling. Models are defined as all methods of abstract representation of the real world, ranging from mental maps to conceptual diagrams to statistical correlations to dynamic computer simulation models. Although a plethora of models exist about various aspects of humans in natural systems, they have only begun to be integrated and adequately tested against our rapidly expanding data bases and for their utility in addressing real world problems.

Integration of the natural and social sciences has lagged, due, in part, to the lack of a common language of discourse. Integrated modeling provides one possible method to bridge this barrier (Costanza et al. 2001). Achieving true "consilience" (Wilson 1998) between the natural and social sciences would be a giant leap forward in our ability to understand the world and manage our activities within it in a sustainable manner. The broader research community has already begun to work together to pose the kinds of questions that will need to be answered to understand humans in natural systems. As an example, a set of 23 system-level questions was posed by the IGBP–GAIM project (now AIMES) in conjunction with the Earth System Science Partnership (ESSP) that will challenge an integrated 21st-century research community (see Box 21.1). This set includes analytical, operational, normative, and strategic questions. These questions can only be fully addressed through close collaboration between the social and natural sciences, as we are proposing in the IHOPE project overall, and especially in the integrated modeling activity.

Furthermore, the importance of considering scale depends on the boundaries set by the model. Whereas global aggregation may be appropriate for the impacts of CO_2 emissions on global climate change, simulation of regional impacts on water quality or quantity would not. A key question to be addressed by the global modeling community is: How can we capture both types of dynamics while maintaining a reasonable level of aggregation so as to keep the driving linkages clear and informative?

IGM DEVELOPMENT

In IHOPE, our approach to models needs to be pluralistic but evaluative. The rationale for any model is the scientific desire to capture the essence and to remove or reduce the redundant aspects of the system under study. What is essential and what is redundant, and thereby what level of reduction is required, to a large degree, depends on the questions being asked. The result is a "model" of reality that is realistic to varying degrees. There is thus no one "right" way to represent reality in models, but we can judge and evaluate the relative quality of different representations for different purposes. Indeed, the level of spatial and temporal aggregation, as well as model assumptions and complexity, should be explicitly driven by the types of questions the model will be used to answer. We must also differentiate between models used for forecasting versus comparative scenario analysis. We recognize that no one model can fulfill all of these needs.

The environment, society, and the economy each represent complex systems characterized by nonlinearities, autocatalysis, time-delayed feedback loops, emergent phenomena, and chaotic behavior (Kauffman 1993; Patten and Jørgensen 1995). Furthermore, these fundamental systems are intimately linked in ways that we are only just beginning to appreciate (Schellnhuber 1998). These complexities pose multiple challenges. Chief among these challenges is the recognition that to achieve the outcomes we desire, it will be necessary to incorporate simultaneously several different perspectives. Clearly it will be

Box 21.1 Earth system questions posed by IGBP–GAIM.

Analytical Questions:

1. What are the vital organs of the ecosphere in view of operation and evolution?
2. What are the major dynamical patterns, teleconnections, and feedback loops in the planetary machinery?
3. What are the critical elements (thresholds, bottlenecks, switches) in the Earth system?
4. What are the characteristic regimes and time scales of natural planetary variability?
5. What are the anthropogenic disturbance regimes and teleperturbations that matter at the Earth system level?
6. Which are the vital ecosphere organs and critical planetary elements that can actually be transformed by human action?
7. Which are the most vulnerable regions under global change?
8. How are abrupt and extreme events processed through nature–society interactions?

Operational Questions:

9. What are the principles for constructing "macroscopes," i.e., representations of the Earth System that aggregate away the details while retaining all systems-order items?
10. What levels of complexity and resolution have to be achieved in Earth system modeling?
11. Is it possible to describe the Earth system as a composition of weakly coupled organs and regions, and to reconstruct the planetary machinery from these parts?
12. What might be the most effective global strategy for generating, processing, and integrating relevant Earth system data sets?
13. What are the best techniques for analyzing and possibly predicting irregular events?
14. What are the most appropriate methodologies for integrating natural-science and social-science knowledge?

Normative Questions:

15. What are the general criteria and principles for distinguishing nonsustainable and sustainable futures?
16. What is the carrying capacity of the Earth?
17. What are the accessible but intolerable domains in the coevolution space of nature and humanity?
18. What kind of nature do modern societies want?
19. What are the equity principles that should govern global environmental management?

Strategic Questions:

20. What is the optimal mix of adaptation and mitigation measures to respond to global change?
21. What is the optimal decomposition of the planetary surface into nature reserves and managed areas?
22. What are the options and caveats for technological fixes like geoengineering and genetic modification?
23. What is the structure of an effective and efficient system of global environment and development institutions?

necessary to incorporate the essential theories, tools, and knowledge of multiple disciplines across the spectrum from social to biological to chemical to physical sciences (Costanza and Jørgensen 2002).

However, it will also be necessary to incorporate the perspectives of non-disciplinary experts who have a strong stake—either directly or indirectly—in the achievement of particular outcomes. These stakeholders include policy makers who must formulate and justify frameworks for future development, resource managers who must interpret and implement those frameworks, and ultimately the communities who will either suffer or benefit from policies and decisions. The core approach of IHOPE will be to attempt to integrate and synthesize perspectives across all of the relevant disciplines and stakeholders.

Simulation modeling will play a pivotal role in helping these disparate groups to incorporate their individual perspectives and visualize a common future (Costanza and Ruth 1998; de Vries, this volume). Clearly, simulation models have the ability to integrate complex dynamics and distill them into readily digestible outputs. They also provide powerful means to communicate and educate, especially when models are used as vehicles to debate assumptions, explore alternative formulations, and discuss responses to different input configurations. In essence, simulation models can be used as "universal translators" that allow individuals with different backgrounds to access a model from their own perspective and then assess the output from this model in a form that can be understood from multiple perspectives. It is thus critical that models are well documented, summarized, and easily accessible to the general modeling community as well as to policy makers. Models also allow us to explore the realm of the possible, setting bounds on what we can realistically achieve with policy.

The IHOPE project will build on integrated modeling experience at multiple scales, from local to global (Leemans 1995; Easterling 1997; Rotmans et al. 1990; Rotmans and van Asselt 2001; Rotmans and de Vries 1997; Krol et al. 2001; Costanza et al. 2002; Ehman et al. 2002; Boumans et al. 2002). IHOPE will allow us to accelerate the development and testing of models of humans in nature that yield better answers to important questions, such as those listed in Box 21.1. Of particular importance are questions that relate to the intertwined futures of human development and environmental quality at scales that include the metropolitan scale, the watershed scale, the country scale, and the global scale. In addition to the general questions outlined in Box 21.1, more specific questions that the IHOPE project can address are:

- What types of demands on natural capital will be generated by land-use change and GDP growth, the provision of clean water, and increases in crops yields?
- In which nations are development trajectories most constrained by natural capital, and where are they most constrained by institutions or systems of governance?

- Can the OECD nations continue to grow at the same time that poverty is reduced in developing nations?

TESTING AND EVALUATION

Significant effort in the IHOPE project needs to be devoted to testing and evaluating model performance relative to observed and reconstructed data and other criteria, as well as to understand and communicate uncertainty. Since the models are complex and the data is of highly variable quality and coverage (see Costanza, Chapter 3, this volume), this is not a trivial task (Perez-Garcia et al. 1995; Tran et al. 2002). We do not wish simply to proliferate new models, but rather we aim to develop a deeper understanding of the many ways in which models relate to data and better ways to judge the performance of models in order to winnow out the best performers for further development. The IHOPE participants have extensive experience in model assessment through intercomparison of models with other models and with widely accepted benchmark datasets. The various intercomparison projects have revealed weaknesses in model formulation, gaps in calibration and validation data, and new approaches that are applicable to other models. In addition IHOPE will contribute to building a "toolbox" of modeling and intercomparison techniques for use by the broader community and participants in the proposed IHOPE activities. In one recent initiative, a catalog of model intercomparisons was developed by IGBP/GAIM in conjunction with the World Climate Research Programme's (WCRP) Working Group on Coupled Modelling (WGCM). Although most of these comparisons strongly focused on physical climate models and a biogeochemical carbon model and rarely included social aspects, this catalog can serve to alert others to what has already been done, and thus eliminate redundancy in future efforts. It can also provide new modelers with an approach and baseline for model development and assessment.

Scenario Generation and Analysis

A key purpose in constructing integrated models of humans in nature is to facilitate quantifying the qualitative scenario narratives of the future under different assumptions about key driving forces, including alternative assumptions about institutions, technology, living standards, and policies (van Notten et al. 2003). Deciding which scenarios to run and interpreting the results are activities that require significant input from a very broad range of stakeholders (see de Vries, this volume). Thus this will be one of the IHOPE project's primary areas for stakeholder input and outreach, and represents a major nexus with communities such as policy makers, nongovernmental organizations, and business. It is, however, also a primary research area, as it inherently addresses the sensitivity of models to changes in underlying assumptions and parameter values.

A SURVEY OF IGMs

Table 21.1 lists and categorizes some historical and existing IGMs according to several relevant criteria. As can be seen, the list is still quite short. This is partly because we have limited it to those models which are both global in their spatial extent and integrated in their inclusion of both human and natural system components as endogenous variables. In the different international assessment processes, many other models have been used (e.g., Harvey et al. 1997), but most of them were not, or only partly, integrated. There are, of course, many additional integrated models at the regional scale (cf. Costanza and Voinov 2003) as well as many additional global models that are not integrated in the sense above because they are assembled from a series of individual disciplinary models (e.g., Prinn et al. 1999). For example, global climate models (GCMs) have become quite popular and useful, but they do not include any endogenous human components. Human activities are usually an outside driving force on these models. The impact of greenhouse gas emissions, for example, is modeled by setting the consequent changes in atmospheric concentrations or radiative forcing as new boundary conditions. Likewise, there are several global economic models that do not include natural systems (e.g., Leontief 1980; Hickman 1983).

Figure 21.1 illustrates the degree to which each IGM included in our survey captures important natural and social systems as well as the level of historic calibration attempted to date for each model. From this figure one can quickly see the range of coverage and complexity of natural and social components of the Earth system. Note that in Table 21.1 we have lumped together (in most cases) models that are simply versions or updates of each other. Of course, this line is a bit fuzzy at times, since models eventually evolve into new forms, but the intention was to list only unique models. Web sites are noted in Table 21.1 for easy access to more detailed information about a model.

Some interesting characteristics of this list, particularly for the purposes of IHOPE, include:

- Although the first models already emerged in the 1970s, integrating spatial and societal dimensions only developed recently.
- The temporal range covered by all models is quite consistent and mostly covers the contemporary history and near future. In fact, all of the models listed are concerned only with the last 100 years (IHOPE's decadal time scale) and 100 years into the future.
- The degree of calibration and testing of the models against real world data is generally low. IHOPE can contribute significantly in this area by making a consistent integrated data base available for the global modeling community.

The degree of human and natural system integration in the models has in general been only moderate. Again, IHOPE can contribute significantly in this area by providing a more integrated data base from which to start.

Table 21.1 Existing and historical IGMs and their characteristics.

Model	Primary references	Temporal extent	Maximum temporal resolution	Spatial extent	Maximum natural system spatial resolution	Maximum socioeconomic system spatial resolution	Major components
World2	Forrester 1971	1900–2060	1 yr	Global	Globally aggregated	Globally aggregated	Population, resources, built capital, pollution, agricultural land, food, industrial output
World3	Meadows et al. 1972, 1992, 2004	1900–2100	1 yr	Global	Globally aggregated	Globally aggregated	Population, resources, built capital, pollution, agricultural land, food, industrial output
WIM	Mesarovic & Pestel 1974; Hughes 1980		1 yr			12 regions	
IWM	Kile & Rabehl 1980		1 yr			24 regions	
IMAGE	Rotmans 1990	1900–2100	0.5 yr	Global (impacts only for the Netherlands)	7 ecosystems	9 regions	Human emissions, atmosphere, biosphere, and oceans
IMAGE 2	Alcamo 1994; Alcamo et al. 1998	1900–2100	1–5 year	Global	.5 degree grid	17 regions	Economy, society, atmosphere, biosphere, and oceans
IFs	Hughes 1996	1960-2100	1 year	Global	182 countries	182 countries	Economy
DICE	Nordhaus 1992, 1994	1990-2100	10 years	Global	Globally aggregated	Globally aggregated	Economy, climate
TARGETS	Rotmans & de Vries 1997	1900-2100	1 year	Global	Globally aggregated	4 regions	Economy, society, atmosphere, biosphere, and oceans
GUMBO	Boumans et al. 2002	1900-2100	1 year	Global	11 biomes	Globally aggregated	Anthroposphere (economy—human, built, social capital), atmosphere, biosphere, hydrosphere, lithosphere, ecosystem services

Table 21.1 (*continued*).

Model	Website	Number of variables and parameters*	Degree of human–natural system integration	Degree of calibration and testing	Use in policy	Modeling language (latest version)
World2		Few	Moderate	Low	Low	DYAMO
World3		Few	Moderate	Low	Moderate	STELLA
WIM	www.inta.gatech.edu/peter/wimg.html	Many	Moderate		Low	Visual Basic
IWM	www.inta.gatech.edu/peter/iwm.html	Many	Moderate	Low	Low	
IMAGE		Many	Low	Low	Moderate	AMSL
IMAGE 2	www.ciesin.org/datasets/rivm/image2.0-home.html	Moderate (but replicated in each grid cell)	Moderate	Medium	High	Fortran, C++, M, and others
IFs	www.du.edu/~bhughes/ifs.html		Moderate	Medium	Low	
DICE	sedac.ciesin.org/mva/DICE/DICEHP.html	Very few	Low	Medium–high	High	GAMS
TARGETS	sedac.ciesin.org/mva/iamcc.tg/TGsec4-2-7.htm	Many	Moderate	Medium	High	M
GUMBO	www.uvm.edu/giee/GUMBO/	Moderate	High	Medium–high	Low (so far)	STELLA

*The classes for the variables correspond to: "very few" up to 10, "few" up to 100; "moderate" up to 1000,"many" up to 10,000 and "very many" more than 10,000. Only the variables for the "unit model" are counted if the model has high spatial resolution. The unit model would be replicated in each cell.

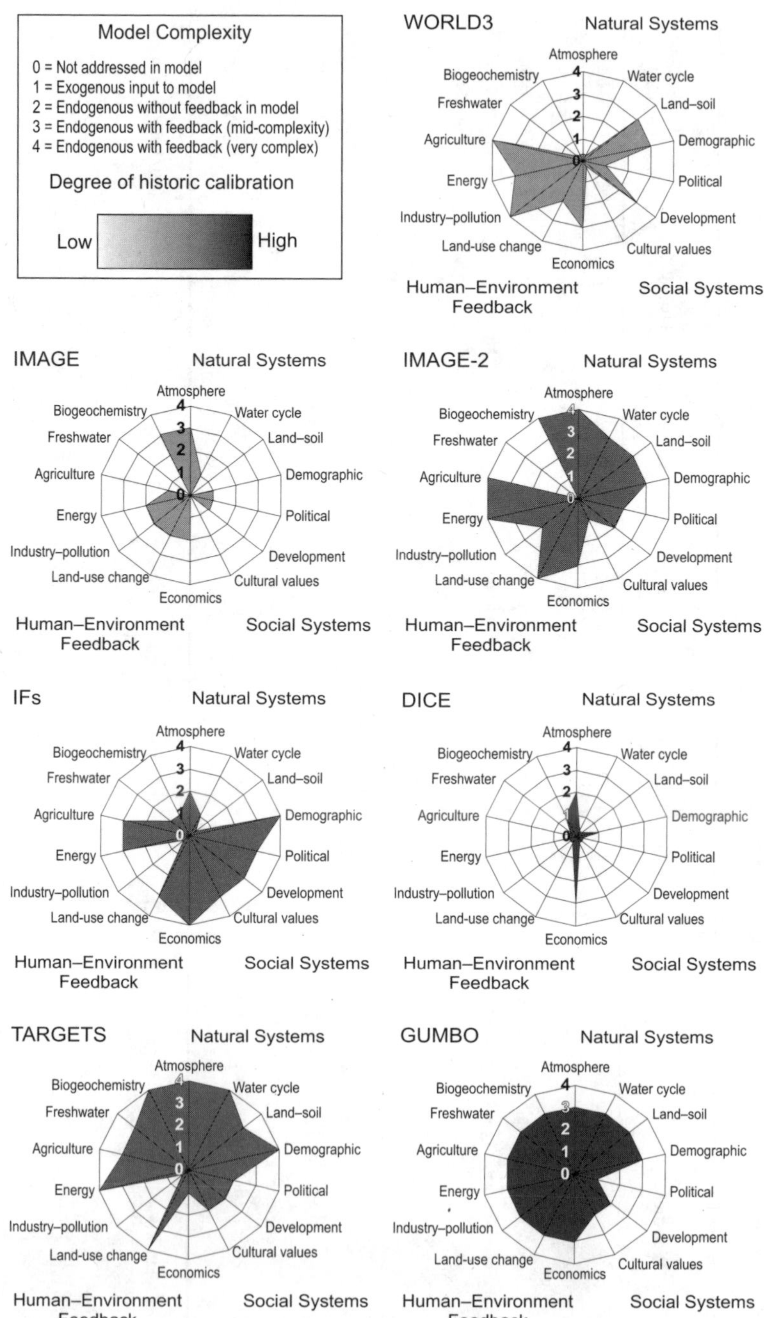

Figure 21.1 Diagram of complexity with which IGMs capture socioeconomic systems, natural systems, and human–environment feedbacks.

Below, some brief observations on each of the major models that are still in active use are given. Thereafter we conclude with some observations on the role of IGMs in IHOPE.

World3

The World3 model has been the subject of three influential books: *The Limits to Growth* (Meadows et al. 1972), *Beyond the Limits* (Meadows et al. 1992), and the recent, *Limits to Growth: The 30-year Update* (Meadows et al. 2004). World3 is a globally aggregated systems dynamics model broken down into 5 sectors (population, capital, agriculture, nonrenewable resources, and persistent pollution) and containing 16 state variables (i.e., population, capital, pollution, and arable land), 100 variables total, and 80 fixed parameters (Table 21.1; Meadows et al. 1974). The latest versions are written in STELLA and are easily runnable on a PC; however, they are not freely available to the public.

Because of the influence of the original book (several million copies were sold), this model has been the topic of intense scrutiny, debate, misunderstanding, and, one could argue, willful misinformation over the years (cf. Meadows, this volume). One interesting bit of misinformation that has been persistently circulating is the idea that the model's "predictions" have been proven totally wrong by subsequent events (*Economist* 1997). In fact, the model's forecasts made in 1972 have been pretty much on target so far.[1] The model's forecasts of collapse under certain scenarios did not start to occur until well past the year 2000. The true tests of this model's forecasts will arrive in the coming decades.

World3 has been criticized on methodological grounds (e.g., Cole et al. 1973). The most often cited difficulties are that it does not explicitly include prices, that it assumes resources are ultimately limited, and that it does not present estimates of the statistical uncertainty on its parameters. If fact, World3 is a viable and effective method to reveal the implications of the primary assumptions about the nature of the world that went into it. That is all that can be claimed for any model. These assumptions or "pre-analytic visions," need to be made clear and placed in direct comparison with the corresponding assumptions of the alternatives, in this case the "unlimited growth model." As Meadows et al. (1992, 2004) have repeatedly stressed, the essential difference in pre-analytic visions centers around the existence and role of limits: thermodynamic limits, natural resource limits, pollution absorption limits, population carrying capacity limits, and, most importantly, the limits of our understanding about where these limits are and how they influence the system. The alternative unlimited growth model, derived from neoclassical economic theory (see, e.g., the DICE model below), assumes there are no limits that cannot be overcome by continued

1 In an interview with the Dutch *Volkskrant* (April 16, 2005), Meadows compares his "1972" scenarios with the outcome of the Millennium Ecosystem Assessment (2005) and states that this assessment actually confirms the scenarios.

technological progress, whereas the limited growth model assumes that there are limits, based on thermodynamic first principles and observations of natural ecosystems. Ultimately, we do not know which pre-analytic vision is correct (they are, after all, assumptions). Thus we have to consider the relative costs of being wrong in each case (Costanza 2000; Costanza et al. 2000).

At the time of its initial release in 1972, World3 was at the cutting edge of computer simulation. Since then, simulation capabilities have increased dramatically, as has the availability of data to calibrate and test global models. The remaining models in this review reveal some of that development. Before leaving this brief discussion of World3, however, we should mention some of the things that could have been done, especially in the recently released 30-year update, but have not yet been. The most important of these has to do with calibration and testing. In all of the books on World3, calibration of the model with historical data is downplayed. This is strange, since the model runs always begin in 1900 and run for 200 years to 2100. Why not show historical data for the variables in the model for which historical data is available in order to demonstrate the model's "skill" at reproducing the past? The reason given is that since the model is only an approximation, one should not put too much emphasis on "precise" calibrations. We think this is ultimately a mistake, since it misses the opportunity to present quantitative tests of the model's performance—tests against which World3 would fare quite well and which would address at least some of the objections of its critics. World3 is also probably the only IGM for which a true "validation" test could be run. One could take the original forecasts made in 1972 of the period from 1972 to 2002 and compare them with the actual data from the 1972–2002 period. Unfortunately, this has not yet been done.

Finally, the discussions of World3 often point to the limited versus unlimited growth assumptions as a key difference with conventional economic models; however, they do not take the opportunity to look at the relative costs and benefits of being right or wrong in those assumptions. When one does this, one can easily see that the cost of assuming no limits and being wrong is the collapse scenarios shown by World3, while the cost of assuming limits and being wrong is only mildly constrained growth (Costanza et al. 2000). Some of the more recent models reviewed below try to elaborate on these scenarios.

IMAGE

IMAGE (Integrated Model to Assess the Greenhouse Effect; Rotmans 1990) signaled a new development in global integrated assessment modeling. Its aim was to create a comprehensive overview of global climate change. IMAGE was one of the first models that implemented OECD's Driver-Pressure-State-Impact-Response (DPSIR) approach (InterFutures Study Team 1979) into a global model by integrating a series of different models for each important subcomponent. The model was global in scope: it simulated the emissions from energy use and tropical deforestation and the consequent climate change in different regions. A few feedback processes (e.g., CO_2 fertilization) were included, mainly to calibrate the model to simulate observed atmospheric CO_2 concentrations. The only impact that was simulated was sea-level rise and its consequences for the Netherlands. One of the conclusions, for example, was that the costs to deal with sea-level rise would amount to more than half a percent of the Dutch national product. Most of the model components simulated the physical system (oceans, atmosphere, and C on land). Societal aspects were not strongly included. They determined emission levels and defined the cost of impacts.

The IMAGE model was developed in the second half of the 1980s at the then new Dutch Institute of Public Health and Environmental Protection (RIVM). RIVM strongly advocated such an integrated modeling approach, which led to presentations in Dutch parliament and helped to raise the awareness of Dutch policy makers to climate change. When the IPCC was established, IMAGE became one of the models that helped to evaluate the earliest IPCC-IS92 scenarios, which defined the input for the climate change modeling and impact community for over a decade. Also, several mitigation scenarios have been developed and published. Scientifically, IMAGE has been instrumental in the debate on global warming potentials. Since not only CO_2 but all greenhouse gases were included in the model, along with a then state-of-the-art atmospheric chemistry model, the advantages and limitations of this concept could be easily determined.

IMAGE has been systematically scrutinized using sensitivity analysis (Rotmans 1990). Automated Monte Carlo analysis was first tested on IMAGE, which led to advanced protocols for experimental design for model testing (Kleijnen et al. 1992). Such analysis resulted in the identification of important variables, which have to be determined accurately, and less sensitive variables, whose values do not matter as much. This analysis helped to advance the modeling research agenda. However, it also showed that variables important to stakeholders rarely were the most sensitive ones, which frustrated the actual agenda-setting process. Additionally, meta-modeling was tested for IMAGE. Meta-modeling searches for simpler relationships between model input and output, so that model experiments can be executed much faster. Meta-modeling techniques proved extremely useful in the development of scenarios, especially for systems whose futures are uncertain. This was especially required earlier, when computing was much less powerful. In addition to the expected costs of dike raising, the meta-modeling of IMAGE revealed that beyond the well-known role of the Antarctic and Greenland ice sheets in determining the impacts of sea-level rise, there must be an additional, major factor. This uncertainty has still not been resolved in subsequent IPCC reports.

IMAGE-2

IMAGE-2 (Integrated Model to Assess the Global Environment; Alcamo et al. 1998) is a further development of IMAGE. The model was initially developed to link important scientific and policy aspects of climate change in a geographically explicit manner to assist decision making (Alcamo 1994). Major emphasis was put on incorporating different components of the Earth system, including oceans, biosphere, atmosphere, and anthroposphere (i.e., society) and all major interactions and feedbacks. Over the years the model has been developed further by incorporating more detail (e.g., water use and land degradation) and improving the underlying datasets. The latest version (IMAGE 2.2) has been used intensively for supporting science–policy dialogues, which has led to the so-called safe-landing approaches (Alcamo and Kreileman 1996) and the scenario developments of IPCC (i.e., SRES; Nakícenovíc et al. 2000), UNEP's Global Environmental Outlook, and the Millennium Ecosystem Assessment. The extensive application of IMAGE-2 is illustrated by the fact that it has been one of the very few models used by of all the different IPCC working groups.

IMAGE-2 is one of the most advanced integrated assessment models currently available. Its major innovative aspect was (and still is) that it simulates energy- and industry-related activities simultaneously with land-use activities for the same set of drivers. This creates a much greater consistency for the different scenarios. Additionally, a spatially explicit global land-use model, based on a few transparent rules, creates highly different dynamics and patterns across different continents. This approach represents a major achievement. The model was calibrated against historic atmospheric CO_2 concentrations to balance oceanic and terrestrial carbon pools. Historic trends in land use, energy use, and industrial activities over the last decades were used to calibrate the socioeconomic models. Data for these trends are derived from large internationally available data bases, compiled by institutions such as FAO, UN, IEA, and the World Bank. IMAGE-2 now includes models for demography and the world economy to provide more detail and consistency for population (e.g., mortality, fecundity, and age structures) and economic drivers (trade, labor forces, resource use, and other economic constraints). The simulations begin in 1995 and run until 2100 with annual time steps. Output of the model is diverse and covers many aspects of the Earth system. Scientific publications have focused on the importance of feedback processes and impacts on ecosystems and agriculture (e.g., Leemans and Eickhout 2004). Policy-oriented applications have focused on scenarios and climate protection targets (e.g., Alcamo and Kreileman 1996).

The philosophy of the IMAGE-2 group has always been to be scientifically sound in order to be accepted by the policy community that needed the scientific advice and/or scenarios. This was achieved by frequently publishing in the international peer-reviewed scientific literature and by installing a scientific advisory board. The strength of IMAGE-2 was therefore its widely available documentation (two books, over 100 papers, and 4 CD-ROMs). IMAGE-2 was especially well accepted by the ecological science community, and scenarios from IMAGE-2 are now widely used by that community. Its weakness is that use of the model requires a well-trained multidisciplinary team, and that it has proven difficult to communicate the detailed results transparently. Some have argued that the presented detail, especially on maps, provides a false impression of precision. IMAGE-2 has consequently been condemned because too little attention was paid to uncertainty (e.g., van der Sluijs 1997). Currently, new methods using aggregated indicators are being developed to communicate results and the inherent uncertainty (e.g., Leemans and Eickhout 2004).

True validation of such a model (with such a forward-looking objective) and its scenarios is only possible by observing the future. As an alternative, the IMAGE-2 team tried to set up a truly independent validation exercise by starting the model in 1900 and simulating the 20th century. The first step of this validation exercise was to develop a historic data base of all relevant drivers. The resulting HYDE data base (documented in Klein Goldewijk 2000 and de Vries and Goudsblom 2002) was then used to initialize the model, after which trends and results were compared with known trends and outcomes. The match was perfect, which created suspicion. A major problem of this validation exercise was the actual coverage of the data. Before World War II much of the agricultural sector was already covered in global summary statistical data bases, but much of forestry and energy data were still part of an informal economy, and only little data with adequate coverage was available. In HYDE, gaps were filled with backward extrapolation of recent trends, model-based reconstructions based on models similar to those used in IMAGE-2, etc. Other available long-term global historic data bases involve similar problems. The validation exercise was thus not independent at all and had little scientific significance. IHOPE could become one of the major activities that will correct this problem.

IFs

Barry Hughes developed the International Futures simulator (IFs) inspired by the following world models: the Mesarovic-Pestel or World Integrated Model (Mesarovic and Pestel 1974; Hughes 1980), the Leontief World Model (Leontief et al. 1977), the Bariloche Foundation's World Model (Herrera et al. 1976), and the Systems Analysis Research Unit Model (SARU 1978). Originally developed for educational purposes, IFs' main function more recently has been that of a policy tool. The IFs model is used by the National Intelligence Council Project 2020, which aims to provide U.S. policy makers with future world developments to inform policy decisions (http://www.cia.gov/nic/ NIC_2020_project.html). It is also being employed in developing and analyzing scenarios for the UNEP GEO-4 report (in conjunction with IMAGE-2 and several other thematic models) and has been used to assess explicitly the attainment of the Millennium Development Goals (MDGs) outlined by the UNDP in 2004.

IFs is a global modeling system based on a data base derived for 182 countries since 1960. The model simulates onward, starting at the initial year 2000 (Hughes 1996; Hughes et al. 2004). The model focuses on capturing trends over the next 10–20 years, although projections out to 2300 are produced for some audiences. Components of the model include a population module, an economic module, an agricultural module, an energy module, a social and international political module, an environmental module, and a technical module. The population module follows 22 age–sex cohorts to old age with cohort-specific fertility and mortality rates of households in response to income and income distribution to simulate average life expectancy at birth, literacy rate, and overall measures of human development (HDI) and physical quality of life. The population model represents migration among the countries and shows the effects of HIV/AIDS. A recent development includes a submodel of formal education.

The economic module is a general equilibrium-seeking algorithm based on a Cobb–Douglas production function, which represents the economy in six sectors: agriculture, materials, energy, industry, services, and technology. It computes and uses input–output matrices that change dynamically with development level. The simulations account for changing consumption patterns and international trade. The model uses a social accounting matrix envelope to tie economic production and consumption to financial flows.

The agricultural module is a partial equilibrium model that represents production, consumption, and trade of crops and meat, ocean fish catch, and aquaculture. This model maintains land use in crops, grazing, forest, urban, and "other" categories dependent on the demand for food, livestock feed, and industrial use of agricultural products.

The energy module is a partial equilibrium module to consider known reserves, consumption and trade of oil, gas, coal, nuclear, hydroelectric, and other renewable energy forms. It portrays changing capital costs of each energy type with technological change as well as with drawdowns of resources.

The political module includes national and international politics. The national politics module computes fiscal policy, six categories of government spending (military, health, education, R&D, foreign aid, and a residual), and computes changes in social conditions, attitudes, and social organization. The national politics module allows for the evolution of democracy and prospects for state failures. The international political module traces changes in power balances across states and region.

The environmental module tracks the remaining resources of fossil fuels, area of forested land, water usage, and atmospheric CO_2 emissions.

Technology solutions are distributed throughout the model and represent the assumptions about rates of technological advance in agriculture, energy, and the broader economy. Technological advances are tied to the extent of electronic networking of individuals in societies and are dependent on the governmental spending model with respect to research and development.

The IFs model has undergone some historic calibration; however, the results are not easily available. Because the relationships driving the model are derived from historic data, the developers of IFs have instead focused on comparative analyses with multiple other model forecasts (Hughes 2004) in each of the issue areas captured by the model. Validation of the model proved difficult in the late 1980s (Liverman 1987). Since then, model assumptions, structure, and resolution have improved. A similar validation exercise would be useful in demonstrating the suitability of the IFs model for global simulation.

The IFs model has the most sophisticated and user-friendly GUI of all the models explored. It has recently been launched on the World Wide Web so that users can run scenarios over the Internet and instantly generate output graphs and comparisons between scenarios, either globally or for selected countries. IFs represents the most highly articulated socioeconomic model in our collection of IGMs, but its natural systems components are rather bare bones. It treats nature as resource sources and sinks, but with no articulation of the internal dynamics of the natural system.

DICE

The Dynamic Integrated Climate and the Economy (DICE) model (Nordhaus 1994) was developed in the early 1990s to investigate the economics of climate change. DICE is the simplest of the models evaluated in this review. The only biophysical process incorporated in the model is a very simple treatment of climate change. The optimization approach used in this model is distinctly different from the systems approach taken in most of the other models considered in this review.

Quoting from Nordhaus (1994, p. 5): "The basic approach of the DICE model is to use a Ramsey model of optimal economic growth with certain adjustments and to calculate the optimal path for both capital accumulation and GHG-emissions reductions." This was done by incorporating a greatly simplified depiction of the global atmosphere to form a set of climate–emissions–damage equations. Although the simplified climate equations might pick up the major features of the emissions–climate link, the link in the model between climate change and economic impact on human and natural systems is by far the weakest one. To pick this up, the DICE model assumes a very simple relationship between global mean temperature (as a proxy for climate change) and damage:

$$\frac{D(t)}{Q(t)} = .00144\,T(t)2,$$

where D is the loss of global output, Q is global output, and T is global mean temperature, all at time t (Nordhaus 1994, Eq. 2.11, p. 18). The missing links are the actual feedbacks between climate change (including the more important features of precipitation change and especially the geographic distribution of changes) and ecosystem changes, as well as between ecosystem changes and economic performance. These links are complex, yet they are the essence of the problem being addressed. Although integrated climate–economy–ecosystem models are still relatively rare (Parson and Fisher-Vanden 1995), several others (see above and below) do a fairly elaborate, spatially explicit job of estimating the climate–ecosystem linkages, and the results are anything but simple. This is why we ranked the "degree of human–natural system integration" in DICE as "low" in Table 21.1.

As Nordhaus (1973) pointed out, any model is only as good as the assumptions that go into it. In the case of the DICE model, a thorough job has been done in analyzing the model's sensitivity to uncertainty about the parameters, but no effort went into analyzing sensitivity to some of the more basic, and more important, assumptions. The Ramsey model of optimal economic growth used as the basis for DICE assumes that economic growth is not limited by natural resources or environmental changes. Economic output in DICE is estimated using a production function that includes only reproducible capital, labor, and technology in its arguments. Population growth and technological change are exogenous, and natural capital is completely missing. These are rather strong assumptions, given that one of the purposes of the DICE model is to integrate economic models with the rest of the natural world. In DICE, the economy goes on its merry way with no real feedback from the natural world. There is only the one-way flow of impacts on climate, and only through that on agriculture and ecosystems. Other work on an economic growth model with natural resources in the production function and endogenous population growth shows some very different results (Brown and Roughgarden 1995), so we can assume that adopting something other than the standard neoclassical growth model would make a big difference to the conclusions.

In addition, both the spatial and temporal resolution of the DICE model are very low, given the problem at hand. DICE is globally averaged and uses a time increment of ten years. Given that most of the underlying relationships are probably highly nonlinear and spatially discontinuous, this level of aggregation has got to cause some serious problems. Nordhaus stated: "The main result of aggregation theory is that aggregation is generally possible only when the underlying micro relations are linear" (Nordhaus 1973, p. 1160). This, combined with the simple basic structure of DICE, means that there are no real possibilities for "surprises" in DICE like the kind we have come to expect in the real world, and which can emerge from some of the other models reviewed here. Yet, there is no discussion of the possibly huge impacts of aggregation error other than Nordhaus's contention that the level of aggregation used was necessary so that "the theoretical model is transparent and the optimization model is empirically tractable." These are good goals, but hardly justification for a model intended to be used to set realistic global policies on greenhouse warming. Some of these critiques have been addressed in RICE, a recent extension of the DICE model. RICE improves spatial resolution by modeling 6–10 regions separately (Nordhaus 1973; Nordhaus and Yang 1995). Further improvements on the model, including an assessment of uncertainty, are currently under development in another version called PRICE.

Finally, DICE assumes that consumption equals welfare: "We assume that the purpose of our policies is to improve the living standards or consumption of humans now and in the future" (Nordhaus 1994, p. 10). This is one purpose, but consumption is not always correlated with overall human well-being or welfare, more broadly defined (e.g., Easterling 1974, 2003; Daly and Cobb 1989; Ekins and Max-Neef 1992). There is some attempt to broaden the concept of consumption beyond conventional GNP by stating that consumption "includes not only traditional market purchases of goods and services like food and shelter but also nonmarket items such as leisure, cultural amenities, and enjoyment of the environment" (Nordhaus 1994, p. 10). After saying this, these nontraditional components are quickly forgotten, and the productive values of natural capital (which are probably more important) are never even considered. The problem is that material growth in the economy can become "anti-economic" if the many uncounted costs of additional growth begin to outweigh the counted benefits (Daly and Cobb 1989). The DICE model, through its simple damage function, includes only a very crude estimate of some of these costs, but it has no way of picking up any nonconsumption welfare effects or feedback effects from the environment to the economy. Yet Nordhaus blithely talks about the "welfare" effects of various policy scenarios. What DICE actually models is (at best) the marketed, and some small piece of the nonmarketed, consumption effects, and these may in fact be opposite to the true welfare effects as the planet's natural capital base continues to erode.

TARGETS

TARGETS (Tool to Assess Regional and Global Environmental and health Targets for Sustainability; Rotmans and de Vries 1997) is also a direct descendent of IMAGE. It aims to redirect and accelerate the discussions from climate change toward global change. Its main innovation was that the model assumed that changes in drivers were a direct function of different perspectives on how the world system functions and is managed. The drivers thus do not define boundary conditions but are an integral part of the model itself. A future, as defined by TARGETS, provides the different implications of such a perspective in terms of population and health, energy, use of land and water, and biogeochemical cycles. Another new aspect is not to list the absolute impacts but to indicate them in terms of risks for unsustainable developments.

The model is thus strongly based on the concept of cultural perspectives to address the apparent uncertainty in the interactions between humans and their natural environment (van Asselt and Rotmans 2002). The natural and socioeconomic dimensions are highly integrated in TARGETS. As such it has departed from the simpler causal chain or DPSIR models used earlier. TARGETS is based on the basic notion that in the absence of complete knowledge and in order to guide choices and actions, people use stylized and simplified images of the world around them. These images are based on experiential trends interpreted by implicit rules. This complex represents human values and beliefs. These images further form the bases of different world views and also determine the behavior of people and thus their interactions in the Earth system.

TARGETS consists of five submodels: population and health, energy, land, food, and water. Each of those submodels is a DPSIR model, but they are linked through a socioeconomic scenario generator, in which policy responses are explicitly incorporated. All submodels can be used in a stand-alone mode as well, to allow model and data comparisons, comparisons with other models, and targeted validation and sensitivity analysis. The TARGETS team argues that having adequate insight into the behavior of the submodels promotes an understanding of the behavior of interactions in the full model. The level of and approach toward integration achieved in TARGETS could well be an example for IGMs to be developed in IHOPE.

The TARGETS approach does not provide simple answers and again stresses that the future is highly uncertain. It generates insights in the accelerating influence of the human race and, as such, strongly supports the notion of the Anthropocene (Crutzen and Steffen 2003). It also provides a new insight (not appreciated by many modelers) that individual people perceive changes in their environment differently, which influences the rationale for selecting appropriate responses. TARGETS is not used to develop scenarios, but utopias: worlds dominated by a particular world view or perspective. Although innovative and challenging, the stakeholders evaluating the model results had large difficulties in understanding the role of all these perspectives (e.g., the Ulysses project: http://zit1.zit.tu-darmstadt.de/ulysses). Additionally, the lack of spatial detail for the land, water, and biogeochemical simulations was also seen as a major drawback. TARGETS, however, was quite influential in setting the stage for the acceptance of the narrative scenarios approach, later also adopted by the IPCC SRES scenarios (Nakícenovíc et al. 2000) and by the Millennium Ecosystem Assessment (2005).

GUMBO

The Global Unified Metamodel of the BiOsphere (GUMBO; Boumans et al. 2002) was developed as part of a working group at the National Center for Ecological Analysis and Synthesis (NCEAS) in Santa Barbara, CA. Its goal was to simulate the integrated Earth system and assess the dynamics and values of ecosystem services. It is a "meta-model" in that it represents a synthesis and a simplification of several existing dynamic global models in both the natural and social sciences at an intermediate level of complexity. The current version of the model contains 234 state variables, 930 variables total, and 1715 parameters (Table 21.1). GUMBO is the first global model to include the dynamic feedbacks among human technology, economic production and welfare, and ecosystem goods and services within the dynamic Earth system. We rated its degree of human–natural system integration as "high" in Table 21.1. GUMBO includes five distinct modules or "spheres": the atmosphere, lithosphere, hydrosphere, biosphere, and anthroposphere. The Earth's surface is further divided into eleven biomes or ecosystem types, which encompass the entire surface area of the planet: open ocean, coastal ocean, forests, grasslands, wetlands, lakes/rivers, deserts, tundra, ice/rock, croplands, and urban. The relative areas of each biome change in response to urban and rural population growth, gross world product (GWP), as well as changes in global temperature. Among the spheres and biomes, there are exchanges of energy, carbon, nutrients, water, and mineral matter. In GUMBO, ecosystem services are aggregated to seven major types, while ecosystem goods are aggregated into four major types. Ecosystem services, in contrast to ecosystem goods, cannot accumulate or be used at a specified rate of depletion. Ecosystem services include: soil formation, gas regulation, climate regulation, nutrient cycling, disturbance regulation, recreation and culture, and waste assimilation. Ecosystem goods include: water, harvested organic matter, mined ores, and extracted fossil fuel. These 11 goods and services represent the output from natural capital, which combines with built capital, human capital, and social capital to produce economic goods and services and social welfare. The model calculates the marginal product of ecosystem services in both the production and welfare functions as estimates of the prices of each service.

Historical calibrations from 1900 to 2000 for 14 key variables, for which quantitative time-series data was available, produced an average R2 of 0.922. A range of future scenarios to the year 2100, representing different assumptions about future technological change, investment strategies, and other factors, have been simulated. The scenarios include a base case (using the "best fit" values of the model parameters over the historical period) and four initial alternative scenarios. These four alternatives are the result of two variations (a technologically optimistic and skeptical set) concerning assumptions about key parameters in the model, arrayed against two variations (a technologically optimistic and skeptical set) of policy settings concerning the rates of investment in the four types of capital (natural, social, human, and built). They correspond to the four scenarios laid out in Costanza (2000) and are very similar to the four scenarios used in the recent Millennium Ecosystem Assessment (2005).

Although this is an early version of GUMBO, some preliminary results and conclusions include:

- A high level of dynamic integration between the biophysical Earth system and the human socioeconomic system is important if we are to develop integrated models with predictive capabilities. The IHOPE project will be extremely important in fostering better integration.
- Preliminary calibration results across a broad range of variables show very good agreement with historical data. This builds confidence in the

GUMBO model and also constrains future scenarios. The model produced a range of scenarios that represent reasonable rates of change of key parameters and investment policies, and these bracketed a range of future possibilities that can serve as a basis for further discussions, assessments, and improvements. Any user is able to change these parameters further and observe the results.

- Assessing global sustainability can only be done using a dynamic integrated model. However, one is still left with decisions about what to sustain (e.g., GWP, welfare, welfare per capita). GUMBO allows these decisions to be made explicitly and in the context of the complex world system. It allows both desirable and sustainable futures to be examined.
- Ecosystem services are an important link between the biophysical functioning of the Earth system and the provision of sustainable human welfare. We have found that their physical and value dynamics are quite complex.
- The overall value of ecosystem services, in terms of their relative contribution to both the production and welfare functions, is shown to be significantly higher than GWP (4.5 times in this preliminary version).

"Technologically skeptical" investment policies are shown to have the best chance (given uncertainty about key parameters) of achieving high and sustainable welfare per capita. This means increased relative rates of investment in knowledge, social capital, and natural capital, as well as reduced relative rates of consumption and investment in built capital.

The GUMBO model is available over the Internet but requires the STELLA software for the user to run. The GUI developed for GUMBO is built into the STELLA software, but does not come with instructions or guidance.

MODEL INTERCOMPARISONS

The natural and social processes important to capture in IGMs include human demographics and development, economic production in multiple sectors (energy, food production, industry, and forestry), and the response of atmospheric, terrestrial, and aquatic ecosystems to increasing degradation. Perhaps the most challenging component of an IGM is to link these impacts on ecosystem health back to human demographics and economic projections (Figure 21.1). In so doing, we must especially pay attention to issues of spatial and temporal scale. The models considered in this chapter range in complexity, spatial aggregation (from globally aggregated models to highly spatially explicit models), and temporal resolution (decadal vs. annual time steps).

We have attempted a concise comparison of model resolution, complexity, feedback, and empiricism. Table 21.1 summarizes the time step, spatial and temporal resolution, and the degree of human–natural system integration found in each model. Model complexity and empiricism is captured in Figure 21.1, which gives a quick graphic overview of the relative coverage and complexity of each model. Model components included as exogenous inputs are given a rank of 1. For example, most of the IGMs use policy scenarios as an exogenous input. Model components that are calculated endogenously but are not included in feedbacks to other model components were assigned a rank of 2. For example, the IMAGE model calculates sea-level rise in the water cycle but this variable does not feed back into other model variables. Endogenous model components that are included in intercompartmental dynamics are ranked 3 and 4 depending on the level of internal complexity captured and spatial scale. For example, the GUMBO model is a globally aggregated meta-model which was given a rank of 3 for most endogenous feedback components. Table 21.2 documents the justification for each of the model component rankings. A gray-scale ramp is included in Figure 21.1 to illustrate the relative level of historic model calibration.

CONCLUSIONS: IHOPE AND IGMs

IHOPE can benefit from explicit connections with IGMs and the IHOPE project can spur significant advances in IGMs through several mechanisms, including:

- Allowing much more elaborate quantitative calibration and testing of IGMs. IHOPE's data bases will be a shared community resource that will be extremely useful for this purpose. We might consider the implementation of an evolutionary selection cycle among models of humans in nature, whereby the IHOPE project would oversee the testing of alternative models against the common data base and rewarding of the most successful models to encourage their further development.
- Extending the temporal extent of IGMs outside the decadal time scale. This will be extremely important for calibration and testing, since the period of 1900 to the present has been one of steady exponential growth in most human system variables. One needs to go further in the past to observe societal collapses (Diamond 2005). Forecasting these collapses (and lack of collapses) will be an important way to test the skill of integrated models.
- Spurring a much higher degree of human and natural system integration in model structure. By assembling integrated teams of researchers from across the natural and social sciences and the humanities to construct an integrated data base, IHOPE will build bridges that will also allow more integrated thinking about model construction and testing.

Table 21.2 Documentation of complexity with which IGMs capture socioeconomic systems, natural systems, and human–environment feedbacks.

	WORLD3	IMAGE	IMAGE-2	IFs	DICE	TARGETS	GUMBO
Atmosphere	None	CO_2, climate change feedback	CO_2 endogenous, climate change feedback	CO_2 endogenous; no feedback	Exogenous	Six energy-related gases calculated; feedback	Carbon, water, nutrient cycles
Water cycle	None	Sea-level rise calculated	Sea-level rise calculated	Exogenous	None	Tracks ten water reservoirs (surface, ground, soil, and ocean)	Calculates surface, ground, soil, and ocean water stocks
Land–soil	Land erosion, fertility	None	None	None	None	Erosion	Weathering, erosion, fertility
Demographic	Feedbacks from all other modules	Exogenous	Demographic module	Complex population predictions	Exogenous	Population and health submodel with feedback	Total population (urban and rural) with feedback
Political	Scenario inputs	Scenario inputs	Scenario inputs	Predicts government spending by sector, democracy, and state failure	None	Scenario inputs	Scenario inputs
Development	Simple	None	None	Predicts HDI and Millennium Development Goals	None	Development indices predicted	Calculates sustainable social welfare and human capital
Culural values	None	None	None	Predicts value change between traditional/ secular rationalism to survival/self-expression	None	Driver of responses modules	Technological optimist vs. technological skeptic
Economics	Industry, service, agricultural output	Calculates impacts of climate change; no feedback	Calculates impacts of climate change; no feedback	Complex dynamics by sector	Damage costs calculated; feedback to other economic components	Economic scenarios drive model	Calculates built capital and ecosystem goods and services

Table 21.2 (*continued*)

	WORLD3	IMAGE	IMAGE-2	IFs	DICE	TARGETS	GUMBO
Land-use change	Agriculture/industry	Agriculture/forest	Spatially explicit global land-use module	Urban/forest/ agriculture	None	Forest/grass/ agriculture	Forest/wetland/ grass/urban/desert
Industry–pollution	Endogenous with feedback	Exogenous	Exogenous	None	None	None	Waste calculated from built capital
Energy	Incorporated into nonrenewable resource sector	Calculates change in energy use; no feedback	Calculates change in energy use; no feedback	Energy demand and fossil fuel reserves computed; feedback with economic sector	None	Demand for five sectors and energy supply; feedback	Calculates energy use and proportion from fossil fuels; feedback
Agricultural	Driven by food per capita and land-use change	Simple; no feedback	Simple endogenous; no feedback	Production driven by population; feedback to land use and economics	None	Food demand driven by population and economics	Calculates agricultural impacts; feedback to economics
Freshwater	None	None	Calculates water use	Calculates water demand but not availability	None	Water demand and availability calculated; feedbacks	Calculates water demand and availability; feedback to economics
Biogeochemistry	None	CH_4, N_2O, CFC, CO_2	CH_4, N_2O, CFC, CO_2	CO_2, simple	None	Tracks cycles of C, N, P, and S	Carbon and nutrient budgets
Reference	Meadows et al. 1974	Rotmans 1990	Alcamo et al. 1998	Hughes 2004	Nordhaus 1973	Rotmans and de Vries 1997	Boumans et al. 2002

REFERENCES

Alcamo, J. 1994. Image2.0 Integrated Modeling of Global Climate Change. Amsterdam: Kluwer.

Alcamo, J., and G.J.J. Kreileman. 1996. Emission scenarios and global climate protection. *Glob. Env. Change: Hum. Pol. Dim.* **6**:305–334.

Alcamo, J., R. Leemans, and G.J.J. Kreileman. 1998. Global change scenarios of the 21st century: Results from the IMAGE 2.1 model. Oxford: Pergamon and London: Elsevier.

Boumans, R., R. Costanza, J. Farley et al. 2002. Modeling the dynamics of the Integrated Earth System and the value of global ecosystem services using the GUMBO Model. *Ecol. Econ.* **41**:529–560.

Brown, G., and J. Roughgarden. 1995. An ecological economy: Notes on harvest and growth. In: Biodiversity Loss: Economic and Ecological Issues, ed. C.A. Perrings, K.-G. Mäler, C. Folke, C.S. Holling, and B.-O. Jansson, pp. 150–189. Cambridge: Cambridge Univ. Press.

Cole, H.S.D., C. Freeman, M. Jahoda, and K.L.R. Pavitt, eds. 1973. Models of Doom: A Critique of the Limits to Growth. New York: Universe.

Costanza, R. 2000. Visions of alternative (unpredictable) futures and their use in policy analysis. *Conserv. Ecol.* **4**:5. http://www.consecol.org/vol4/iss1/art5

Costanza, R., H. Daly, C. Folke et al. 2000. Managing our environmental portfolio. *BioScience* **50**:149–155.

Costanza, R., and S.E. Jørgensen, eds. 2002. Understanding and Solving Environmental Problems in the 21st Century: Toward a New, Integrated "Hard Problem Science." Amsterdam: Elsevier.

Costanza, R., B. Low, E. Ostrom, and J. Wilson, eds. 2001. Institutions, Ecosystems, and Sustainability. Boca Raton, FL: Lewis/CRC.

Costanza, R., and M. Ruth. 1998. Using dynamic modeling to scope environmental problems and build consensus. *Env. Manag.* **22**:183–195.

Costanza, R., and A. Voinov, eds. 2003. Landscape Simulation Modeling: A Spatially Explicit, Dynamic Approach. New York: Springer.

Costanza, R., A. Voinov, R. Boumans et al. 2002. Integrated ecological economic modeling of the Patuxent River watershed, Maryland. *Ecol. Mono.* **72**:203–231.

Crutzen, P.J., and W. Steffen. 2003. How long have we been in the Anthropocene era? *Clim. Change* **61**:251–257.

Daly, H.E., and J. Cobb. 1989. For the Common Good: Redirecting the Economy towards Community, the Environment, and a Sustainable Future. Boston: Beacon.

de Vries, B., and J. Goudsblom, eds. 2002. Mappae Mundi: Humans and Their Habitats in a Long-term Socio-ecological Perspective. Myths, Maps and Models. Amsterdam: Amsterdam Univ. Press.

Diamond, J.M. 2005. Collapse: How Societies Choose to Fail or Succeed. New York: Viking.

Easterling, R.A. 1974. Does economic growth improve the human lot? In: Nations and Households in Economic Growth, ed. P.A. David and M.W. Reder, pp. 89–125. New York: Academic.

Easterling, R.A. 2003. Explaining happiness. *Proc. Natl. Acad. Sci.* **100**:11,176–11,183.

Easterling, W.E. 1997. Why regional studies are needed in the development of full-scale integrated assessment modelling of global change processes. *Glob. Env. Change* **7**:337–356.

Economist. 1997. Plenty of gloom. Dec 20. **345**:19–20.

Ehman, J.L., W.H. Fan, J.C. Randolph et al. 2002. An integrated GIS and modeling approach for assessing the transient response of forests of the southern Great Lakes region to a doubled CO_2 climate. *Forest Ecol. Manag.* **155**:237–255.

Ekins, P., and M. Max-Neef, eds. 1992. Real-life Economics: Understanding Wealth Creation. London: Routledge.

Forrester, J. 1971. World Dynamics. Cambridge, MA: MIT Press.

Gallagher, R., and S. Carpenter. 1997. Human-dominated ecosystems. *Science* **277**: 485–490.

Harvey, D., J. Gregory, M. Hoffert et al. 1997. An introduction to simple climate models used in the IPCC Second Assessment Report. IPCC Technical Paper Intergovernmental Panel on Climate Change. Geneva: IPPC.

Herrera, A.O., H.D. Scholnik et al. 1976. Catastrophe or New Society? A Latin American World Model. Ottawa: Intl. Development Research Center.

Hickman, B.G., ed. 1983. Global International Economic Models. Amsterdam: Elsevier.

Hughes, B.B. 1980. World Modeling: The Mesarovic-Pestel World Model in the Context of Its Contemporaries. Lexington, MA: Lexington Books.

Hughes, B.B. 1996. International Futures: Choices in the Creation of a New World Order. 2d ed. Boulder, CO: Westview.

Hughes, B.B. 2004. The Base Case of International Futures (IFs): Comparison with Other Forecasts. Report prepared for the NIC Project 2020. http://www.du.edu/~bhughes/ ifsreports.html

Hughes, B.B., A. Hossain, and M. Irfan. 2004. The structure of International Futures (IFs). http://www.du.edu/~bhughes/ifsreports.html

InterFutures Study Team. 1979. Mastering the Probable and Managing the Unpredictable. Paris: OECD

IPCC. 1992. Climate Change 1992: The Supplementary Report to the IPCC Scientific Assessment. Cambridge: Cambridge Univ. Press.

IPCC. 1995. Impacts, Adaptations, and Mitigation of Climate Change: Scientific-technical Analyses. Contribution of Working Group II to the Second Assessment Report of the Intergovernmental Panel on Climate Change. Cambridge: Cambridge Univ. Press.

IPCC. 2001. Summary for Policymakers: A Report of Working Group I of the Intergovernmental Panel on Climate Change. http://www.ipcc.ch (2001)

Kauffman, S. 1993. The Origins of Order: Self-Organization and Selection in Evolution. New York: Oxford Univ. Press.

Kile, F., and A. Rabehl. 1980. Structure and use of the Integrated World Model. *Technol. Forecst. Soc. Change* **17**:73–87.

Kleijnen, J.P.C., G. van Ham, and J. Rotmans. 1992. Techniques for sensitivity analysis of simulation models: A case study of the CO_2 greenhouse effect. *Simulation* **58**: 410–417.

Klein Goldewijk, C.G.M. 2000. Estimating global land-use change over the past 300 years: The HYDE 2.0 data base. *Glob. Biogeochem. Cyc.* **15**:417–434.

Krol, M.S., A. Jaeger, A. Bronstert et al. 2001. The semi-arid integrated model (SIM), a regional integrated model assessing water availability, vulnerability of ecosystems and society in NE-Brazil. *Phys. Chem. Earth* **B 26**:529–533.

Leemans, R. 1995. Determining the global significance of local and regional mitigation strategies: Setting the scene with global integrated assessment models. *Env. Monit. Assess.* **38**:205–216.

Leemans, R., and B. Eickhout. 2004. Another reason for concern: Regional and global impacts on ecosystems for different levels of climate change. *Glob. Env. Change* **14**:219–228.

Leontief, W. 1980. The world economy of the year 2000. *Sci. Am.* **243**:207–231.

Leontief, W., A.P. Carter, and P. Petri. 1977. The Future of the World Economy: A United Nations Study. New York: Oxford Univ. Press.

Liverman, D.M. 1987. Forecasting the impact of climate on food systems: Model testing and model linkage. *Clim. Change* **11**:267–285.

Meadows, D.H. 1985. Charting the way the world works. *Technol. Rev.* **88**:55–56.

Meadows, D.H., D.L. Meadows, J. Randers, and W.W. Behrens. 1972. The Limits to Growth. New York: Universe.

Meadows, D.H., D.L. Meadows, and J. Randers. 1992. Beyond the Limits: Confronting Global Collapse, Envisioning a Sustainable Future. Post Mills, VT: Chelsea Green.

Meadows, D.H., J. Randers, and D.L. Meadows. 2004. Limits to Growth: The 30-year Update. Post Mills, VT: Chelsea Green.

Meadows, D.L., W.W. Behrens III, D.H. Meadows et al. 1974. Dynamics of Growth in a Finite World. Cambridge, MA: Wright-Allen.

Meadows, D.L., and D.H. Meadows. 1975. A summary of limits to growth: Its critics and its challenge. In: Beyond Growth: Essays on Alternative Futures, ed. D.L. Meadows et al., pp. 3–23. New Haven, CT: Yale Univ.

Millennium Ecosystem Assessment. 2005. Millennium Ecosystem Assessment Synthesis Report: A Report of the Millennium Ecosystem Assessment. Washington, D.C.: Island. http://www.millenniumassessment.org/

Mesarovic, M., and E. Pestel. 1974. Mankind at the Turning Point. New York: Dutton.

Nakícenovíc, N., J. Alcamo, G. Davis et al. 2000. Special Report on Emissions Scenarios. Cambridge: Cambridge Univ. Press.

Nordhaus, W.D. 1973. World dynamics: Measurement without data. *Econ. J.* **83**:1156–1183.

Nordhaus, W.D. 1992. An optimal transition path for controlling greenhouse gases. *Science* **258**:1315–1319.

Nordhaus, W.D. 1994. Managing the Global Commons: The Economics of Climate Change. Cambridge, MA: MIT Press.

Nordhaus, W.D., and Z. Yang. 1995. RICE: A Regional Dynamic General Equilibrium Model of Optimal Climate-change Policy. New Haven, CT: Yale Univ. and Cambridge, MA: MIT.

Parson, E.A., and K. Fisher-Vanden. 1995. Searching for integrated assessment: A preliminary investigation of methods, models, and projects in the integrated assessment of global climatic change. University Center, MI: Consortium for Intl. Earth Science Information Network (CIESIN).

Patten, B.C., and S.E. Jørgensen. 1995. Complex Ecology: The Part-Whole Relationship in Ecosystems. Englewood Cliffs, NJ: Prentice Hall.

Perez-Garcia, J., L.A. Joyce, and A.D. McGuire. 1995. Temporal uncertainties of integrated ecological/economic assessments at the global and regional scales. *Forest Ecol. Manag.* **162**:105–115.

Pfirman, S., and the AC-ERE. 2003. Complex Environmental Systems: Synthesis for Earth, Life, and Society in the 21st Century. A Report Summarizing a 10-year Outlook in Environmental Research and Education for the National Science Foundation. Arlington, VA: NSF.

Prinn, R., H. Jacoby, A. Sokolov et al. 1999. Integrated global system model for climate policy assessment: Feedbacks and sensitivity studies. *Clim. Change* **41**:469–546.

Rotmans, J. 1990. Image: An Integrated Model to Assess the Greenhouse Effect. Amsterdam: Kluwer.

Rotmans, J., H. Deboois, and R.J. Swart. 1990. An integrated model for the assessment of the greenhouse effect: The Dutch approach. *Clim. Change* **16**:331–356.

Rotmans, J., and B. de Vries. 1997. Perspectives on Global Change: The TARGETS Approach. Cambridge: Cambridge Univ. Press.

Rotmans, J., and M.B.A. van Asselt. 2001. Uncertainty management in integrated assessment modeling: Towards a pluralistic approach. *Env. Monit. Assess.* **69**:101–130.

SARU (Systems Analysis Research Unit). 1978. SARUM Handbook. London: U.K. Depts. of Environment and Transport.

Schellnhuber, H.-J. 1998. Global change: Quantity turns into quality. In: Earth System Analysis, ed. H.-J. Schellnhuber and V. Wenzel, pp. 12–195. Berlin: Springer.

Tran, L.T., C.G. Knight, R.V. O'Neill et al. 2002. Fuzzy decision analysis for integrated environmental vulnerability assessment of the Mid-Atlantic region. *Env. Manag.* **29**:845–859.

van Asselt, M.B.A., and J. Rotmans. 2002. Uncertainty in integrated assessment modeling: From positivism to pluralism. *Clim. Change* **54**:75–105.

van der Sluijs, J.P. 1997. Anchoring amid Uncertainty: On the Management of Uncertainties in Risk Assessment of Anthropogenic Climate Change. Ph.D. diss. Univ. Utrecht, Utrecht.

van Notten, P.W.F., J. Rotmans, M.B.A van Asselt, and D.S. Rothman. 2003. An updated scenario typology. *Futures* **35**:423–443.

Wilson, E.O. 1998. Consilience: The Unity of Knowledge. New York: Knopf.

Left to right: Dennis Meadows, John Finnigan, Bob Costanza, Marianne Young, Michael Young, Rik Leemans, Bert de Vries, Roelof Boumans, and Uno Svedin

22

Group Report: Future Scenarios of Human–Environment Systems

Marianne N. Young and Rik Leemans, Rapporteurs
Roelof M. J. Boumans, Robert Costanza, Bert J. M. de Vries, John Finnigan, Uno Svedin, and Michael D. Young

INTRODUCTION

The fundamental question that our working group addressed was: "How can we use historical narratives and models of human and environmental systems to generate plausible insights about the future in such a way as to advance policy, decision making, and stakeholders?" This question emphasizes the need to understand the dynamics of human–environment interactions. Properly organized, such an understanding can then be used to inform and constructively influence future actions.

In this chapter we first discuss properties of human–environment systems, the questions that need to be asked, and ways in which models can be used to answer them. Historical analogues are central to this discussion. Modeling techniques and approaches are then discussed and data needs explored.

Mirroring the investigation undertaken by the other working groups (see Redman et al., Dearing et al., and Hibbard et al., all this volume), we distinguish three future time periods for analysis:

1. the next generation, up to 30 to 50 years from the present;
2. the next century, with a focus on slower changes, the consequences of which will become apparent only toward the end of the period, and
3. the next millennium, seeking to identify consequences over very long time periods and, in particular, circumstances that may only become a problem some hundreds of years from now but which are triggered by events closer to the present.

The first time period is obviously relevant for planners in public life and business. It is the planning horizon for major societal systems such as transport, energy, and land-use infrastructure. The second period is of interest to national governments and international organizations who are farsighted enough to take a long view of the consequences of their collective actions, while the longer period is academic in terms of action but nevertheless of scientific interest.

In attempting to draw lessons from the past to help plot a course into the future, we assume implicitly that there are regularities and patterns in the dynamics of the Earth system that we can discover and capture by observations and mimic in models. This is an uncontentious idea when applied to the biophysical workings of the Earth. Global climate models, for example, are based on mathematical expressions of such regularities, some of which qualify as "laws of nature" (e.g., conservation of mass and energy) and so form a solid basis for calculation and prediction. It is harder to find such regularities in social dynamics. In economics, however, sets of such rules are well understood and incorporated in different schools of economic modeling, as we discuss later. When we come to the paths traced through time by whole societies and smaller groups within them, there are few general principles upon which all can agree. Nevertheless, the search for recurrent themes and patterns in past events forms a tradition in historical studies (exemplified by Toynbee 1961). The other working groups in this volume focused on the similarities and drivers of events that punctuated the steady courses of civilizations in the past—the societal collapses that we discuss below—to try to draw universal lessons from collapses widely separated in space, time, and circumstances. In this report we ask how we would fare in describing these events using the tools of dynamic modeling.

To make sense of the past we construct narratives that describe the course of events at some place and time. These "stories" must have an internal consistency and incorporate the best knowledge and understanding that we can bring to bear, from radiochemical dating to textual analysis, from contemporary reports to cultural classification of artifacts. In the context of IHOPE, we strive to pay particular attention to less traditional streams of data such as the paleorecords of past climates trapped in ice and sediment cores, the records of resource availability written in ancient pollens, in accounts of commerce and trade, and even exotic archives like the traces left in the genomes of diseases and their hosts. Nevertheless, when we construct plausible narratives, we unavoidably bring together the regularities that constrain the Earth system and the contingent events that play out within these constraints. These contingencies or chance events range from the qualities of dominant individuals, the choices they and groups in society make, the environmental forcings (e.g., climate and its haphazard extreme events), and the essentially chaotic conjunctions of circumstances that depend on the paths that societies have already traced through time.

When modeling the future, all we can hope to do is to capture the regularities that place bounds on the working of these contingent events. Hence we cannot talk about predicting the exact future of a single society or many societies. Instead, we must allow sets of contingencies to play out in the regularities of our models and to trace different paths through the future. We call these paths "scenarios" (de Vries, this volume). Scenarios are "narratives of possible futures" where we specify the regularities (i.e., our models) and combine them with the set of contingent events to describe a particular future trajectory. The probability that we can attach to any scenario is tied to the likelihood of the set of contingent circumstances that form its initial conditions (but quantifying such probability is exceptionally difficult). Thus when we ask what we can learn from the historical events dissected by the other working groups, we need to dissect their narratives to distinguish the universal themes from the contingent events that are entwined around them. When we consider building scenarios for the future we need to pay equal attention to the dynamics captured in the models and to the range of initial conditions from which we start the models and the choices of parameters within them. One obvious lesson from Hibbard et al. (this volume) is that many of the contingent events that have proved critical in shaping the course of the 20th century would not have been recognized as the most obvious or probable choices for a modeling effort at the onset of the century.

SOCIETAL DYNAMICS, EVOLUTION, OR COLLAPSE

The millennial, centennial, and decadal working groups identified and described system collapses in the far, medium, and recent past as a major focus for IHOPE in describing societal dynamics. While "collapse," "successful evolution," or even "sustainable development" should be seen as merely points on a continuum, any focus on collapse dictates the need for agreement on a generic definition, which should become part of the common vocabulary of IHOPE. This is especially important if one wants to understand the relationships between the events leading up to collapses, which are needed to model them effectively, rather than just describing their courses. Up to now, most integrated assessment models and Earth systems models only mimicked gradually evolving societies. Only few showed collapses (see Meadows as well as Costanza et al., both this volume). Here, we do not provide a precise definition but sketch some likely properties. The presentation, however, is not exhaustive.

According to the MSN Encarta dictionary (http://encarta.msn.com/), *collapse* represents a failure or sudden end to something or a sudden reduction or decrease in value. Collapses in human–environment systems are often triggered by events or trends that have occurred long before, and thus the underlying processes can involve long time lags. In the context of future-oriented studies, collapse can also denote a sudden, uncontrollable change away from desired ends. The rate of change during a collapse is more rapid and usually in an opposite direction to that preferred by at least some members of society. Often the path being followed is one that some members of a society find intolerable. During a collapse, system connections break down and no longer respond in ways previously understood (see, e.g., Meadows, this volume).

In most cases, the causes of collapse result from a combination of external and internal processes. Partial collapse of some elements of the system or even partial collapse of an entire system may be necessary for social learning. Even when a collapse is known to be imminent, it often cannot be controlled, and recovery cannot occur until considerable reconfiguration and downsizing has taken place. Although most collapses happen unexpectedly and are usually deemed, at least by some, as socially intolerable, *some collapses are actually planned or desired*. Examples of unanticipated collapses are the destruction of Newfoundland's cod fishery and the collapse of the Soviet Union and subsequent decline of the Russian Empire. However, societal collapses can be deliberately planned as part of serious social conflicts and wars. The collapse of the Taliban regime by international forces could exemplify this (although from the Taliban's standpoint, it certainly was not desired).

Let us define *collapse* as any situation where the rate of change to a system:

- has negative effects on human welfare, which, in the short or long term, are socially intolerable;
- will result in a fundamental downsizing, a loss of coherence, and/or significant restructuring of the constellation of arrangements that characterize the system; and
- cannot be stopped or controlled via an incremental change in behavior, resource allocation, or institutional values.

In addition, collapses can be characterized according to whether they are (a) capricious or predictable, linear or nonlinear and chaotic, unexpected or expected; (b) irreversible or reversible; or (c) have local, regional, or global consequences. Local collapses and even regional ones might have negligible consequences globally; alternatively, they could become globally significant through their cumulative impacts. For the purpose of developing models and scenarios, we define the opposite of collapse as a set of arrangements that retain system coherence, are accepted by humans as evolving at acceptable rates, and do not diminish perceived quality of life. In many cases, considerable reconfiguration (e.g., in the development of new institutional arrangements or the construction of new infrastructure) is possible without collapse.

MAJOR QUESTIONS

Successful models are often guided by sharp questions. Hence, the choice of questions plays a critical role in framing the models we build. In general the questions we ask of the future will have a different character to those we ask of the past, but the two are linked by the necessity of distinguishing the regularities in past events. In framing our questions, the successes and failures of earlier attempts to model the future provide an invaluable guide.

We classify our questions in four categories along the lines of IGBP–GAIM (cf. Box 21.1 in Costanza et al., this volume): normative, strategic, analytical, and operational. *Normative questions* delimit the domains of application of the models and define their structure, scope, and interfaces. *Strategic questions* directly address policy issues. They reveal the purposes for which IHOPE models will be used. *Analytical questions* focus on how to analyze and model the behavior of the human–environment system. Major interactions (e.g., feedbacks, trade-offs, synergies, and time lags) and the complexity of the components of the system have to be quantified. Finally, *operational questions* address how we develop and integrate the different models. Answers to these questions can be expected to catalyze improvements in modeling of the human–environment interface in many ways.

Normative and strategic questions need to be developed through stakeholder involvement. In the Millennium Ecosystem Assessment (MEA 2003, 2005), for example, strong interactions with its users were cultivated as a means of responding to changes in their needs. This resulted in a timely evaluation of the Millennium Development Goals and the internationally accepted 2010 biodiversity target. Another example is the climate change debate in the Netherlands, which was facilitated by a dialogue between science and the public (Berk et al. 2002). When issues are addressed that involve value judgments, scientists must be clear as they communicate/interact with stakeholders. Many issues associated with the future, including those associated with the rise and collapse of states, fall into this category. In such situations, scientists are primarily problem recognizers, not problem solvers.

Normative Questions

Normative questions reflect our desire to sustain and develop certain qualities of life for human beings. Thus, formulating these questions necessarily involves value judgments about what human futures are desirable. Examples might be:

1. How can we measure quality of life and describe inequalities?
2. Can the rich be expected or forced to share wealth and/or economic opportunities more equitably with the poor? Under what conditions can societies collectively be expected to make trade-offs that do not offer win–win outcomes for all?
3. For the purposes of making trade-offs, should one value people's interests according to their anticipated income, value all people's interests equally, or use some other measure? Should the interests of future people be valued equally or discounted?
4. To what extent have human aspirations overloaded the human–environment system and brought the system to the brink of collapse?

Strategic Questions

Strategic questions require the combined conceptual skills of scientists, social scientists, and stakeholders. They assume action and presuppose that model output is likely to be of value for those whose actions affect the future. A strong dialogue with stakeholders is necessary to ensure that their strategic needs are met.

Model type and structure are strongly dictated by these strategic questions. Examples include:

1. What future trajectory of societal development is compatible with the sustainability of a human society where individuals have resources and choices similar to those enjoyed by western Europeans and North Americans in the last two decades of the 20th century?
2. How can the diffusion of new knowledge, science, and technology help to shift the carrying capacity of the Earth, and what are the trade-offs between services provided by ecosystems and the immediate needs of society and the longer-term sustainability of the human–environment system?
3. What mechanisms cause greater or lesser wealth disparity, and what does the past teach us about them?
4. What proactive change or adaptation could we imagine that would reverse the current trend toward greater such disparity?
5. How could societies collectively organize themselves to make trade-offs that affect relative wealth among groups of societies in a particular region, across regions, and through time? How can we adapt, increase resilience, and reduce vulnerability of society?
6. What resources and choices for individuals are compatible with avoiding damaging changes to the planet's life support systems over the next century, given likely demographic trends and foreseeable technologies?

Analytical Questions

These questions address the structure and behavior of our models and thus determine their complexity, the number of components, the nature and direction of interactions, and the processes that determine lags in the system's responses. The analysis used to answer these questions should identify the most important components and processes. As we consider analytical questions we begin to come closer to the regularities of the Earth system that we hope studies of the past will help reveal. Typical questions are:

1. What is the historical life cycle of different cultures or civilizations, and what determines their longevity and relative success?
2. Are there historical insights about interactions among societies that would let us anticipate the future course of corporate regimes, nongovernmental organizations, and governments?
3. Which sets of events and conditions have led to concerted governance actions for "broader public gains," such as the Marshall plan?
4. What governance and other institutional arrangements have enabled some societies to perform better than others?
5. What can we learn by looking at the life cycle of past societies in relation to the environment? What are the critical characteristics of historical

societies that successfully adapted to sudden declines in resources? What are the critical components, processes, and interactions in the human–environment system, and can we identify dangerous thresholds, bottlenecks, and abysses?

6. What have been the actual local, regional, and global trends in the social, economic, and environmental costs and benefits of key resources, activities, and sectors?
7. Which institutional arrangements and policy designs have proved to be most robust in the allocation, use, and management of the environment and natural resources?
8. What are the relative contributions of governments, business hierarchies, and markets in causing and mitigating problems?
9. Does the extent of globalization, increased trade, and rapid communication mean that the traditional approach of organizing economic activity by country or society is now irrelevant? Consequently, does the idea of groups of societies or countries interacting with one another need to be replaced with notions of interconnected networks of business and governmental arrangements?
10. What may happen to the relative power of different entities if resource availability declines rapidly in the near future and this scarcity is not offset by the development of alternative resources and alternative technology?
11. What will determine the nature and direction of technology and the emergence of new institutional arrangements?

Operational Questions

Operational questions relate to the "art of modeling" and thus have a specific relevance to the modeling community. The answers to operational questions largely define the application domain of a model. For example, operational questions concern issues like the data sets necessary for initializing, testing, and validating models; for setting parameters; and for scaling processes:

1. How do we couple the different components of the human–environment system? How does one model trade-offs and synergies in a system-dynamic framework? How do we integrate the social, economic, environmental, and energy dimensions in human–environment system models?
2. What is the best strategy to generate consistent data sets from the various historical studies?
3. What are the best techniques to use historical analogues to test and refine models?
4. Are there appropriate techniques to recognize imminent and potential collapses as well as other irregular or extreme events in the near future in models and in reality?

5. How do we effectively communicate to stakeholders the enhanced understanding of positive and negative trends obtained by human–environment system models?

A POSSIBLE FRAMEWORK FOR MODELING THE FUTURE

Now that we have outlined the kinds of questions we wish to answer, we can ask which aspects of the past Earth system and societal dynamics should be incorporated into models of the future. We present a generalized framework in Figure 22.1 to define aspects of future human–environment interactions that require careful consideration:

- the size and quality of natural resources or natural capital, the ecosystems services that flow from them, and the processes that maintain their productivity;
- human population size, birth and death rates, and changes in the structure of age cohorts;
- investments in infrastructure and other forms of built capital that facilitate the production of goods and the provision of services to humankind, with attention to the rate at which these investments depreciate and options to increase their longevity;
- human knowledge and capacity to innovate so as to increase productivity and reduce adverse impacts on natural capital and infrastructure;
- the capacity of institutional rules to guide decision making and, in particular, facilitate the redistribution of income in ways that improve system performance and equity; and
- the nature of human aspirations, preferences, and value systems.

Particular attention needs to be paid to the processes that work to increase or decrease overall capacity to provide ecosystem services, as these influence human behavior and define human–environment interactions. A number of these processes tend to work in opposite directions. In sustainable systems, such opposing effects should provide a balance. When they are not in balance, an outcome could ultimately be an environmental collapse or decline in the well-being of a significant number of people. Not all changes, however, need to be negative. Increasing knowledge, science, and technology, for example, may redress negative forces and allow previously unused elements or concepts to become feasible in combination with new technologies. Within such a perspective, it could be argued that new technologies can be expected to produce large efficiency gains, significant opportunities for de-materialization, and dramatic shifts in resource use. These drivers change the underlying dynamics of the processes. At the same time, one lesson from the past shows clearly that the solution to one problem may bring with it unforeseeable side effects. An example can be found in

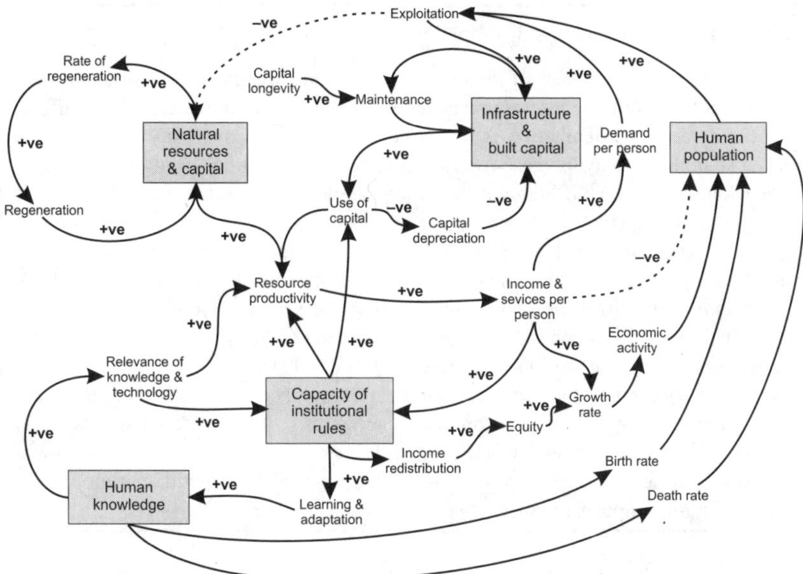

Figure 22.1 An illustrative wiring diagram for possible IHOPE models. +ve indicates positive feedback; –ve a negative feedback. Figure drafted by Dennis Meadows.

20th-century improvement in the quality of life and longevity based on better nutrition, hygiene, and medical techniques, which has driven the increase in global human population (see Hibbard et al., this volume).

Narratives of the Past and the Role of Historical Analogues

The millennial, centennial, and decadal working groups were comprised of archaeologists and historians, who have collected data and (re-)constructed narratives of past events, as well as scientists from ecology, anthropology, and climatology. Overall, these scholars have studied humanity at times and locations where human groups had less connection to each other and less control over nature than we have now or can expect to exert in the future. Both technology and the degree of connectedness in human society appear to have increased almost exponentially (Box 22.1). What limits does this place upon what we can learn from the past about the future? Is human society moving into a "no-analogue state" (Steffen et al. 2004) similar to that which now applies to the global climate, and which would invalidate the central assumption of IHOPE? To answer this we need to be much clearer about what we mean by analogues and how they differ from metaphors and narratives.

An analogue is a thing, an idea, or institution that is similar to or has the same function as another. For our purposes, two events are analogous if their initial conditions and dynamics are similar enough so that by studying one we can

Box 22.1 Globalization and complexity.

Depending upon the nature of institutional arrangements, increasing global system connectedness could be interpreted as having a stabilizing or a destabilizing effect. Erosion of diversity of practice, for example, may reduce food security but access to a wider spatial area may increase food security by smoothing out temporal variations. The result could be a world that is both more stable in the short term but, because of this, increasingly vulnerable to sudden change or collapse.

The fraction of natural systems being operated by, imposed upon, or managed for human purposes has been expanding quickly. These are the characteristics of the Anthropocene (Crutzen 2002) or anthroposphere (de Vries and Goudsblom 2002). As noted in the recent Millennium Ecosystem Assessment, highly modified systems now dominate much of the world's environment (MEA 2005).

The emerging world system seems to be moving outside the extremely long historical record of swings in environmental variables (e.g., climate variations). The question has been posed as to whether this is human induced and connected to global industrialization of the last century. The answer may well have hidden time lags in terms of future impacts of entrenched processes.

reasonably predict how the other will unfold. The other working groups compared and contrasted past societal collapses to see which can function as analogues and which cannot.

In natural science, analogues have been very helpful. In the social sciences, however, there seems to be a tendency to drift from strict analogues to metaphors. Metaphors are figures of speech that can be useful in bringing color to a description or in communicating a concept. They are often used to describe some essential feature of a complex phenomenon in the terms of simpler, better-known situations; however, they must not be used as a basis for prediction. An example of a metaphor is the use of tectonic plate movements to describe structural change in an economy (e.g., Krugman 2002). Contrast this with the analogy between chemical homeostasis and the broad-scale dynamics of the Earth system envisaged by James Lovelock in his Gaia hypothesis (Lovelock 2003). In the first example, we would rightly think it ludicrous to expect that the economy should follow the same dynamic laws as continental drift, whereas in the case of Gaia, the suggestion is precisely that homeostasis governs the Earth's dynamics. Krugman offers a metaphor; Lovelock an analogue.

Narratives are descriptions of the trajectory of past events that combine the regularities and contingent factors of the Earth system in a logical and consistent way. They are reconstructed from data that we can access, but these are often inadequate to constrain the narrative completely, especially when events are in the distant past. In such cases, theory of some kind is used to fill in the gaps. Such theories may be views of social drivers or "cyclic" theories of history, as discussed earlier, or they may be formal mathematical models, analogous to those used to describe the biophysical world (see, e.g., chapters by Hassan, Hole, and Tainter and Crumley, all this volume). Even the description of recent events can be colored by the subjective views we bring to the construction of the narrative, as the postmodern school of social theorists has so strongly emphasized.

As we build models to describe the future, our only hope is to capture the regularities of the Earth system in mathematical formulations.[1] Contingent events form the initial conditions of models of the future and also appear in critical parameters. A good example of the latter is the rate and kind of technological innovation that we might include in a model describing the evolution of world society over the next fifty years. The choice of these contingent factors is ultimately subjective but can be done in an informed way and probabilities assigned to the choices. In fact, formal methodologies, such as Bayesian belief networks, are available to distill expert views into such probabilities.

By running a model into the future and constructing a future narrative based on a set of contingent factors, we create a scenario. The plausibility of any scenario depends both on the confidence we can place on the processes incorporated in the model and the likelihood we assign to the set of contingent factors that set the initial conditions and parameters of the model. From this point of view, when we look at the narratives of the past, we need to try to separate the contingent factors from the regularities in order to extract the maximum learning from history. We expect that the biophysical processes operating outside the human sphere are controlled by laws of nature that will not change in the future so that these allow us to build some regularities into our future models with confidence. With slightly less confidence, we can build in the constraints of resource availability, conscious that these can be flawed by inadequate present knowledge or removed by technological advances. Nevertheless, these can provide powerful constraints (e.g., Meadows et al. 1972). The processes that we know least well are those of economics and social dynamics, in that order.

The simplest models of the future are found by taking past narratives as analogues. Given what we have just said, we should expect that past events which, upon analysis, do not appear to be shaped by extreme contingencies might form the best clue as to the future, at least to a future where extreme contingencies will not dominate either. This is likely to be the only future that we can predict anyway.

We have, therefore, a broad prescription for how we should analyze the past to inform future models, including the simplest models, analogues. This approach separates, as far as we can, the roles of regularities and contingencies in shaping past events and is especially critical of the assumptions we have made in filling in the data gaps of the past. In particular we need to determine the extent that those processes we expect to continue unchanged (e.g., planetary biophysics and resource availabilities) have played in controlling and constraining past events. Finally, we hope that comparative studies of superficially similar events might give us solid clues as to the dynamics of universal social processes.

---

[1] It is worth emphasizing that these regularities need not be perfectly deterministic. Stochastic processes can be valid components of Earth system models and such random processes themselves place limits on the confidence we can assign to prediction.

MODELS FOR UNDERSTANDING THE PAST AND ANTICIPATING THE FUTURE

The Nature of Models

Let us now discuss in much more detail what we mean by models, the different kinds of models we might construct, and their domains of application. Models are abstract representations of reality that help us to understand and act. They range from straightforward analogues based on past events to computer simulation models. To answer the kind of questions set above, the models we have in mind are usually dynamic computer simulations at different space and time scales.

In the pursuit of a better understanding of human–environment systems, there are several reasons for building models. First, from a Popperian stance we can frame hypotheses for testing by comparison with data. In the complex human–environment systems we are considering, a dynamic computer model is often the simplest way of framing a hypothesis satisfactorily. This approach is aimed at better process understanding. Second, we can build scenarios of the future that are linked to choices and contingencies as a tool for risk management and wise government. Third, we can build models in collaboration with the stakeholders and use the model-building process itself as a tool for education and community transformation. Finally, we can build models to support gaming and role-playing exercises to allow users to see the implications of their (and other's) decisions. For these purposes, three general types of models are recognized: conceptual, optimization, and dynamic models (Figure 22.2).

Conceptual models attempt to describe relationships in a robust manner. Typically, they focus on cause and effect and the mathematical form of responses to a stimulus. They are particularly useful in specifying the most appropriate way to build a model. Network models and agent-based modeling (presented below),

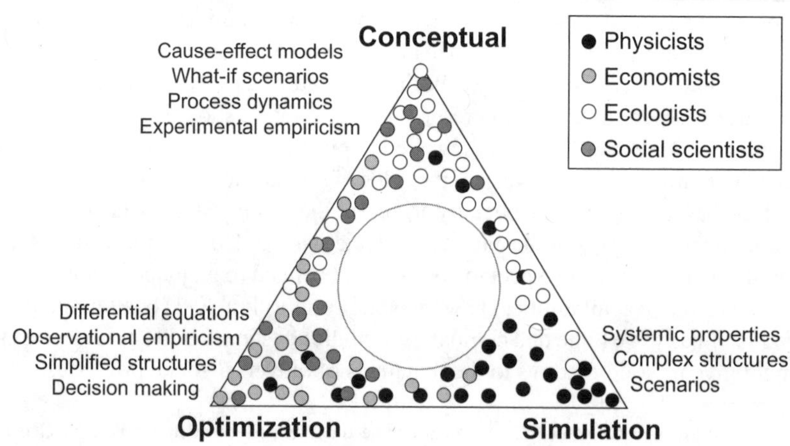

Figure 22.2 Caricature of different modeling approaches and disciplinary preferences.

but also experimental economics techniques that use people to simulate human behavior in a controlled environment, can be useful in the development and testing of conceptual models. Here the primary aim is to assist model development by attending to the development of relationships upon which scenarios for the future are critically dependent.

Optimization models are most commonly designed to assist decision makers in making trade-offs. Value laden, they make trade-offs according to pre-specified normative rules. Many economic models fit into this category. Typically, economic models search for the most efficient outcome given a preference for profit maximization in a world where inter-temporal trade-offs are made in the same way that financiers tend to discount future income. Some optimization models are dynamic in the sense that they seek a solution using processes that account for the effects of one process on another. A computable general equilibrium (CGE) model is one example of such dynamic optimization models. Typically, CGE models explore the effects of a shock to one part of an economy on all the others. Inter-temporal optimization is possible but because of the computational time necessary to find a solution, it is usually only undertaken when knowledge about an adjustment pathway is necessary.

Simulation models differ from optimization models in the sense that they attempt to describe interactions in an "if X and Y occur then the result will be Z" type of scenario. Both approaches are valid and optimization processes are often included within simulation models. The main advantage of simulation models is that they facilitate a much richer and more complex exploration of consequences across space and through time. Choice of the approach to take depends heavily on the nature of the question being asked, and as such each profession tends to have greater experience in using different types of models. Physicists, for example, tend to prefer simulation models whereas neoclassical economists tend to prefer optimization models.

Earth System Models in the Anthropocene

Natural scientists are familiar with models that capture the dynamics of the biophysical parts of the Earth system. Global climate models are perhaps the best known contemporary example. In IHOPE, we wish to generate models that will integrate the biophysical and societal dynamics of the Earth system to enable us to understand and mitigate human impacts on the system during the Anthropocene. Modeling social dynamics and the consequences of human choice has a much shorter pedigree than biophysical modeling and presents the greatest hurdle to integrated models (Chapter 21, Costanza et al., this volume). The aspect of societal dynamics that has been modeled quantitatively for the longest time is the economy so, although this is only one aspect of social dynamics, it has become conventional to treat the economy separately. Hence, while one can align modeling components along a continuum from biophysical through economic to social, it is convenient to describe these integrated Earth system models separately as interdependent biophysical, economic, and societal models.[2]

The major processes of biophysical models are based on a series of well-established components, behaving through processes that follow fundamental laws of nature, and their other components are often represented by regression models based on large data sets. Economic models are based on some general principles, upon which many economists agree; they also rely on large data sets. These economic rules are not, however, universal laws of nature, and there is a good deal of controversy about their universality. At best they use functional relationships that apply within restricted spatial and temporal domains. For example, simple linear relationships between inflation, wages, and employment can obtain for some years and then go through very rapid nonlinear adjustments (Ormerod 1994). At worst, the economic rules are really expressions of normative statements about concepts like market primacy with poor connection to empirical observations (see, e.g., the comments on the DICE model of Nordhaus in Chapter 21, Costanza et al., this volume).

In general, the social processes included in integrated global models (reviewed by Costanza, this volume) incorporate functional relationships between population and resource use with the drivers of demographic change such as health, fecundity, and migration represented explicitly. The more recent models include social dynamics in quite some detail. For example, in TARGETS (Rotmans and de Vries 1997), human decision making is modulated through the implementation of parameter changes reflecting slow changes in dominant world views. However, very few, if any, universal laws or theories exist to guide parameterizations of social processes. For example, whereas the modulation of human decision making by widely shared world views is relatively uncontentious, it is far from clear at what scale it is appropriate to apply this principle. In general, it is fair to describe the modeling of social processes at the global scale as based on far fewer data than the corresponding functions employed in economic models (see below for more discussion).

Models to Reproduce or Anticipate Rapid Changes in the Earth System

The examples of collapse that have been used as bases of comparison of societal dynamics in the past by the other working groups involve rapid, abrupt, or irreversible changes. Experience in modeling such transitions in natural systems (e.g. Scheffer et al. 2001) has revealed how difficult it can be to predict the magnitude and location of such changes, if they are a result of bifurcations or higher-order discontinuities intrinsic to the dynamics of the system. Mathematicians call such discontinuous changes "catastrophes." They have classifications for catastrophes of different kinds and can make general deductions about

2 Note that the agent-based models of societal–biosphere interaction, discussed later in this chapter, more often than not treat economics as an integral societal process.

whether a given system of equations can support catastrophes of a given order. In contrast, rapid but locally smooth changes that result from nonlinear combinations of forcing and system dynamics can be located with much more confidence. It is important for modelers to distinguish between the two kinds of change, if they can, and also to realize that it may be impossible to make that distinction from data records. The conventional way to model such phenomena is in a systems dynamics framework (see below), where the models consist of sets of coupled nonlinear differential equations. Even in relatively simple model systems, nonlinearity means that finding analytic solutions is usually difficult or impossible, and we quickly resort to numerical solutions on a computer.

When the chosen set of governing equations contains further complications, such as time-delayed feedbacks or integrals that accumulate the results of earlier dynamics, even finding stable numerical solutions can be difficult. The problems in capturing such complex dynamics emphasize the importance of building, where we can, meta-models (i.e., simpler models of the more complex models), which mimic the dynamics and character of abrupt changes from the more comprehensive models (Liljenström and Svedin 2005). With such tools, we can hope to capture and analyze the critical dynamics of the real system (e.g., Claussen et al. 2002). To do this may require a set of steps where parts of the total system are isolated from the whole so that their individual dynamics may be properly understood. It is perilous indeed to assume that the complex numerical simulation of a whole Earth system will accurately indicate and locate collapses or catastrophes if this kind of careful analysis has not been performed. This is especially so when our confidence in the parameterization of component processes ranges from good to poor, as discussed above.

Below, we discuss in detail some different examples of the kinds of models that we have been discussing.

Economic Optimization Models

Regional, national, and global optimization models are frequently used in economics and decision support (cf. description of DICE in Chapter 21, Costanza et al., this volume). These models come in many forms but usually have some type of an input–output table at their core. These tables show how changes in one sector or region affect other sectors and other regions. In our complex world, CGE models are used to explore how market processes are likely, under various constraints, to interact with each other and how this is likely to affect the welfare of people. The capacity of these models to recognize economic constraints means that it is possible to link them with other models that focus on infrastructure and ecological constraints without loss of economic detail. Changes in resource stocks induced by overexploitation and/or the loss of ecosystem services can be accommodated in a similar manner.

Dynamic CGE models are typically used to examine the consequences of path-dependent changes and temporal consequences over long periods of time.

Embedded in integrated dynamic simulation models and/or linked to simulation models, they have the potential to help produce scenarios that recognize and account for economic, social, and ecological considerations in much more detail than is normally available.

Agent-based Models

A powerful technique for modeling social processes that has grown rapidly in popularity in the last decade is agent-based modeling, also called multi-agent simulation. Agent-based models specify attributes of agents (individuals, family groups, village communities) and the rules of local interactions between agents and their environment and then let these agents interact within a computer simulation. The global social behavior is not specified; it "emerges" from multiple local interactions between the agents. Agent-based models are now widely used to simulate and investigate a wide range of social processes to develop an understanding of social dynamics (Bonebeau 1999). So far they have only been applied at regional scales.

Of particular interest to the IHOPE human–environment systems debate are agent-based models that capture the interactions between societies and their environment. Models such as Catchscape (Becu et al. 1999), for example, combine local hydrology and crop models, climate forcing, and human decision making at the village level to model interacting and antagonistic village farming systems in Thailand. Enkimdu (Christiansen 2000) is another model that focuses on a large community in ancient Mesopotamia. In these models, economic processes are subsumed in the agent behavior rules and regarded as part of the social process.

The key distinguishing ideas of the agent-based modeling approach are that:

- Agents have bounded cognition, i.e., they are only aware of, respond to, and affect their "neighborhoods" (and this may be a necessary condition for emergent behavior) (Finnigan 2005).
- It is easier to describe the rules of social interaction at an agent–agent level than at a group level.
- Getting the rules of agent decision making precisely correct is often less critical than might have been supposed. Simple decision rules often lead to the same sensible group behavior as more subtle rules.

Agent-based models have been shown to be capable of reproducing the fine-scale societal behavior observed and described by historians and observed historical events on the societal, regional, and global scale using robust and agreed upon rules of agent behavior.

Network Models

Recent developments in network theory (cf. Box 22.2) have shown that the total trajectory of a dynamical system is determined partly by the nature of the interactions between agents and partly by the global topology of those connections. This explains why simple decision rules in agent-based models often produce the same societal behavior as subtle ones. In network models, slow societal changes are described by repeatedly applying local rules governing the formation or reformation of links between agents (nodes), resulting in slow changes in network topology. Network properties such as connectivity (the ability to get from any node in the network to any other), however, do not change smoothly as links are added or removed but go through sudden "phase changes." Any property that depends on connectivity, like the ability of ideas, goods, diseases, or energy to spread through the network, changes rapidly. A particularly important property that changes with connectivity is the stability of the dynamical system underlying the network (Brede and Finnigan 2004).

In this sense, network models can be good models of rapid societal change, if we can map rules of societal behavior into rules of network evolution. Several examples of such mappings have been proposed at a qualitative level by Barabasi et al. (2001). In particular, network approaches suggest that contingent events in history occur within a landscape of possibilities whose character is determined by the (slowly changing) network topology. Network models might be good candidates for understanding recurrent historical patterns and distinguishing the roles of contingency and determinacy in shaping the historical record. The corollary, however, is that the functional rules we would build for predictive models of global change would be probabilistic.

Gaming and Interactive Simulation Models

Interactive models are used for a number of research, educational, and entertainment purposes. These include the provision of learning experiences about the consequences of complex interactions for an individual and for society as a whole. The search is for unexpected insights about system interaction and the testing of likely human responses to different structures and institutional arrangements. Simulation gaming is the name for the activity in which one uses a model in an interactive fashion within a well-specified context. The user(s) of the model make decisions about how they would like to steer the system along a certain desired trajectory. Such a gaming setup can be used for entertainment (there are many single- and multiplayer games available for this purpose) or education purposes. Examples of such games in the domain of human–environment interactions are Stratagem and FishBanks (Meadows 1990, 1996).

One can also use a gaming context as a research tool by letting (many) users interact with the model in a structured way to find out about their behavior. Examples are the way in which students from MIT attempted to control simple nonlinear feedback systems (Sterman and Sweeney 2002) and the use by experimental economists who replace the code in their models that describes human decision making with a person who, in a gaming environment, is asked to make decisions in response to inputs from other parts of a model. It can be used, for

Box 22.2 Networks, complexity, and diversity.

In the past, it has been traditional to model complexity without attention to the nature of networks and the way network structures influence stability and prospects for avoiding collapse. One of the emerging issues well suited to network analysis is globalization (cf. Box 22.1). This new phenomena is rapidly connecting human societal and environmental interaction at a planetary scale. [It is true that "semiglobal" connections, for example, international trade, have been operating at various levels for a few hundred years but the degree of spatial and temporal connectivity is now totally different from that which occurred in the past.] Strong product and value set homogenization is occurring. As a result, there may be less room, but greater opportunity, for conflict about inequality and opportunity.

Complexity is different from diversity. May (1977) showed that as one randomly increased the complexity or connectedness of a dynamic system it became less stable. The challenge, well understood by ecologists, is to find ways of increasing complexity *but* simultaneously increasing stability. One of the emerging hypotheses for IHOPE to explore is the merits associated with globalization, trade, and other international connections (see inset for a conceptualization; after Brede and Finnigan 2004).

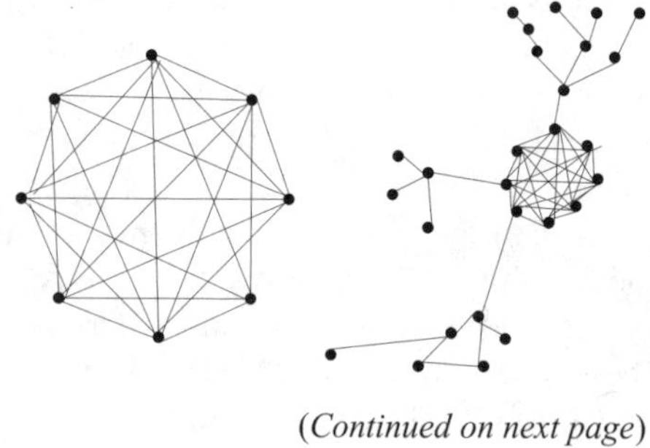

(*Continued on next page*)

instance, to test the capacity of different cohorts of people to learn about decision making through time.

A third context for gaming simulation is to assist in policy development and decision support. In this case, a model is constructed in close interaction with relevant stakeholders. The resulting model may actually be less important than the process of its construction, but it can play an important role in communicating lessons and insights to other similar situations (van den Belt 2004). An interesting example is the management of a natural area north of Montpellier in France. A model was constructed in dialogue with local farmers, shepherds, and environmentalists to learn about and appreciate each other's view of the situation and thus create a platform for cooperative policy strategies (Etienne and Le Page 2002). Similarly, it has been used in combination with agent-based models to develop new institutional rules in partnership with local communities in Mali, Kiribati, and other developing nations.

Mathematical formulations for computer games do not have to be different from research models. It is more the focus on user interface packaging that separates gaming models from research models. In gaming models, the packaging requires a higher degree of reality recognition for the players, with the main purpose being to make abstracted mental models available for teaching while providing fun. Although games are a well-proven methodology to teach about

Box 22.2 (*continued*) A related issue to explore is the flow-on effects that increasing connectedness has for global economies and global ecosystems. If societies fail to reveal the cost of externalities on the environment then costs will be wrongly signaled. In an unconnected world, it may only be necessary to track costs that are "close" to goods being produced. In a globally connected world, where one is seeking an optimal solution, *all* costs need to be signaled correctly, including those distant to the locus of the good or service being produced:

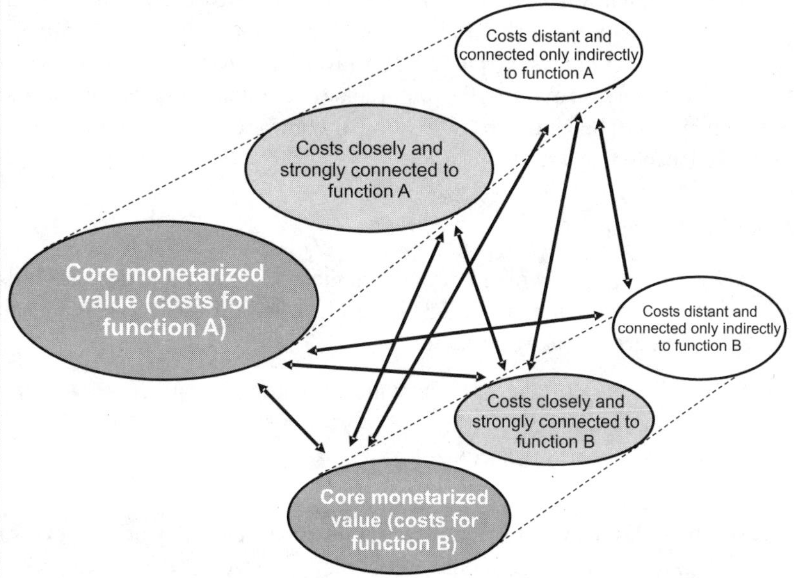

Consideration of such relationships is particularly important when one considers, for example, the pros and cons of pursuing a regional or global free trade agreement. If some but not all costs are included in the model very erroneous recommendations could be made. The historical evolution of trade, in particular in Europe, may provide some interesting data and insights (Jones 2003; Elgin 2000).

societal dynamics, computer games, and in particular the emergence of online gaming worlds, create a new and unexplored avenue to disseminate and carry out global change research. Modern play routines mimic the development and execution of scenario runs of research models, where players can explore and sort through hypotheses to make sense of a game's environment. In doing so, players are required to manage a dizzying array of information and options, not unlike researchers who are mining data bases and other scientific works for clues that would help in forecasting global futures. Modern gaming is an opportunity for the global change scientific community to utilize the creativity of a large group of people in designing and thinking through scenarios while disseminating information.

These kinds of approachs, where human agents form part of the model software (liveware), offer insights on human decision making that can then be incorporated as algorithms in dynamic models. Agent-based models are particularly well adapted to use such data.

Applying Models

Once we have run our models, or even before we have designed them, it is necessary to think about how we will use the results. Recent research, in which the authors are involved or of which they are aware (van Daalen et al. 1998; Costanza and Jørgensen 2002; Schröter et al. 2005), draws particular attention to the role of scenarios in decision making, the importance of community engagement, and the new insights to be gained from looking at human–environment systems in terms of connected networks.

The Role of Scenarios

The use of scenarios has evolved throughout the last decades and de Vries (this volume) and van Notten et al. (2003) have identified different roles they can play. Scenarios are nowadays used to:

- facilitate consistent, shared understanding of important processes and systemic behavior;
- make abstract narratives concrete;
- reinforce interdisciplinary discovery and learning.

Scenarios thus play a critical unifying role at several levels. They bring a level of transparency to systems by providing a coherent "real-world" description of the motivations and events that trigger developments. Finally, they provide a common ground for communicating and conveying the needs of the users to the modelers. However, their ultimate effectiveness can still be undermined by the quality of models and data.

Community Engagement and Outreach

As IHOPE models begin to capture the causes and (combinations of) events that potentially lead to collapse and/or successful evolution of societies, one of the major challenges will be to communicate this knowledge to the user community and to build support for appropriate responses. Many of the choices associated with the future direction of humankind require consideration of value sets that are not universally accepted. Just as the normative questions that guide the building of models incorporate value sets, so judgment about the merits or otherwise of any future scenario may also be value laden. These judgments involve trade-offs that depend upon the user's values. For example, a preference for sustainability over wealth creation leads to a quite different set of policy choices.

Given this, insofar as it is practical, the full range of stakeholders likely to be affected should be involved in setting the normative questions and in developing the sets of contingent factors that we inject into our models to generate scenarios. Finally, we must organize our data and model outputs in a way that enables these stakeholders to explore and understand consequences from their various perspectives. In other words, the full range of stakeholders likely to be affected by a model needs to be engaged. Although full engagement with all global stakeholders is impossible, we are strongly of the view that engagement should cover the full spectrum of views, including those from developing countries.

RELEVANT DATA AND ITS QUALITY AND COLLECTION

IHOPE will require the integration and synthesis of data about human and natural systems from a huge range of sources. For example, Table 22.1 lists a minimum set of *core variables* that would be of interest for almost any integrated investigation of historical systems. We do not imply that these are the only variables of interest, but this core set must include social, economic, demographic, and environmental variables.

Given the range of variables required, these data will be of highly variable quality. While experts in a field of study usually have a good working understanding of the quality constraints on their data, this understanding is not often or easily communicated across fields. What we need for the IHOPE effort is a

Table 22.1 Core variables.

Economics Prices Markets	*Population* Health Labor Birth/death rates
Scale/scaling	*Land use*
Technology Agriculture Industrial Information technology *Culture* Social organization/governance Politics/nation-states/elites Trade Warfare Knowledge/information/wisdom	*Natural systems* Climates Water Soils Nutrients Biodiversity Geo-resources/energy (energy return on investment) Biomass/productivity
Information feedbacks Property rights Distribution of wealth/resources Language Religion	*Physical infrastructure* Buildings/cities Roads Canals Business

system to communicate the full range of data quality—from statistically valid estimates to informed guesses—from historical narratives to detailed quantitative measurements. Communicating data quality is a prerequisite to integrating effectively the full range of information we hope to assemble. One can think of this process as *grading* data (cf. Costanza, this volume). A grading scheme for communicating the "degree of goodness" associated with data has high potential utility. If consistently applied, it can provide nonexperts with greater competence in interpreting the degree of uncertainty associated with complex estimates. In modeling and analysis, it will provide a much needed input to help assess the overall uncertainty of results, based on the quality of the input data, combined with information on the structure and quality of the models used to process the data. We can look to the example of natural resource measurement and modeling such as that carried out in the Millennium Ecosystem Assessment for guidance here. The data sets used in such studies are now routinely accompanied by "meta data" that conveys information on the quality and reliability of the primary data.

THE WAY FORWARD: THE NEED FOR HYBRID APPROACHES

In this chapter we set out to answer the question: How can past events and human–environment histories be used to understand future human–environment interactions and test models that simulate these interactions? This question is ambitious and broad.

To move to the next level of understanding of complex human–environmental systems, it will be necessary to take advantage of the strengths of each of the modeling approaches we have enumerated. The major questions that we identified can best be addressed by comparing the results of different modeling approaches. Different model options thus provide complementary approaches.

An alternative strategy would be to integrate the approaches into hybrid models. For example, one could build a systems dynamics submodel to deal with the natural system components of a system and an agent-based submodel to address the human system components. One could first implement this model with real humans playing the model as a game and observe their behavior to structure the rules and parameterize the behavior of the computer agents subsequently used in the hybrid model. We suspect that this hybrid approach to modeling would produce the kinds of results needed to address the complex, integrated questions noted above better than any one approach used independently.

ACKNOWLEDGMENT

We thank especially Dennis Meadows for his insight and contributions to this report.

REFERENCES

Barabasi, A.L., R. Erzsebet, and T. Vicsek. 2001. Deterministic scale-free networks. *Physica. A* **299**:559–564.

Becu, N., P. Perez, A. Walker, O. Barreteau, and C. Le Page. 1999. Agent-based simulation of a small catchment water management in northern Thailand: Description of the Catchscape model. *Ecol. Mod.* **170**:319–331.

Berk, M.M., J.G. van Minnen, B. Metz, and W. Moomaw, eds. 2002. Keeping Our Options Open: A Strategic Vision on Near-term Implications of Long-term Climate Policy Options. Bilthoven: RIVM and NRP.

Bonebeau, E. 1999. Swarm Intelligence: From Natural to Artificial Systems. Oxford: Oxford Univ. Press.

Brede, M., and J.J. Finnigan. 2004. Constructing scale-free networks from a matrix stability criterion. In: ICCS2004 Abstract book. Boston: New England Complex Institute. http://necsi.net/events/iccs/openconf/author/abstractbook.php

Christiansen, J.H. 2000. A flexible object-based software framework for modeling complex systems with interacting natural and societal processes. In: 4th Intl. Conf. on Integrating GIS and Environmental Modeling (GIS/EM4): Problems, Prospects and Research Needs, Banff, Alberta, Canada, pp. 1–10.

Claussen, M., L.A. Mysak, A.J. Waever et al. 2002. Earth system models of intermediate complexity: Closing the gap in the spectrum of climate system models. *Clim. Dyn.* **18**:579–586.

Costanza, R., and S.E. Jørgensen, eds. 2002. Understanding and Solving Environmental Problems in the 21st Century: Toward a New, Integrated Hard Problem Science. Amsterdam: Elsevier.

Crutzen, P.J. 2002. Geology of mankind: The Anthropocene. *Nature* **415**:23.

de Vries, B., and J. Goudsblom, eds. 2002. Mappae Mundi: Humans and their Habitats in a Long-term Socio-ecological Perspective. Myths, Maps and Models. Amsterdam: Amsterdam Univ. Press.

Elgin, D. 2000. Promise Ahead: A Vision of Hope and Action for Humanity's Future. New York: HarperCollins.

Etienne, M., and C. Le Page. 2002. Modelling contrasted management behaviours of stakeholders facing a pine encroachment process: An agent-based simulation process. In: Proc. Intl. Environmental Modelling and Software Soc. Conf., Lugano, June 2002, pp. 208–213. Avignon/Montpellier: INRA/CIRAD.

Finnigan, J.J. 2005. The science of complex systems. *Australasian Sci.* June:32–34.

Jones, E.L. 2003. The European Miracle: Environments, Economies and Geopolitics in the History of Europe and Asia. 3d ed. Cambridge: Cambridge Univ. Press.

Krugman, P. 2002. For richer: How the permissive capitalism of the boom destroyed American equality. *New York Times Mag*. Oct. 20.

Liljenström, H., and U. Svedin, eds. 2005. Micro, Meso, Macro: Addressing Complex Systems Coupling. Hackensack, NJ: World Scientific.

Lovelock, J. 2003. The living Earth. *Nature* **426**:769–770.

May, R.M. 1977. Thresholds and breakpoints in ecosystems with a multiplicity of stable states. *Nature* **269**:471–477.

MEA (Millennium Ecosystem Assessment). 2003. Ecosystems and Human Well-being: A Framework for Assessment. Washington, D.C.: Island. Millennium Ecosystem Assessment. 2005. Millennium Ecosystem Assessment Synthesis Report. Washington, D.C.: Island. http://www.millenniumassessment.org/

MEA. 2005. Ecosystems and Human Well-being: Synthesis. Washington, D.C.: Island.

Meadows, D. 1990. Stratagem-I User's Manual. Durham, NH: Institute for Policy and Social Science Research (IPSSR), Univ. of New Hampshire.

Meadows, D. 1996. Fish Banks, Ltd. News. http://www.unh.edu/ipssr/FishBank.html

Meadows, D.H., D.L. Meadows, J. Randers, and W.W. Behrens. 1972. The Limits to Growth. New York: Universe.

Ormerod. 1994. The Death of Economics. London: Faber and Faber.

Rotmans, J., and B. de Vries, eds. 1997. Perspectives on Global Change: The TARGETS Approach. London: Cambridge Univ. Press.

Scheffer, M., S.R. Carpenter, J.A. Foley, C. Folke, and B. Walker. 2001. Stochastic events can trigger large state shifts in ecosystems with reduced resilience. *Nature* **413**:591–596.

Schröter, D., W. Cramer, R. Leemans et al. 2005. Ecosystem service supply and vulnerability to global change in Europe. *Science* **310**:1333–1337.

Steffen, W., A. Sanderson, P.D. Tyson et al. 2004. Global Change and the Earth System: A Planet under Pressure. The IGBP Book Series. Berlin: Springer.

Sterman, J., and L.B. Sweeney. 2002. Cloudy skies: Assessing public understanding of global warming. *Syst. Dyn. Rev.* **18**:207–240.

Toynbee, A.J. 1961. A Study of History. London: Oxford Univ. Press.

van Daalen, C.E., W.A.H. Thissen, and M.M. Berk. 1998. The Delft process: Experiences with a dialogue between policy makers and global modellers. In: Global Change Scenarios of the 21st Century: Results from the IMAGE 2.1 model, ed. J. Alcamo, R. Leemans, and G. J. J. Kreileman, pp. 267–285. London: Elsevier.

van den Belt, M. 2004. Mediated Modeling: A System Dynamics Approach to Environmental Consensus Building. Washington, D.C.: Island.

van Notten, P.W.F., J. Rotmans, M.B.A. van Asselt, and D.S. Rothman. 2003. An updated scenario typology. *Futures* **35**:423–443.

List of Acronyms

AAO	AntArctic Oscillation
ABMs	Agent-Based Models
AC-ERE	Advisory Committee for Environmental Research and Education
ACIA	Arctic Climate Impact Assessment
AIDS	Acquired ImmunoDeficiency Syndrome
AIMES	Analysis, Integration and Modelling of the Earth System
AL	Aleutian Low
AMO	Atlantic Multidecadal Oscillation
AO	Arctic Oscillation
ARIMA	AutoRegressive Integrated Moving Average modeling
BCIF	Biophysical Climate Impact Factors
BECC	Border Environment Cooperation Commission
CA	Cellular Automata
CCD	Convention to Combat Desertification
CEC	Commission for Environmental Cooperation
CEMCOS	CEllular Model for COastal Simulation
CFCs	ChloroFluoroCarbons
CGE	Computerized General Equilibrium
CNRS	Centre National de la Recherche Scientifique
COOL	Climate OptiOns for the Long term
CSIRO	Commonwealth Scientific and Industrial Research Organisation
DDP	Dahlem Desertification Paradigm
DDT	DichlorDiphenylTrichlorethan
DICE	Dynamic Integrated Climate and the Economy
DPSIR	Driver-Pressure-State-Impact-Response
EA	Energy Analysis
EC	European Community
ENSO	El Niño-Southern Oscillation
EPICA	European Project for Ice Coring in Antarctica
EROI	Energy Return On Investment
ESR	Electron Spin Resonance
ESSP	Earth System Science Partnership
FAO	Food and Agricultural Organization
GAIM	Global Analysis, Integration and Modelling
GCM	Global Climate Model
GCTE	Global Change and Terrestrial Ecology
GDP	Gross Domestic Product
GHG	GreenHouse Gas
GIS	Geographic Information System
GLOBUS	Generating Long-term Options By Using Simulation
GNP	Gross National Product
GPP	Gross Primary Product
GUI	Graphical User Interface

GUMBO	Global Unified Metamodel of the BiOsphere
GWP	Gross World Product
HDI	Human Development Index
HFCs	HydroFluoroCarbons
HIV	Human Immunodeficiency Virus
HYDE	HistorY Data base of the global Environment
IAM	Integrated Assessment Modeling
ICSU	International Council for Scientific Unions
ICT	Information Communication Technology
IEA	International Energy Agency
IFs	International Futures simulation
IGBP	International Geosphere–Biosphere Programme
IGMs	Integrated Global Models
IHDP	International Human Dimensions Programme
IHOPE	Integrated History and future Of People on Earth
IMAGE	Integrated Model to Assess the Greenhouse Effect
IMAGE-2	Integrated Model to Assess the Global Environment
IMF	International Monetary Fund
IPCC	Intergovernmental Panel on Climate Change
IPO	Interdecadal Pacific Oscillation
ISCOM	Information Society as a COMplex system
ISSC	International Social Science Council
IWM	Integrated World Model
LEK	Local Environmental Knowledge
LGM	Last Glacial Maximum
LIA	Little Ice Age
LIATE	Little Ice Age-Type Events
LIATIMP	Little Ice Age-Type IMPacts
LUCC	Land-Use and Cover Change
MAB	Man And Biosphere
MDG	Millenium Development Goals
MEA	Millennium Ecosystem Assessment
mer	market exchange rate
MIT	Massachusetts Institute of Technology
MNP	Milieu en Natuur Planbureau (Netherlands Environmental Assessment Agency)
MODIS	MODerate-resolution Imaging Spectroradiometer
MSA	Metropolitan Statistical Area
MSN	MicroSoft Network
NAFTA	North American Free Trade Agreement
NAO	North Atlantic Oscillation
NASA	National Aeronautics and Space Administration
NBIC	Nanotechnology–Biotechnology–Information–Communications
NCCR	National Centre of Competence in Research
NCEAS	National Center for Ecological Analysis and Synthesis
NGO	Non Governmental Organization
NOAA	National Oceanic and Atmospheric Administration
NP Index	North Pacific Index

NPP	Net Primary Production
NRC	National Research Council
NUSAP	Numeral, Unit, Spread, Assessment, Pedigree
OECD	Organisation for Economic Co-operation and Development
OSL	Optically Stimulated Luminescence
PAGES	PAst Global changES
PDO	Pacific Decadal Oscillation
PFCs	PerFluoroCarbons
ppp	purchasing power parity
PSIR	Pressure–State–Impact–Response
R&D	Research and Development
RICE	Regional Integrated model of Climate and the Economy
RIVM	Rijksinstituut voor Volksgezondheid en Milieu (Dutch National Institute for Public Health and the Environment)
SARU	Systems Analysis Research Unit
SRES	Special Report on Emission Scenarios
T21 model	Millennium Institute's Threshold 21 model
TARGETS	Tool to Assess Regional and Global Environmental and health Targets for Sustainability
TOE	Tons Oil Equivalent
UN	United Nations
UNCED	United Nations Conference on Environment and Development
UNDP	United Nations Development Programme
UNEP	United Nations Environment Programme
UNESCO	United Nations Educational, Scientific and Cultural Organization
UNFCCC	United Nations Framework Convention on Climate Change
UNFPA	United Nations Population Fund
WCED	World Commission on Environment and Development
WCRP	World Climate Research Programme
WGCM	Working Group on Climate Modelling
WIM	World Integrated Model
WMO	World Meteorological Organization
WOMP	World Order Models Project
WRI	World Resources Institute
WTP	Willingness To Pay
WWI	World War I
WWII	World War II

Author Index

Name Index

Subject Index